# Stability Problems for Stochastic Models

# Stability Problems for Stochastic Models

## Theory and Applications

Editors

**Alexander Zeifman**
**Victor Korolev**
**Alexander Sipin**

MDPI • Basel • Beijing • Wuhan • Barcelona • Belgrade • Manchester • Tokyo • Cluj • Tianjin

*Editors*

Alexander Zeifman
Vologda State University
Russia

Victor Korolev
Lomonosov Moscow State University
Russia
Russian Academy of Sciences
Russia

Alexander Sipin
Vologda State University
Russia

*Editorial Office*
MDPI
St. Alban-Anlage 66
4052 Basel, Switzerland

This is a reprint of articles from the Special Issue published online in the open access journal *Mathematics* (ISSN 2227-7390) (available at: https://www.mdpi.com/journal/mathematics/special_issues/Stochastic_Processes_Theory_Applications_2020).

For citation purposes, cite each article independently as indicated on the article page online and as indicated below:

LastName, A.A.; LastName, B.B.; LastName, C.C. Article Title. *Journal Name* **Year**, *Volume Number*, Page Range.

**ISBN 978-3-0365-0452-0 (Hbk)**
**ISBN 978-3-0365-0453-7 (PDF)**

# Contents

**About the Editors** . . . . . . . . . . . . . . . . . . . . . . . . . . . . . . . . . . . . . . . . . . . . . . . . **vii**

**Preface to "Stability Problems for Stochastic Models"** . . . . . . . . . . . . . . . . . . . . . . . . **ix**

**Irina Shevtsova and Mikhail Tselishchev**
A Generalized Equilibrium Transform with Application to Error Boundsin the Rényi Theorem with No Support Constraints
Reprinted from: *Mathematics* **2020**, *8*, 577, doi:10.3390/math8040577 . . . . . . . . . . . . . . . . . **1**

**Dmitry Efrosinin, Andreas PlankNatalia Stepanova and Janos Sztrik**
Approximations in Performance Analysis of a Controllable Queueing System with Heterogeneous Servers
Reprinted from: *Mathematics* **2020**, *8*, 1803, doi:10.3390/math8101803 . . . . . . . . . . . . . . . . **23**

**Mohammed Al-Nator and Sofya Al-Nator**
Accumulative Pension Schemes with Various Decrement Factors
Reprinted from: *Mathematics* **2020**, *8*, 2081, doi:10.3390/math8112081 . . . . . . . . . . . . . . . . **41**

**Seokjun Lee, Sergei Dudin, Olga Dudina, Chesoong Kim and Valentina Klimenok**
A Priority Queue with Many Customer Types, Correlated Arrivals and Changing Priorities
Reprinted from: *Mathematics* **2020**, *8*, 1292, doi:10.3390/math8081292 . . . . . . . . . . . . . . . . **57**

**Georgy Shevlyakov**
Highly Efficient Robust and Stable *M*-Estimates of Location
Reprinted from: *Mathematics* **2021**, *9*, 105, doi:10.3390/math9010105 . . . . . . . . . . . . . . . . . **77**

**Arsen L. Yakymiv**
Local Limit Theorem for the Multiple Power Series Distributions
Reprinted from: *Mathematics* **2020**, *8*, 2067, doi:10.3390/math8112067 . . . . . . . . . . . . . . . . **87**

**Yury Khokhlov, Victor Korolev and Alexander Zeifman**
Multivariate Scale-Mixed Stable Distributions and Related Limit Theorems
Reprinted from: *Mathematics* **2020**, *8*, 749, doi:10.3390/math8050749 . . . . . . . . . . . . . . . . . **95**

**Edward Omey and Meitner Cadena**
On Convergence Rates of Some Limits
Reprinted from: *Mathematics* **2020**, *8*, 634, doi:10.3390/math8040634 . . . . . . . . . . . . . . . . . **125**

**Andrey Borisov and Igor Sokolov**
Optimal Filtering of Markov Jump Processes Given Observations with State-Dependent Noises: Exact Solution and Stable Numerical Schemes
Reprinted from: *Mathematics* **2020**, *8*, 506, doi:10.3390/math8040506 . . . . . . . . . . . . . . . . . **143**

**Francesco Iafrate and Enzo Orsingher**
On the Fractional Wave Equation
Reprinted from: *Mathematics* **2020**, *8*, 874, doi:10.3390/math8060874 . . . . . . . . . . . . . . . . . **165**

**Victor Korolev and Andrey Gorshenin**
Probability Models and Statistical Tests for Extreme Precipitation Based on Generalized Negative Binomial Distributions
Reprinted from: *Mathematics* **2020**, *8*, 604, doi:10.3390/math8040604 . . . . . . . . . . . . . . . . . **179**

# About the Editors

**Alexander Zeifman** Professor, Head of Department of Applied Mathematics, Vologda State University, Vologda, Russia; Senior Researcher, Institute of Informatics Problems, Federal Research Center "Computer Sciences and Control" of the Russian Academy of Sciences, Russia; Chief Researcher, Vologda Research Center of the Russian Academy of Sciences, Russia.

Graduate of Vologda State Pedagogical Institute, 1976. Candidate of Science in Physics and Mathematics (Ph.D.), 1981. Doctor of Science in Physics and Mathematics (1994, Institute of Control Sciences, Russian Academy of Sciences). Main research interests: stochastic models, continuous-time Markov chains, bounds on the rate of convergence, perturbation bounds, queueing models, biological models, and queueing theory.

**Victor Korolev** Professor, Head of Department of Mathematical Statistics, Faculty of Computational Mathematics and Cybernetics, Lomonosov Moscow State University, Moscow, Russia; Leading researcher, Institute of Informatics Problems, Federal Research Center "Computer Sciences and Control" of the Russian Academy of Sciences, Moscow, Russia; Professor, Hangzhou Dianzi University, Hangzhou, China.

Graduate of Faculty of Computational Mathematics, Lomonosov Moscow State University, 1977. Candidate of Science in Physics and Mathematics (Ph.D.), 1981. Doctor of Science in Physics and Mathematics (1994, Lomonosov Moscow State University).

Main research interests: limit theorems of probability theory and their applications in distribution theory, statistics, risk theory, and reliability theory; probability models of real processes in physics, meteorology, financial mathematics, and other fields.

**Alexander Sipin** Professor at the Department of Applied Mathematics, Vologda State University, Institute of Mathematics, Natural and Computer Sciences, Russia.

Graduate of Faculty of Mathematics and Mechanics, Leningrad State University, now St. Petersburg State University, Russia in 1975. Candidate of Science in Physics and Mathematics (Ph.D.), 1979. Doctor of Science in Physics and Mathematics (2016, St. Petersburg State University, Russia).

Research interests: Monte Carlo and quasi-Monte Carlo methods, Markov chains, and meshless numerical methods for solving boundary value problems.

# Preface to "Stability Problems for Stochastic Models"

The aim of this Special Issue of *Mathematics* is to commemorate the outstanding Russian mathematician Vladimir Zolotarev whose 90th birthday will be celebrated on February 27th, 2021.

Through his mathematical maturation, Zolotarev took much from distinguished mathematicians E. B. Dynkin and A.N. Kolmogorov, who were his direct teachers during his studies at Lomonosov Moscow State University, in addition to B.V. Gnedenko and Yu. V. Linnik during his graduate studies. In 1958, he defended his candidate (Ph.D.) thesis "Analytic Properties of Infinitely Divisible Distribution Laws" prepared under the supervision of academician Andrey Kolmogorov. In 1966, Vladimir Zolotarev defended his second thesis "Distribution of Sums of Independent Random Variables and Stochastic Processes with Independent Increments" and obtained the degree of Doctor of Sciences. One of his main interests was study of the properties of stable distributions. Zolotarev extended the concept of stable law to the schemes of maximum and multiplication of random variables. Zolotarev's studies on stable laws were summarized in the book "One-Dimensional Stable Distributions" published in 1983, which was soon translated into English (1985) and quickly gained widespread recognition. As this book saw the light, concepts such as Zolotarev's theorem, Zolotarev's formula, and the Zolotarev transformation became quite conventional. Contemporaneously with the study of stable laws, Zolotarev began to work in the field of limit theorems for sums of independent random variables. He made a substantial contribution to the so-called nonclassical theory of summation. The cornerstone of this scheme was the break of the habitual approach, in which an individual summand does not have an effect on the form of the limit distribution (in nonclassical summation theory, an individual summand may play a prominent role). It is fair to say that Vladimir Zolotarev is one of the fathers of this direction in probability theory. He generalized the results of his predecessors, P. Lévy and Yu. V. Linnik who, on the heuristic level, pointed out the possibility of a new approach to limit theorems for sums of independent random variables. The key point of this approach is that limit theorems of probability theory are treated as special stability theorems. Zolotarev created the theoretical foundation of the key method used within this approach, the theory of probability metrics. This approach assumes that statements establishing convergence must be accompanied by statements establishing the convergence rate. Zolotarev called the conditions of convergence that simultaneously serve as convergence rate estimates "natural", In the 1970s, the annual International Seminar on Stability Problems for Stochastic Models were launched, with wide participation of mathematicians from many countries. Today, this seminar is internationally recognized for the originality and relevance of the considered problems and presented results. The seminar formed and developed a breakthrough approach to limit theorems of probability theory as stability theorems. Below is the complete list of the sessions of the International Seminar on Stability Problems for Stochastic Models:

I: May 1974, Leningrad (now St. Petersburg), USSR
II: November 1974, Vilnius, Lithuanian SSR
III: 5–9 December 1977, Moscow, USSR
IV: 11–16 April 1979, Palanga, Lithuanian SSR
V: 17–22 November 1980, Panevezys, Lithuanian SSR
VI: 19–27 April 1982, Moscow, USSR
VII: 18–24 June 1984, Saratov, USSR
VIII: 23–29 September 1984, Uzhgorod, Ukrainian SSR

IX: 13–19 May 1985, Varna, Bulgaria
X: October 1986, Kuybyshev (now Samara), USSR
XI: 4–11 October 1987, Sukhumi, Abkhasian ASSR
XII: October 1988, Kharkov, Ukrainian SSR
XIII: October 1989, Kirillov, Vologda Region, USSR
XIV: 27 January–2 February 1991, Suzdal, USSR
XV: 1–6 June 1992, Perm, Russia
XVI: 29 August–3 September 1994, Eger, Hungary
XVII: 19–26 June 1995, Kazan', Russia
XVIII: 26 January–1 February 1997, Hajdúszoboszló, Hungary
XIX: 6–12 September 1998, Vologda, Russia
XX: 5–11 September 1999, Nałeczów, Poland
XXI: 28 January–3 February 2001, Eger, Hungary
XXII: 25–31 May 2002, Varna, Bulgaria
XXIII: 12–17 May 2003, Pamplona, Spain
XXIV: 10–17 September 2004, Majori (Jurmala), Latvia
XXV: 20–24 September 2005, Maiori (Salerno), Italy
XXVI: 27 August–2 September 2006, Sovata-Bai, Romania
XXVII: 22–26 October 2007, Nahariya, Israel
XXVIII: 31 May–5 June 2009, Zakopane, Poland
XXIX: 10–16 October 2011, Svetlogorsk, Russia
XXX: 24–30 September 2012, Svetlogorsk, Russia
XXXI: 23–27 April 2013, Moscow, Russia
XXXII: 15–21 June 2014, Trondheim, Norway
XXXIII: 13–18 June 2016, Svetlogorsk, Russia
XXXIV: 24–28 August 2017, Debrecen, Hungary
XXXV: 24–28 September 2018, Perm, Russia
XXXVI: 22–26 June 2020, Petrozavodsk, Russia (online session), 21–25 June 2021, Petrozavodsk, Russia (offline session).

Devoting his heart and soul to science, he had always demanded the same from his colleagues and numerous students. Vladimir Zolotarev had to spend the last years of his life away from Russia in California (USA), where the local equable climate helped him overcome the consequences of a stroke suffered in 1995. On November 7, 2019, the outstanding mathematician Vladimir Mikhailovich Zolotarev passed away.

The present Special Issue contains a collection of new papers by the colleagues and followers of Vladimir Zolotarev, who were participants in sessions of the International Seminar on Stability Problems for Stochastic Models.

**Alexander Zeifman, Victor Korolev, Alexander Sipin**
*Editors*

mathematics

MDPI

Article

# A Generalized Equilibrium Transform with Application to Error Bounds in the Rényi Theorem with No Support Constraints

**Irina Shevtsova** [1,2,3,4,*] **and Mikhail Tselishchev** [2,4*]

[1] Department of Mathematics, School of Science, Hangzhou Dianzi University, Hangzhou 310018, China
[2] Department of Mathematical Statistics, Faculty of Computational Mathematics and Cybernetics, Lomonosov Moscow State University, GSP-1, 1-52 Leninskiye Gory, Moscow 119991, Russia
[3] Federal Research Center "Informatics and Control" of the Russian Academy of Sciences, Moscow 119333, Russia
[4] Moscow Center for Fundamental and Applied Mathematics, Moscow 119991, Russia
* Correspondence: ishevtsova@cs.msu.ru (I.S.); m.tselishchev@cs.msu.ru (M.T.)

Received: 18 March 2020; Accepted: 8 April 2020; Published: 13 April 2020

**Abstract:** We introduce a generalized stationary renewal distribution (also called the equilibrium transform) for arbitrary distributions with finite nonzero first moment and study its properties. In particular, we prove an optimal moment-type inequality for the Kantorovich distance between a distribution and its equilibrium transform. Using the introduced transform and Stein's method, we investigate the rate of convergence in the Rényi theorem for the distributions of geometric sums of independent random variables with identical nonzero means and finite second moments without any constraints on their supports. We derive an upper bound for the Kantorovich distance between the normalized geometric random sum and the exponential distribution which has exact order of smallness as the expectation of the geometric number of summands tends to infinity. Moreover, we introduce the so-called asymptotically best constant and present its lower bound yielding the one for the Kantorovich distance under consideration. As a concluding remark, we provide an extension of the obtained estimates of the accuracy of the exponential approximation to non-geometric random sums of independent random variables with non-identical nonzero means.

**Keywords:** Rényi theorem; Kantorovich distance; zeta-metrics; Stein's method; stationary renewal distribution; equilibrium transform; geometric random sum; characteristic function

---

## 1. Introduction

Let $X_1, X_2, \ldots$ be a sequence of independent and, for simplicity in this Introduction, identically distributed (i.i.d.) random variables (r.v.s) with $a := \mathbf{E}X_1 \neq 0$. Let $N$ be a random variable independent of $\{X_1, X_2, \ldots\}$ and having the geometric distribution $\mathrm{Geom}(p)$ with parameter $p \in (0,1)$, i.e., $\mathbf{P}(N = n) = p(1-p)^{n-1}$ for $n \in \mathbb{N}$. Denote also $N_0 := N - 1$ the shifted geometric r.v. Let $S_n := \sum_{k=1}^{n} X_k$, $n \in \mathbb{N}$, $S_0 := 0$. The well-known Rényi theorem states that the distribution of a properly normalized geometric random sum $S_N$ converges weakly to the exponential law as $p$ tends to zero. More precisely,

$$W := \frac{S_N}{\mathbf{E}S_N} \xrightarrow{d} \mathscr{E} \text{ as } p \downarrow 0, \quad \text{where } \mathscr{E} \sim \mathrm{Exp}(1) \text{ and } \mathbf{E}S_N = \mathbf{E}N\,\mathbf{E}X_1 = a/p. \tag{1}$$

Here, the notation $\mathrm{Exp}(\lambda)$ stands for the exponential distribution with density $\lambda e^{-\lambda x}\mathbb{1}_{(0,\infty)}(x)$, $\lambda > 0$. Originally, Rényi proved Equation (1) under the additional assumption of nonnegativeness of

$\{X_n\}$. However, it can be made sure that Equation (1) holds also: (i) for alternating $\{X_n\}$ (by alternating r.v. we mean a r.v. that may take values of both signs); and (ii) for

$$W_0 := \frac{S_{N_0}}{\mathbf{E}S_{N_0}} = \frac{pS_{N_0}}{a(1-p)},$$

in place of $W$ (still without any support assumptions on the distribution of $\{X_k\}$). This can be done, for example, by showing that the characteristic function (ch.f.) of $W$ (and also of $W_0$) converges pointwisely to that of the exponential distribution.

The importance of every limit theorem only increases if it is accompanied by the corresponding estimates of the rate of convergence. There are several bounds on the accuracy of approximation in Equation (1), mainly w.r.t. the Kolmogorov (uniform) and $\zeta$-metrics, which are cited below. All of them assume additional conditions on the distribution of random summands including the finiteness of higher-order moments.

Recall that both the Kolmogorov and $\zeta_s$-metrics are defined as simple probability metrics with $\zeta$-structure (see Section 2 of [1]) between probability distributions (d.f.s $F$, $G$) of r.v.s $X$, $Y$:

$$\zeta_{\mathcal{H}}(F,G) \equiv \zeta_{\mathcal{H}}(\mathscr{L}(X),\mathscr{L}(Y)) \equiv \zeta_{\mathcal{H}}(X,Y) := \sup_{h\in\mathcal{H}} \left| \int_{\mathbb{R}} h\,dF - \int_{\mathbb{R}} h\,dG \right| \tag{2}$$

for specific classes $\mathcal{H}$ of real Borel functions on $\mathbb{R}$ (to simplify the notation, here and in what follows, we use r.v.s as well as their distributions and d.f.s in the arguments of simple probability metrics interchangeably; this should not cause any misunderstanding). The Kolmogorov metric $\rho$ is obtained with $\mathcal{H} = \{\mathbb{1}_{(-\infty,a)}(x) \mid a \in \mathbb{R}\}$, the class of indicators of all open intervals with unbounded left endpoint:

$$\rho(F,G) := \sup_{x\in\mathbb{R}} |F(x) - G(x)|,$$

while $\zeta$-metric of order $s > 0$, originally introduced by Zolotarev [2] (see also [3]) as an example of an ideal metric with $\zeta$-structure, is defined as $\zeta_{\mathcal{H}}$ with $\mathcal{H} = \mathcal{F}_s^\infty$, where

$$\mathcal{F}_s^\infty := \left\{ h \in \mathcal{F}_s \colon h \text{ is bounded} \right\},$$
$$\mathcal{F}_s := \left\{ h\colon \mathbb{R} \to \mathbb{R}\colon \left| h^{(m)}(x) - h^{(m)}(y) \right| \le |x-y|^{s-m} \ \forall x,y \in \mathbb{R} \text{ with } m := \lceil s-1 \rceil \in \mathbb{N}_0 \right\},\ s > 0,$$

that is,

$$\zeta_s(F,G) := \sup_{h\in\mathcal{F}_s^\infty} \left| \int_{\mathbb{R}} h\,dF - \int_{\mathbb{R}} h\,dG \right|. \tag{3}$$

Observe that $h \in \mathcal{F}_s$ iff $h' \in \mathcal{F}_{s-1}$, $s > 1$. If $\mathbf{E}|X|^s < \infty$ and $\mathbf{E}|Y|^s < \infty$, then $\zeta_s(F,G) < \infty$ and the least upper bound w.r.t. to $h \in \mathcal{F}_s^\infty$ in Equation (3) may be replaced with that over a wider class $\mathcal{F}_s$. For further properties of $\zeta_s$-metrics, we refer to the works in [3,4] and Section 4 of [5].

In the present paper, we focus mostly on $\zeta_1$-metrics between distributions with finite first moments; under this assumption, the definition of $\zeta_1$-metric can be rewritten as

$$\zeta_1(F,G) = \sup_{h\in\mathrm{Lip}_1} \left| \int_{\mathbb{R}} h\,dF - \int_{\mathbb{R}} h\,dG \right|, \tag{4}$$

where

$$\mathrm{Lip}_c := \left\{ h\colon \mathbb{R} \to \mathbb{R} \,\middle|\, |h(x) - h(y)| \le c\,|x-y| \quad \forall x,y \in \mathbb{R} \right\}, \quad c > 0,$$

so that $\mathrm{Lip}_1 = \mathcal{F}_1$. It is worth noting that $\zeta_1$ has several alternative representations. The Kantorovich–Rubinstein theorem states that $\zeta_1(X,Y)$ is minimal with respect to the compound metric $\mathbf{E}|X-Y|$, while the results in [6] imply that the optimal coupling is attained at the comonotonic pair (that is, with $(X,Y) = (F^{-1}(U), G^{-1}(U))$, $U$ having the uniform distribution on $(0,1)$, $F^{-1}$, $G^{-1}$ being generalized inverse d.f.s):

$$\zeta_1(F,G) = \min_{\mathscr{L}(X',Y'):\, X' \overset{d}{=} X,\, Y' \overset{d}{=} Y} \mathbf{E}|X'-Y'| = \int_0^1 \left|F^{-1}(u) - G^{-1}(u)\right| du = \int_{-\infty}^{\infty} |F(x) - G(x)|\, dx. \quad (5)$$

The rightmost representation in Equation (5), as the mean metric between the d.f.s $F$ and $G$, follows from the geometrical interpretation. The metric $\zeta_1$ is also called the Kantorovich, or the Wasserstein distance.

Thus, coming back to the convergence rate estimates in Equation (1), we first mention the paper by Solovyev [7], which gives the following uniform bound for nonnegative $\{X_k\}$, as pointed out in [8]:

$$\rho(W_0, \mathscr{E}) \le 24p\,\frac{\gamma_r}{r-2}, \quad 2 < r \le 3, \quad (6)$$

where $\gamma_r = \left(\mathbf{E}X_1^r / a^r\right)^{1/(r-1)}$.

Kalashnikov and Vsekhsvyatskii [9] proved a uniform upper bound for nonnegative summands in terms of their moments of order $s \in (1,2]$:

$$\rho(W, \mathscr{E}) \le Cp^{s-1}\frac{\mathbf{E}X_1^s}{a^s}, \quad (7)$$

where $C$ is an absolute constant.

Kruglov and Korolev [10] gave the following nonuniform bound of the accuracy of the exponential approximation to the normalized geometric distribution (i.e., for degenerate $\{X_n\}$):

$$\begin{aligned} \left|\mathbf{P}(pN < x) - (1 - e^{-x})\right| &\le x\,\mathbb{1}_{\{x<p\}} + \left(e^{-x} - e^{-Q(p)x}\right)\mathbb{1}_{\{x \ge p\}} \le \\ &\le x\left[\mathbb{1}_{\{x<p\}} + \frac{p}{2(1-p)}e^{-x}\,\mathbb{1}_{\{x \ge p\}}\right], \end{aligned} \quad (8)$$

where $Q(p) = (1 - p/2)/(1-p)$.

Brown [8] proved an asymptotically exact (as $p \to 0$) upper bound for nonnegative summands, which does not require moments of order greater than two:

$$\rho(W_0, \mathscr{E}) \le p\,\frac{\mathbf{E}X_1^2}{a^2}\max\left(1, \frac{1}{2(1-p)}\right). \quad (9)$$

Brown also showed that Equation (9) is tighter than Equation (6) for all $2 < r \le 3$ and $p \in (0, 0.5]$. Moreover, Equation (9) can be treated as a specification of Equation (7) for $s = 2$ with a concrete value of $C$.

Sugakova [11] presented some bounds for the d.f. $F_{S_{N_0}}(t)$ for $t > 1$ using the characteristics of the renewal process built on top of independent and not necessary identically distributed alternating $\{X_n\}$ with identical means.

Kalashnikov [12] provided estimates of the rate of convergence in the Rényi theorem for i.i.d. alternating $\{X_n\}$ w.r.t. $\zeta_s$-metrics of order $s \in [1,2]$ and the uniform metric (the latter is done under the additional assumption of bounded density), in particular, for any $s \in (1,2]$,

$$\zeta_s(W, \mathscr{E}) \le p^{s-1}\zeta_s(X_1, \mathscr{E}), \quad (10)$$

$$\zeta_1(W, \mathscr{E}) \le p\,\zeta_1(X_1, \mathscr{E}) + 2(1-p)p^{s-1}\zeta_s(X_1, \mathscr{E}), \quad (11)$$

provided that $\mathbf{E}X_1 = 1$.

Among other valuable things, Peköz and Röllin [13] exploited Stein's method and equilibrium (stationary renewal) distributions (see Section 3) to estimate the Kantorovich distance between the exponential distribution and that of a normalized geometric random sum $W$ of square integrable independent and not necessary identically distributed nonnegative random summands $\{X_n\}$ with identical positive means under the technical assumption $\mathbf{E}X_k = 1$:

$$\zeta_1(W,\mathscr{E}) \le 2p \sum_{n=1}^{\infty} \mathbf{P}(N=n)\, \zeta_1(X_n, X_n^e), \tag{12}$$

where $X_n^e$ has an equilibrium distribution w.r.t. $X_n$, $n \in \mathbb{N}$. Using the trivial bound $\zeta_1(X,Y) \le \mathbf{E}|X| + \mathbf{E}|Y|$ that follows from representation (5) and holds true for arbitrary r.v.s $X, Y$ with finite first moments, the inequality in Equation (12) can be naturally extended to

$$\zeta_1(W,\mathscr{E}) \le 2p \sup_n \zeta_1(X_n, X_n^e) \le p \sup_n \left(\mathbf{E}X_n^2 + 2\right), \tag{13}$$

as done in [14].

Equation (22) of Hung [15] gives the following bound for the Trotter distance between $W$ and $\mathscr{E}$ in the case of i.i.d. nonnegative summands $\{X_n\}$ with $\mathbf{E}X_1 = 1$:

$$d_T(W,\mathscr{E};h) := \sup_{t\in\mathbb{R}} |\mathbf{E}h(W+t) - \mathbf{E}h(\mathscr{E}+t)| \le p^{s-1}\left(\mathbf{E}X_1^2 + 3\right), \quad h \in \mathcal{F}_s^\infty,\ s \in (1,2]. \tag{14}$$

Given that $\zeta_s(W,\mathscr{E}) = \sup_{h\in\mathcal{F}_s^\infty} d_T(W,\mathscr{E};h)$, the estimate in Equation (14) may be rewritten as

$$\zeta_s(W,\mathscr{E}) \le p^{s-1}\left(\mathbf{E}X_1^2 + 3\right) \quad \text{for} \quad s \in (1,2]. \tag{15}$$

To compare Equation (15) with Kalashnikov's bound in Equation (10), observe that, by Theorem 1(i,c) below, the dual representation of $\zeta_s(X,Y)$-metric as the minimal w.r.t. the compound metric $\mathbf{E}|X-Y|^s$ for $s \in (0,1]$ (see, e.g., Corollary 5.2.2 of [4]), and, finally, Theorem 1(g) below, for $s \in (1,2]$, we have

$$\begin{aligned}\zeta_s(X_1,\mathscr{E}) = \zeta_{s-1}(X_1^e,\mathscr{E}^e) = \zeta_{s-1}(X_1^e,\mathscr{E}) = \inf_{\mathscr{L}(X,Y):\ X \overset{d}{=} X_1^e,\ Y \overset{d}{=} \mathscr{E}} \mathbf{E}\,|X-Y|^{s-1} \le \\ \le \mathbf{E}\,|X_1^e - \mathscr{E}| + 1 \le \mathbf{E}X_1^e + \mathbf{E}\,\mathscr{E} + 1 = \mathbf{E}X_1^2/2 + 2 < \mathbf{E}X_1^2 + 3,\end{aligned}$$

hence, Kalashnikov's bound in Equation (10) is tighter than Equation (15).

Thus, most existing estimates of the rate of convergence in the Rényi theorem were obtained under the additional assumption of nonnegativeness of random summands $\{X_n\}$. However, there are many applications where geometric random sums appear with alternating random summands, for example, as profit-or-losses in financial mathematics, risk theory, queuing theory, etc. Hence, extensions of such sharp and natural estimates as Equations (9), (12), and (13), say, to the alternating random summands, would not only represent a theoretical interest, but can also be in great demand by various applications of probability theory.

In the present paper, we focus on $\zeta_1$-estimates, in particular, we extend bounds in Equations (12) and (13) to the alternating case. More precisely, in Theorem 4 below, we prove that, for square integrable independent and not necessarily identically distributed random summands $\{X_n\}$ with identical nonzero means (for simplicity, equal to one), the following estimates hold:

$$\zeta_1(W,\mathscr{E}) \le 2p\sum_{n=1}^{\infty}\mathbf{P}(N=n)\,\zeta_1\big(\mathscr{L}(X_n),\mathscr{L}^e(X_n)\big) \le p\left(\mathbf{E}X_N^2 - 2\,\mathbf{P}(X_N\le 0)\right), \tag{16}$$

$$\zeta_1(W_0,\mathscr{E}) \le \frac{2p}{1-p}\,\zeta_1\left(\delta_0,\mathscr{L}^e(X_N)\right) = \frac{p}{1-p}\,\mathbf{E}X_N^2, \tag{17}$$

where $\delta_0$ is the Dirac measure concentrated in zero and $\mathscr{L}^e(X_n)$ is the equilibrium transform of $\mathscr{L}(X_n)$, which is a generalization of the equilibrium distribution introduced in Section 3 below and, generally speaking, is no more a probability measure (therefore, we write $\mathscr{L}^e(X_n)$ instead of $\mathscr{L}(X_n^e)$), but allows eliminating the support constraints on the distribution of $X_n$. The notion of the $\zeta_1$-metric between signed measures is introduced in Section 2 below and coincides with that of the ordinary $\zeta_1$-metric in case of probability measures. Thus, the intermediate estimate in Equation (16) coincides with estimate (12), but now also holds true for alternating random summands $\{X_n\}$. Furthermore, it can easily be seen that the right-hand side of Equation (16) does not exceed

$$p\sup_n\left(\mathbf{E}X_n^2 - 2\,\mathbf{P}(X_n\le 0)\right)$$

and, hence, is tighter than estimate (13) and does not require that $\{X_n\}$'s take only positive values. The comparison of estimates (16) and Kalashnikov's bound in Equation (11) with $s=2$

$$\begin{aligned}\zeta_1(W,\mathscr{E}) &\le p\,\zeta_1(X_1,\mathscr{E}) + 2p(1-p)\,\zeta_2(X_1,\mathscr{E}) = \\ &= p\,\zeta_1(X_1,\mathscr{E}) + 2p(1-p)\,\zeta_1\big(\mathscr{L}^e(X_1),\mathrm{Exp}(1)\big)\end{aligned} \tag{18}$$

(for the equality here, see Theorem 1(i) below) is complicated in the general case, since, due to Theorem 3 below, the rightmost expression does not exceed

$$2p(2-p)\,\zeta_1\big(\mathscr{L}(X_1),\mathscr{L}^e(X_1)\big),$$

which is asymptotically twice greater than the intermediate expression in Equation (16), while the intermediate estimate in Equation (16), by the triangle inequality, yields the bound

$$\zeta_1(W,\mathscr{E}) \le 2p\,\zeta_1(X_1,\mathscr{E}) + 2p\,\zeta_1\big(\mathscr{L}^e(X_1),\mathrm{Exp}(1)\big)$$

with the first term twice larger than that in Equation (18).

We use the same techniques and recipes as in [13]. First, we bound the left-hand side of Equation (16) from above with $\zeta_1\left(\mathscr{L}(W),\mathscr{L}^e(W)\right)$ using Stein's method (see Theorem 3 in Section 4). Second, we estimate $\zeta_1\left(\mathscr{L}(W),\mathscr{L}^e(W)\right)$ by the $\zeta_1$-distances between $X_n$ and their equilibrium transforms $\mathscr{L}^e(X_n)$, $n\in\mathbb{N}$. Third, we construct an optimal upper bound for $\zeta_1\left(\mathscr{L}(X_n),\mathscr{L}^e(X_n)\right)$ in terms of the second moments of $X_n$ and $\mathbf{P}(X_n\le 0)$, $n\in\mathbb{N}$ (see Theorem 2 in Section 3). The resulting upper bounds for $\zeta_1(W,\mathscr{E})$ and $\zeta_1(W_0,\mathscr{E})$ are given in Theorem 4 of Section 5. Furthermore, we provide asymptotic lower bounds for $\zeta_1(W,\mathscr{E})$ and $\zeta_1(W_0,\mathscr{E})$ (see Theorem 5 in Section 5) in terms of the so-called *asymptotically best constants* introduced in Section 5. The constructed lower bounds turn out to be asymptotically four times smaller than the upper ones. Finally, we extend the obtained estimates of the accuracy of the exponential approximation to non-geometric random sums of independent random variables with non-identical nonzero means of identical signs (see Theorem 6 in Section 5).

## 2. The Kantorovich Distance between Signed Measures

In the next sections, we need to calculate the Kantorovich (or $\zeta_1$-) distance between measures on $(\mathbb{R}, \mathcal{B})$ that are no longer probabilities, but still have unit mass on $\mathbb{R}$. Denote by $\mathcal{M}^1$ the linear space of signed measures on $(\mathbb{R}, \mathcal{B})$ with finite total variations and finite first moments, and by $\mathcal{M}_0^1$ the subspace of measures $\sigma \in \mathcal{M}^1$ with $\sigma(\mathbb{R}) = 0$.

The Kantorovich norm on $\mathcal{M}_0^1$ is defined as (see Section 3.2 of [16])

$$\|\sigma\|_K := \sup_{f \in \mathrm{Lip}_1} \left| \int_{\mathbb{R}} f \, d\sigma \right|.$$

Now let $\mu, \nu \in \mathcal{M}^1$ and $\mu(\mathbb{R}) = \nu(\mathbb{R})$, so that $\mu - \nu \in \mathcal{M}_0^1$. The *induced Kantorovich distance* $\zeta_1$ between $\mu$ and $\nu$ is

$$\zeta_1(\mu, \nu) := \|\mu - \nu\|_K = \sup_{f \in \mathrm{Lip}_1} \left| \int_{\mathbb{R}} f \, d\mu - \int_{\mathbb{R}} f \, d\nu \right|. \tag{19}$$

It is easy to see that in the case of probability measures $\mu$ and $\nu$ Equation (19) coincides with the definition of $\zeta_1$-distance given in Equation (4).

Using the Jordan decompositions $\mu = \mu^+ - \mu^-$ and $\nu = \nu^+ - \nu^-$, as well as the alternative representation in Equation (5) of the $\zeta_1$-distance between nonnegative measures $\lambda = \mu^+ + \nu^-$ and $\pi = \nu^+ + \mu^-$ with $\lambda(\mathbb{R}) = \pi(\mathbb{R})$ in terms of their d.f.s, after a proper normalization, one can rewrite Equation (19) as

$$\zeta_1(\mu, \nu) = \zeta_1(\lambda, \pi) = \int_{\mathbb{R}} |F_\lambda(x) - F_\pi(x)| \, dx = \int_{\mathbb{R}} |F_\mu(x) - F_\nu(x)| \, dx, \tag{20}$$

where $F_\mu(x) = \mu\left((-\infty, x)\right)$, $F_\nu(x) = \nu\left((-\infty, x)\right)$, $x \in \mathbb{R}$, are the d.f.s of the signed measures $\mu$ and $\nu$, respectively. In other words, the alternative representation of Zolotarev's $\zeta_1$-distance in terms of d.f.s in Equation (5) is preserved for signed measures with identical masses of $\mathbb{R}$.

We also use the convolution of signed measures $\mu * \lambda$, which is defined word-for-word as that of probability distributions. The uniqueness and multiplication theorems (see, e.g., Chapter 6 of [17] or Section 3.8 of [18]) state that the characteristic function of $\mu$ (the Fourier–Stieltjes transform of $F_\mu$)

$$\widehat{\mu}(t) := \int_{\mathbb{R}} e^{itx} \mu(dx) = \int_{\mathbb{R}} e^{itx} dF_\mu(x), \quad t \in \mathbb{R},$$

defines the signed measure $\mu$ as well as its d.f. $F_\mu$ uniquely and

$$\widehat{\mu * \nu} = \widehat{\mu} \cdot \widehat{\nu}.$$

The following lemma, which is a simple corollary to representation (20), shows that the well-known properties of homogeneity and regularity of the Kantorovich distance between probability distributions are preserved for signed measures, but with a slight correction.

**Lemma 1.** *The Kantorovich distance $\zeta_1$ on the space $\mathcal{M}_D^1$ of finite signed Borel measures on the real line with the masses of $\mathbb{R}$ equal to $D \in \mathbb{R}$ and finite first moments possesses the following properties:*

**(a) Homogeneity of order 1.** *For every $\mu, \nu \in \mathcal{M}_D^1$ and $c \neq 0$, with $\mu_c(B) := \mu(cB)$, $\nu_c(B) := \nu(cB)$ and $cB := \{cx \mid x \in B\}$, $B \in \mathcal{B}$, we have*

$$\zeta_1(\mu_c, \nu_c) = \frac{1}{|c|} \zeta_1(\mu, \nu).$$

**(b) Regularity.** *For all* $\mu, \nu \in \mathcal{M}_D^1$ *and* $\lambda \in \mathcal{M}^1$, *we have*

$$\zeta_1(\mu * \lambda, \nu * \lambda) \leq |\lambda|(\mathbb{R}) \cdot \zeta_1(\mu, \nu),$$

*where* $|\lambda| := \lambda^+ + \lambda^-$ *is the total variation of* $\lambda$.

To avoid abusing the notation, in what follows, we also use $\zeta_1(F, G)$ for the Kantorovich distance between (signed) measures uniquely restored (Section 3.5, Theorem 3.29 of [19]) from distribution functions $F$ and $G$.

## 3. The Equilibrium Transform of Probability Distributions

The notion of *equilibrium distribution* w.r.t. nonnegative r.v.s with finite positive means originally arises in the renewal theory as the distribution of the initial delay of a renewal process which makes its renewal rate constant (Chapter 11, §4 of [20]) and, more generally, the renewal process stationary (Chapter 5, §4 of [21]), which is why it is also called the *stationary renewal distribution*. Equilibrium distribution appears also as the limit distribution of the residual waiting times, or hitting probabilities (Chapter 11, §4 of [20]) and in the celebrated Pollaczek–Khinchin–Beekman formula which expresses the ruin probability in the classical risk process in terms of geometric random sum of i.i.d. r.v.s whose common distribution is the equilibrium transform of the distributions of claims. Due to the definition given in a more general form in Equation (21) below, equilibrium distribution is also called the *integrated tail* one ([12], p. 37, [22]). Concerning the equilibrium transform, we would also like to mention the work of Harkness and Shantaram [23] who considered the *iterated equilibrium transform* for d.f.s with nonnegative support and investigated limit theorems for normalized iterations, the description of limit laws being given in [24]. In particular, the authors of [23] calculated the ch.f. of the equilibrium transform that can be used as the definition of the equilibrium transform in the general case and hence, with the inverse formula, can give a hint to definition in Equation (21) of the equilibrium d.f. with no support constraints.

We introduce an extension of the equilibrium distribution that is applicable for alternating random variables with finite nonzero first moments, but leads out of the class of probability distributions.

Let $P$ be a probability measure with the d.f. $F(x) = P((-\infty, x))$, $x \in \mathbb{R}$, ch.f. $f(t) = \int e^{itx} P(dx) = \int_{\mathbb{R}} e^{itx} dF(x)$, $t \in \mathbb{R}$, and a finite first moment $a := \int x P(dx) = \int_{\mathbb{R}} x dF(x)$. If a r.v. $X$ (on some probability space $(\Omega, \Sigma, \mathbf{P})$) has the distribution $P$, we also write $P = \mathscr{L}(X)$, $f(t) = \mathbf{E}e^{itX} =: f_X(t)$, $F(x) = \mathbf{P}(X < x) =: F_X(x)$, $a = \mathbf{E}X$.

**Definition 1.** *The equilibrium d.f. (distribution) w.r.t. the d.f.* $F$ *(probability distribution* $P$ */ law* $\mathscr{L}(X)$*) with* $a \neq 0$ *is a function of bounded variation (a (signed) measure* $P^e$ */* $\mathscr{L}^e(X)$ *on* $\mathcal{B}(\mathbb{R})$ *with the d.f.)*

$$F^e(x) := \begin{cases} -\dfrac{1}{a} \displaystyle\int_{-\infty}^{x} F(y)\, dy, & \text{if } x \leq 0, \\ -\dfrac{\mathbf{E}X^-}{a} + \dfrac{1}{a} \displaystyle\int_0^x (1 - F(y))\, dy, & \text{if } x > 0, \end{cases} \tag{21}$$

$$= \frac{1}{a}\left(x^+ - \int_{-\infty}^{x} F(y)\, dy\right), \quad x \in \mathbb{R}. \tag{22}$$

In Theorem 1(a) below, it is proved that $F^e$, indeed, has bounded variation and some useful properties of the equilibrium transform are stated as well.

We call $F^e / P^e / \mathscr{L}^e(X)$ the equilibrium transform (d.f./distribution) w.r.t. $F/P/\mathscr{L}(X)/X$ correspondingly, although it may not be a probability d.f./distribution at all. At the same time, it can be easily seen that $\mathscr{L}^e(X)$ is a probability measure if and only if $X$ does not change sign (that is, if and only if $P$ is concentrated either on $(-\infty, 0]$ or on $[0, \infty)$), in which case one might construct a random variable $X^e$ with the distribution $\mathscr{L}(X^e) = \mathscr{L}^e(X)$ and such that $X$ and $X^e$ are either both nonnegative or both nonpositive.

In what follows, to indicate the r.v. whose equilibrium transform is considered, we use the corresponding lower index and write $F^e_X$ and $f^e_X$ for $(F_X)^e$ and $(f_X)^e$, respectively.

**Theorem 1.** *Let $X$ be a r.v. with the d.f. $F$ and $a \neq 0$, and $F^e$ be the equilibrium d.f. w.r.t. $F$ defined in Equation (21). Then:*

**(a)** **Absolute continuity.** *The function $F^e$ has bounded variation on $\mathbb{R}$ with*

$$|\mathscr{L}^e(X)|(\mathbb{R}) = \mathbf{E}|X|/|\mathbf{E}X|, \quad F^e(-\infty) = 0, \quad F^e(+\infty) = 1,$$

*and, hence, $\mathscr{L}^e(X)$ is a Borel measure with unit on $\mathbb{R}$; moreover, $F^e$ is a.c. with the Lebesgue derivative*

$$p^e(x) = \begin{cases} -\frac{1}{a}F(x), & \text{if } x \leq 0, \\ \frac{1}{a}(1 - F(x)), & \text{if } x > 0, \end{cases} \tag{23}$$

*and* $\operatorname{supp}\mathscr{L}^e(X)$ *coincides with the convex hull of* $\operatorname{supp}\mathscr{L}(X)$.

**(b)** **Characteristic function.** *The ch.f. (Fourier–Stieltjes transform) of $F^e$ has the form*

$$f^e(t) := \int_{\mathbb{R}} e^{itx} dF^e(x) = \frac{f(t) - 1}{tf'(0)} = \frac{f(t) - 1}{ita}, \quad \text{if } t \neq 0, \quad \text{and} \quad f^e(0) = 1. \tag{24}$$

**(c)** **Fixed points.** *$\mathscr{L}^e(X) = \mathscr{L}(X)$ iff $X \sim$ Exp$(1/a)$, that is, if and only if $F(x) = (1 - e^{-x/a})\mathbb{1}_{(0,\infty)}(x)$ for some $a > 0$.*

**(d)** **Test functions.** *$F^e$ is the equilibrium d.f. w.r.t. $X$ if and only if*

$$\mathbf{E}g(X) - g(0) = \mathbf{E}X \cdot \int_{\mathbb{R}} g'(x)\, dF^e(x) \tag{25}$$

*for all Lipschitz functions $g\colon \mathbb{R} \to \mathbb{R}$.*

**(e)** **Mixture preservation.** *For arbitrary d.f.s $F_1, F_2, \ldots$ with identical nonzero expectations and a discrete probability distribution $p_n \geq 0$, $n \in \mathbb{N}$, $\sum_{n=1}^{\infty} p_n = 1$, we have*

$$\left(\sum_{n=1}^{\infty} p_n F_n\right)^e = \sum_{n=1}^{\infty} p_n F^e_n. \tag{26}$$

**(f)** **Homogeneity.** *For all $c \in \mathbb{R} \setminus \{0\}$, we have*

$$(F_{cX})^e(x) = F^e_X(x/c), \quad x \in \mathbb{R}, \tag{27}$$

*or, in terms of (constant-sign) r.v.s, $(cX)^e \stackrel{d}{=} cX^e$, $c \in \mathbb{R} \setminus \{0\}$. In other words, equilibrium transform respects scaling.*

**(g)** **Moments.** *If $\mathbf{E}|X|^{r+1} < \infty$ for some $r > 0$, then for all $k \in \mathbb{N} \cap [1, r]$ we have*

$$\int_{\mathbb{R}} x^k\, dF^e(x) = \frac{\mathbf{E}X^{k+1}}{(k+1)\mathbf{E}X}, \quad \int_{\mathbb{R}} |x|^r\, dF^e(x) = \frac{\mathbf{E}X|X|^r}{(r+1)\mathbf{E}X}, \tag{28}$$

$$\int_{\mathbb{R}} x^k\, |dF^e|(x) = \frac{\mathbf{E}|X|X^k}{(k+1)|\mathbf{E}X|}, \quad \int_{\mathbb{R}} |x|^r\, |dF^e|(x) = \frac{\mathbf{E}|X|^{r+1}}{(k+1)|\mathbf{E}X|}. \tag{29}$$

**(h)** **Single summand property.** *Let* $N, X_1, X_2, \ldots$ *be independent r.v.s, such that* $a_n := \mathbf{E}X_n \in (0,\infty)$, $n \in \mathbb{N}$, $\mathbf{P}(N \in \mathbb{N}_0) = 1$, $S_N := X_1 + \ldots + X_N$, $S_0 := 0$, $A := \mathbf{E}S_N = \sum\limits_{n=1}^{\infty} a_n \mathbf{P}(N \geq n)$ *be finite, and* $M$ *be a* $\mathbb{N}$*-valued r.v. with the distribution*

$$\mathbf{P}(M = m) = \frac{a_m}{A}\mathbf{P}(N \geq m), \quad m \in \mathbb{N}.$$

*Then,*

$$\mathscr{L}^e(S_N) = \sum_{m=1}^{\infty} \mathbf{P}(M = m)\, \mathscr{L}(S_{m-1}) * \mathscr{L}^e(X_m), \tag{30}$$

*where* $*$ *denotes the convolution of two Borel measures, or, in terms of (constant-sign) r.v.s,*

$$S_N^e \stackrel{d}{=} S_{M-1} + X_M^e,$$

*where all the r.v.s are independent. In particular, if* $N \sim \mathrm{Geom}(p)$ *and all* $X_k$*'s have identical nonzero expectations, then* $M \stackrel{d}{=} N$ *and*

$$\mathscr{L}^e(S_N) = \mathscr{L}^e(S_{N-1}) = \sum_{n=1}^{\infty} p(1-p)^{n-1} \mathscr{L}(S_{n-1}) * \mathscr{L}^e(X_n), \tag{31}$$

*which can be also rewritten, in the case of i.i.d.* $\{X_k\}$, *in the form*

$$\mathscr{L}^e(S_N) = \mathscr{L}^e(S_{N-1}) = \mathscr{L}(S_{N-1}) * \mathscr{L}^e(X_1).$$

**(i)** **Relation between** $\zeta$**-distances.** *For arbitrary d.f.s* $F$ *and* $G$ *with finite moments of order* $s > 1$ *and identical expectations* $a \neq 0$, *we have*

$$\zeta_s(F, G) = |a|\; \zeta_{s-1}(F^e, G^e). \tag{32}$$

Theorem 2 below provides also an optimal upper bound for $\zeta_1(F, F^e)$ given $F(0+)$ and the second-order moment of $F$.

**Remark 1.** *Theorem 1(h) shows that the equilibrium transform of the geometric random sum of independent r.v.s with identical nonzero means does not depend on whether or not one takes the geometric distribution starting from zero.*

Let us make several historical remarks. Some of the properties of the equilibrium distribution stated in Theorem 1 were known for a nonnegative r.v. $X$. Thus, the characteristic function of $X^e$ given in Equation (24) was found in [23], Equation (25) was taken as the definition of (the distribution of) $X^e$ in [13,14]. In Theorem 2.1 of [13], it was proved that the exponential distribution is the only fixed point of the equilibrium transform; this fact is proved directly also in Lemma 5.2 of [14]. In [14] (p. 268), it is observed that $(cX)^e \stackrel{d}{=} cX^e$ for $c > 0$. Some moment calculations were given in [22]. Single summand property for $S_N$ was demonstrated in the proof of Theorem 3.1 of [13] for nonnegative, but not necessarily independent $\{X_k\}$. The fact that $\mathscr{L}^e(S_N) = \mathscr{L}^e(S_{N-1})$ for i.i.d. nonnegative $\{X_k\}$ was observed in [8] (p. 1394). The equality in Equation (32) for $F(0) = G(0) = 0$ and $s = 2$ was stated in [12] (p. 37).

To prove Theorem 1, we require the following auxiliary statement.

**Lemma 2.** *For every $n \in \mathbb{N}$ and $z_1, \dots, z_n \in \mathbb{C}$, we have*

$$\prod_{k=1}^{n} z_k - 1 = \sum_{k=1}^{n} (z_k - 1) \prod_{j=1}^{k-1} z_j = \sum_{k=1}^{n} (z_k - 1) \prod_{j=k+1}^{n} z_j, \tag{33}$$

*where $\prod_{j=a}^{b}(\cdot) := 1$ for $b < a$.*

**Proof.** We use the induction w.r.t. $n$. For $n = 1$ Equation (33) is trivial. Let Equation (33) hold for $n = 1, \dots, m-1$; let us prove it for $n = m$. Using the inductive transition in the second equality below, we get

$$\prod_{k=1}^{m} z_k - 1 = (z_m - 1) \prod_{k=1}^{m-1} z_k + \prod_{k=1}^{m-1} z_k - 1 = (z_m - 1) \prod_{k=1}^{m-1} z_k + \sum_{k=1}^{m-1} (z_k - 1) \prod_{j=1}^{k-1} z_j = \sum_{k=1}^{m} (z_k - 1) \prod_{j=1}^{k-1} z_j.$$

The second equality in Equation (33) can be deduced from the first one just by the re-numeration of $\{z_k\}_{k=1}^{n} : z_k \leftarrow z_{n-k+1}, k = 1, \dots, n$. □

**Proof of Theorem 1.** (a) It follows immediately from the definition in Equation (21) of $F^e$ that $F^e$ is a.c. with the density given in Equation (23). In turn, Equation (23) implies that $\operatorname{supp} \mathscr{L}^e(X)$ is the convex hull of $\operatorname{supp} \mathscr{L}(X)$ and, accounting for $|\mathscr{L}^e(X)|(\mathbb{R}) = \int |p^e(x)| dx = \mathbf{E}|X|/|\mathbf{E}X| < \infty$, also that $F^e$ has bounded variation. The limiting values $F^e(\pm\infty)$ can be found directly using the definition of $F^e$.

(b) Using the density of $F^e$ (see Equation (23)) and integrating by parts, we have

$$\begin{aligned} f^e(t) &= \frac{1}{a} \int_{\mathbb{R}} e^{itx} p^e(x)\, dx = \frac{1}{a} \int_{\mathbb{R}} e^{itx} \left(\mathbb{1}_{(0,\infty)}(x) - F(x)\right) dx = \frac{1}{ita} \int_{\mathbb{R}} \left(\mathbb{1}_{(0,\infty)}(x) - F(x)\right) de^{itx} = \\ &= \frac{1}{ita} \left[ - e^{itx} F(x) \Big|_{-\infty}^{0} + e^{itx}(1 - F(x)) \Big|_{0}^{\infty} + \int_{\mathbb{R}} e^{itx}\, dF(x) \right] = \frac{f(t) - 1}{ita}, \end{aligned}$$

which coincides with Equation (24).

(c) This statement follows immediately due to the uniqueness of the solution to the linear equation

$$f^e(t) \equiv \frac{f(t) - 1}{ita} = f(t) \quad \Leftrightarrow \quad f(t) = \frac{1}{1 - ita} \sim \mathrm{Exp}(1/a)\,.$$

(d)–(g) These statements follow from the definition and integration by parts for (d) and (g) or the linearity of the Lebesgue–Stieltjes integral for (e).

(h) Let us denote $f_0(t) \equiv 1$, $f_k(t) = \mathbf{E}e^{itX_k}$, $k \in \mathbb{N}$, $t \in \mathbb{R}$. Using the fact that

$$f_{S_N}(t) = \sum_{n=0}^{\infty} \mathbf{P}(N = n) \mathbf{E} e^{itS_n} = \sum_{n=0}^{\infty} \mathbf{P}(N = n) \prod_{k=0}^{n} f_k(t),$$

together with the equation for the equilibrium ch.f. in Equations (24) and (33), we get

$$\begin{aligned} f^e_{S_N}(t) &= \frac{f_{S_N}(t) - 1}{t f'_{S_N}(0)} = \frac{1}{itA} \sum_{n=1}^{\infty} \mathbf{P}(N = n) \left( \prod_{k=1}^{n} f_k(t) - 1 \right) = \\ &= \sum_{n=1}^{\infty} \mathbf{P}(N = n) \sum_{k=1}^{n} \frac{f_k(t) - 1}{itA} \prod_{j=1}^{k-1} f_j(t) = \\ &= \sum_{n=1}^{\infty} \mathbf{P}(N = n) \sum_{k=1}^{n} \frac{a_k}{A} f^e_k(t)\, f_{S_{k-1}}(t). \end{aligned}$$

Changing the order of summation, which is possible by virtue of the absolute convergence of the above series, and recalling the definition of $\mathscr{L}(M)$, we obtain

$$f^e_{S_N}(t) = \sum_{k=1}^{\infty} f^e_k(t)\, f_{S_{k-1}}(t) \cdot \frac{a_k}{A} \sum_{n=k}^{\infty} \mathbf{P}(N=n) = \sum_{k=1}^{\infty} f^e_k(t)\, f_{S_{k-1}}(t)\, \mathbf{P}(M=k),$$

which is equivalent to Equation (30) by virtue of the uniqueness theorem.

If now $N \sim \mathrm{Geom}(p)$ and $a_1 = a_2 = \ldots = a$, then $A = a\mathbf{E}N = a/p$, $\mathbf{P}(M=k) = p(1-p)^{k-1} = \mathbf{P}(N=k)$, $k \in \mathbb{N}$. Denoting by $M_0$ a r.v. corresponding to $N_0 := N-1$ with the distribution

$$\mathbf{P}(M_0 = k) := a_k \mathbf{P}(N_0 \geq k) \Big/ \sum_{k=1}^{\infty} a_k \mathbf{P}(N_0 \geq k) = \mathbf{P}(N_0 \geq k)/\mathbf{E}N_0 = p(1-p)^{k-1}, \quad k \in \mathbb{N},$$

we observe that $M_0 \stackrel{d}{=} N \stackrel{d}{=} M$. This proves Equation (31).

(i) This statement follows from Theorem 4.2(a), Equation (4.20) of [5]. It can also be proved independently, namely, by virtue of (d) we have

$$\begin{aligned}\zeta_s(F,G) &= \sup_{h\in\mathcal{F}_s} \left| \int_{\mathbb{R}} h\, dF - \int_{\mathbb{R}} h\, dG \right| = |a| \sup_{h\in\mathcal{F}_s} \left| \int_{\mathbb{R}} h'\, dF^e - \int_{\mathbb{R}} h'\, dG^e \right| = \\ &= |a| \sup_{h\in\mathcal{F}_{s-1}} \left| \int_{\mathbb{R}} h\, dF^e - \int_{\mathbb{R}} h\, dG^e \right| = |a|\, \zeta_{s-1}(F,G). \quad \square\end{aligned}$$

To conclude this section, we construct an optimal upper bound for the Kantorovich distance between an arbitrary probability distribution with nonzero mean and its equilibrium transform given its second moment and the mass of nonpositive axis. Before formulating the corresponding result, we have to note that Cantelli's (one-sided Chebyshev's) inequality yields $\mathbf{P}(X \leq 0) \leq 1 - 1/\mathbf{E}X^2$ for an arbitrary r.v. $X$ with $0 < \mathbf{E}X^2 < \infty$, and, hence,

$$\mathbf{E}X^2 \geq \frac{1}{1 - \mathbf{P}(X \leq 0)}.$$

This remark explains the choice of the domain of parameters $q$ and $b$ in the following Theorem 2.

**Theorem 2.** *Take any $q \in [0,1)$ and $b \geq \frac{1}{\sqrt{1-q}}$ and let $X$ be a square integrable r.v. with $\mathbf{E}X = 1$, $\mathbf{E}X^2 = b^2$, and $\mathbf{P}(X \leq 0) = q$. Then,*

$$\zeta_1\left(\mathscr{L}(X), \mathscr{L}^e(X)\right) \leq \frac{b^2}{2} - q, \tag{34}$$

*where $\mathscr{L}^e(X)$ is the equilibrium transform of $\mathscr{L}(X)$. The equality in Equation (34) is attained for every $q \in (0,1)$ and $b \geq \frac{1}{\sqrt{1-q}}$ on the two-point distribution $\mathscr{L}(X) = q\delta_u + (1-q)\delta_v$ with*

$$u = 1 - \sqrt{\frac{1-q}{q}(b^2-1)}, \quad v = 1 + \sqrt{\frac{q}{1-q}(b^2-1)}, \tag{35}$$

*and for $q = 0$ and $b = 1$ on the degenerate distribution $\mathscr{L}(X) = \delta_1$.*

**Remark 2.** *With the account of Theorem 1(f) and Lemma 1(a), for arbitrary $\mathbf{E}X \neq 0$, Equation (34) takes the form*

$$\zeta_1\left(\mathscr{L}(X), \mathscr{L}^e(X)\right) \leq \frac{1}{2} \cdot \frac{\mathbf{E}X^2}{|\mathbf{E}X|} - |\mathbf{E}X| \cdot \mathbf{P}(X \leq 0).$$

**Proof of Theorem 2.** Let $F$ be the d.f. of $X$ and $F^e$ be its equilibrium transform. Consider the following functional on the space $\mathbf{F}$ of probability d.f.s with unit mean and finite second moment:

$$J(F) = \zeta_1(F, F^e) - \frac{1}{2}\int_{\mathbb{R}} x^2\, dF(x) + F(0+), \quad F \in \mathbf{F}. \tag{36}$$

Then, Equation (34) would follow from

$$\sup_{F\in\mathbf{F}} J(F) \le 0. \tag{37}$$

Let us prove Equation (37).

Since $h \in \mathrm{Lip}_1$ if and only if $(-h) \in \mathrm{Lip}_1$, the modulus sign in the definition of $\zeta_1(F, F^e)$ (see Equation (19)) may be omitted. Hence, we can rewrite

$$J(F) = \sup_{h\in\mathrm{Lip}_1} J_1(F,h), \quad \text{where} \quad J_1(F,h) = \int_{\mathbb{R}} h\, dF - \int_{\mathbb{R}} h\, dF^e - \frac{1}{2}\int_{\mathbb{R}} x^2\, dF(x) + F(+0), \quad F \in \mathbf{F}.$$

Note that $J_1(F,h)$ is linear w.r.t. $F \in \mathbf{F}$ for every $h \in \mathrm{Lip}_1$, by definition. According to Theorems 2 and 3 of [25], for any fixed $h \in \mathrm{Lip}_1$, the least upper bound $\sup_{F\in\mathbf{F}} J_1(F,h)$ w.r.t. probability d.f $F$ satisfying two linear conditions (we can also fix the value $b^2 \ge 1$ of the second moment and then take the least upper bound w.r.t. all $b \ge 1$) coincides with that over the set of three-point distributions from $\mathbf{F}$. Since every three-point distribution has finite moments of all orders, the condition of finiteness of the second-order moments may be eliminated, so that

$$\sup_{F\in\mathbf{F}} J(F) = \sup_{h\in\mathrm{Lip}_1} \sup_{F\in\mathbf{F}_3} J_1(F,h),$$

where $\mathbf{F}_3$ is the space of all discrete probability d.f.s with at most three jumps and unit first moment. Furthermore, according to Hoeffding [26], the least upper bound $\sup_{F\in\mathbf{F}_3} J_1(F,h)$ w.r.t. discrete probability d.f.s $F$ with finite number of jumps and satisfying one moment condition is attained on two-point distributions, hence,

$$\sup_{F\in\mathbf{F}} J(F) = \sup_{h\in\mathrm{Lip}_1} \sup_{F\in\mathbf{F}_2} J_1(F,h) = \sup_{F\in\mathbf{F}_2} J(F),$$

where $\mathbf{F}_2$ is the space of all discrete probability d.f.s with at most two jumps and unit first moment. Therefore, to prove Equation (37), it suffices to show that $J(F) \le 0$ for every $F \in \mathbf{F}_2$

Let $F$ correspond to a two-point distribution $p\,\delta_u + (1-p)\,\delta_v$ with $u < v$ and $p \in [0,1)$. The condition $\int_{\mathbb{R}} x\, dF(x) = 1$ yields $u < 1 \le v$ and $v = (1-pu)/(1-p)$, so that there are only three possibilities:

**Case 1:** $u \le 0 < 1 \le v$ and $p \in [0,1)$. Then,

$$q = \mathbf{P}(X \le 0) = p, \quad b^2 = \mathbf{E}X^2 = \frac{pu^2 - 2pu + 1}{1-p}, \tag{38}$$

and, by definition of $F^e$ given in Equation (21), we have

$$F^e(x) = \begin{cases} 0, & \text{for } x \le u, \\ pu - px, & \text{for } u < x \le 0, \\ pu + (1-p)x, & \text{for } 0 < x \le v, \\ 1, & \text{for } x > v. \end{cases}$$

Observing that the difference $F(x) - F^e(x)$ has exactly one sign change at $x = p(1-u)/(1-p) = v-1 \in [0, v)$ and using Equation (20), after some elementary calculations, we get

$$\zeta_1(F, F^e) = \frac{1}{2}u^2p - up + \frac{1}{2}(1-u)^2\frac{p^2}{1-p} + \frac{1}{2}(1-p)\cdot 1,$$

and, hence,

$$J(F) = \zeta_1(F, F^e) - \frac{pu^2 - 2pu + 1}{2(1-p)} + p = 0,$$

which means that $J(F) = 0$ for arbitrary two-point probability distribution with unit first moment and a nonpositive atom. Expressing $u$ and $v$ in terms of $q$ and $b^2$ (see Equation (38)), we get Equation (35).

**Case 2:** $0 < u < 1 \le v$ and $p \in [0, u]$. Then, $q = \mathbf{P}(X \le 0) = 0$,

$$F^e(x) = \begin{cases} 0, & \text{for } x \le 0, \\ x, & \text{for } 0 < x \le u, \\ u + (1-p)(x-u), & \text{for } u < x \le v, \\ 1, & \text{for } x > v, \end{cases}$$

and by $F^e(x) - F(x) \ge 0$ for all $x \in \mathbb{R}$, we get $\zeta_1(F, F^e) = \frac{1}{2}u^2 + \frac{1}{2}(v-u)(u+1-2p) = 1 - \frac{1}{2}\mathbf{E}X^2$. Hence,

$$J(F) = \zeta_1(F, F^e) - \frac{1}{2}\mathbf{E}X^2 + q = 1 - \mathbf{E}X^2 \le 0,$$

since $\mathbf{E}X^2 \ge (\mathbf{E}X)^2 = 1$ by Jensen's inequality. The equality here and, hence, in Equation (34) is attained in the case of degenerate distribution $\delta_1$.

**Case 3:** $0 < u < 1 < v$ and $p \in (u, 1)$. Then, $q = 0$ and $F^e$ has the same form as in the previous case, but the function $F^e(x) - F(x)$ now has exactly one sign change at $x = p(1-u)/(1-p) = v - 1 \in (u, v)$, and, hence, $\zeta_1(F, F^e) = \frac{1}{2}u^2 + \frac{1}{2}(p-u)^2\frac{1}{1-p} + \frac{1}{2}(1-p)\cdot 1$. Thus,

$$J(F) = \zeta_1(F, F^e) - \frac{1}{2}\mathbf{E}X^2 + q = u^2 - p < 0,$$

since $u^2 < u < p$ in this case, and the equality in Equation (37) (and, hence, in Equation (34)) is not attained. □

**Remark 3.** *Analyzing the proof, one can make sure that Equation (34) admits a slight improvement:*

$$\zeta_1(\mathscr{L}(X), \mathscr{L}^e(X)) \le \frac{\mathbf{E}X^2}{2} - \mathbf{P}(X \le 0) - \mathbf{E}(1-X)^2\mathbb{1}_{(0,1]}(X)$$

*for any r.v. $X$ with $\mathbf{E}X = 1$ and finite second moment. The proof differs only by the appearance (subtraction) of an additional term $\int_{(0,1]}(1-x)^2\,dF(x)$ in definition in Equation (36) of $J(F)$, which is still linear w.r.t. $F$, and, hence, does not change the logic. One has only to check that the new $J(F)$ is nonpositive for two-point distributions. In Case 1, $J(F)$ is retained. In Cases 2 and 3, the additional term is of the form $p(1-u)^2$ and it can be made sure that this term does not affect the sign of $J(F)$.*

## 4. Stein's Method

Stein's method, first introduced in [27] for normal approximation, is a powerful technique that allows to estimate distances with $\zeta$-structure (see Equation (2)) between probability distributions and a fixed target distribution (of a r.v.) $Z$. A complete survey on Stein's method may be found, e.g., in [14]. Suppose that the distance $\zeta_{\mathcal{H}}$ is of the form given in Equation (2) for a specific class $\mathcal{H}$ of real-valued functions. As mentioned in the Introduction, this is the case for both uniform (Kolmogorov) and Kantorovich distances with $\mathcal{H} = \{\mathbb{1}_{(-\infty,a)}(\cdot) \mid a \in \mathbb{R}\}$ and $\mathcal{H} = \mathrm{Lip}_1$, respectively.

The first step of Stein's method is to construct the so-called *Stein operator* $\mathcal{A}$ in some space $\mathcal{F}$ of real functions, such that

$$\mathbf{E}\mathcal{A}f(Z) = 0 \quad \forall f \in \mathcal{F}. \tag{39}$$

The second step is to find the solution $f_h$ to the *Stein equation*

$$\mathcal{A}f_h(x) = h(x) - \mathbf{E}h(Z) \tag{40}$$

for every $h \in \mathcal{H}$. Once the solution is found, it becomes possible to estimate the distance between the distributions of $X$ and $Z$ as

$$\begin{aligned}\zeta_{\mathcal{H}}(X,Z) &= \sup_{h\in\mathcal{H}}\left|\int_{\mathbb{R}} h\,dF_X - \int_{\mathbb{R}} h\,dF_Z\right| = \sup_{h\in\mathcal{H}}\left|\int_{\mathbb{R}} h\,dF_X - \mathbf{E}h(Z)\right| = \\ &= \sup_{h\in\mathcal{H}}\left|\int_{\mathbb{R}} (h - \mathbf{E}h(Z)\,)\,dF_X\right| = \sup_{h\in\mathcal{H}}\left|\int_{\mathbb{R}} \mathcal{A}f_h\,dF_X\right| = \sup_{h\in\mathcal{H}}\left|\mathbf{E}\mathcal{A}f_h(X)\right|.\end{aligned} \tag{41}$$

The final estimate for $\zeta_{\mathcal{H}}(X,Z)$ is usually derived by bounding the latest expression in Equation (41) from above using the properties of the Stein operator $\mathcal{A}$ and those of the solutions $f_h$ to the Stein Equation (40).

It can be made sure that for $Z \overset{d}{=} \mathscr{E} \sim \mathrm{Exp}(1)$ the following operator satisfies Equation (39) on the space $\mathcal{F}$ of absolutely continuous functions with $\mathbf{E}|f'(\mathscr{E})| < +\infty$ and thus appears to be the Stein operator:

$$\mathcal{A}f(x) = f'(x) - f(x) + f(0). \tag{42}$$

Peköz and Röllin [13] found an explicit solution to Stein Equation (40) in this case:

$$f_h(x) = -e^x \int_x^{+\infty} \widetilde{h}(t)e^{-t}\,dt, \quad \text{where} \quad \widetilde{h}(t) = h(t) - \mathbf{E}h(\mathscr{E}), \tag{43}$$

for every $h$ with $\mathbf{E}|h(\mathscr{E})| < \infty$. Note that $f_h(0) = 0$.

The following theorem extends results of Peköz and Röllin [13] in Theorem 2.1 to distributions with no support constraints and provides estimates of the accuracy of the exponential approximation in terms of the Kantorovich distance characterizing the proximity to the equilibrium transform.

**Theorem 3.** *Let $X$ be a square integrable r.v. with $\mathbf{E}X = 1$ and $\mathscr{E} \sim \mathrm{Exp}(1)$. Then,*

$$\begin{aligned}\zeta_1(X,\mathscr{E}) &\le 2\,\zeta_1(\mathscr{L}(X),\mathscr{L}^e(X)),\\ \zeta_1(\mathscr{L}^e(X),\mathrm{Exp}(1)) &\le \zeta_1(\mathscr{L}(X),\mathscr{L}^e(X)),\end{aligned}$$

*where $\mathscr{L}^e(X)$ is the equilibrium transform of $\mathscr{L}(X)$.*

**Proof.** Let $f_h$ be defined by Equation (43). Then, by Equations (41), (42), and (25), we have

$$\zeta_1(X,\mathscr{E}) = \sup_{h\in\mathrm{Lip}_1}\left|\mathbf{E}\mathcal{A}f_h(X)\right| = \sup_{h\in\mathrm{Lip}_1}\left|\mathbf{E}f_h'(X) - \mathbf{E}f_h(X)\right| = \sup_{h\in\mathrm{Lip}_1}\left|\int_{\mathbb{R}} f_h'\,dF_X - \int_{\mathbb{R}} f_h'\,dF_X^e\right|$$

and

$$\begin{aligned}\zeta_1\left(\mathscr{L}^e(X),\mathrm{Exp}(1)\right) &= \sup_{h\in \mathrm{Lip}_1}\left|\int_{\mathbb{R}} h(x)\,dF_X^e(x) - \mathbf{E}h(\mathscr{E})\right| = \sup_{h\in \mathrm{Lip}_1}\left|\int_{\mathbb{R}} \tilde{h}(x)\,dF_X^e(x)\right| = \\ &= \sup_{h\in \mathrm{Lip}_1}\left|\int_{\mathbb{R}} \mathcal{A}f_h(x)\,dF_X^e(x)\right| = \sup_{h\in \mathrm{Lip}_1}\left|\int_{\mathbb{R}} f_h'(x)\,dF_X^e(x) - \int_{\mathbb{R}} f_h(x)\,dF_X^e(x)\right| = \\ &= \sup_{h\in \mathrm{Lip}_1}\left|\int_{\mathbb{R}} f_h(x)\,dF_X(x) - \int_{\mathbb{R}} f_h(x)\,dF_X^e(x)\right|.\end{aligned}$$

In Lemma 4.1 of [13] (see also Lemma 5.3 of [14]), it is proved that $f_h \in \mathrm{Lip}_1$ and $f_h' \in \mathrm{Lip}_2$ for $h \in \mathrm{Lip}_1$. This remark together with the observation that $\mathscr{L}(X)$ and $\mathscr{L}^e(X)$ have finite first moments immediately leads to the statement of the theorem. □

Less formally, Theorem 3 states that, if $\mathscr{L}(X)$ and $\mathscr{L}^e(X)$ are close, then so are $\mathscr{L}(X)$ and $\mathrm{Exp}(1)$, and, hence, may be regarded as the continuity theorem to the fixed-point property stated in Theorem 1(c).

## 5. Main Results

**Theorem 4.** *Let $X_1, X_2, \ldots$ be a sequence of independent square integrable random variables with $\mathbf{E}X_n = a \neq 0$ and $S_n := \sum_{i=1}^n X_i$ for $n \in \mathbb{N}$, $S_0 := 0$. Let $p \in (0,1)$, $N \sim \mathrm{Geom}(p)$, be independent of all $\{X_n\}$, $N_0 := N-1$, and $W := S_N/\mathbf{E}S_N = pS_N/a$, $W_0 := S_{N_0}/\mathbf{E}S_{N_0} = pS_{N_0}/(a(1-p))$ be normalized geometric random sums, $\mathscr{E} \sim \mathrm{Exp}(1)$. Then,*

$$\zeta_1(W,\mathscr{E}) \le \frac{2p}{|a|}\sum_{n=1}^{\infty}\mathbf{P}(N=n)\,\zeta_1(\mathscr{L}(X_n),\mathscr{L}^e(X_n)) \le p\left(\frac{\mathbf{E}X_N^2}{a^2} - 2\,\mathbf{P}(X_N \le 0)\right), \tag{44}$$

$$\zeta_1(W_0,\mathscr{E}) \le \frac{p}{1-p}\cdot\frac{\mathbf{E}X_N^2}{a^2}. \tag{45}$$

Before proceeding to the proof, we need the following auxiliary statement.

**Lemma 3.** *Under the conditions of Theorem 4, we have*

$$\zeta_1\left(\mathscr{L}(S_N),\mathscr{L}^e(S_N)\right) \le \sum_{n=1}^{\infty} p(1-p)^{n-1}\zeta_1\left(\mathscr{L}(X_n),\mathscr{L}^e(X_n)\right),$$

$$\zeta_1\left(\mathscr{L}(S_{N_0}),\mathscr{L}^e(S_{N_0})\right) \le \frac{\mathbf{E}X_N^2}{2|a|}.$$

**Proof.** Let $F_n$ be the d.f. of $X_n$, $n \in \mathbb{N}$. Then, according to Equation (20), Theorem 1(h), Tonelli's theorem, and an obvious fact that $\mathscr{L}(S_n) = \mathscr{L}(S_{n-1}) * \mathscr{L}(X_n)$, we have

$$\begin{aligned}\zeta_1(\mathscr{L}(S_N),\mathscr{L}^e(S_N)) &= \int_{\mathbb{R}}\left|F_{S_N}(x) - F_{S_N}^e(x)\right|dx \le \\ &\le \sum_{n=1}^{\infty} p(1-p)^{n-1}\int_{\mathbb{R}}\int_{\mathbb{R}}\left|F_n(x-s) - F_n^e(x-s)\right|dF_{S_{n-1}}(s)\,dx = \\ &= \sum_{n=1}^{\infty} p(1-p)^{n-1}\int_{\mathbb{R}}\int_{\mathbb{R}}\left|F_n(x-s) - F_n^e(x-s)\right|dx\,dF_{S_{n-1}}(s) = \\ &= \sum_{n=1}^{\infty} p(1-p)^{n-1}\zeta_1(\mathscr{L}(X_n),\mathscr{L}^e(X_n)),\end{aligned}$$

which proves the first claim of the lemma, and, similarly,

$$\zeta_1\left(\mathscr{L}(S_{N_0}),\mathscr{L}^e(S_{N_0})\right) \le \sum_{n=1}^{\infty} p(1-p)^{n-1}\int_{\mathbb{R}}\int_{\mathbb{R}}\left|\mathbb{1}_{(0,+\infty)}(x-s)-F_n^e(x-s)\right| dF_{S_{n-1}}(s)\,dx =$$
$$= \sum_{n=1}^{\infty} p(1-p)^{n-1}\int_{\mathbb{R}}\left|\mathbb{1}_{(0,+\infty)}(x)-F_n^e(x)\right| dx = \sum_{n=1}^{\infty} p(1-p)^{n-1}\zeta_1(\delta_0,\mathscr{L}^e(X_n)),$$

where $\delta_0$ denotes the Dirac delta-measure concentrated in 0. As can easily be seen from the definition of the equilibrium transform given in Equation (21),

$$\begin{aligned} &\text{if } a>0, \quad \text{then} \quad F^e(x)\le 0,\ x\le 0, \quad F(x)\le 1,\ x\ge 0,\\ &\text{if } a<0, \quad \text{then} \quad F^e(x)\ge 0,\ x\le 0, \quad F(x)\ge 1,\ x\ge 0, \end{aligned}$$

hence, we write

$$|\mathbb{1}_{(0,+\infty)}(x)-F_n^e(x)| = \begin{cases} F_n^e(x)\,\mathrm{sign}\,a, & x\le 0,\\ (1-F_n^e(x))\,\mathrm{sign}\,a, & x\ge 0, \end{cases}$$

and also using Equation (28), we obtain

$$\zeta_1\left(\delta_0,\mathscr{L}^e(X_n)\right) = \mathrm{sign}\,a\cdot\left(-\int_{-\infty}^{0}F_n^e(x)\,dx + \int_0^{+\infty}(1-F_n^e(x))\,dx\right) = \mathrm{sign}\,a\cdot\int_{\mathbb{R}} x\,dF_n^e(x) = \frac{\mathbf{E}X_n^2}{2|a|}.$$

The second claim of the lemma follows now by the total probability formula and independence conditions. □

**Proof of Theorem 4.** Due to the homogeneity of both the Kantorovich metric (Lemma 1(a)) and the equilibrium transform (Theorem 1(f)), without loss of generality, we can assume that $a=1$. The second inequality in Equation (44) is the implication of Theorem 2, thus it remains only to prove the first inequality in Equation (44) and the inequality in Equation (45). Indeed, by Theorems 3 and 1(f) and Lemmas 1 and 3, we have

$$\zeta_1(W,\mathscr{E}) \le 2\zeta_1(\mathscr{L}(W),\mathscr{L}^e(W)) =$$
$$= 2p\,\zeta_1(\mathscr{L}(S_N),\mathscr{L}^e(S_N)) \le 2p\sum_{n=1}^{\infty}\mathbf{P}(N=n)\,\zeta_1(\mathscr{L}(X_n),\mathscr{L}^e(X_n)),$$

and

$$\zeta_1(W_0,\mathscr{E}) \le 2\zeta_1\left(\mathscr{L}(W_0),\mathscr{L}^e(W_0)\right) =$$
$$= \frac{2p}{1-p}\zeta_1\left(\mathscr{L}(S_{N_0}),\mathscr{L}^e(S_{N_0})\right) \le \frac{p}{1-p}\mathbf{E}X_N^2. \quad \square$$

**Corollary 1.** *Under the conditions of Theorem 4 and* $\sup_n \mathbf{E}X_n^2 < \infty$, *we have*

$$\zeta_1(W,\mathscr{E}) \le \frac{2p}{|a|}\sup_n \zeta_1\left(\mathscr{L}(X_n),\mathscr{L}^e(X_n)\right) \le p\sup_n\left(\frac{\mathbf{E}X_n^2}{a^2} - 2\mathbf{P}(X_n\le 0)\right), \tag{46}$$
$$\zeta_1(W_0,\mathscr{E}) \le \frac{p}{(1-p)a^2}\sup_n \mathbf{E}X_n^2. \tag{47}$$

**Remark 4.** *The right-hand side of Equation (47) is no less than that of Equation (46) because of the factor* $\frac{1}{1-p} > 1$ *and the absence of the nonpositive term* $-2\mathbf{P}(X_n \le 0)$. *This result agrees with the intuition that* $W$ *may be closer to* $\mathscr{E}$ *than* $W_0$, *because* $S_N$ *contains a.s. one summand more than* $S_{N_0}$.

**Corollary 2.** *Under the conditions of Theorem 4, we have*

$$\zeta_2(W,\mathscr{E}) \le \frac{3p}{|a|}\sum_{n=1}^{\infty}\mathbf{P}(N=n)\,\zeta_1\big(\mathscr{L}(X_n),\mathscr{L}^e(X_n)\big) \le \frac{3p}{2}\left(\frac{\mathbf{E}X_N^2}{a^2} - 2\,\mathbf{P}(X_N\le 0)\right), \tag{48}$$

$$\zeta_2(W_0,\mathscr{E}) \le \frac{p}{1-p}\cdot\frac{3\,\mathbf{E}X_N^2}{2a^2}. \tag{49}$$

Recently, Korolev and Zeifman [28] obtained a bound similar to Equation (49), but with the constant factor of $1/2$ on the right-hand side instead of $3/2$, i.e., three times smaller. The estimate in Equation (48) is also worse than Kalashnikov's bound in Equation (10) obtained in the i.i.d. case and $\mathbf{E}X_1 = 1$, since Equation (10) with $s = 2$, by Theorem 3, yields

$$\zeta_2(W,\mathscr{E}) \le p\zeta_1(X_1,\mathscr{E}) \le 2p\zeta_1\big(\mathscr{L}(X_1),\mathscr{L}^e(X_1)\big),$$

while Equation (48) in the i.i.d. case with $\mathbf{E}X_1 = 1$ reduces to

$$\zeta_2(W,\mathscr{E}) \le 3p\zeta_1\big(\mathscr{L}(X_1),\mathscr{L}^e(X_1)\big),$$

which is 1.5 times greater.

**Proof.** Using subsequently Theorem 1(i,c), the triangle inequality for the Kantorovich metric, Theorem 3, and Lemma 3 together with the homogeneity of the Kantorovich distance and the equilibrium transform, we obtain

$$\zeta_2(W,\mathscr{E}) = \zeta_1\big(\mathscr{L}^e(W),\mathscr{L}^e(\mathscr{E})\big) = \zeta_1\big(\mathscr{L}^e(W),\mathscr{L}(\mathscr{E})\big) \le \zeta_1\big(\mathscr{L}^e(W),\mathscr{L}(W)\big) + \zeta_1(W,\mathscr{E}) \le$$
$$\le 3\,\zeta_1\big(\mathscr{L}^e(W),\mathscr{L}(W)\big) \le \frac{3p}{|a|}\sum_{n=1}^{\infty}\mathbf{P}(N=n)\,\zeta_1\big(\mathscr{L}(X_n),\mathscr{L}^e(X_n)\big).$$

Similarly,

$$\zeta_2(W_0,\mathscr{E}) \le 3\,\zeta_1\big(\mathscr{L}^e(W_0),\mathscr{L}(W_0)\big) \le \frac{3}{2}\cdot\frac{p}{1-p}\cdot\frac{\mathbf{E}X_N^2}{a^2}. \quad \square$$

To study the problem of the accuracy of the estimates obtained above in Equations (46) and (47), let us introduce the *asymptotically best constant* for the Kantorovich distance in the Rényi theorem for geometric random sums of i.i.d. r.v.s in a way similar to the definition of the asymptotically best constant [29] in the classical Berry–Esseen inequality (see also [3,30–35]):

$$C_{\text{AB}} := \sup_{\{X_n\}\sim\text{i.i.d.}:\ \mathbf{E}X_1\ne 0,\ \mathbf{E}X_1^2<\infty}\ \lim_{p\to+0}\zeta_1(W,\mathscr{E})\frac{(\mathbf{E}X_1)^2}{p\mathbf{E}X_1^2}, \tag{50}$$

which serves as a lower bound to the constant $C$ in the inequality

$$\zeta_1(W,\mathscr{E}) \le Cp\,\mathbf{E}X_1^2/(\mathbf{E}X_1)^2, \tag{51}$$

still if it is supposed to hold only for sufficiently small $p$. Similarly, define $C_{\text{AB}}^0$ for $W_0$. The inequality in Equation (46) (similarly, Equation (47)) trivially yields the validity of Equation (51) with $C = 1$ for all $p \in (0,1)$. Since

$$C \ge C_{\text{AB}},$$

it is easy to conclude that $C_{\text{AB}} \le 1$.

**Theorem 5.** *For the asymptotically best constants* $C_{\rm AB}$, $C^0_{\rm AB}$ *defined in Equation (50), for* $W$ *and* $W_0$ *we have*

$$C_{\rm AB} \geq 1/4, \quad C^0_{\rm AB} \geq 1/4.$$

**Proof.** Taking all $X_n := 1$, we get $\mathbf{E}X_n = \mathbf{E}X_n^2 = 1$ and $W = pN$, $W_0 := pN_0/(1-p)$, where $N \sim \mathrm{Geom}(p)$ and $N_0 := N-1$. To estimate $\zeta_1(W, \mathscr{E})$, we use the definition of the Kantorovich distance in Equation (19) and take $h(x) = \frac{1}{t}\sin(tx) \in \mathrm{Lip}_1$ as a test function, where $t \in \mathbb{R} \setminus \{0\}$ is the free parameter to be chosen later. Recalling the ch.f.s of the exponential and the geometric distributions, we obtain

$$\mathbf{E}h(\mathscr{E}) = \frac{1}{t}\Im\mathbf{E}e^{it\mathscr{E}} = \Im\frac{1}{t(1-it)} = \frac{1}{1+t^2},$$

$$\mathbf{E}h(W) = \mathbf{E}h(pN) = \frac{1}{t}\Im\mathbf{E}e^{itpN} = \frac{1}{t}\Im\left[\frac{pe^{itp}}{1-(1-p)e^{itp}}\right] =$$
$$= \frac{1}{t}\Im\left[\frac{pe^{itp}\left(1-(1-p)e^{-itp}\right)}{1+(1-p)^2-2(1-p)\cos(tp)}\right] = \frac{p\sin(tp)}{tp^2+2t(1-p)\left(1-\cos(tp)\right)},$$
$$\mathbf{E}h(W_0) = \mathbf{E}h\left(\frac{pN_0}{1-p}\right) = \frac{1}{t}\Im\mathbf{E}e^{itpN_0/(1-p)} = \frac{p(1-p)\sin\left(\frac{tp}{1-p}\right)}{tp^2+2t(1-p)\left(1-\cos\left(\frac{tp}{1-p}\right)\right)}.$$

Thus,

$$C_{\rm AB} \geq \limsup_{p\to+0}\sup_{t\neq 0}\frac{|\mathbf{E}h(W)-\mathbf{E}h(\mathscr{E})|}{p} \geq \sup_{t\neq0}\lim_{p\to+0}\left|\frac{\mathbf{E}h(W)-\mathbf{E}h(\mathscr{E})}{p}\right| =$$
$$= \sup_{t\neq0}\lim_{p\to+0}\left|\frac{p^3t^3+o(p^3)}{p^3t(t^2+1)^2+o(p^3)}\right| = \sup_{t\neq0}\frac{t^2}{(t^2+1)^2} = 1/4,$$

and, similarly,

$$C^0_{\rm AB} \geq \sup_{t\neq0}\lim_{p\to+0}\left|\frac{\mathbf{E}h(W_0)-\mathbf{E}h(\mathscr{E})}{p}\right| = \sup_{t\neq0}\frac{t^2}{(t^2+1)^2} = 1/4. \quad \square$$

Theorem 1(h) allows extending Theorem 4 to non-geometric random sums of independent random variables with arbitrary means of identical signs. Namely, the following statement holds.

**Theorem 6.** *Let* $X_1, X_2, \ldots$ *be a sequence of independent random variables, independent of all else, with*

$$a_n := \mathbf{E}X_n > 0, \quad b_n := \mathbf{E}X_n^2 < \infty, \quad n \in \mathbb{N},$$

*and* $S_n := \sum_{i=1}^n X_i$ *for* $n \in \mathbb{N}$, $S_0 := 0$. *Let* $N$ *be a* $\mathbb{N}_0$*-valued r.v.,*

$$A := \mathbf{E}S_N = \sum_{n=1}^{\infty} a_n\mathbf{P}(N \geq n) < \infty,$$

*and* $M$ *be a* $\mathbb{N}$*-valued r.v. with the distribution*

$$\mathbf{P}(M = m) = \frac{a_m}{A}\mathbf{P}(N \geq m), \quad m \in \mathbb{N}.$$

*Assume also that* $\mathbf{E}S_M < \infty$. *Then, with* $W := S_N/\mathbf{E}S_N = A^{-1}S_N$, *for any joint distribution* $\mathscr{L}(N,M)$, *we have*

$$\zeta_1(W,\mathscr{E}) \le 2A^{-1}\Big(\sup_n \mathbf{E}|X_n|\cdot\mathbf{E}|N-M| + \sum_{m\in\mathbb{N}} \mathbf{P}(M=m)\,\zeta_1\big(\mathscr{L}(X_m),\mathscr{L}^e(X_m)\big)\Big) \le \tag{52}$$

$$\le 2A^{-1}\left(\sup_n \mathbf{E}|X_n|\cdot\mathbf{E}|N-M| + \mathbf{E}\left(\frac{b_M}{2a_M} - a_M\cdot\mathbf{P}(X_M\le 0|M)\right)\right). \tag{53}$$

**Remark 5.** *If both expectations* $\mathbf{E}N$ *and* $\mathbf{E}M$ *are finite, then* $\mathbf{E}|N-M|$ *in Equations (52) and (53) can be replaced with* $\zeta_1(N,M)$.

**Remark 6.** *Theorem 6 reduces to ([13], Theorem 3.1) in the case of nonnegative* $\{X_n\}$ *and to Theorem 4, Equation (44), in the case of* $N\sim\mathrm{Geom}(p)$ *and identical* $a := \mathbf{E}X_n \ne 0$, $n\in\mathbb{N}$. *For shifted geometric* $N$, *i.e.,* $\mathbf{P}(N=n) = p(1-p)^n$, $n\in\mathbb{N}_0$, *under the assumptions of Theorem 4, Theorem 6 yields a bound*

$$\zeta_1(W_0,\mathscr{E}) \le \frac{p}{1-p}\left(2\sup_n\frac{\mathbf{E}|X_n|}{|a|} + \sum_{n\in\mathbb{N}}\mathbf{P}(N=n-1)\,\zeta_1\big(\mathscr{L}(X_n),\mathscr{L}^e(X_n)\big)\right) \le$$

$$\le \frac{p}{1-p}\left(\frac{\mathbf{E}X_{N+1}^2}{a^2} + 2\sup_n\frac{\mathbf{E}|X_n|}{|a|} - \mathbf{P}(X_{N+1}\le 0)\right),$$

*whose rightmost part is worse than the estimate in Equation (45), generally speaking (for example, in the i.i.d. case), since* $\mathbf{E}|X_n| \ge |a|$ *for all* $n\in\mathbb{N}$ *and* $\mathbf{P}(X_{N+1}\le 0)\le 1$.

**Proof of Theorem 6.** By Theorem 3 and homogeneity of the Kantorovich distance and the equilibrium transform (see Lemma 1(a) and Theorem 1(f)), we have

$$\zeta_1(W,\mathscr{E}) \le 2\zeta_1\big(\mathscr{L}(W),\mathscr{L}(W^e)\big) = 2A^{-1}\zeta_1\big(\mathscr{L}(S_N),\mathscr{L}^e(S_N)\big). \tag{54}$$

Let us bound $\zeta_1\big(\mathscr{L}(S_N),\mathscr{L}^e(S_N)\big)$ from above.

For a given joint distribution $\mathscr{L}(N,M)$, let $p_{nm} := \mathbf{P}(N=n, M=m)$, $n\in\mathbb{N}_0$, $m\in\mathbb{N}$. Denoting $S_{j,k} := \sum_{i=j}^k X_i$ for $j\le k$ and using the representation in Equation (20) and Theorem 1(h), we have

$$\zeta_1\big(\mathscr{L}(S_N),\mathscr{L}^e(S_N)\big) = \int_{\mathbb{R}}\left|F_{S_N}(x) - F^e_{S_N}(x)\right|dx = \int_{\mathbb{R}}\left|F_{S_N}(x) - F_{S_{M-1}} * F^e_{X_M}(x)\right|dx =$$

$$= \int_{\mathbb{R}}\left|\sum_{n\in\mathbb{N}_0, m\in\mathbb{N}} p_{nm}\left(F_{S_n}(x) - F_{S_{m-1}} * F^e_{X_m}(x)\right)\right|dx \le$$

$$\le \sum_{n,m} p_{nm}\int_{\mathbb{R}}\left|F_{S_n}(x) - F_{S_{m-1}} * F^e_{X_m}(x)\right|dx \le$$

$$\le \sum_{n<m} p_{nm}\int_{\mathbb{R}}\left|\mathbb{1}_{(0,+\infty)}(x) - F_{S_{n+1,m-1}} * F^e_{X_m}(x)\right|dx +$$

$$+ \sum_{n\ge m} p_{nm}\int_{\mathbb{R}}\left|F_{S_{m,n}}(x) - F^e_{X_m}(x)\right|dx.$$

Adding and subtracting $F_{S_{n+1,m}}(x)$ under the modulus sign in the integrands in the first sum (w.r.t. $n<m$) and $F_{X_m}(x)$ in the second one (w.r.t. $n\ge m$) and using further the triangle inequality and Lemma 1(b), we obtain

$$
\begin{aligned}
\zeta_1\big(\mathscr{L}(S_N),\mathscr{L}^e(S_N)\big) &\leq \sum_{n<m} p_{nm}\,\zeta_1(\delta_0,S_{n+1,m}) + \sum_{n\geq m} p_{nm}\,\zeta_1(S_{m+1,n},\delta_0)+\\
&\quad+\sum_{n,m} p_{nm}\,\zeta_1\big(\mathscr{L}(X_m),\mathscr{L}^e(X_m)\big) =\\
&= \sum_{n,m} p_{nm}\mathbf{E}\left|\sum_{i=(n\wedge m)+1}^{n\vee m} X_i\right| + \sum_{m\in\mathbb{N}} \mathbf{P}(M=m)\,\zeta_1\big(\mathscr{L}(X_m),\mathscr{L}^e(X_m)\big) \leq\\
&\leq \sup_i \mathbf{E}|X_i|\cdot\sum_{n,m} p_{nm}|n-m| + \sum_{m\in\mathbb{N}} \mathbf{P}(M=m)\,\zeta_1\big(\mathscr{L}(X_m),\mathscr{L}^e(X_m)\big)\\
&= \sup_i \mathbf{E}|X_i|\cdot\mathbf{E}|N-M| + \sum_{m\in\mathbb{N}} \mathbf{P}(M=m)\,\zeta_1\big(\mathscr{L}(X_m),\mathscr{L}^e(X_m)\big)\,.
\end{aligned}
$$

Substituting the latter bound into Equation (54) yields Equation (52). The bound in Equation (53) follows from Equation (52) by Theorem 2 (see also Remark 2). □

**Author Contributions:** Conceptualization, I.S.; methodology, I.S. and M.T.; formal analysis, I.S. and M.T.; investigation, I.S. and M.T.; writing—original draft preparation, I.S. and M.T.; writing—review and editing, I.S. and M.T.; supervision, I.S.; funding acquisition, and I.S. All authors have read and agreed to the published version of the manuscript.

**Funding:** The results of Sections 1–3 (including Theorem 1) were obtained under support by the Russian Science Foundation, project No. 18-11-00155. The rest of the study was funded by RFBR, project number 20-31-70054, and by the grant of the President of Russia No. MD–189.2019.1.

**Acknowledgments:** The authors would like to thank Professor Victor Korolev for the careful editing of the manuscript and to anonymous referee for a suggestion resulting in Theorem 6.

**Conflicts of Interest:** The authors declare no conflict of interest.

## Abbreviations

The following abbreviations are used in this manuscript:

| | |
|---|---|
| r.v. | random variable |
| i.i.d. | independent identically distributed |
| d.f. | distribution function |
| ch.f. | characteristic function |
| a.s. | almost sure |
| a.c. | absolute continuity, absolutely continuous |
| w.r.t. | with respect to |

## References

1. Zolotarev, V.M. Probability metrics. *Theory Probab. Appl.* **1984**, *28*, 278–302. [CrossRef]
2. Zolotarev, V.M. Ideal metrics in the problems of probability theory and mathematical statistics. *Austral. J. Statist* **1979**, *21*, 193–208. [CrossRef]
3. Zolotarev, V.M. *Modern Theory of Summation of Random Variables*; VSP: Utrecht, The Netherlands, 1997.
4. Rachev, S. *Probability Metrics and the Stability of Stochastic Models*; John Wiley ans Sons: Chichester, UK, 1991.
5. Mattner, L.; Shevtsova, I.G. An optimal Berry–Esseen type theorem for integrals of smooth functions. *Lat. Am. J. Probab. Math. Stat.* **2019**, *16*, 487–530. [CrossRef]
6. Cambanis, S.; Simons, G.; Stout, W. Inequalities for $Ek(X,Y)$ when the marginals are fixed. *Z. Wahrsch. Verw. Geb.* **1976**, *36*, 285–294. [CrossRef]
7. Solovyev, A.D. Asymptotic behaviour of the time of first occurrence of a rare event. *Engrgy Cybern.* **1971**, *9*, 1038–1048.
8. Brown, M. Error Bounds for Exponential Approximations of Geometric Convolutions. *Ann. Probab.* **1990**, *18*, 1388–1402. [CrossRef]

9. Kalashnikov, V.V.; Vsekhsvyatskii, S.Y. Metric estimates of the first occurrence time in regenerative processes. In *Stability Problems for Stochastic Models*; Lecture Notes in Mathematics; Springer: Berlin/Heidelberg, Germany, 1985; Volume 1155, pp. 102–130. [CrossRef]
10. Kruglov, V.M.; Korolev, V.Y. *Limit Theorems for Random Sums*; Moscow State University: Moscow, Russia, 1990. (In Russian)
11. Sugakova, E.V. Estimates in the Rényi theorem for differently distributed terms. *Ukr. Math. J.* **1995**, *47*, 1128–1134. [CrossRef]
12. Kalashnikov, V.V. *Geometric Sums: Bounds for Rare Events with Applications: Risk Analysis, Reliability, Queueing*; Mathematics and Its Applications; Springer: Dordrecht, The Netherlands, 1997.
13. Peköz, E.A.; Röllin, A. New rates for exponential approximation and the theorems of Rényi and Yaglom. *Ann. Probab.* **2011**, *39*, 587–608. [CrossRef]
14. Ross, N. Fundamentals of Stein's method. *Probab. Surv.* **2011**, *8*, 210–293, doi:10.1214/11-PS182. [CrossRef]
15. Hung, T.L. On the rate of convergence in limit theorems for geometric sums. *Southeast Asian J. Sci.* **2013**, *2*, 117–130.
16. Bogachev, V.I. Weak Convergence of Measures. *Am. Math. Soc.* **2018**, *234*.
17. Glivenko, V.I. *Stieltjes Integral*, 2nd ed.; URSS: Moscow, Russia, 2007.
18. Bogachev, V.I. *Measure Theory. Vol. I, II.*; Springer: Berlin, Germany, 2007.
19. Folland, G.B. *Real Analysis*, 2nd ed.; John Wiley & Sons: New York, NY, USA, 1999.
20. Feller, W. *An Introduction to the Probability Theory and Its Applications. Vol. II.*, 2nd ed.; John Wiley: New York, NY, USA, 1971.
21. Asmussen, W. *Applied Probability and Queues*; Springer: New York, NY, USA, 2003.
22. Lin, X.S., Integrated Tail Distribution. Encyclopedia of Actuarial Science. *Am. Cancer Soc.* **2006**. [CrossRef]
23. Harkness, W.L.; Shantaram, R. Convergence of a sequence of transformations of distribution functions. *Pacific J. Math.* **1969**, *31*, 403–415. [CrossRef]
24. Shantaram, R.; Harkness, W.L. On a certain class of limit distributions. *Ann. Math. Stat.* **1972**, *43*, 2067–2071. [CrossRef]
25. Mulholland, H.P.; Rogers, C.A. Representation Theorems for Distribution Functions. *Proc. Lond. Math. Soc.* **1958**, *s3-8*, 177–223. [CrossRef]
26. Hoeffding, W. The extrema of the expected value of a function of independent random variables. *Ann. Math. Stat.* **1955**, *26*, 268–275. [CrossRef]
27. Stein, C. A bound for the error in the normal approximation to the distribution of a sum of dependent random variables. In *Proceedings of the Sixth Berkeley Symposium on Mathematical Statistics and Probability, Volume 2: Probability Theory*; University of California Press: Berkeley, CA, USA, 1972; pp. 583–602.
28. Korolev, V.; Zeifman, A. Bounds for convergence rate in laws of large numbers for mixed Poisson random sums. *arXiv* **2020**, arXiv:2003.12495.
29. Esseen, C.G. A moment inequality with an application to the central limit theorem. *Skand. Aktuarietidskr.* **1956**, *39*, 160–170. [CrossRef]
30. Kolmogorov, A.N. Some recent works in the field of limit theorems of probability theory. *Bull. Mosc. Univ.* **1953**, *10*, 29–38. (In Russian)
31. Chistyakov, G.P. Asymptotically proper constants in the Lyapunov theorem. *J. Math. Sci.* **1999**, *93*, 480–483. [CrossRef]
32. Chistyakov, G.P. A new asymptotic expansion and asymptotically best constants in Lyapunov's theorem. I. *Theory Probab. Appl.* **2002**, *46*, 226–242. [CrossRef]
33. Chistyakov, G.P. A new asymptotic expansion and asymptotically best constants in Lyapunov's theorem. II. *Theory Probab. Appl.* **2002**, *46*, 516–522. [CrossRef]
34. Chistyakov, G.P. A new asymptotic expansion and asymptotically best constants in Lyapunov's theorem. III. *Theory Probab. Appl.* **2003**, *47*, 395–414. [CrossRef]
35. Shevtsova, I.G. On the asymptotically exact constants in the Berry–Esseen–Katz inequality. *Theory Probab. Appl.* **2011**, *55*, 225–252. [CrossRef]

mathematics

MDPI

Article

# Approximations in Performance Analysis of a Controllable Queueing System with Heterogeneous Servers

**Dmitry Efrosinin [1,2], Natalia Stepanova [3], Janos Sztrik [4,*] and Andreas Plank [1]**

1 Insitute for Stochastics, Johannes Kepler University, 4040 Linz, Austria; dmitry.efrosinin@jku.at (D.E.); andreasplank@gmx.net (A.P.)

2 Department of Information Technologies, Faculty of Mathematics and Natural Sciences, Peoples' Friendship University of Russia (RUDN University), 117198 Moscow, Russia

3 Laboratory N17, Trapeznikov Institute of Control Sciences of RAS, 117997 Moscow, Russia; natalia0410@rambler.ru

4 Department of Informatics and Networks, Faculty of Informatics, University of Debrecen, 4032 Debrecen, Hungary

* Correspondence: sztrik.janos@inf.unideb.hu

Received: 21 September 2020; Accepted: 9 October 2020; Published: 16 October 2020

**Abstract:** The paper studies a controllable multi-server heterogeneous queueing system where servers operate at different service rates without preemption, i.e., the service times are uninterrupted. The optimal control policy allocates the customers between the servers in such a way that the mean number of customers in the system reaches its minimal value. The Markov decision model and the policy-iteration algorithm are used to calculate the optimal allocation policy and corresponding mean performance characteristics. The optimal policy, when neglecting the weak influence of slow servers, is of threshold type defined as a sequence of threshold levels which specifies the queue lengths for the usage of any slower server. To avoid time-consuming calculations for systems with a large number of servers, we focus here on a heuristic evaluation of the optimal thresholds and compare this solution with the real values. We develop also the simple lower and upper bound methods based on approximation by an equivalent heterogeneous queueing system with a preemption to measure the mean number of customers in the system operating under the optimal policy. Finally, the simulation technique is used to provide sensitivity analysis of the heuristic solution to changes in the form of inter-arrival and service time distributions.

**Keywords:** heterogeneous servers; Markov decision process; policy-iteration algorithm; mean number of customers; decomposable semi-regenerative process

## 1. Introduction

The study of multi-server queueing systems in most cases assumes the servers to be homogeneous when the individual service rates are the same for all the servers in the system. However, in many real applications, the assumption of the homogeneity cannot be valid, e.g., a group of servers with different types of processors as a consequence of irregular system updates, nodes in telecommunication networks with links of different unequal capacities and availability, nodes in wireless systems serving different mobile users, peer-to-peer services for data streaming, file sharing and storage, where heterogeneous servers arrive and depart randomly, multi-processor systems with heterogeneous processor's attributes like a throughput and an electric energy consumption, etc. Moreover, in many cases the heterogeneous server system outperforms its homogeneous server counterpart. This reality leads to necessity to analyse multi-server queueing systems with heterogeneous servers. The assumption of the heterogeneity of servers does not automatically mean that the stochastic modelling of such a queueing system becomes more complex. If the customers can change the server to a faster one during a service, in other words, the service is with a preemption, this is a classic one-dimensional birth-and-death process, that can be analysed in a standard way. The task of analysing a heterogeneous system becomes much more complex with the assumption that the customer cannot change the server during a service time, i.e., service without any preemption. In this case, on the one hand, the dimension of the corresponding random process increases and on the other hand, a mechanism for allocation of customers between the servers must be introduced.

The systems with heterogeneous servers are mostly investigated with respect to heuristic allocation policies, e.g., allocation according to the fastest server first (FSF) policy or the randomly chosen server (RCS). The results dedicated to the heterogeneous systems operating under these policies and some approximations of such models can be found in papers of Alves et al. [1], Bilgen and Altintas [2], Melikov et al. [3], The question of how to allocate the customers between the heterogeneous servers in order to minimize the mean number of customers in the system was studied for the queueing system with two servers in terms of a Markov decision process (MDP), e.g., by Larsen [4], Larsen and Agrawala [5], who conjectured the optimality of threshold policy that functions as follows: the fastest server must be used whenever it is idle and the slower one must be used only if the number of customers in the queue exceeds some prespecified threshold level $q \geq 1$. Based on the MDP model, Lin and Kumar in [6] considered a similar problem and proved the optimality of a threshold policy. Simple proofs of corresponding results have later been given by Koole [7], Luh and Viniotis [8], Walrand [9] and Weber [10]. The problem of an optimal control of a two-server queueing system with failures was studied by Özkan and Kharoufeh [11]. The problem of the optimal control allocation in the systems with more than two servers were investigated by Armony and Ward [12], Efrosinin [13], Rosberg and Makowski [14], Viniotis and Ephremides [15]. Rykov in [16] gave evidence for certain monotonicity properties of an optimal policy in case of the mean number of customers minimization. The techniques to prove such results are based on monotonicity properties of the dynamic programming relative value function. The case of infinitely many servers was proposed by Shenker and Weinrib [17], where an asymptotic analysis of large heterogeneous queueing systems is performed.

As it was shown in [18,19], also taking into account the incompleteness of the theoretical proof noticed by Vericourt and Zhou in [20], the optimal allocation policy, which minimizes the mean number of customers in heterogeneous queueing system without preemption, belongs to a set of structural policies. According to this policy for the servers' enumeration (1) the first server is used whenever it is free and there is a waiting customer in the queue, while the empty server with a number $k+1$ must be occupied only if the first $k$ faster servers are busy and the number of customers in the queue reaches some threshold level $q_{k+1} \geq 1$. Numerical analysis shows that the threshold level $q_k$ in general case can have a very weak dependence of slower servers' states. Due to our observations, the optimal threshold may vary by at most 1 when the state of a slower server changes. Moreover, since this deviation has no influence on the mean number of customers in the system, it can be neglected. Hence the optimal allocation policy can be defined as a classic threshold one through a sequence of threshold levels $1 = q_1 \leq q_2 \leq \cdots \leq q_K < \infty$, that is the first $k$ servers must be occupied whenever there are $q$ customers in the queue and $q_k \leq q \leq q_{k+1} - 1$.

While there is a certain amount of work being done on heterogeneous systems, there are still many open questions related to the accurate and quick calculation of the optimal control policy and the resulting performance characteristics. Searching for optimal values $q_2, \ldots, q_K$ by a direct minimizing the mean number of customers in the system can be performed only for small $K$ by solving the system of difference equations for the steady-state probabilities or by means of a matrix geometric approach introduced by Neuts [21]. However, when $K$ is large, these methods become too complicated. For example, the involved in computation matrix sizes become infeasible large even for the moderate numbers of servers like $K \geq 4$, see e.g., [22]. To calculate the optimal threshold levels the MDP model and a policy iteration algorithm [23–25], which constructs a sequence of improved policies that converges to optimal one, can also be used. While this approach is a powerful tool for solving many optimization tasks, it has significant limitations on dimension of the model, number of states, convergence in a heavy traffic case due to processing time and memory requirements. The contribution of the paper is three-fold. First, we provide a simple heuristic solution (HS) for a sub-optimal policy in order to avoid the time-consuming search for the optimal one in case of an arbitrary number of servers. Second, we investigate the possibility to use the equivalent queueing system with a preemption and a threshold-based policy to evaluate the lower and upper boundaries for the optimal mean number of customers in the system without preemption. Third, we check by means of a simulation, whether the proposed heuristic solution for the optimal thresholds is insensitive to changes in the form of inter-arrival and service time distributions.

This paper is organized as follows. In Section 2 we discuss a queueing model, formulate the corresponding MDP and specify a policy-iteration algorithm used for evaluation the optimal threshold policy. Section 3 introduces a heuristic solution for the optimal threshold levels based on a simple discrete fluid approximation, that turn out to be nearly optimal. In Section 4 we propose approximations to calculate the lower and upper bounds for the mean number of customers in the system under the optimal allocation policy. In Section 5 the simulation is used to provide sensitivity analysis of the heuristic solution to changes in inter-arrival and service time distributions. Finally, we make some conclusions and remarks.

## 2. Mathematical Model and MDP Formulation

Consider an infinite-capacity $M/M/K$ queueing system with $K$ heterogeneous servers and one common queue, see Figure 1. The customers arrive to the system according to a homogeneous Poisson process with a rate $\lambda$. The $j$th server has an exponentially distributed service time with a rate $\mu_j$. The server $j$ is called an available server if it is idle. The service of customers is has no preemption, i.e., a customer being served on a server cannot change it. In this case a threshold-based policy defined below which is used for the customer allocation has sense. The inter-arrival and service times are assumed to be mutually independent. Assume that the servers are enumerated in a way

$$\mu_1 \geq \cdots \geq \mu_K. \tag{1}$$

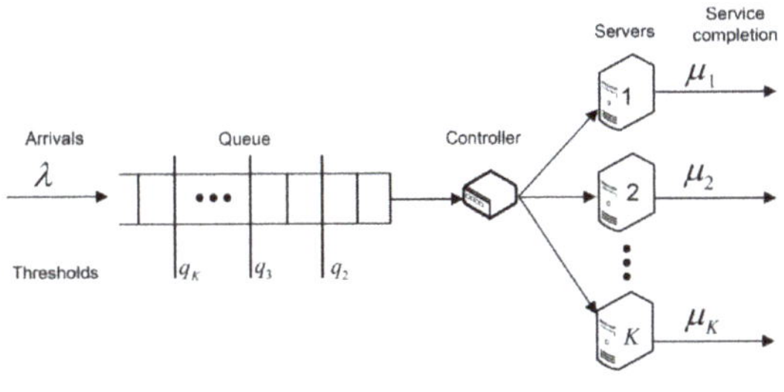


**Figure 1.** Controllable multi-server queueing system with heterogeneous servers.

The stability condition is obviously defined through the inequality

$$\lambda < \sum_{j=1}^{K} \mu_j. \tag{2}$$

The controller or decision maker has a full information about system states. It allocates customers between servers according to a control policy $f$ either to one of available servers or to queue at a new arrival and service completion epoch if it occurs with a nonempty queue. The system dynamics is common for the systems with one common queue and heterogeneous servers. At each arrival epoch the customer joins the queue and the controller can allocate the customer staying at the head of the queue to an available server $j$. At service completion epochs the controller may decide to allocate the customer from the head of nonempty queue to an available server or leave the customer in the queue. As it was mentioned above, the optimal control policy $f$, which minimizes the mean number of customers in the system with servers ordered according to (1), belongs to a set of threshold policies defined as a sequence of threshold levels

$$1 = q_1 \leq q_2 \leq \cdots \leq q_K < \infty. \tag{3}$$

According to this policy the first $k$ servers must be occupied whenever there are $q$ customers in the queue and $q_k \leq q \leq q_{k+1} - 1$ for $k = 1, \ldots, K-1$, and $q_k \leq q < \infty$ for $k = K$. For example, the policy $f$ for the system $M/M/5$ with $K = 5$ servers and thresholds $(q_1, q_2, \ldots, q_5) = (1, 3, 4, 5, 12)$ means that the fastest server is used whenever upon arrival of a customer it is free and there are $q$ customers in the queue with $1 \leq q \leq 2$. The first two servers are used when $q = 3$. The first three servers must be occupied whenever there are $q = 4$ customers in the queue, the first four customers are used when $5 \leq q \leq 11$. All servers must be used when the queue length exceeds the level $q \geq 12$. When the queue length drops below a specific threshold level, then the corresponding busy server remains idle after a service completion. As thresholds can take on different values, there are a huge number of admissible threshold policies. Hence the main goal is to calculate the optimal values for threshold levels $q_k$ and the minimized mean number of customers in the system.

We formulate the above optimization problem as a Markov decision process associated with a multi-dimensional continuous-time Markov chain

$$\{X(t)\}_{t \geq 0} = \{Q(t), D_1(t), \ldots, D_K(t)\}_{t \geq 0} \tag{4}$$

with a set of admissible actions $A = \{0, 1, \ldots, K\}$ with elements $a$, where $a = 0$ means the allocation of the customer to the queue and $a = j \neq 0$ – to the $j$th server. The term $Q(t) \in \mathbb{N}_0$ in (4) denotes the number of customers in the queue at time $t$, $D_j(t) \in \{0, 1\}$ – the state of server $j$ at time $t$, where

$$D_j(t) = \begin{cases} 0 & \text{if server } j \text{ is idle} \\ 1 & \text{if server } j \text{ is busy.} \end{cases}$$

For any fixed allocation policy $f$ we wish to guarantee that the process $\{X(t)\}_{t \geq 0}$ is an irreducible, positive recurrent Markov chain with a state space $E = \{x = (q(x), d_1(x), \ldots, d_K(x))\} \equiv \mathbb{N}_0 \times \{0, 1\}^K$ and infinitesimal generator $\Lambda^f$ which depend on the policy $f$. The notations $q(x)$ and $d_j(x)$ will be used further in the paper to specify the components of the vector state $x \in E$, where $q(x)$ denotes the queue length in state $x$ and $d_j(x)$ – the state of the $j$th server in state $x$. We use next the notations

$$J_0(x) = \{j : d_j(x) = 0\}, \ J_1(x) = \{j : d_j(x) = 1\}$$

to specify respectively a set of idle and busy servers in state $x$, $A(x) = J_0(x) \cup \{0\} \subseteq A$ the subset of admissible actions in state $x$ and $\mathbf{e}_j$ stands for a vector of dimension $K + 1$ with 1 in the $j$th position ($j = 0, 1, \ldots, K$) and 0 elsewhere.

For the ergodic Markov decision process a long-run average cost in the system per unit of time for the policy $f$ coincides with the corresponding assemble average, i.e.,

$$g^f = \limsup_{t \to \infty} \frac{1}{t} V^f(x, t) = \sum_{y \in E^f} l(y) \pi_y^f < \infty, \tag{5}$$

where $l(y) = q(y) + \sum_{j=1}^{K} d_j(y)$ in our model is a number of customers in state $y \in E$,

$$V^f(x,t) = \mathbb{E}^f\Big[\int_0^t \Big(Q(t) + \sum_{j=1}^{K} D_j(t)\Big)dt|X(0) = x\Big]$$

denotes the total average number of customers up to time $t$ given initial state is $x$ and $\pi_y^f = \mathbb{P}^f[X(t) = y]$ is a stationary state probability of the process under given policy $f$. The policy $f^*$ is said to be optimal when for $g^f$ defined in (5) we evaluate

$$g^* = \inf_f g^f = \min_{q_2,\ldots,q_K} g(q_2,\ldots,q_K). \tag{6}$$

One fruitful approach to finding optimal policy $f^*$ is through solving the Bellman's optimality equation, which in our case is of the form

$$Bv(x) = (\lambda + \sum_{j\in J_1(x)} \mu_j)v(x) + g, \tag{7}$$

where $B$ is a dynamic programming operator acting on a relative value function $v : E \to \mathbb{R}$ which indicates a transient effect of an initial state $x$ to the total average cost, and, according to Howard [23], the following asymptotic relation for the function $V^f(x,t)$ in case of a Markov-chain with one ergodic class holds,

$$V^f(x,t) = g^f t + v^f(x) + o(1),\ x \in E,\ t \to \infty. \tag{8}$$

The functions $v^f$ and $g^f$ further in the paper will be denoted by $v$ and $g$ without upper index $f$.

**Proposition 1.** *The Bellman's optimality Equation (7) is defined as follows*

$$\begin{aligned} Bv(x) = l(x) + \lambda \min_{a\in A(x)} v(x+\mathbf{e}_a) + \sum_{j\in J_1(x)} \mu_j v(x-\mathbf{e}_j)1_{\{q(x)=0\}} + \\ + \sum_{j\in J_1(x)} \mu_j \min_{a\in A(x-\mathbf{e}_j-\mathbf{e}_0)} v(x-\mathbf{e}_j-\mathbf{e}_0+\mathbf{e}_a)1_{\{q(x)>0\}}, \end{aligned} \tag{9}$$

*where the notation $1_{\{A\}}$ specifies the indicator function, which takes the value 1 if the event A holds, and 0 otherwise.*

**Proof.** According to [26], the behaviour of the function $V(x,t)$ in the interval $[t, t+dt)$ by letting $t \to \infty$ and taking into account the asymptotic relation (8) can be represented as a system of linear equations, which in general case is of the form

$$v(x) = \min_a \Big\{\frac{1}{\lambda_x(a)}\Big[c(x) + \sum_{y\neq x} \lambda_{xy}(a)v(y) - g\Big]\Big\}.$$

Evaluating these equations for analyzed queueing system and taking into account the transition rates of the specified Markov decision model we get

$$v(x) = \frac{1}{\lambda + \sum_{j\in J_1(x)} \mu_j}[Bv(x) - g].$$

The relation for $Bv(x)$ contains the term $l(x)$ specifying a number of customers in state $x \in E_X$, the second term represents the changing of the state accompanying with a new arrival which occurs with a rate $\lambda$. The third and the fourth terms represent transitions due to service completions at server $j$ with a rate $\mu_j$ by en empty and non-empty queue respectively. □

To generate a data-set for the queueing system under study which includes optimal threshold levels and corresponding values of the system parameters the policy-iteration Algorithm 1 is used. For numerical results the truncated equivalent system with a buffer size $W$ is considered. The algorithm consists of two main steps: policy evaluation and policy improvement. In the first step, for a given control policy $f$ the system of linear equations for the relative value function $v(x), x \in E \setminus \{(0,0,\ldots,0)\}$ must be solved together with an equation $g = \lambda v(\mathbf{e}_1)$. In the second step, the obtained in the first step relative function is used to improve the current policy. The algorithm stops when a new policy coincides with a previous one. As an initial policy the FSF allocation policy is used.

**Algorithm 1** Policy-iteration algorithm

1: **procedure** PIA($K, W, \lambda, \mu_j, j = 1,2,\ldots,K$)

2: $\quad f^{(0)}(x) = \text{argmax}_{j \in J_0(x)} \{\mu_j\}$ ▷ Initial policy

3: $\quad n \leftarrow 0$

4: $\quad g^{(n)} \leftarrow \lambda v^{(n)}(\mathbf{e}_1)$ ▷ Policy evaluation

5: $\quad$ **for** $x = (0,1,0,\ldots,0)$ **to** $(N,1,1,\ldots,1)$ **do**

6:

$$\begin{aligned} v^{(n)}(x) \leftarrow \frac{1}{\lambda + \sum_{j \in J_1(x)} \mu_j} \Big[ & l(x) - g^{(n)} + \lambda v^{(n)}(x + \mathbf{e}_{f^{(n)}(x)}) \\ & + \sum_{j \in J_1(x)} \mu_j v^{(n)}(x - \mathbf{e}_j) 1_{\{q(x)=0\}} \\ & + \sum_{j \in J_1(x)} \mu_j v^{(n)}(x - \mathbf{e}_j - \mathbf{e}_0 + \mathbf{e}_{f^{(n)}(x - \mathbf{e}_j - \mathbf{e}_0)}) 1_{\{q(x)>0\}} \Big] \end{aligned}$$

7: $\quad$ **end for**

8: ▷ Policy improvement

$$f^{(n+1)}(x) \leftarrow \text{argmin}_{a \in A(x)} v^{(n)}(x + \mathbf{e}_a)$$

9: $\quad$ **if** $f^{(n+1)}(x) \leftarrow f^{(n)}(x), x \in E$ **then return** $f^{(n+1)}(x), v^{(n)}(x), g^{(n)}$

10: $\quad$ **else** $n \leftarrow n+1$, **go to step 4**

11: $\quad$ **end if**

12: **end procedure**

We convert by implementing the Algorithm 1 the $K+1$-dimensional state space $E$ of the Markov decision process ordered in a certain way to a one-dimensional equivalent state space $\mathbb{N}_0$, $\Delta : E \to \mathbb{N}_0$, for state $x = (q(x), d_1(x), \ldots, d_K(x)) \in E$,

$$\Delta(x) = q(x)2^K + \sum_{i=1}^{K} d_i(x) 2^{i-1}. \tag{10}$$

Therefore, in one-dimensional case the changing of the state $x$ due to adding or removing a customer from the queue and due to occupation or departure of a customer from the $j$th server can be respectively represented in the form,

$$\begin{aligned} \Delta(x \pm \mathbf{e}_0) &= (q(x) \pm 1)2^K + \sum_{i=1}^{K} d_i(x) 2^{i-1} = \Delta(x) \pm 2^K, \\ \Delta(x \pm \mathbf{e}_j) &= q(x)2^K + \sum_{i=1}^{K} d_i(x) 2^{i-1} \pm 2^{j-1} = \Delta(x) \pm 2^{j-1}, \; j = 1,2,\ldots,K. \end{aligned}$$

For further details about derivation of the dynamic programming equation needed to evaluate the optimal policy the interested readers are referred to [13]. The infinite buffer queueing system is approximated by a finite buffer equivalent system in such a way that the loss probability does not exceed some specified small number $\varepsilon > 0$.

**Remark 1.** *For the bounded buffer size $W$ the number of states is*

$$|E| = 2^K(W+1).$$

*If the queue length $q \geq q_K$, all servers must be busy and the system behaves like a $M/M/1$ queueing system with a service rate $\sum_{j=1}^{K} \mu_j$. The stationary state probabilities $\pi_{(q,1,\dots,1)}, q \geq q_K$, satisfy the difference equation*

$$\lambda \pi_{(q-1,1,\dots,1)} - \Big(\lambda + \sum_{j=1}^{K} \mu_j\Big) \pi_{(q,1,\dots,1)} + \sum_{j=1}^{K} \mu_j \pi_{(q+1,1,\dots,1)} = 0,$$

*which has a solution in a geometric form, $\pi_{(q,1,\dots,1)} = \pi_{(q_K,1,\dots,1)} \rho^{q-q_K}, q \geq q_K$. For details and theoretical substantiation see e.g., [27]. The threshold level $q_K$ can be estimated using HS (11). The buffer size $W$ is chosen in such a way that it satisfies the condition for the loss probability*

$$\sum_{q=W}^{\infty} \pi_{(q,1,\dots,1)} = \pi_{q_K} \sum_{q=W}^{\infty} \rho^{q-q_K} \leq \sum_{q=W}^{\infty} \rho^{q-q_K} = \frac{\rho^{W-q_K}}{1-\rho} < \varepsilon,$$

*where $\rho = \frac{\lambda}{\sum_{j=1}^{K} \mu_j}$. After simple algebra it implies*

$$W > \frac{\log \varepsilon(1-\rho)}{\log(\rho)} + q_K.$$

The algorithm was implemented in C++ and tested for model problems up to 10 servers and a queue of size $W = 100$. It shows matching results to the proposed heuristic solution but is only viable for relative small number of servers. For system with 100 servers the maximum number of states would be in the order of $2^{100}$ which makes a reasonable usage of the policy-iteration algorithm impossible.

**Example 1.** *Consider the system $M/M/5$ with $K = 5$ and $\lambda = 15$. The service rates take the following values: $(\mu_1, \mu_2, \mu_3, \mu_4, \mu_5) = (20, 8, 4, 2, 1)$. The buffer size is $W = 80$ which for $\varepsilon = 0.0001$ guaranties that $W > \frac{\log 0.0001(1-14/36)}{\log(14/36)} + q_5 = 22.2734$, where $q_5 = 12$ is evaluated by (11). The table of evaluated control actions $f(x)$ for selected system states $x$ is of the form:*

| *System state x* | *Queue length q(x)* | | | | | | | | | | | | | |
|---|---|---|---|---|---|---|---|---|---|---|---|---|---|---|
| $d = (d_1, d_2, d_3, d_4, d_5)$ | 0 | 1 | 2 | 3 | 4 | 5 | 6 | 7 | 8 | 9 | 10 | 11 | 12 | ... |
| *(0,*,*,*,*)* | **1** | 1 | 1 | 1 | 1 | 1 | 1 | 1 | 1 | 1 | 1 | 1 | 1 | 1 |
| *(1,0,*,*,*)* | 0 | 0 | **2** | 2 | 2 | 2 | 2 | 2 | 2 | 2 | 2 | 2 | 2 | 2 |
| *(1,1,0,*,*)* | 0 | 0 | 0 | **3** | 3 | 3 | 3 | 3 | 3 | 3 | 3 | 3 | 3 | 3 |
| *(1,1,1,0,*)* | 0 | 0 | 0 | 0 | **4** | 4 | 4 | 4 | 4 | 4 | 4 | 4 | 4 | 4 |
| *(1,1,1,1,0)* | 0 | 0 | 0 | 0 | 0 | 0 | 0 | 0 | 0 | 0 | 0 | **5** | 5 | 5 |
| *(1,1,1,1,1)* | 0 | 0 | 0 | 0 | 0 | 0 | 0 | 0 | 0 | 0 | 0 | 0 | 0 | 0 |

*Threshold levels $q_k$, $k = 1, \dots, K = 5$, can be evaluated by comparing the optimal actions $f(q, \underbrace{1, \dots, 1}_{k-1}, \underbrace{0, \dots, 0}_{K-k+1}) < f(q+1, \underbrace{1, \dots, 1}_{k-1}, \underbrace{0, \dots, 0}_{K-k+1})$ for $q = 0, \dots, W-1$. In this example the optimal policy $f^*$ is defined here through a sequence of threshold levels $(q_2, q_3, q_4, q_5) = (3, 4, 5, 12)$ and $g^* = 4.92897$.*

## 3. Heuristic Solution

As it was mentioned above, the policy iteration algorithm has restrictions on dimension of the model, number of states, convergence in a heavy traffic case. In this section we derive a heuristic solution (HS) to estimate threshold levels $q_k, k = 2, \dots, K$, for the arbitrary $K$ using a simple discrete fluid approximation $Q(t) - Q\left(t + \frac{1}{\sum_{j=1}^{k-1} \mu_j - \lambda}\right) = 1, t = 0, \frac{1}{\sum_{j=1}^{k-1} \mu_j - \lambda}, \dots, \frac{q_k - 1}{\sum_{j=1}^{k-1} \mu_j - \lambda}$, for the queue length at time $t$, as illustrated in Figure 2.

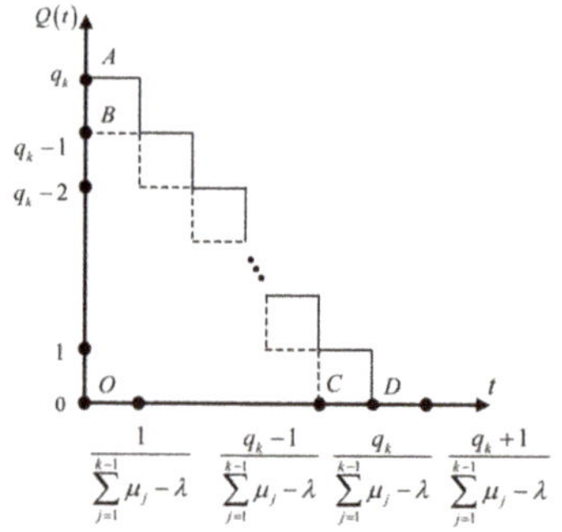


**Figure 2.** Fluid approximation.

We now explain how this fluid model can be employed for our aim. Assume that $q_k$ is an optimal threshold to allocate the customer to server $k$ in state $(q_k - 1, \underbrace{1, \dots, 1}_{k-1}, \underbrace{0, \dots, 0}_{K-k+1})$, where the first $k-1$ servers are busy. Now we compare the queues of the system given initial state is $x_0 = (q_k, \underbrace{1, \dots, 1}_{k-1}, 0, \underbrace{0, \dots, 0}_{K-k})$, where the $k$th server is not used for a new customer, and $y_0 = (q_k - 1, \underbrace{1, \dots, 1}_{k-1}, 1, \underbrace{0, \dots, 0}_{K-k})$, where the $k$th server is occupied by a waiting customer. It is assumed that the stability condition holds. In Figure 2, the queue lengths are labeled by $A = q_k$ and $B = q_k - 1$. If the queue dynamics corresponded to the deterministic fluid, it would decrease at the rate $\sum_{j=1}^{k-1} \mu_j - \lambda$. When this rate is maintained until the queue is empty, it occurs respectively at points $D = \frac{q_k}{\sum_{j=1}^{k-1} \mu_j - \lambda}$ and $C = \frac{q_k - 1}{\sum_{j=1}^{k-1} \mu_j - \lambda}$. The total holding times of customers in a queue with lengths $q_k$ and $q_k - 1$ are equal obviously to the areas

$$F_{AOD} = \frac{q_k(q_k+1)}{2} \cdot \frac{1}{\sum_{j=1}^{k-1} \mu_j - \lambda} \quad \text{and} \quad F_{BOC} = \frac{q_k(q_k-1)}{2} \cdot \frac{1}{\sum_{j=1}^{k-1} \mu_j - \lambda}$$

of triangles $AOD$ and $BOC$. The mean service time of customers by first $k-1$ busy servers until the queue is empty starting from state $x_0$ is equal to

$$q_k \left( \frac{1}{\mu_1} \frac{\mu_1}{\sum_{j=1}^{k-1} \mu_j} + \dots + \frac{1}{\mu_{k-1}} \frac{\mu_{k-1}}{\sum_{j=1}^{k-1} \mu_j} \right) = q_k \frac{k-1}{\sum_{j=1}^{k-1} \mu_j},$$

where $\frac{\mu_i}{\sum_{j=1}^{k-1} \mu_j}$ is a probability to be served by the $i$th server, and starting from state $y_0$—is equal to $(q_k - 1) \frac{k-1}{\sum_{j=1}^{k-1} \mu_j}$.

According to a specified deterministic fluid schema we formulate

**Proposition 2.** *The optimal thresholds $q_k, k = 2, \dots, K$, are defined by*

$$q_k \approx \hat{q}_k = \min\Big\{ \hat{q}_{k-1}, \Big\lfloor \Big( \sum_{j=1}^{k-1} \mu_j - \lambda \Big) \Big( \frac{1}{\mu_k} - \frac{k-1}{\sum_{j=1}^{k-1} \mu_j} \Big) \Big\rfloor + 1 \Big\}. \tag{11}$$

**Proof.** Denote by $V(x)$ the overall average holding time of customers until the system is empty given initial state is $x \in E$. The decision to perform the allocation to the $k$th server in state $(q_k - 1, \underbrace{1, \dots, 1}_{k-1}, \underbrace{0, \dots, 0}_{K-k+1})$ must lead to a reduction of the overall holding time under fluid schema, i.e.,

$$V(x_0) - V(y_0) > 0. \tag{12}$$

where

$$V(x_0) = F_{AOD} + q_k \frac{k-1}{\sum_{j=1}^{k-1} \mu_j} + V(0, \underbrace{1, \dots, 1}_{k-1}, \underbrace{0, \dots, 0}_{K-k+1}), \tag{13}$$

$$\begin{aligned} V(y_0) &= \frac{1}{\mu_k} + V(q_k - 1, \underbrace{1, \dots, 1}_{k-1}, 0, \underbrace{0, \dots, 0}_{K-k}) \\ &= \frac{1}{\mu_k} + F_{BOC} + (q_k - 1) \frac{k-1}{\sum_{j=1}^{k-1} \mu_j} + V(0, \underbrace{1, \dots, 1}_{k-1}, \underbrace{0, \dots, 0}_{K-k+1}). \end{aligned}$$

After substitution of (13) into (12) and some simple manipulations we get that the heuristic solution for the optimal threshold $q_k$ is defined then as the integer larger then 1 and the smallest integer (11) satisfying the inequality (12). □

**Example 2.** *Consider a queueing system from previous example for $K = 5$. We generate a data-set $S$ in form of a list*

$$S = \Big\{ (\lambda, \mu_1, \dots, \mu_K) \to (q_2, \dots, q_K) : \lambda \in [1, 45], \mu_1, \dots, \mu_K \in [1, 40],\ \lambda < \sum_{j=1}^{K} \mu_j, \mu_1 \geq \dots \geq \mu_K \Big\}. \tag{14}$$

*and evaluate with HS for the corresponding thresholds $q_k, k = 1, \dots, K$. Confusion matrices in Figure 3 visualize the performance of proposed heuristics respectively for the threshold levels $(q_2, q_3, q_4, q_5)$. Each row of these matrices represents the instances in a predicted value while each column represents the instances in an actual value. We notice the heavily diagonally dominant matrices that indicates a very good classification. This fact is confirmed also by overall accuracies. Such metrics describe the closeness of the heuristic measurements to a real threshold value and are calculated through the ratio of correct predictions to total predictions. Calculations of the overall accuracies as well as the accuracies for results with an acceptable deviation of threshold values by ±1 from the real value are summarized in Table 1.*

**Table 1.** Accuracy for prediction with heuristic solution (HS).

| **HS** | $q_2$ | $q_3$ | $q_4$ | $q_5$ |
|---|---|---|---|---|
| Accuracy | 0.8430 | 0.8778 | 0.7899 | 0.6282 |
| Accuracy ±1 | 0.9861 | 0.9884 | 0.9871 | 0.9769 |

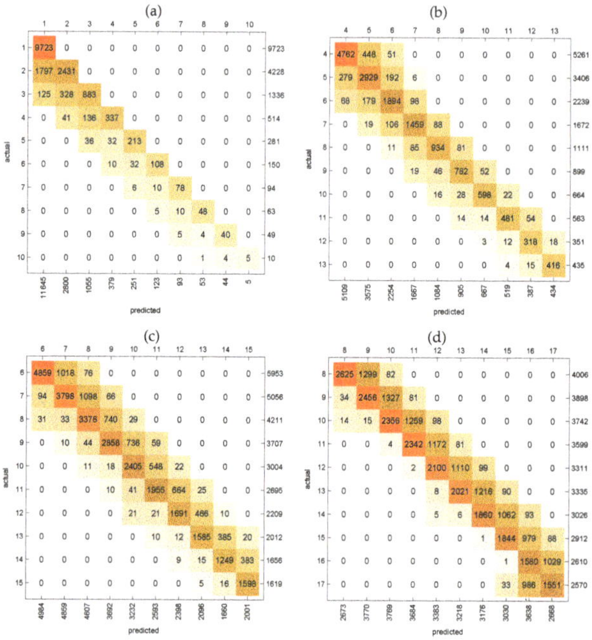


**Figure 3.** Confusion matrices (**a–d**) for prediction of $q_2, q_3, q_4$ and $q_5$ using HS.

**Example 3.** *Consider the queueing system $M/M/K$ with a different number of servers $K = 2, 3, \dots, 8$. Service rates take the values as given in Table 2.*

**Table 2.** Service rates $\mu_k, k = 1, \dots, K$.

| $\mu_k$ \ $K$ | 2 | 3 | 4 | 5 | 6 | 7 | 8 |
|---|---|---|---|---|---|---|---|
| $\mu_1$ | 34 | 32 | 28 | 20 | 18 | 16 | 14 |
| $\mu_2$ | 1 | 2 | 4 | 8 | 8 | 8 | 6 |
| $\mu_3$ | - | 1 | 2 | 4 | 4 | 4 | 5 |
| $\mu_4$ | - | - | 1 | 2 | 2 | 3 | 4 |
| $\mu_5$ | - | - | - | 1 | 2 | 2 | 2 |
| $\mu_6$ | - | - | - | - | 1 | 1 | 2 |
| $\mu_7$ | - | - | - | - | - | 1 | 1 |
| $\mu_8$ | - | - | - | - | - | - | 1 |

*Table 3 lists values of optimal thresholds $q_k$ and corresponding heuristic solutions $\hat{q}_k$. As we see it, the maximum deviation of the optimal thresholds from the heuristic solution is 1 independently of the number of servers.*

**Table 3.** Threshold values $q_k$ and $\hat{q}_k, k = 2, \dots, K$.

| $(q_k, \hat{q}_k)$ \ $K$ | 2 | 3 | 4 | 5 | 6 | 7 | 8 |
|---|---|---|---|---|---|---|---|
| $(q_2, \hat{q}_2)$ | (24,24) | (11,11) | (5,4) | (1,1) | (1,1) | (1,1) | (1,1) |
| $(q_3, \hat{q}_3)$ | - | (23,23) | (10,10) | (4,4) | (3,3) | (3,3) | (2,2) |
| $(q_4, \hat{q}_4)$ | - | - | (22,22) | (9,9) | (8,9) | (4,5) | (2,2) |
| $(q_5, \hat{q}_5)$ | - | - | - | (21,22) | (8,9) | (8,8) | (7,7) |
| $(q_6, \hat{q}_6)$ | - | - | - | - | (20,21) | (19,20) | (7,8) |
| $(q_7, \hat{q}_7)$ | - | - | - | - | - | (19,20) | (19,19) |
| $(q_8, \hat{q}_8)$ | - | - | - | - | - | - | (19,20) |

*The large number of numerical experiments carried out using the policy-iteration algorithm and simulations allows us to conclude that deviations of certain thresholds by 1 have practically no effect on the value of the minimised function. Thus, we believe that the proposed heuristic solution is effective for an arbitrary number of servers.*

## 4. Simple Bounds for the Optimal Mean Number of Customers in the System

As established in previous section, the estimation of the optimal threshold policy is possible by means of a simple heuristic solution. Nevertheless, with this knowledge it is quite complicated to calculate the optimal mean number of customers in the system with a high number of servers. A possible solution for this problem consists in construction a proper approximation of the original system with a preemption by an equivalent system without preemption. In this case a multidimensional Markov-chain can be described by an one-dimensional process. In this section we develop approximations for the low $\bar{L}_l$ and upper $\bar{L}_u$ bounds for the optimal gain function $g = \bar{L}$, $\bar{L}_l \leq \bar{L} \leq \bar{L}_u$.

To calculate the lower bound $\bar{L}_l$ we use a heterogeneous queueing system with a preemption and threshold-based control policy denoted by $S_l$ Further define by $\{Y_l(t)\}_{t\geq 0}$ the corresponding continuous-time Markov chain with a state space $E_l = \{y : y \in \mathbb{N}_0\}$ describing the number of customers in the system. The state transition diagram of this system is illustrated in Figure 4.

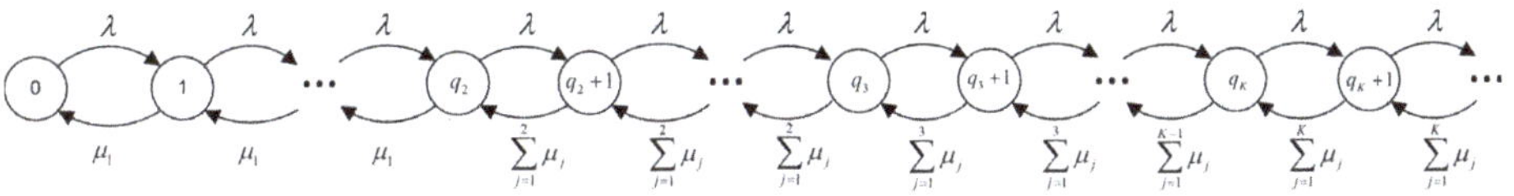

**Figure 4.** The state transition diagram for the queueing system $S_l$.

The optimal threshold levels $q_k, k = 2, \ldots, K$ are calculated using the heuristic solution (11). Obviously, since the customer being served in a slower server can change it as the faster one becomes empty, the mean number of customers in the system must be lower comparing to an original queue. The steady-state probabilities $\pi_y = \lim_{t\to\infty} \mathbb{P}[Y_l(t) = y]$ obviously exist under the stability condition (2).

**Proposition 3.** *The steady-state probabilities $\pi_y$ of the Markov chain $\{Y_l(t)\}_{t\geq 0}$ are given by*

$$\pi_0 = \left[1 + \sum_{y=1}^{q_2}\left(\frac{\lambda}{\mu_1}\right)^y + \sum_{k=3}^{K}\sum_{y=q_{k-1}}^{q_k}\left(\frac{\lambda}{\sum_{j=1}^{k-1}\mu_j}\right)^{y-q_{k-1}} \cdot \prod_{i=1}^{k-2}\left(\frac{\lambda}{\sum_{j=1}^{i}\mu_j}\right)^{q_{i+1}-q_i} + \right.$$

$$\left. + \prod_{i=1}^{K-1}\left(\frac{\lambda}{\sum_{j=1}^{i}\mu_j}\right)^{q_{i+1}-q_i} \cdot \frac{\lambda}{\sum_{j=1}^{K}\mu_j - \lambda}\right]^{-1},$$

$$\pi_y = \begin{cases} \pi_0 \cdot \left(\frac{\lambda}{\mu_1}\right)^y, & 1 \leq y \leq q_2 \\ \pi_0 \cdot \prod_{i=1}^{k-2}\left(\frac{\lambda}{\sum_{j=1}^{i}\mu_j}\right)^{q_{i+1}-q_i} \cdot \left(\frac{\lambda}{\sum_{j=1}^{k-1}\mu_j}\right)^{y-q_{k-1}}, & q_{k-1} \leq y \leq q_k, \; 3 \leq k \leq K \\ \pi_0 \cdot \prod_{i=1}^{K-1}\left(\frac{\lambda}{\sum_{j=1}^{i}\mu_j}\right)^{q_{i+1}-q_i} \cdot \left(\frac{\lambda}{\sum_{j=1}^{K}\mu_j}\right)^{y-q_K}, & y \geq q_K + 1 \end{cases}$$

**Proof.** The proposition follows directly from the properties of the ergodic birth-and-death process $\{Y_l(t)\}_{t\geq 0}$ [28]. □

From the probabilities $\pi_y$ it is possible to derive the performance measures of the system, e.g., the mean number of customers in the system $\bar{L}$ and the mean number of customers in the queue $\bar{Q}$.

**Corollary 1.** *The mean number of customers in the system $S_l$ satisfies the relation*

$$\bar{L}_l = \sum_{y=0}^{\infty} \pi_y = \sum_{y=0}^{q_K} \pi_y + \frac{\lambda(\sum_{j=1}^{K} \mu_j + (\sum_{j=1}^{K} \mu_j - \lambda) q_K)}{(\sum_{j=1}^{K} \mu_j - \lambda)^2} \pi_{q_K}. \tag{15}$$

The upper bound $\bar{L}_u$ for the optimal mean number of customers in the system can be obtained from an equivalent system under the FSF policy, see a state transition diagram in Figure 5, where $q_k = 1$ for $k = 1, \ldots, K$.

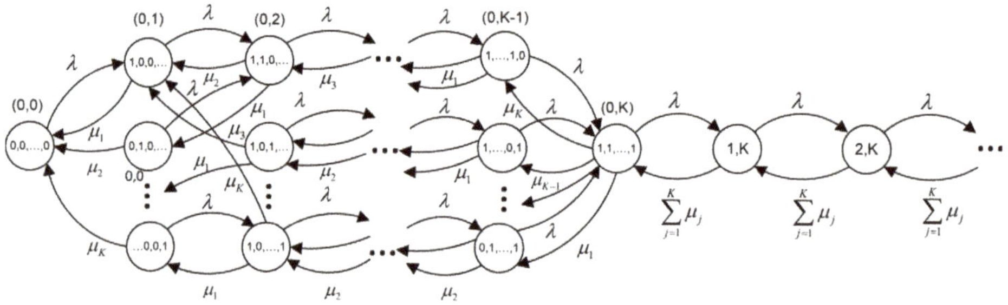

**Figure 5.** The state transition diagram for the heterogeneous queueing system with the fastest server first (FSF) policy.

In this diagram the group of states with a certain number of busy servers are labeled by $(q, \sum_{j=1}^{K} d_j)$ according to the number of busy servers in a state. An analytical solution for the heterogeneous queueing system with the FSF policy, where all states of servers are taken into account, although possible, but it is limited by the number of servers in the system. The latter system can be approximated in turn by a heterogeneous system $S_u$ with a preemption with appropriate evaluated service rates $m_j, j = 1, \ldots, K$. The dynamics of the system $S_u$ is described by the continuous-time Markov-chain $\{Y_u(t)\}$ with a state space $E_u = \{y : y \in \mathbb{N}_0\}$, where $Y_u(t)$ specifies the number of customers in the system at time $t$. The state transition diagram for this Markov-chain is presented in Figure 6.

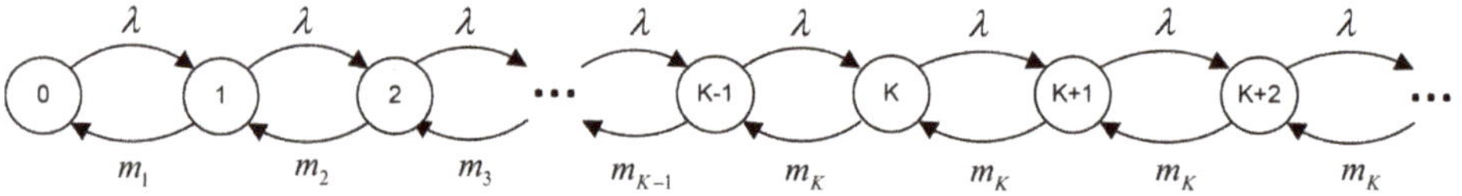

**Figure 6.** The state transition diagram for the queueing system $S_u$.

The approximations for $m_j$ are based on the observation that the incentive to occupy the slower servers is getting higher as arrival rate increases.

**Proposition 4.** *The service rates $m_j, j = 1, \ldots, K-1$, of the queueing model $\{Y_u(t)\}_{t \geq 0}$ can be approximated by the following relations*

$$m_j = \begin{cases} \sum_{i=1}^{j} \mu_i, & 0 < \lambda \leq \sum_{i=0}^{j-1} \mu_{K-i} \\ \sum_{i=0}^{j-1} \frac{\mu_{K-i}}{\lambda} \sum_{i=1}^{j} \mu_i + \sum_{i=1}^{k-j} \left( \frac{\mu_{K-i-j+1}}{\lambda} \sum_{n=1}^{k} \mu_{n+i} \right) & \\ \qquad + \left(1 - \sum_{i=0}^{k-1} \frac{\mu_{K-i}}{\lambda}\right) \sum_{i=k-j+2}^{k+1} \mu_j, & \sum_{i=0}^{k-1} \mu_{K-i} < \lambda \leq \sum_{i=0}^{k} \mu_{K-i}, j \leq k \leq K-1, \end{cases}$$

$$m_K = \sum_{i=1}^{K} \mu_i. \tag{16}$$

**Proof.** For small arrival rate, e.g., $0 < \lambda \leq \mu_K$, most probably that only the first server which is the fastest one will be occupied and hence it will have the main contribution to the service rate $m_1$. When the values $\lambda$ are larger, e.g., $\mu_K < \lambda \leq \mu_{K-1} + \mu_K$, the first server will have a contribution to $\mu_1$ with a probability $\frac{\mu_K}{\lambda}$ and the second server—with a complementary probability $(1 - \frac{\mu_K}{\lambda})$. For larger values of $\lambda$, $\mu_{K-1} + \mu_K < \lambda \leq \mu_{K-2} + \mu_{K-1} + \mu_K$ the first three servers will contribute to $\mu_1$ with probabilities $\frac{\mu_K}{\lambda}$, $\frac{\mu_{K-1}}{\lambda}$ and $\left(1 - \frac{\mu_{K-1}+\mu_K}{\lambda}\right)$. Similarly we may derive the contribution of the servers larger values of $\lambda$ up to the condition $\mu_2 + \ldots, +\mu_K < \lambda \leq \mu_1 + \mu_2 + \cdots + \mu_K$. To evaluate the contribution to the service rate $m_2$ in a state with two busy servers the same schema can be used. When $\lambda$ is small, $0 < \lambda \leq \mu_{K-1} + \mu_K$, the first two servers will form the service rate $\mu_2$. If $\mu_{K-1} + \mu_K < \lambda \leq \mu_{K-2} + \mu_{K-1} + \mu_K$, the first three servers will have a contribution to $\mu_2$, the first and second servers contribute with a probability $\frac{\mu_1+\mu_2}{\lambda}$, the second and fourth – with a probability $\left(1 - \frac{\mu_2+\mu_3}{\lambda}\right)$. When $\sum_{j=0}^{2} \mu_{K-j} < \lambda \leq \sum_{j=0}^{3} \mu_{K-j}$, the four faster servers will serve the customers, the first and second server with probability $\frac{\mu_{K-1}+\mu_K}{\lambda}$, the second and third server with probability $\frac{\mu_3}{\lambda}$ and the third and fourth with probability $\left(1 - \frac{\mu_{K-2}+\mu_{K-1}+\mu_K}{\lambda}\right)$. The procedure can be continued for larger values of $\lambda$ in a similar way as before. The proposed arguments can be summarized for all service rates $m_j, j = 1, \ldots, K$, and the arbitrary number of servers $K$ in form of the approximation (16). □

It can be verified that for any $j$ the quotient $\frac{\lambda}{m_j} < 1$ and $\lambda < m_K = \sum_{j=1}^{K} \mu_j$. Now we can use the approximation (16) to derive the steady-state distribution.

**Proposition 5.** *The steady-state probabilities $\pi_y$ of the Markov-chain $\{Y_u(t)\}_{t\geq 0}$ are given by*

$$\pi_0 = \left[1 + \sum_{y=1}^{K-1} \frac{\lambda^y}{\prod_{j=1}^{y} m_j} + \frac{\lambda^{K+1}}{(m_K - \lambda)\prod_{j=1}^{K} m_j}\right]^{-1},$$

$$\pi_y = \begin{cases} \pi_0 \cdot \frac{\lambda^y}{\prod_{j=1}^{y} m_j} & 1 \leq y \leq K, \\ \pi_0 \cdot \frac{\lambda^y}{m_K^{y-K} \prod_{j=1}^{K} m_j} & y \geq K+1. \end{cases}$$

**Proof.** The proposition follows from the properties of the ergodic birth-and-death process $\{Y_u(t)\}_{t\geq 0}$ [28]. □

**Corollary 2.** *The mean number of customers in the system $S_u$ satisfies the relation*

$$\bar{L}_u = \sum_{y=0}^{\infty} \pi_y = \sum_{y=0}^{K} \pi_y + \frac{\lambda(m_K + (m_K - \lambda)K)}{(m_K - \lambda)^2}. \tag{17}$$

**Example 4.** *Consider the $M/M/K$ queueing system with a total service intensity equal to $\sum_{j=1}^{K} \mu_j = 35$. Here we analyse the systems with different number of servers and their heterogeneity.*

*A Gini's index $G(\mu), 0 \leq G(\mu) \leq 1$, can be used to measure the inequality for individual data $\mu$, see for details [29], and hence is quite appropriate as a metric for the heterogeneity of servers. This index can be obtained by computing the moments of the data set $\mu = \{\mu_K, \mu_{K-1}, \ldots, \mu_1\}$ with $\mu_j$ sorted in increasing order,*

$$G(\mu) = \frac{2Cov[\mu, n_K]}{K\bar{\mu}}, \ \bar{\mu} = \frac{1}{K}\sum_{j=1}^{K} \mu_j, \ n_K = \{1, 2, \ldots, K\}.$$

*The Gini's index ranges from a minimum value of zero, when all individuals are equal, e.g., for the homogeneous servers $G(\mu) = 0$, to a theoretical maximum of one when every individual except one has a value zero. Two different values of heterogeneity are studied within this example, namely $G(\mu) = 0.63$ and*

*$G(\mu) = 0.40$. The corresponding values of service intensities for three types of systems with $K = 3$, $K = 5$ and $K = 8$ are presented in Table 4.*

**Table 4.** Service intensities versus Gini's index.

| $K$ | $\mu$ | $G(\mu)$ | $K$ | $\mu$ | $G(\mu)$ |
|---|---|---|---|---|---|
| 3 | (1,11,23) | 0.63 | 3 | (5,11,19) | 0.40 |
| 5 | (1,2,4,8,20) | 0.63 | 5 | (2,4,6,10,13) | 0.40 |
| 8 | (0.5,1,1.5,2,2.5,3,7,17.5) | 0.63 | 8 | (1.5,1.5,2,3,4,6,8,9) | 0.40 |

*In Figures 7–9 we display the values $\bar{L}$ with bounds $\bar{L}_l$ and $\bar{L}_u$ calculated respectively by the policy-iteration Algorithm 1 and by expressions (15) and (17) as functions of $\lambda$ and number of servers $K = 3, 5, 8$. The Gini's index $G(\mu) = 0.63$ in a figures labeled by (a) and $G(\mu) = 0.40$—by (b). The curves in figures show, that the mean number of customers as well as the size of the gap between the lower and upper bounds increases with increasing values of K. As expected, the low and upper bounds must coincide with a mean value $\bar{L}$ for the system with homogeneous servers, where $G(\mu) = 0$. Indeed, in figures with less heterogeneity of servers the curves for $\bar{L}$ $\bar{L}_u$ and $\bar{L}_l$ are getting closer, as the Gini's index decreases. Moreover, we notice that the functions take similar values in a light traffic case when $\lambda << \sum_{j=1}^{K} \mu_j$ and tend to the same values as the traffic becomes heavier, i.e., if $\lambda \to \sum_{j=1}^{K} \mu_j$.*

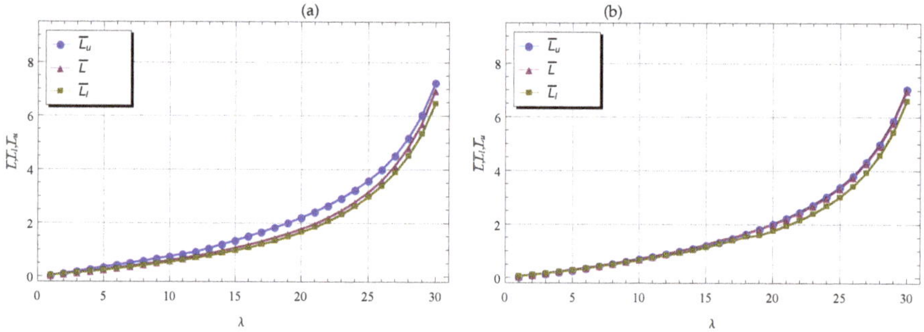


**Figure 7.** Mean value $\bar{L}$ with the bounds versus $\lambda$.

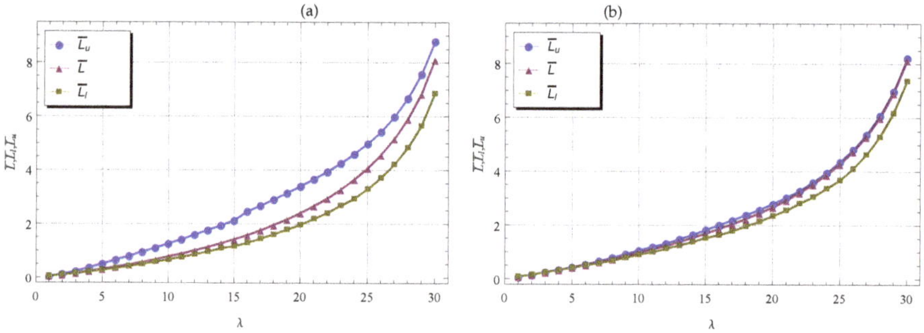


**Figure 8.** Mean value $\bar{L}$ with the bounds versus $\lambda$.

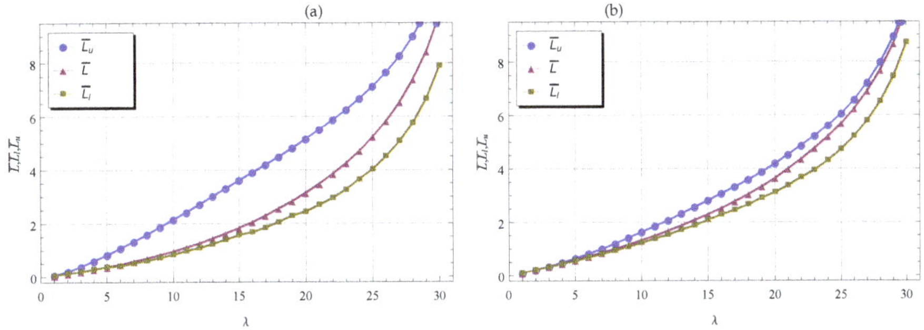


**Figure 9.** Mean value $\bar{L}$ with the bounds versus $\lambda$.

## 5. Sensitivity Analysis of the Heuristic Solution to Changes in Distribution

Another natural method to calculate the mean number of customers in the system and to check whether a certain policy leads to a reduction of this value is a simulation. This approach, while time-consuming, also makes it possible to examine the sensitivity of the optimal control policy $f$ and the corresponding mean performance characteristics to changes in distribution types other than exponential. An implementation of a simulation model is shown in the Figure 10 bellow.

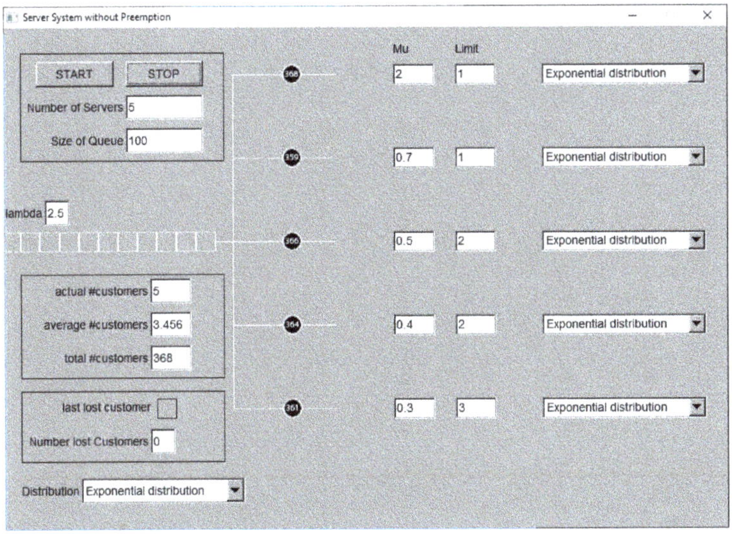


**Figure 10.** Simulation of the heterogeneous queueing system without preemption.

For this specific implementation it is possible to set the number of servers, the buffer capacity, threshold levels (limits), the arrival and service rates. The customers are indicated by a black circle, and are numbered accordingly to their arrival times. On the graphical interface there are also fields that show the actual amount of customers in the system, the average number and the total number of customers in the system including the already processed customers, the number of lost customers due to the truncated buffer capacity. The stability condition is taken into account and the buffer size is big enough so there should be hardly any lost customers. Hence the results with a truncated queue are comparable to systems with infinite queue lengths. Unfortunately, simulations are also unfit to solve systems with a large number of servers and states, as one would need to simulate a large number of different configurations with thousands of customers to get acceptable results. This fact further confirms the relevance of the results obtained in the previous sections.

The inter-arrival $A$ and service times $B_j, j = 1,2,\dots,K$, of customers follow exponential, gamma, Pareto, log-normal and hyper-exponential distributions. To get comparable results the parameters of the distributions are chosen to have the same means $\mathbb{E}[A] = \frac{1}{\lambda}$, $\mathbb{E}[B_j] = \frac{1}{\mu_j}, j = 1,2,\dots,K$, and variances $\mathbb{V}[A] = \frac{1}{\lambda^2}$, $\mathbb{V}[B_j] = \frac{1}{\mu_j^2}, j = 1,2,\dots,K$, as the system driven by exponential distribution. For this purpose we use formulas describing the parameters in terms of the mean and variance given by Toth et al. [30]. The main goal of the simulation experiments consists in understanding weather the heuristic solution (11) for $\lambda = \frac{1}{\mathbb{E}[A]}$ and $\mu_j = \frac{1}{\mathbb{E}[B_j]}$ is insensitive to changes in forms of distributions.

**Example 5.** *As a reference, we first simulate the system $M/M/5$ with an arrival rate $\lambda = 25$, $(\mu_1,\mu_2,\mu_3,\mu_4,\mu_5) = (20,8,4,2,1)$. Table 6 lists the mean number of customers in the system the optima, heuristic, FSF policies as well for other threshold policies with lower and higher values of thresholds.*

*We now simulate the systems like $GI/M/5$, $M/G/5$ as well as $GI/G/5$ with heterogeneous servers and threshold-based allocation policy where either the inter-arrival times, the service time or both follow one of the distributions mentioned above. For all the following simulation results we hereby want to find the mean number of customers in the system $\bar{L}$ for the policies specified in the preceding table for the markovian queueing system $M/M/5$. Table 6 provide a sensitivity and comparative analysis of the system performance obtained by employing different inter-arrival and service time distributions.*

*Of course, finding the optimal control policy through a simulation modelling is not an easy task. But in our example, we do not want to find the real values of the optimum thresholds, but rather to understand whether the optimum control and heuristic solution changes drastically when the distribution of the corresponding random values characterising the behaviour of a queueing system changes. Note that $\bar{L}$ for the optimal and heuristic policy takes always the values between those corresponding to the policies with lower and higher thresholds. The results of this example, as well as numerous other results carried out for systems with other parameters, show that while the absolute values of the mean number of customers vary as distributions change, the values of the optimal and heuristic thresholds are concentrated sufficiently close to the respective thresholds for markovian systems. Thus, we strongly believe, it is possible to use a heuristic solution with the replacement of exponential intensities by intensities of arbitrary distributions as a quasi-optimal solution in the problem of minimising the mean number of customers in the system with non-exponential inter-arrival and service time distributions.*

**Table 5.** Simulation results for the $M/M/5$ queueing system.

| **Exponential Distribution** | | | | |
|---|---|---|---|---|
| **Optimal Solution** | **Heuristic Solution** | **FSF** | **Lower Thresholds** | **Higher Thresholds** |
| $q_2 = 1$ | $q_2 = 1$ | $q_2 = 1$ | $q_2 = 1$ | $q_2 = 2$ |
| $q_3 = 2$ | $q_3 = 2$ | $q_3 = 1$ | $q_3 = 1$ | $q_3 = 3$ |
| $q_4 = 4$ | $q_4 = 3$ | $q_4 = 1$ | $q_4 = 2$ | $q_4 = 4$ |
| $q_5 = 9$ | $q_5 = 8$ | $q_5 = 1$ | $q_5 = 7$ | $q_5 = 9$ |
| $L = 4.082$ | $L = 4.189$ | $L = 4.860$ | $L = 4.213$ | $L = 4.674$ |

**Table 6.** Simulation results for the $GI/M/5$, $M/G/5$ and $GI/G/5$ queueing systems.

| **gamma distribution** | | | | |
|---|---|---|---|---|
| **optimal solution** | **heuristic solution** | **FSF** | **lower thresholds** | **higher thresholds** |
| GI/M/5: $L = 4.491$ | GI/M/5: $L = 4.499$ | GI/M/5: $L = 5.230$ | GI/M/5: $L = 4.375$ | GI/M/5: $L = 5.002$ |
| M/G/5: $L = 4.527$ | M/G/5: $L = 4.646$ | M/G/5: $L = 5.011$ | M/G/5: $L = 4.742$ | M/G/5: $L = 5.223$ |
| GI/G/5: $L = 4.048$ | GI/G/5: $L = 4.154$ | GI/G/5: $L = 4.827$ | GI/G/5: $L = 4.352$ | GI/G/5: $L = 4.719$ |
| **Pareto distribution** | | | | |
| **optimal solution** | **heuristic solution** | **FSF** | **lower thresholds** | **higher thresholds** |
| GI/M/5: $L = 3.857$ | GI/M/5: $L = 3.958$ | GI/M/5: $L = 4.426$ | GI/M/5: $L = 3.889$ | GI/M/5: $L = 4.561$ |
| M/G/5: $L = 4.211$ | M/G/5: $L = 4.321$ | M/G/5: $L = 4.870$ | M/G/5: $L = 4.477$ | M/G/5: $L = 4.913$ |
| GI/G/5: $L = 3.385$ | GI/G/5: $L = 3.473$ | GI/G/5: $L = 3.837$ | GI/G/5: $L = 3.461$ | GI/G/5: $L = 4.051$ |

**Table 6.** *Cont.*

| log-normal distribution | | | | |
|---|---|---|---|---|
| **optimal solution** | **heuristic solution** | **FSF** | **lower thresholds** | **higher thresholds** |
| GI/M/5: $L = 4.366$ | GI/M/5: $L = 4.479$ | GI/M/5: $L = 4.911$ | GI/M/5: $L = 4.509$ | GI/M/5: $L = 5.037$ |
| M/G/5: $L = 4.429$ | M/G/5: $L = 4.545$ | M/G/5: $L = 4.870$ | M/G/5: $L = 4.824$ | M/G/5: $L = 5.139$ |
| GI/G/5: $L = 3.821$ | GI/G/5: $L = 3.921$ | GI/G/5: $L = 4.636$ | GI/G/5: $L = 3.975$ | GI/G/5: $L = 4.593$ |
| **hyper-exponential distribution** | | | | |
| **optimal solution** | **heuristic solution** | **FSF** | **lower thresholds** | **higher thresholds** |
| GI/M/5: $L = 4.043$ | GI/M/5: $L = 4.148$ | GI/M/5: $L = 4.771$ | GI/M/5: $L = 4.129$ | GI/M/5: $L = 4.645$ |
| M/G/5: $L = 4.024$ | M/G/5: $L = 4.129$ | M/G/5: $L = 4.707$ | M/G/5: $L = 4.167$ | M/G/5: $L = 4.801$ |
| GI/G/5: $L = 4.021$ | GI/G/5: $L = 4.126$ | GI/G/5: $L = 4.709$ | GI/G/5: $L = 4.233$ | GI/G/5: $L = 4.768$ |

## 6. Conclusions

The queueing systems with heterogeneous servers have many real applications. The optimal control policy which minimizes the mean number of customers in the system without preemption under certain assumptions belongs to a threshold policy. Classical methods, such as the solution of difference equations, matrix-analytic and dynamic-programming approach, have significant restrictions due to the dimension of the random processes involved. A heuristic solution is obtained for the optimal threshold levels in a system with an arbitrary number of servers. The simple lower and upper bounds for the minimal mean number of customers in the system are derived using one dimensional processes for the equivalent heterogeneous queues with a preemption. The gap between the bounds increases with increasing of the servers' heterogeneity and the number of servers in the system. We have further conducted simulation to provide sensitivity analysis of the obtained HS to changes in inter-arrival and service time distributions. Simulation results showed that the optimal thresholds are likely to depend on the mean inter-arrival and service times and hence the proposed heuristic solution can be used as a quasi-optimal in systems with arbitrary distributions.

**Author Contributions:** Conceptualization, D.E.; Formal analysis, D.E., N.S.; Investigation, D.E., N.S., A.P.; Methodology, D.E., J.S.; Software, D.E., N.S.; Writing—original draft, D.E., N.S.; Writing—review & editing, D.E., J.S. All authors have read and agreed to the published version of the manuscript.

**Funding:** This research was funded by RUND University Program grant number 5-100.

**Acknowledgments:** The authors are very grateful to the reviewers for their valuable comments and suggestions which improved the quality and the presentation of the paper.

**Conflicts of Interest:** The authors declare no conflict of interest.

## References

1. Alves, F.S.Q.; Yehia, H.C.; Cruz, F.R.B.; Pedrosa, L.A.C. Upper bounds on performance measures of heterogeneous $M/M/c$ queues. *Math. Probl. Eng.* **2011**, *2011*, 702834. [CrossRef]
2. Bilgen, S.; Altintas, O. An approximate solution for the resequencing problem in packet-switching networks. *IEEE Trans. Commun.* **1994**, *42*, 229–232. [CrossRef]
3. Melikov, A.Z.; Ponomarenko, L.A.; Mekhbaliyeva, E.V. Analyzing the models of systems with heterogeneous servers. *Cybern. Syst. Anal.* **2020**, *56*, 89–99. [CrossRef]
4. Larsen, R.L. Control of Multiple Exponential Servers with Application to Computer Systems. Ph.D. Thesis, University of MD, Maryland, America, 1981.
5. Larsen R.L.; Agrawala, A.K. Control of a heterogeneous two-server exponential queueing system. *IEEE Trans. Softw. Eng.* **1983**, *9*, 522–526. [CrossRef]
6. Lin, W.; Kumar, P.R. Optimal control of a queueing system with two heterogeneous servers. *IEEE Trans. Autom. Control.* **1984**, *29*, 696–703. [CrossRef]
7. Koole, G. A simple proof of the optimality of a threshold policy in a two-server queueing system. *Syst. Control Lett.* **1995**, *26*, 301–303. [CrossRef]

8. Luh, H.P.; Viniotis, I. Threshold control policies for heterogeneous server systems. *Math. Methods Oper. Res.* **2002**, *55*, 121–142. [CrossRef]
9. Walrand, J. A note on 'Optimal control of a queueing system with two heterogeneous servers'. *Syst. Control Lett.* **1984**, *4*, 131–134. [CrossRef]
10. Weber, R. On a conjecture about assigning jobs to processors of different speeds. *IEEE Trans. Autom. Control* **1993**, *38*, 166–170. [CrossRef]
11. Özkan, E.; Kharoufeh, J.P. Optimal control of a two-server queueing system with failures. *Probab. Eng. Informational Sci.* **2014**, *28*, 489–527. [CrossRef]
12. Armony, M.; Ward, A.R. Fair dynamic routing in large-scale heterogeneous-server systems. *Oper. Res.* **2010**, *58*, 624–637. [CrossRef]
13. Efrosinin, D. *Controlled Queueing Systems with Heterogeneous Servers: Dynamic Optimization and Monotonicity Properties of Optimal Control Policies in Multiserver Heterogeneous Queues;* VDM Verlag: Saarbrücken, Germany, 2008.
14. Rosberg, Z.; Makowski A.M. Optimal routing to parallel heterogeneous servers—Small arrival rates. *Trans. Autom. Control* **1990**, *35*, 789–796. [CrossRef]
15. Viniotis, I.; Ephremides, A. Extension of the optimality of a threshold policy in heterogeneous multi-server queueing systems. *IEEE Trans. Autom. Control* **1988**, *33*, 104–109. [CrossRef]
16. Rykov, V. Monotone Control of Queueing Systems with Heterogeneous Servers. *QUESTA* **2001**, *37*, 391–403.
17. Shenker, S.; Weinrib, A. *Asymptotic Analysis of Large Heterogeneous Queueing Systems;* Bell Communication Research: Murray Hill, NJ, USA 1988.
18. Efrosinin, D. Queueing model of a hybrid channel with faster link subject to partial and complete failures. *Ann. Oper. Res.* **2013**, *202*, 75–102. [CrossRef]
19. Rykov, V.; Efrosinin, D. On the slow server problem. *Autom. Remote Control* **2010**, *70*, 2013–2023. [CrossRef]
20. de Vericourt, F.; Zhou, Y.P. On the incomplete results for the heterogeneous server problem. *Queueing Syst.* **2006**, *52*, 189–191. [CrossRef]
21. Neuts, M.F. *Matrix-Geometric Solutions in Stochastic Models;* The John Hopkins Univ. Press: Baltimore, MA, USA, 1981.
22. Efrosinin, D.; Rykov, V. On performance characteristics for queueing systems with heterogeneous servers. *Autom. Remote Control* **2008**, *69*, 61–75. [CrossRef]
23. Howard, R.A. *Dynamic Programming and Markov Processes;* Wiley Series; Wiley: New York, NY, USA, 1960. .
24. Puterman, M.L. *Markov Decision Process;* Wiley Series in Probability and Mathematical Statistics, John Wiley and Sons: Hoboken, NJ, USA, 1994..
25. Tijms, H.C. *Stochastic Models. An Algorithmic Approach;* John Wiley and Sons: Hoboken, NJ, USA, 1994.
26. Rykov, V.V. Controllable queueing systems. *Itogi Nauk. I Techniki. Teor. Verojatnostey I Mat. Stat. Kibern.* **1975**, *12*, 45–152. (In Russian)
27. Efrosinin, D.; Sztrik, J. An algorithmic approach to analyzing the reliability of a controllable unreliable queue with two heterogeneous servers. *Eur. J. Oper. Res.* **2018**, *271*, 934–952. [CrossRef]
28. Karlin, S.; McGregor, J. The classification of birth and death processes. *Trans. Am. Math. Soc.* **1957**, *86*, 366–400. [CrossRef]
29. Shalit, H. Calculating the Gini index of inequality for individual data. *Oxf. Bull. Econ. Stat.* **1985**, *47*, 185–189. [CrossRef]
30. Toth, A.; Sztrik, J.; Kuki, A.; Berczes, T.; Efosinin, D. Reliability analysis of finite-source retrial queues with outgoing calls using simulation. In Proceedings of the 2019 International Conference on Information and Digital Technologies (IDT), Zilina, Slovakia, 25–27 June 2019; pp. 504–511.

mathematics

MDPI

Article

# Accumulative Pension Schemes with Various Decrement Factors

**Mohammed S. Al-Nator** *,† **and Sofya V. Al-Nator** †

Department of Mathematics, Financial University under the Government of the Russian Federation, 49 Leningradsky Prospekt, Moscow 125993, Russia; salnator@yandex.ru
* Correspondence: malnator@yandex.ru
† These authors contributed equally to this work.

Received: 18 October 2020; Accepted: 16 November 2020; Published: 22 November 2020

**Abstract:** We consider accumulative defined contribution pension schemes with a lump sum payment on retirement. These schemes differ in relation to inheritance and provide various decrement factors. For each scheme, we construct the balance equation and obtain an expression for calculation of gross premium. Payments are made at the end of the insurance event period (survival to retirement age or death or retirement for disability within the accumulation interval). A simulation model was developed to analyze the constructed schemes.

**Keywords:** pension schemes; balance equation; gross premium; premium load; lump sum; defined contribution pension schemes; decrement tables

## 1. Introduction

At present, the accumulative pension schemes with a lump sum payment are not used in the Russian Federation. Such schemes may be of interest to many people for many reasons. For example, the pensioner plans to make large purchases, such as purchase of housing, purchase of household appliances and/or purchase of expensive treatment. In addition, such schemes allow elderly parents to ensure the future of their minor children.

At the end of October 2019, the Central Bank and the Ministry of Finance of the Russian Federation announced a draft law for a new system of voluntary retirement savings, called Guaranteed Retirement Plan. This draft law inspired the authors to write this article.

Pension schemes with the possibility of inheritance are popular in many countries. For example, in the United Kingdom, in many circumstances, one can inherit a pension. Examples of such pension schemes are defined contribution pension funds, joint life annuities and annuities with guarantee periods. In the UK, it has now become easier to inherit a pension thanks to the 2015 pension freedoms and the introduction of flexi-access drawdown, which is a newer, more flexible version of pension drawdown (see [1–4]).

In this work, we used standard methods of actuarial calculations. However, the authors would like to emphasize that since the actuarial calculations associated with a specific insurance contract are made by the actuaries of the insurance companies and pension funds; the details of these calculations are confidential.

Let us now describe the main results of this work. In this work, we consider accumulative defined contribution pension schemes with a lump sum payment on retirement. These schemes differ in relation to inheritance and provide various decrement factors: mortality or mortality and disability (recall that the disability retirement is a plan of retirement which is invoked when person covered is disabled from working to normal retirement age). It is assumed that contributions are paid regularly at the beginning of each period (monthly, quarterly, yearly, etc.). In other words, the contributions (without premium load) form an annuity due (for more details about basic actuarial definitions and terminology, see [5,6]). Recall that the premium load is a percentage of insurance premium deducted from the premium payments to cover policy administrative expenses, including the agent's sales commissions and a return on investment.

For each scheme, we construct the balance equation and obtain an expression for calculation of gross premium (Propositions 1–4). The balance equation is based on the equivalence principle of financial liabilities of the pension fund and the insured person. In general, the equivalence principle states that the (expected) present values of premiums and benefits should be equal. In other words, the equivalence principle ensures that, at any time, the net contributions (i.e., the contributions balance after taking into account the premium load) in the past provide precisely the amount needed to meet net liabilities in the future. Note that the premium load leads to a decrease of contributions when applying the principle of equivalence.

For all schemes with the possibility of inheritance (see Sections 2.2, 2.3 and 3), we assume that if the death or disability of the insured person occurs before reaching the retirement age then the payments will be made not at the time of death or disability, but at the end of the last contribution period in which the death or disability occurs (in other words, we consider the so-called discrete models of pension insurance). It is assumed that the net contributions are not refunded if the death occurs within the last period of the insurance contract. In the event of disability within the last period of the insurance contract, the insured receives the lump sum payment at the end of this period.

Note that for long-term insurance, the investment income of the collected premiums should be taken into account in the balance equation. This income is related to the changing value of money over time. In particular, when deriving the balance equation, it is necessary to find the (present) value of liabilities of the pension fund and the insured person (i.e., the value of contributions and payments) relative to one time point [5,6]. For our schemes, we derive the balance equation relative to the moment of concluding the pension insurance contract.

To analyze the constructed schemes, in Section 4 we discuss the results of a simulation model developed by the authors for calculating the gross premium.

This work is dedicated to Zolotarev V.M. the founder of the International Seminar on Stability Problems for Stochastic Models and to Kalashnikov V.V., whose works have important applications in actuarial mathematics and mathematical risk theory (see [7–11]. See also [12]).

Some results of this work were announced in Russian in [13,14].

## 2. Cumulative Models of Pension Insurance Based on one Decrement Factor (Mortality)

In this section, we consider accumulative defined contribution pension schemes with a lump sum payment on retirement. These schemes differ in relation to inheritance and provide one decrement factor (mortality). It is assumed that contributions are paid regularly at the beginning of each period (monthly, quarterly, yearly, etc.).

In this section, we use the following notation:

- $x$ is the age of the insured person at the time of concluding the pension insurance contract.
- $y$ is the retirement age (for example, $y = 55$, 60 or 65).
- $B$ is the gross premium paid by the policyholder at the beginning of each period (monthly, quarterly, yearly, etc.) until retirement.
- $P_l$ is the lump sum payment on retirement if the insured person survives to the retirement age.
- $\alpha_1, \ldots, \alpha_{y-x}$ are the annual premium loads under the insurance pension contract.
- $i$ is the effective annual interest rate, $v = \frac{1}{1+i}$ is the annual discount factor and $a = 1 + i$ is the annual growth rate.
- $P_r$ is the present value of the net contributions refunded to the inheritors if the insured dies before surviving to the retirement age. It is assumed that the net contributions are not refunded if the death occurs within the last period of the insurance contract.

Let us recall the definitions of some actuarial symbols (life table notation) that are used in this section (for more details see [5,6]).

With a starting group of $l_0$ newborns $l_z$ denotes the expected number of survivors at age $z$ from the original group.

Some definitions and relationships involving life table functions are:

- $d_z = l_z - l_{z+1}$ is the number of deaths between ages $z$ and $z+1$.
- The number of deaths between ages $z$ and $z+t$ is

$$ {}_t d_z = l_z - l_{z+t}. $$

  If $t = n$ is an integer, then

$$ {}_n d_z = d_z + d_{z+1} + \ldots + d_{z+n-1}. $$

  Note that ${}_1 d_z = d_z$.
- The notation $(z)$ refers to an individual alive at age $z$.
- ${}_t q_z = \frac{{}_t d_z}{l_z}$ is the probability that $(z)$ will die within the next $t$ years (by age $z+t$).
- The complement of ${}_t q_z$ is denoted by ${}_t p_z = 1 - {}_t q_z = \frac{l_{z+t}}{l_z}$ which is the probability that $(z)$ survives at least to time $t$ (and dies some time after $t$).
- Recall that for discrete $n$-year temporary life annuity due the first payment (contribution) is made at age $z$ and the latest possible payment is made at age $z+n-1$ (a maximum of $n$ payments, with last occurring at the beginning of the $n$-th year). The actuarial present value of this annuity is denoted by $\ddot{a}_{z:\overline{n}|}$. We have (see [5,6]).

$$ \ddot{a}_{z:\overline{n}|} = \sum_{k=0}^{n-1} v^k {}_k p_z. $$

### 2.1. Scheme 1. Accumulative Pension Scheme with Annual Contributions and a Lump Sum Payment on Retirement and without the Possibility of Inheritance

Consider an accumulative defined contribution pension scheme with a lump sum payment on retirement and without the possibility of inheritance (i.e., $P_r = 0$). Suppose that a group of individuals at age $x$ start to pay the pension contributions of size $B$ at the beginning of each year until retirement, i.e., until the age $y$. A schematic drawing of this pension model is presented in Figure 1.

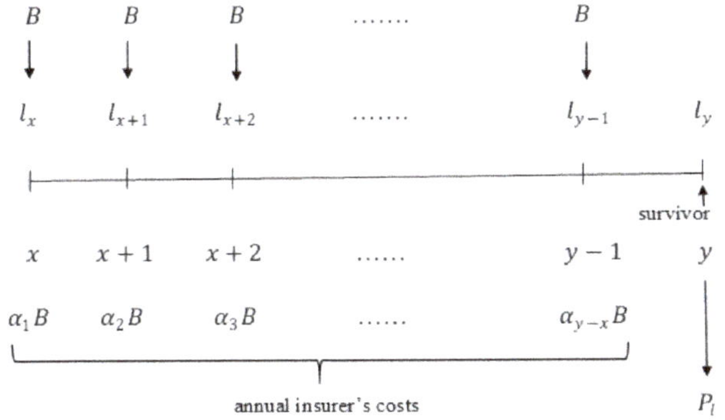


**Figure 1.** Schematic drawing of the accumulative pension scheme with annual contributions and a lump sum payment on retirement and without the possibility of inheritance.

**Proposition 1.** *Consider the pension Scheme 1. Then, the balance equation has the form*

$$B\ddot{a}_{x:\overline{y-x}|} = B\sum_{k=0}^{y-x-1} \alpha_{k+1}v^k\, {}_kp_x + P_l\, {}_{y-x}p_x\, v^{y-x}. \quad (1)$$

*In particular, the annual gross premium B is equal to*

$$B = \frac{P_l\, {}_{y-x}p_x\, v^{y-x}}{\ddot{a}_{x:\overline{y-x}|} - \sum\limits_{k=0}^{y-x-1} \alpha_{k+1}v^k\, {}_kp_x}. \quad (2)$$

The proof of Proposition 1 will be given in Section 2.3 as a special case of more general pension scheme (Scheme 3).

### 2.2. Scheme 2. Accumulative Pension Scheme with Annual Contributions and a Lump Sum Payment on Retirement and with Possibility of Inheritance

Consider Scheme 1 with possibility of inheritance. In other words, assume that if the insured dies before surviving to the retirement age, then all the net contributions are paid before the deaths are refunded to the inheritors at the end of the death year.

Recall that $P_r$ is the present value of the net contributions refunded to the inheritors and $P_l$ is the lump sum payment on retirement. Recall also that the net contributions are not refunded if the death occurs within the last period of the insurance contract (i.e., within $(y-1, y)$). A schematic drawing of this pension model is presented in Figure 2.

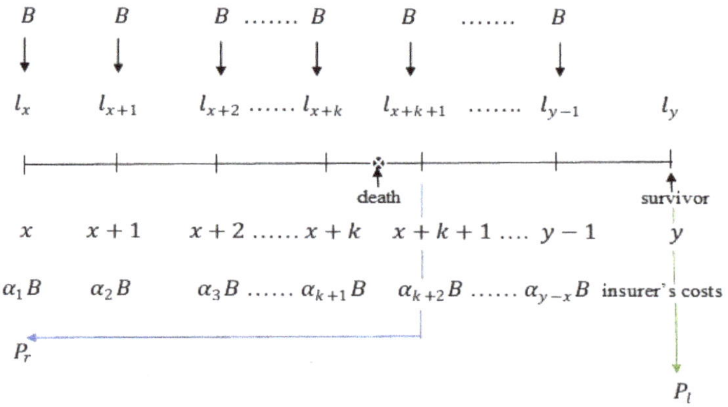


**Figure 2.** Schematic drawing of the accumulative pension scheme with annual contributions and a lump sum payment on retirement and with the possibility of inheritance.

**Proposition 2.** *Consider the pension Scheme 2. Then, the balance equation has the form*

$$B\ddot{a}_{x:\overline{y-x}|} = B\sum_{k=0}^{y-x-1} \alpha_{k+1}v^k\, {}_kp_x + P_l\, {}_{y-x}p_x\, v^{y-x} + P_r', \quad (3)$$

*where*

$$P_r' = \frac{P_r}{l_x} = B\sum_{j=1}^{y-x-1} {}_{j-1|}q_x \sum_{k=1}^{j}(1-\alpha_k)v^{k-1} \quad (4)$$

*and*

$$ {}_{s|}q_x = \frac{l_{x+s} - l_{x+s+1}}{l_x} \tag{5} $$

*is the probability that* $(x)$ *will die in a year, deferred s years; that is, that he will die in the* $(s+1)th$ *year (see [5,6] for more details about this actuarial symbol).*

*In particular, the annual gross premium B is equal to*

$$ B = \frac{P_l\, {}_{y-x}p_x\, v^{y-x}}{\ddot{a}_{x:\overline{y-x}|} - \sum\limits_{k=0}^{y-x-1} \alpha_{k+1} v^k\, {}_k p_x - \sum\limits_{j=1}^{y-x-1} {}_{j-1|}q_x \sum\limits_{k=1}^{j} (1-\alpha_k) v^{k-1}}. \tag{6} $$

The proof of Proposition 2 will be given in Section 2.3 as a special case of a more general pension scheme (Scheme 3).

*2.3. Scheme 3. Accumulative Pension Scheme with m-thly Payable Contributions and a Lump Sum Payment on Retirement and with Possibility of Inheritance*

Assume in Scheme 2 that the contributions form an $m$-thly payable annuity due, in which each year is broken into $m$ equal fractions, and the annuity pays $B$ at the start of each fraction of a year, as long as the annuitant survives. The first payment is made at age $x$. If the insured dies before surviving to the retirement age, then all the net contributions paid before the death are refunded to the inheritors at the end of the death period. Recall that the net contributions are not refunded if the death occurs within the last period of the insurance contract (i.e., within $(y - 1/m, y)$). A schematic drawing of this pension model is presented in Figure 3.

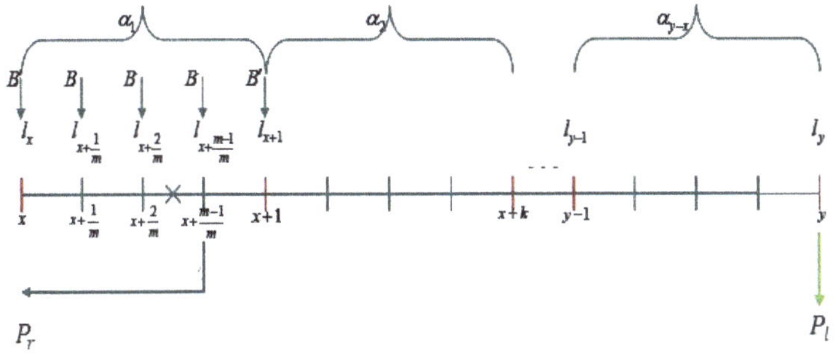

**Figure 3.** Schematic drawing of the accumulative pension scheme with $m$-thly payable contributions and a lump sum payment on retirement and with possibility of inheritance. The cross indicates the moment of death.

Since the life table functions in the published mortality tables are given for exact integer ages, actuaries make fractional age assumptions when insurance event occurs at non-integer ages. A fractional age assumption is an interpolation of life table functions between integer age values which are accepted as given. Three fractional age assumptions have been widely used by actuaries: the uniform distribution of death (UDD) assumption, the constant force assumption and the hyperbolic or Balducci assumption (see [5,6]). For Scheme 3, we apply the uniform distribution of death assumption. UDD assumption is equivalent to the linear interpolation of $l_z$:

$$ l_{z+t} = (1-t)l_z + t l_{z+1} \text{ for integer } z \text{ and } t \in [0,1]. \tag{7} $$

**Proposition 3.** *Consider the pension Scheme 3. Then, under UDD assumption, the balance equation has the form*

$$B\sum_{k=0}^{y-x-1}(1-\alpha_{k+1})\sum_{j=0}^{m-1}v^{k+\frac{j}{m}}\left({}_kp_x-\frac{j}{m}\,{}_{k|}q_x\right)=P_r'+P_l\,{}_{y-x}p_xv^{y-x}, \tag{8}$$

*where*

$$P_r'=\frac{P_r}{l_x}=\frac{B}{m}\sum_{k=1}^{m(y-x)-1}\,{}_{\lfloor\frac{k-1}{m}\rfloor|}q_x\sum_{j=1}^{k}\left(1-\alpha_{\lfloor\frac{j-1}{m}\rfloor+1}\right)v^{\lfloor\frac{j-1}{m}\rfloor+\frac{(j-1)\bmod m}{m}} \tag{9}$$

*and* $\lfloor\cdot\rfloor$ *is the floor function.*

*In particular, the m-thly payable gross premium B is equal to*

$$B=\frac{P_l\,{}_{y-x}p_xv^{y-x}}{\sum\limits_{k=0}^{y-x-1}(1-\alpha_{k+1})\sum\limits_{j=0}^{m-1}v^{k+\frac{j}{m}}\left({}_kp_x-\frac{j}{m}\,{}_{k|}q_x\right)-P_r''}, \tag{10}$$

*where* $P_r''=P_r'/B$.

**Proof of Proposition 3.** For the convenience of calculation of the balance equation, we give in Table 1 the present values of net contributions (PVNCs) and payments (PVPs) to the inheritors for each period. The last line consists of the present value of the lump sum payments (PVLS).

Denote respectively by

$$PVNC=\sum PVNCs,\;\; PVP=\sum PVPs$$

the present values of all net contributions and payments to the inheritors. Then, the balance equation has the form

$$PVNC=PVP+PVLS. \tag{11}$$

According to Table 1, the present value of the net contributions is

$$PVNC=l_xB(1-\alpha_1)v^0+l_{x+\frac{1}{m}}B(1-\alpha_1)v^{\frac{1}{m}}+l_{x+\frac{2}{m}}B(1-\alpha_1)v^{\frac{2}{m}}+\ldots+l_{x+1}B(1-\alpha_2)v^1+$$
$$+l_{x+1+\frac{1}{m}}B(1-\alpha_2)v^{1+\frac{1}{m}}+\ldots+l_{y-1+\frac{1}{m}}B(1-\alpha_{y-x})v^{y-x-1+\frac{1}{m}}+\ldots+$$
$$+l_{y-1+\frac{m-1}{m}}B(1-\alpha_{y-x})v^{y-x-1+\frac{m-1}{m}}. \tag{12}$$

According to Table 1, the present value of the payments is

$$PVP=\left(l_x-l_{x+\frac{1}{m}}\right)B(1-\alpha_1)v^0+\left(l_{x+\frac{1}{m}}-l_{x+\frac{2}{m}}\right)B\left[(1-\alpha_1)v^0+(1-\alpha_1)v^{\frac{1}{m}}\right]+\ldots+$$
$$+\left(l_{x+\frac{m-1}{m}}-l_{x+1}\right)B\left[(1-\alpha_1)v^0+(1-\alpha_1)v^{\frac{1}{m}}+(1-\alpha_1)v^{\frac{2}{m}}+\ldots+(1-\alpha_1)v^{\frac{m-1}{m}}\right]+\ldots+$$
$$+\left(l_{y-1}-l_{y-1+\frac{1}{m}}\right)B\left[(1-\alpha_{y-x})v^{y-x-1}+(1-\alpha_{y-x-1})v^{y-x-1-\frac{1}{m}}+\ldots+\right.$$
$$\left.+(1-\alpha_1)v^0\right]+\ldots+\left(l_{y-1+\frac{m-2}{m}}-l_{y-1+\frac{m-1}{m}}\right)B\left[(1-\alpha_{y-x})v^{y-x-1+\frac{2}{m}}+\right.$$
$$\left.+(1-\alpha_{y-x})v^{y-x-1+\frac{1}{m}}+(1-\alpha_{y-x})v^{y-x-1}+\ldots+(1-\alpha_1)v^0\right]. \tag{13}$$

**Table 1.** Present values of individual net contributions and payments to the inheritors by period (except the last line). The last line consists of the present value of the lump sum payments.

| Age | PVNCs | PVPs |
|---|---|---|
| $x$ | $l_xB(1-\alpha_1)v^0$ | $0$ |
| $x+\frac{1}{m}$ | $l_{x+\frac{1}{m}}B(1-\alpha_1)v^{\frac{1}{m}}$ | $\left(l_x-l_{x+\frac{1}{m}}\right)B(1-\alpha_1)(i+1)^{\frac{1}{m}}v^{\frac{1}{m}} = \left(l_x-l_{x+\frac{1}{m}}\right)B(1-\alpha_1)v^0$ |
| $x+\frac{2}{m}$ | $l_{x+\frac{2}{m}}B(1-\alpha_1)v^{\frac{2}{m}}$ | $\left(l_{x+\frac{1}{m}}-l_{x+\frac{2}{m}}\right)B\left[(1-\alpha_1)(i+1)^{\frac{1}{m}}+(1-\alpha_1)(i+1)^{\frac{2}{m}}\right]v^{\frac{2}{m}} = \left(l_{x+\frac{1}{m}}-l_{x+\frac{2}{m}}\right)B\left[(1-\alpha_1)v^0+(1-\alpha_1)v^{\frac{1}{m}}\right]$ |
| $\vdots$ | $\vdots$ | $\vdots$ |
| $x+1$ | $l_{x+1}B(1-\alpha_2)v^1$ | $\left(l_{x+\frac{m-1}{m}}-l_{x+1}\right)B\left[(1-\alpha_1)(i+1)^{\frac{1}{m}}+(1-\alpha_1)(i+1)^{\frac{2}{m}}+\ldots+(1-\alpha_1)(i+1)^{\frac{m-1}{m}}+(1-\alpha_1)(i+1)^1\right]v^1 = \left(l_{x+\frac{m-1}{m}}-l_{x+1}\right)B\left[(1-\alpha_1)v^0+(1-\alpha_1)v^{\frac{1}{m}}+(1-\alpha_1)v^{\frac{2}{m}}+\ldots+(1-\alpha_1)v^{\frac{m-1}{m}}\right]$ |
| $x+1+\frac{1}{m}$ | $l_{x+1+\frac{1}{m}}B(1-\alpha_2)v^{1+\frac{1}{m}}$ | $\left(l_{x+1}-l_{x+1+\frac{1}{m}}\right)B\left[(1-\alpha_2)(i+1)^{\frac{1}{m}}+(1-\alpha_1)(i+1)^{\frac{2}{m}}+\ldots(1-\alpha_1)(i+1)^{\frac{m-1}{m}}+(1-\alpha_1)(i+1)^1+(1-\alpha_1)(i+1)^{1+\frac{1}{m}}\right]v^{1+\frac{1}{m}} = \left(l_{x+1}-l_{x+1+\frac{1}{m}}\right)B\left[(1-\alpha_2)v^1+(1-\alpha_1)v^{\frac{m-1}{m}}+\ldots+(1-\alpha_1)v^{\frac{2}{m}}+(1-\alpha_1)v^{\frac{1}{m}}+(1-\alpha_1)v^0\right]$ |
| $\vdots$ | $\vdots$ | $\vdots$ |
| $y-1+\frac{1}{m}$ | $l_{y-1+\frac{1}{m}}B\left(1-\alpha_{y-x}\right)v^{y-x-1+\frac{1}{m}}$ | $\left(l_{y-1}-l_{y-1+\frac{1}{m}}\right)B\left[\left(1-\alpha_{y-x}\right)(i+1)^{\frac{1}{m}}+\left(1-\alpha_{y-x-1}\right)(i+1)^{\frac{2}{m}}+\ldots+\left(1-\alpha_{y-x-1}\right)(i+1)^{1+\frac{1}{m}}+\ldots+(1-\alpha_1)(i+1)^{y-x-2+\frac{1}{m}}+\ldots+\ldots+(1-\alpha_1)(i+1)^{y-x-1+\frac{1}{m}}\right]v^{y-x-1+\frac{1}{m}} = \left(l_{y-1}-l_{y-1+\frac{1}{m}}\right)B\left[\left(1-\alpha_{y-x}\right)v^{y-x-1}+\left(1-\alpha_{y-x-1}\right)v^{y-x-1-\frac{1}{m}}+\ldots+(1-\alpha_1)v^0\right]$ |
| $\vdots$ | $\vdots$ | $\vdots$ |
| $y-1+\frac{m-1}{m}$ | $l_{y-1+\frac{m-1}{m}}B\left(1-\alpha_{y-x}\right)v^{y-x-1+\frac{m-1}{m}}$ | $\left(l_{y-1+\frac{m-2}{m}}-l_{y-1+\frac{m-1}{m}}\right)B\left[\left(1-\alpha_{y-x}\right)(i+1)^{\frac{1}{m}}+\left(1-\alpha_{y-x}\right)(i+1)^{\frac{2}{m}}+\ldots+\left(1-\alpha_{y-x}\right)(i+1)^{\frac{m-1}{m}}+\ldots+(1-\alpha_1)(i+1)^{y-x-1}+\ldots+(1-\alpha_1)(i+1)^{y-x-1+\frac{m-1}{m}}\right]v^{y-x-1+\frac{m-1}{m}} = \left(l_{y-1+\frac{m-2}{m}}-l_{y-1+\frac{m-1}{m}}\right)B\left[\left(1-\alpha_{y-x}\right)v^{y-x-1+\frac{2}{m}}+\left(1-\alpha_{y-x}\right)v^{y-x-1+\frac{1}{m}}+\left(1-\alpha_{y-x}\right)v^{y-x-1}+\ldots+(1-\alpha_1)v^0\right]$ |
| | | **PVLS** |
| $y$ | $0$ | $P_l l_y v^{y-x}$ |

The present value of the lump sum payments (PVLS) is

$$PVLS = P_l l_y v^{y-x}. \tag{14}$$

Under the assumption of uniform distribution of deaths, we have (see (7))

$$l_{x+k+\frac{j}{m}} = \left(1-\frac{j}{m}\right)l_{x+k}+\frac{j}{m}l_{x+k+1} = l_{x+k}-\frac{j}{m}\left(l_{x+k}-l_{x+k+1}\right), \tag{15}$$

where $x, k$ are integers and $j/m \in [0,1]$. Hence, $PVNC$ can be rewritten as

$$PVNC = B\Big[(1-\alpha_1)\Big\{l_x v^0 + \Big(l_x - \frac{1}{m}(l_x - l_{x+1})\Big)v^{\frac{1}{m}} + \Big(l_x - \frac{2}{m}(l_x - l_{x+1})\Big)v^{\frac{2}{m}} + + \ldots + \Big(l_x - \frac{m-1}{m}(l_x - l_{x+1})\Big)v^{\frac{m-1}{m}}\Big\} + + (1-\alpha_2)\Big\{l_{x+1}v^1 + \Big(l_{x+1} - \frac{1}{m}(l_{x+1} - l_{x+2})\Big)v^{1+\frac{1}{m}} + \ldots + + \Big(l_{x+1} - \frac{m-1}{m}(l_{x+1} - l_{x+2})\Big)v^{1+\frac{m-1}{m}}\Big\} + \ldots + (1-\alpha_{y-x})\Big\{l_{y-1}v^{y-x-1} + + \Big(l_{y-1} - \frac{1}{m}(l_{y-1} - l_y)\Big)v^{y-x-1+\frac{1}{m}} + \ldots + \Big(l_{y-1} - \frac{m-1}{m}(l_{y-1} - l_y)\Big)v^{y-x-1+\frac{m-1}{m}}\Big\}\Big]. \quad (16)$$

Consider now $PVP$. From (15), it follows that

$$l_{x+k+\frac{j}{m}} - l_{x+k+\frac{j+1}{m}} = \frac{1}{m}(l_{x+k} - l_{x+k+1}), \quad (17)$$

where $x, k$ are integers and $j/m,\ (j+1)/m \in [0,1]$. Hence, $PVP$ can be rewritten as

$$PVP = \frac{1}{m}(l_x - l_{x+1})B(1-\alpha_1)v^0 + \frac{1}{m}(l_x - l_{x+1})B\Big[(1-\alpha_1)v^0 + (1-\alpha_1)v^{\frac{1}{m}}\Big] + \ldots + +\frac{1}{m}(l_x - l_{x+1})B\Big[(1-\alpha_1)v^0 + (1-\alpha_1)v^{\frac{1}{m}} + (1-\alpha_1)v^{\frac{2}{m}} + \ldots + (1-\alpha_1)v^{\frac{m-1}{m}}\Big] + + \ldots + \frac{1}{m}(l_{y-1} - l_y)B\Big[(1-\alpha_{y-x})v^{y-x-1} + + (1-\alpha_{y-x-1})v^{y-x-1-\frac{1}{m}} + \ldots + (1-\alpha_1)v^0\Big] + +\frac{1}{m}(l_{y-1} - l_y)B\Big[(1-\alpha_{y-x})v^{y-x-1+\frac{2}{m}} + (1-\alpha_{y-x})v^{y-x-1+\frac{1}{m}} + + (1-\alpha_{y-x})v^{y-x-1} + \ldots + (1-\alpha_1)v^0\Big]. \quad (18)$$

Multiplying both sides of the balance Equation (11) by $1/l_x$ and taking into account (14), (16) and (18), we obtain

$$\frac{1}{l_x}PVNC = B\Big[(1-\alpha_1)\Big\{{}_0p_x v^0 + \Big({}_0p_x - \frac{1}{m}{}_{0|}q_x\Big)v^{\frac{1}{m}} + \Big({}_0p_x - \frac{2}{m}{}_{0|}q_x\Big)v^{\frac{2}{m}} + \ldots + + \Big({}_0p_x - \frac{m-1}{m}{}_{0|}q_x\Big)v^{\frac{m-1}{m}}\Big\} + (1-\alpha_2)\Big\{{}_1p_x v^1 + \Big({}_1p_x - \frac{1}{m}{}_{1|}q_x\Big)v^{1+\frac{1}{m}} + \ldots + + \Big({}_1p_x - \frac{m-1}{m}{}_{1|}q_x\Big)v^{1+\frac{m-1}{m}}\Big\} + \ldots + (1-\alpha_{y-x})\Big\{{}_{y-x-1}p_x v^{y-x-1} + + \Big({}_{y-x-1}p_x - \frac{1}{m}{}_{y-x-1|}q_x\Big)v^{y-x-1+\frac{1}{m}} + \ldots + + \Big({}_{y-x-1}p_x - \frac{m-1}{m}{}_{y-x-1|}q_x\Big)v^{y-x-1+\frac{m-1}{m}}\Big\}\Big] = = B\sum_{k=0}^{y-x-1}(1-\alpha_{k+1})\sum_{j=0}^{m-1}v^{k+\frac{j}{m}}\Big({}_kp_x - \frac{j}{m}{}_{k|}q_x\Big), \quad (19)$$

$$\frac{1}{l_x}PVP = \frac{1}{m}\,{}_{0|}q_x B\left(1-\alpha_1\right)v^0 + \frac{1}{m}\,{}_{0|}q_x B\left[\left(1-\alpha_1\right)v^0 + \left(1-\alpha_1\right)v^{\frac{1}{m}}\right] + \ldots +$$
$$+\frac{1}{m}\,{}_{0|}q_x B\left[\left(1-\alpha_1\right)v^0 + \left(1-\alpha_1\right)v^{\frac{1}{m}} + \left(1-\alpha_1\right)v^{\frac{2}{m}} + \ldots +\right.$$
$$\left.+\left(1-\alpha_1\right)v^{\frac{m-1}{m}}\right] + \ldots + \frac{1}{m}\,{}_{y-x-1|}q_x B\left[\left(1-\alpha_{y-x}\right)v^{y-x-1} +\right.$$
$$+\left(1-\alpha_{y-x-1}\right)v^{y-x-1-\frac{1}{m}} + \ldots + \left.\left(1-\alpha_1\right)v^0\right] + \frac{1}{m}\,{}_{y-x-1|}q_x B\left[\left(1-\alpha_{y-x}\right)v^{y-x-1+\frac{2}{m}} +\right.$$
$$+\left(1-\alpha_{y-x}\right)v^{y-x-1+\frac{1}{m}} + \left(1-\alpha_{y-x}\right)v^{y-x-1} + \ldots + \left.\left(1-\alpha_1\right)v^0\right] =$$
$$= \frac{B}{m}\sum_{k=1}^{m(y-x)-1} {}_{\lfloor\frac{k-1}{m}\rfloor|}q_x \sum_{j=1}^{k}\left(1-\alpha_{\lfloor\frac{j-1}{m}\rfloor+1}\right)v^{\lfloor\frac{j-1}{m}\rfloor+\frac{(j-1)\ \mathrm{mod}\ m}{m}}, \tag{20}$$

$$\frac{1}{l_x}PVLS = P_l\,{}_{y-x}p_x v^{y-x}. \tag{21}$$

Relations (19)–(21) prove Proposition 3. □

**Proof of Proposition 1.** Relations (1), (2) can be obtained from Proposition 3 if we set in (8)–(10) $P_r = 0$ and $m = 1$. □

**Proof of Proposition 2.** Relations (3), (4), (6) can be obtained from Proposition 3 if we set in (8)–(10) $m = 1$. □

## 3. Scheme 4. Cumulative Model of Pension Insurance with Possibility of Inheritance Based on Two Decrement Factors (Mortality and Disability) and with Annual Contributions and a Lump Sum Payment on Retirement

This scheme differs from Scheme 2 (Section 2.2) in that it provides for exit from the pension fund for reasons of death, old-age retirement or disability. A schematic drawing of this pension model is presented in Figure 4.

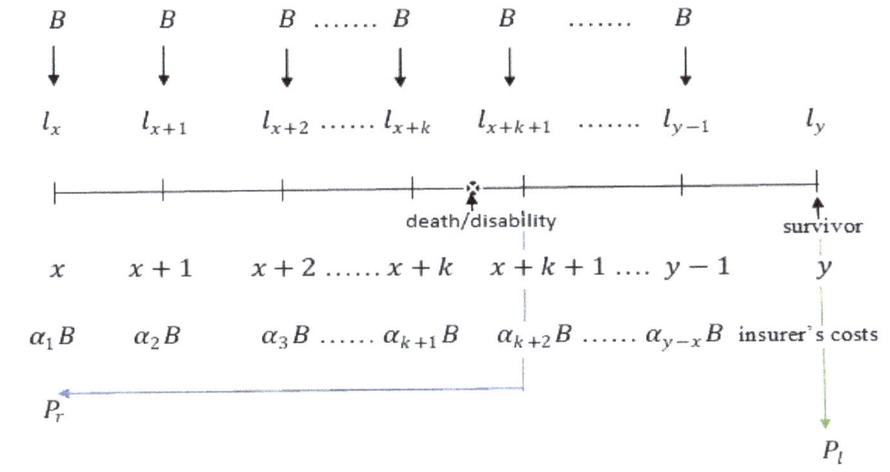


**Figure 4.** Schematic drawing of the accumulative pension scheme with possibility of inheritance based on two decrement factors and with annual contributions and a lump sum payment on retirement.

For this scheme, some of the actuarial symbols introduced in Section 2 may have another interpretation (the symbols $x$, $y$ and $d_z$ have the same meaning as in Section 2):

- $l_z$ is the expected number of active insured persons of the pension scheme at age $z$.
- $P_r$ is the present value of the net contributions refunded to the inheritors/insured if the death/disability occurs before the retirement age. The net contributions are not refunded if the death occurs within the last period of the insurance contract (i.e., within $(y-1, y)$). If the insured person has become disabled within $(y-1, y)$ and reaches the retirement age $y$, then the pension fund pays him the lump sum at age $y$.
- $i_z$ is the number of insured persons who leave the pension scheme due to disability between ages $z$ and $z+1$.
- ${}_n p_z$ is the probability for an active insured aged $z$, of being active at age $z+n$. This probability is equal to

$$ {}_n p_z = \frac{l_{z+n}}{l_z} = 1 - \frac{\sum\limits_{j=0}^{n-1} (d_{z+j} + i_{z+j})}{l_z}. $$

  If $z = x$ and $n = y - x$, then according to our assumptions, we have

$$ {}_{y-x} p_x = \frac{l_y}{l_x} = 1 - \frac{\sum\limits_{j=0}^{y-x-2} (d_{x+j} + i_{x+j}) + d_{y-1}}{l_x}. $$

- ${}_{n|}q_z$ is the probability for an active insured aged $z$, of being active at age $z+n$ and leaving the pension scheme during the next year (i.e., between ages $z+n$ and $z+n+1$). This probability is equal to

$$ {}_{n|}q_z = \frac{l_{z+n} - l_{z+n+1}}{l_z} = \frac{d_{z+n} + i_{z+n}}{l_z}. \tag{22} $$

**Proposition 4.** *Consider the pension scheme 4. Then, the balance equation has the form*

$$ B\ddot{a}_{x:\overline{y-x}|} = B \sum_{k=0}^{y-x-1} \alpha_{k+1} v^k \, {}_k p_x + P_l \, {}_{y-x} p_x \, v^{y-x} + P_r', \tag{23} $$

*where*

$$ P_r' = \frac{P_r}{l_x} = B \sum_{j=1}^{y-x-1} {}_{j-1|}q_x \sum_{k=1}^{j} (1 - \alpha_k) v^{k-1}. \tag{24} $$

*In particular, the annual gross premium $B$ is equal to*

$$ B = \frac{P_l \, {}_{y-x} p_x \, v^{y-x}}{\ddot{a}_{x:\overline{y-x}|} - \sum\limits_{k=0}^{y-x-1} \alpha_{k+1} v^k \, {}_k p_x - \sum\limits_{j=1}^{y-x-1} {}_{j-1|}q_x \sum\limits_{k=1}^{j} (1 - \alpha_k) v^{k-1}}. \tag{25} $$

**Remark 1.** *Although the relations in Propositions 2 and 4 seem to coincide, they have different meanings, since the relations in Proposition 4 contain decrement functions and probabilities. In particular, $\ddot{a}_{x:\overline{n}|}$ is the decrement n-year temporary life annuity due.*

The proof of this proposition is similar to the proof of Proposition 3 with $m = 1$. One can do this with the help of Table 2.

**Table 2.** Present values of individual net contributions and payments to the inheritors by period (except the last line). The last line consists of the present value of the lump sum payments.

| Age | PVNCs | PVPs |
|---|---|---|
| $x$ | $l_x B(1-\alpha_1)v^0$ | 0 |
| $x+1$ | $l_{x+1}B(1-\alpha_2)v^1$ | $(d_x+i_x)B(1-\alpha_1)(i+1)v^1 = $ $= (d_x+i_x)B(1-\alpha_1)v^0$ |
| $x+2$ | $l_{x+2}B(1-\alpha_3)v^2$ | $(d_{x+1}+i_{x+1})B\left[(1-\alpha_1)(i+1)^2+ \right.$ $\left.+(1-\alpha_2)(i+1)\right]v^2 =$ $= (d_{x+1}+i_{x+1})B\left[(1-\alpha_1)v^0+(1-\alpha_2)v^1\right]$ |
| $\vdots$ | $\vdots$ | $\vdots$ |
| $y-1$ | $l_{y-1}B(1-\alpha_{y-x})v^{y-x-1}$ | $(d_{y-2}+i_{y-2})B\left[(1-\alpha_1)v^0+(1-\alpha_2)v^1+\right.$ $\left.+\ldots+(1-\alpha_{y-x-1})v^{y-x-2}\right]$ |
| | | **PVLS** |
| $y$ | 0 | $P_l l_y v^{y-x}$ |

## 4. Simulation Results and Discussion

A simulation model was developed to analyze the constructed schemes. The annual gross premium was calculated on the base of this simulation model for the schemes with one decrement factor (Sections 2.1–2.3) and for the scheme with two decrement factors (Section 3). In the calculations, the authors used the male/female mortality table of the Russian Federation for schemes with one decrement factor and the decrement table for the scheme with two factors. A comparative analysis of these four schemes was also carried out.

For calculation purposes, an Excel macro code was written in the VBA programming language (Visual Basic for Applications).

In our calculations, we assumed (we keep the notation of Sections 2 and 3)

1. The retirement age $y = 55$ or 60 for females and 60 or 65 for males.
2. The age $x$ of the insured person ranges from 18 to $y$.
3. The annual premium loads $\alpha_1, \ldots, \alpha_{y-x}$ are selected by two rules (deterministic and random rules). Under the deterministic rule, the premium load increases every year by a fixed value, for example by 1%: $\alpha_{k+1} = \alpha_k + 0.01$, $k = 1, \ldots, y-x-1$ for a given $\alpha_1$ (for example $\alpha_1 = 0.05$). Under the random rule, the premium loads are generated by the Excel built-in function RAND. Recall that RAND returns an evenly distributed random real number greater than or equal to 0 and less than 1.
4. The effective annual interest rate $i = 3\%$, 4% or 5%.
5. The lump sum payment may take any non-negative value.

The calculation results show that all the constructed schemes are qualitatively adequate, namely:

- All other things equal, when the accumulation period decreases, the contribution size $B$ increases. The results of some scenarios are shown in Figure 5. In particular, since females reach retirement age earlier than males, the possible accumulation period for females is always shorter. Hence, the contribution size for females is always higher than for males (see Figure 6).
- All other things equal, when the interest rate $i$ increases, the contribution size $B$ decreases. The results of some scenarios are shown in Figure 7.

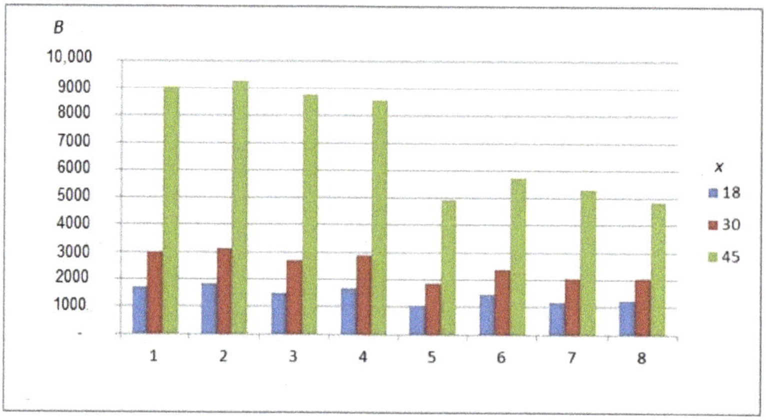


**Figure 5.** Gross premium size as a function of the accumulation period.

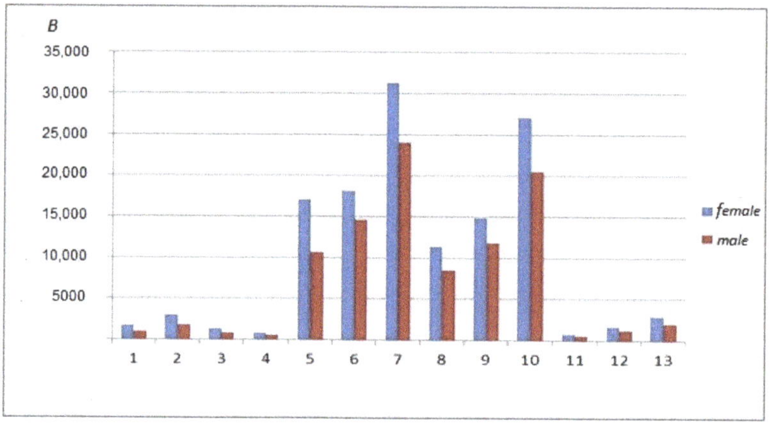


**Figure 6.** Gross premium size for males and females.

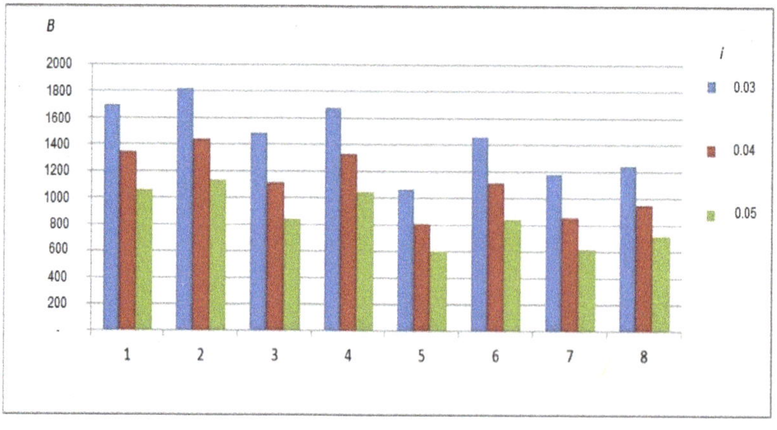


**Figure 7.** Gross premium size as a function of the interest rate.

- All other things equal, when the lump sum payment $P_l$ decreases the contribution size $B$ also decreases. Namely if we replace $P_l$ by $rP_l$ then $B$ is replaced by $rB$ and vice versa. This immediately follows from (2), (6), (10) and (25). The results of some scenarios are shown in Figure 8.
- For Scheme 3: the lower number of contributions $m$ per year, the higher contribution size $B$.

Let us now give a comparative analysis of the constructed schemes (see Figure 9).

- Consider schemes with one decrement factor (mortality) and annual contributions (Schemes 1 and 2). Then for both genders, all other things equal, the scheme without inheritance (Scheme 1) is the cheapest. This immediately follows from (2) and (6).
- Consider schemes with inheritance and one decrement factor (Schemes 2 and 3). Then for both genders the scheme with at least two contributions per year is more expensive than the scheme with annual contributions (Scheme 2).
- Consider schemes with inheritance and annual contributions (Schemes 2 and 4). Then the scheme with two decrement factors (Scheme 4) is the cheapest.

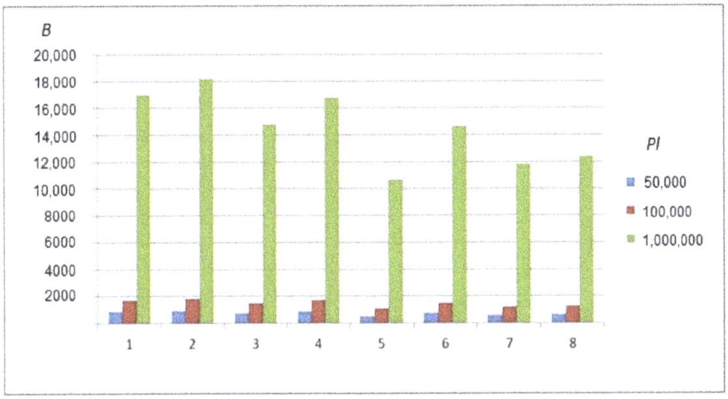


**Figure 8.** Gross premium size as a function of the lump sum payment.

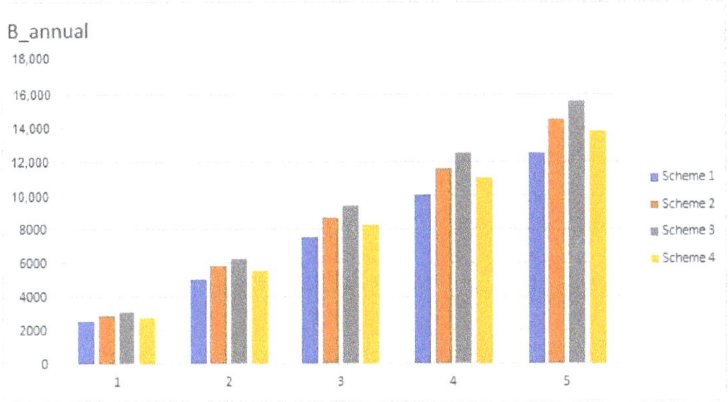


**Figure 9.** Comparative analysis of Schemes 1–4.

## 5. Conclusions

In this work, four accumulative defined contribution pension schemes with a lump sum payment on retirement are proposed. These schemes differ in relation to inheritance and provide various decrement factors. The availability of various schemes allows the insured to choose the most suitable scheme depending on the size of the contribution or the size of the lump sum payment. When choosing a pension scheme, the insured person may have health or family reasons, for example, elderly parents would like to ensure the future of their minor children.

The results of this work are directly related to the Russian draft law of a new system of voluntary retirement savings, called Guaranteed Retirement Plan. In particular, these results are of practical importance for the Pension Fund of the Russian Federation, as well as for private pension funds and pension actuaries in general.

Let us summarize the main results of this work. For each scheme, we use standard methods of actuarial calculations to construct the balance equation and to obtain an expression for the gross premium. A simulation model was developed to analyze the constructed schemes. The gross premium was calculated on the base of this simulation model. A comparative analysis of various schemes was also carried out. According to our calculation, we obtained

- Scheme 1 is the cheapest for the insured person, since it does not provide for the possibility of inheritance and takes into account only one decrement factor (mortality).
- For schemes with inheritance and annual contributions, the scheme with the two decrement factors is the cheapest.
- Consider schemes with one decrement factor (mortality). Then, Scheme 3 with $m > 1$ is the most expensive in terms of total annual contribution.

This work can be developed in two directions. It is interesting to consider schemes with a limited inheritance period. This allows one to reduce the size of the gross premium as well as the obligations of the insurer. Since some policyholders may lose jobs and/or do not have funds to pay contributions in full and on time for a certain period, it is interesting to consider schemes that take this into account.

**Author Contributions:** Conceptualization, M.S.A.-N. and S.V.A.-N.; methodology, M.S.A.-N.; software, S.V.A.-N.; validation, M.S.A.-N. and S.V.A.-N.; formal analysis, M.S.A.-N. and S.V.A.-N.; investigation, M.S.A.-N. and S.V.A.-N.; resources, S.V.A.-N.; data curation, S.V.A.-N.; writing—original draft preparation, M.S.A.-N. and S.V.A.-N.; writing—review and editing, M.S.A.-N. and S.V.A.-N.; visualization, S.V.A.-N.; supervision, M.S.A.-N. All authors have read and agreed to the published version of the manuscript.

**Funding:** This research received no external funding. It was done within the framework of the Department of Mathematics and the Financial University under the Government of the Russian Federation Basic Research Programs.

**Acknowledgments:** This work is supported by the Financial University under the Government of the Russian Federation. The authors are grateful to the reviewers for their valuable comments that helped us to improve this work. We would like to thank Solovyev A.K., the head of the Department of Actuarial Calculations and Strategic Planning of the Pension Fund of the Russian Federation for useful discussions.

**Conflicts of Interest:** The authors declare no conflict of interest.

## Abbreviations

The following abbreviations are used in this manuscript:

| | |
|---|---|
| PVNCs | Present value of individual net contributions (by period) |
| PVNC | Present value of all net contributions |
| PVPs | Present value of individual payments (by period) |
| PVP | Present value of all payments |
| PVLS | Present value of the lump sum payments |
| UDD | Uniform distribution of death |

## References


1. Pension Freedoms and DWP Benefits (Guidance). Published 27 March 2015. Available online: http://gov.uk/government/publications/pension-freedoms-and-dwp-benefits (accessed on 13 November 2020).
2. Pension Flexibility: New Options from 6 April 2015 (Guidance). Published 12 February 2015. Available online: https://www.gov.uk/government/publications/pension-flexibility-new-options-from-6-april-2015 (accessed on 13 November 2020).
3. Nick Green. What Happens to My Pension when I Die? Published 3 September 2020. Available online: https://www.unbiased.co.uk/life/pensions-retirement/pensions-and-inheritance (accessed on 12 November 2020).
4. Tom Conner. Can I Inherit a Pension? Published 10 October 2019. Available online: https://www.drewberryinsurance.co.uk/knowledge/pensions/can-i-inherit-a-pension (accessed on 12 November 2020).
5. Bowers, N.L.; Gerber H.U.; Hickman J.C.; Jones D.A.; Nesbitt C.J. *Actuarial Mathematics*; The Society of Actuaries: Schaumburg, IL, USA, 1986; 624p.
6. Gerber, H.U. *Life Lnsurance Mathematics*, 3rd ed.; Springer: Berlin/Heidelberg, Germany, 1997; 217p.
7. Zolotarev, V.M. *One-Dimensional Stable Distributions*; American Mathematical Society: Providence, RI, USA, 1986.
8. Uchaikin, V.V.; Zolotarev, V.M. *Chance and Stability Stable Distributions and Their Applications*; VSP: Utrecht, The Netherlands, 1999; p. 569.
9. Kalashnikov, V.V. *Geometric Sums: Bounds for Rare Events with Applications: Risk Analysis, Reliability, Queueing*; Kluwer Academic Publishers: Dordrecht, The Netherlands, 1997.
10. Kalashnikov, V.V. Two-sided bounds of ruin probabilities. *Scand. Actuar. J.* **1996**, *1*, 1–18. [CrossRef]
11. Kalashnikov, V.V.; Tsitsiashvili, G.S. *Asymptotically Correct Bounds of Geometric Convolutions with Subexponential Components*; Working Paper No. 151; Lab. of Actuarial Math., Univ. of Copenhagen: Copenhagen, Denmark, 1998.
12. Korolev, V.; Bening, V.; Shorgin, S. *Mathematical Foundations of Risk Theory*, 2nd ed.; FIZMATLIT: Moscow, Russia, 2011.
13. Al-Nator, M.S.; Al-Nator S.V.; Olenchenko N.A. Accumulative models of pension insurance. In *Materials of Conference on Applied Probability Theory and Theoretical Computer Science*; IPI RAN: Moscow, Russia, 2012; pp. 5–7.
14. Al-Nator, M.S.; Al-Nator S.V. Accumulative pension scheme with two decrement factors and possibility of inheritance. In *Materials of Conference on Information and Telecommunication Technologies and Mathematical Simulation of Hi-Tech Systems*; Peoples' Friendship University of Russia: Moscow, Russia, 2020; pp. 229–232.

mathematics

MDPI

Article

# A Priority Queue with Many Customer Types, Correlated Arrivals and Changing Priorities

**Seokjun Lee [1], Sergei Dudin [2], Olga Dudina [2], Chesoong Kim [3,*] and Valentina Klimenok [2]**

1 Department of Management Information Systems, Sangji University, Wonju 26339, Korea; digitaldesign@sangji.ac.kr
2 Laboratory of Applied Probabilistic Analysis, Belarusian State University, 4, Nezavisimosti Ave., 220030 Minsk, Belarus; dudin85@mail.ru (S.D.); dudina@bsu.by (O.D.); vklimenok@yandex.ru (V.K.)
3 Department of Business Administration, Sangji University, Wonju 26339, Korea
* Correspondence: dowoo@sangji.ac.kr

Received: 15 July 2020; Accepted: 3 August 2020; Published: 5 August 2020

**Abstract:** A single-server queueing system with a finite buffer, several types of impatient customers, and non-preemptive priorities is analyzed. The initial priority of a customer can increase during its waiting time in the queue. The behavior of the system is described by a multi-dimensional Markov chain. The generator of this chain, having essential dependencies between the components, is derived and formulas for computation of the most important performance indicators of the system are presented. The dependence of some of these indicators on the capacity of the buffer space is illustrated. The profound effect of the phenomenon of correlation of successive inter-arrival times and variance of the service time is numerically demonstrated. Results can be used for the optimization of dispatching various types of customers in information transmission systems, emergency departments and first aid stations, perishable foods supply chains, etc.

**Keywords:** priority system; marked Markov arrival process; phase-type distribution; change of the priority; dispatching

## 1. Introduction

Queueing theory is successfully applied in various fields of human activity for optimization of the consumption and scheduling certain restricted resources and provisioning the high quality of service. The overwhelming majority of the existing literature in this theory is devoted to the systems with homogeneous customers; see, e.g., [1]. Because real-world customers are very often heterogeneous in many respects, new developments in the analysis of queues with heterogeneous customers are of great importance. The heterogeneity of the customers with respect to the required resources, level of service, and their economical or social value causes the necessity of the optimal management of their service. Such management can be implemented, e.g., in various generalizations of polling disciplines, processor sharing, applying versatile priority schemes. For some references, see, e.g., [2]. Priority schemes assume the assignment of a certain priority to each class of customers and providing the advantage of access to the restricted resource (we will call this resource as a server) to available customers having the highest priority. Static priorities suggest that once the priorities are assigned, a low priority customer does not have any chance to start service until the server finishes service of all high priority customers presenting in the system. This may cause a low priority customer to wait in the queue much longer than the just arrived high priority customer. To avoid this evident unfairness to the low priority customers, dynamic priorities were taken into consideration. The dynamic priority assumes, e.g., that the low priority customers obtain the chance to start service in presence of high priority customers when: (i) the queue of the low priority customers exceeds some threshold values, see, e.g., [3–6]; or (ii) some relation between the queue lengths of priority and non-priority customers is fulfilled, see, e.g., [7]; or (iii) a certain limit of the number of high priority customers that can overtake the low priority customers is exceeded, see, e.g., [8]. The use of dynamic priorities allows to essentially improve the quality of the system operation. The shortcomings of such priorities are: (i) the necessity to permanently monitor the values of the queue length of different classes of customers what is not always possible (or costly) in some real-world systems and (ii) dependence of the waiting time of a concrete low priority customer on the rate of future arrival of other low priority customers. Another opportunity of providing more fair access to low priority customers is assumed in the models where a low priority customer can become higher priority customer after a certain period of waiting in the buffer. A currently popular model assumes that the low priority customers accumulate a priority during the stay in the queue. The accumulation of the priority may be described as some function, e.g., linear or piece-wise linear function, of the time spent by the customer in a queue. The rate of the increase of the priority may depend on the class to which the customer belongs. Such a type of model was considered, e.g., in the papers [9–14]. The main interest to the queues with accumulating priorities stems from their applicability to modeling operation of emergency departments of hospitals. Arriving customers (patients) are preliminarily sorted (triaged) into several groups according to the severity of the patient's condition. However, during the waiting for treatment by the doctors, a state of health of some patient, which was initially classified as not requiring very urgent treatment, can become essentially worse and this patient has to be transferred to the group of very urgent patients. Because in the described situation the increase of the priority of a customer is not defined by some deterministic function of the elapsed waiting time, another type of model, with the randomized change of a priority, exists in the literature. This type of model was considered, e.g., in [15,16] and the recent paper [17]. The table presenting the state of art in the analysis of queues with priority change after some random amount of time is presented in [17]. It follows from that table that only a few papers consider the models where the arrival processes of customers of different types are not defined by the stationary Poisson arrival process, while it is already well recognized that the flows in many real systems and networks are poorly described by the stationary Poisson arrival process. The rare exceptions, when a more complicated arrival process is considered, are the papers [18–20]. In all these papers, an arbitrary number of priority classes is suggested. In [18], it is assumed that all the flows, except the flow having the highest priority, are described by the stationary Poisson arrival process. The arrival flow of customers having the highest priority is described by a much more general Markov arrival process ($MAP$); see, e.g., [21–23] for more details. In [19,20], the arrival flow is described by even more general marked Markov arrival process ($MMAP$). The $MMAP$, as the essential generalization of the $MAP$ to the case of heterogeneous customers, was introduced in [24]. The models with the $MAP$ or $MMAP$ are much more difficult for analysis than the models with the stationary Poisson arrival process. This explains why only some bounds and tail distributions were obtained in [18] and only the problem of establishing the ergodicity condition (but not the problem of computation of the stationary distribution of the system states and performance measures) is solved in [19,20]. The problem of computation of the stationary distribution of the system states is successfully solved in [17] but only for two classes of customers. The advantage of our paper over [17] is that we suggest any finite number $R$ of priority classes. The arrival process is described by the $MMAP$. The system has a finite buffer and any arriving customer is admitted to the buffer if it is not full. If the buffer is full while some waiting customers have lower priority than the arriving customer, the arriving customer pushes out from the buffer a customer having the lowest priority among the presenting ones. During the stay in the buffer, after an exponentially distributed time, any customer can increase its priority. The service time has a phase-type distribution. After the service completion, the next service is provided for a customer with the highest priority among the presented in the buffer.

It is worth mentioning that the problem of assigning the priorities to different classes of customers is often closely related to the problem of the account of possible impatience of customers from different classes, e.g., if customers of two types are almost equally valuable for the system, the more impatient customers should be given higher priority (and the possibility to increase the priority during the waiting time in a buffer) to avoid the loss of the customer and possible starvation (and poor utilization) of the server in the future. In our model, we pay significant attention to the account of impatience.

Besides the above-mentioned popular model of treatment of patients in a hospital emergency department, we mention the following examples of potential applications of the considered model to the analysis and optimization of real-world systems.

(1) Let us consider the operation of an information transmission channel. Several kinds of information having approximately the same transmission times, but having different importance for the system and different tolerance to the delay are transmitted through this channel. Initially, the priorities can be assigned to the different types of information depending on their importance. However, to avoid the loss of low priority and delay-sensitive information units (and possible under-utilization of the channel in the future), it makes sense to allow a low priority information unit whose obsolescence time is almost expired to become a high priority information unit and receive the service soon.

(2) Let us consider the operation of a first aid station. The station has to accept the calls for help, categorize the urgency of the required help, and to manage the assignment of the necessary ambulance car for providing help, e.g., in the Republic of Belarus (as of 1 January 2020), there are three possible categories of the urgency of the required help.

    (a) An emergency call—when a patient suddenly has diseases, conditions and (or) exacerbation of chronic diseases that pose a threat to the patient's life and (or) others requiring emergency medical intervention;
    (b) An urgent call is associated with a sharp deterioration in the patient's health status when it is not possible to clarify the reasons for treatment;
    (c) A less urgent call—when the patient suddenly has diseases, conditions, and/or exacerbation of chronic diseases without obvious signs of a threat to the patient's life, requiring urgent medical intervention.

Accordingly, the emergency call has the highest priority, the urgent call has the middle priority, and the less urgent calls have the lowest priority. However, along with this categorization and establishing the priority in service, there exist strict standards for starting the provisioning of help. A dispatcher has to assign an ambulance car for providing help to patients before the fixed deadlines. In Minsk, the capital of the Republic of Belarus, these standards are fixed as four minutes for the emergency call, fifteen minutes for the urgent call, and sixty minutes for the less urgent call. Violation of this standard is punished. In this example, the service time can be interpreted as a time between the sequential release of ambulance cars. The service time essentially depends on the number of available cars and medical teams. The results of the analysis of the model given in our paper can be useful for the optimization of the work of the described first aid station via a proper choice of the number of ambulance teams to guarantee the required quality of service.

Methodological value of the paper consists of presenting a way for analysis of various transitions of a set of interacting Markov processes, which define the dynamics of the number of customers of several types in the system, caused by new customers of various types arrival, service completion, departure due to impatience, changing the priority, and pushing out the low priority customers in the case of the buffer overflow.

The organization of the text is as follows. In Section 2, the mathematical model is described and graphically illustrated. The multi-dimensional Markov chain including as components the total number of customers in the system, the states of the underlying processes of customers arrival and service, and the number of customers of all types presenting in the system is defined in Section 3. The set of matrices defining the probabilities or intensities of transitions of the number of customers of all types are given and the generator of the Markov chain is written down. Formulas for computation of the main performance measures of the system are presented in Section 4. The numerical example illustrating the dependence of performance measures of the system on the capacity of the buffer is presented in Section 5. The importance of account of a complicated pattern of arrival process and variance of the service time is demonstrated there. Section 6 concludes the paper.

## 2. Mathematical Model

We consider a single-server queuing system where service is provided to $R$ types of customers. The structure of the system is presented in Figure 1.

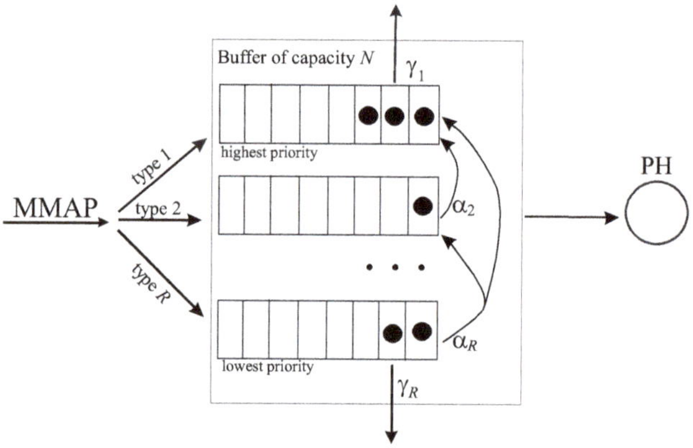


**Figure 1.** Structure of the system.

The customer arrival process is assumed to be defined by the $MMAP$ (see, e.g., [24]). As the recent papers where the queuing models with the $MMAP$ are analyzed, we can mention, e.g., [25–27].

Customer arrivals in the $MMAP$ can occur at the moments of the transitions of the irreducible continuous-time Markov chain $\nu_t$, $t \geq 0$, having a state space $\{1, 2, ..., W\}$. The $MMAP$ is completely described by the square matrices $D_0$, $D_r$, $r = \overline{1, R}$. Hereinafter, the denotation like $r = \overline{1, R}$ means that the parameter $r$ takes values $\{1, \ldots, R\}$.

The matrix $D_r$ defines the transition intensities of the underlying process $\nu_t$ that lead to arrival of a type-$r$ customer, $r = \overline{1, R}$. The non-diagonal entries of the matrix $D_0$ define the transition intensities of the underlying process that do not lead to any arrival. The moduli of the diagonal entries of the matrix $D_0$ define the intensity of the the process $\nu_t$ departure of from its states. The matrix $D(1) = D_0 + D$ where $D = \sum\limits_{r=1}^{R} D_r$ is the generator of the underlying process.

The mean arrival rate $\lambda$ is defined by $\lambda = \boldsymbol{\theta} D \mathbf{e}$ where $\boldsymbol{\theta}$ is the invariant probability row vector of the underlying process. This vector is computed as the unique solution for the finite system $\boldsymbol{\theta} D(1) = \mathbf{0}$, $\boldsymbol{\theta}\mathbf{e} = 1$. Hereinafter, $\mathbf{e}$ denotes a column vector of appropriate size consisting of 1s and $\mathbf{0}$ denotes a row vector consisting of zeroes.

The mean rate $\lambda_r$ of type-$r$ customers arrival is computed as $\lambda_r = \boldsymbol{\theta} D_r \mathbf{e}$, $r = \overline{1, R}$. The squared coefficient of variation $c_{var}^2$ of the intervals between successive arrivals is given by $c_{var}^2 = 2\lambda\boldsymbol{\theta}(-D_0)^{-1}\mathbf{e} - 1$. The coefficient of correlation $c_{cor}$ of two successive intervals between arrivals is given by

$$c_{cor} = (\lambda\boldsymbol{\theta}(-D_0)^{-1}D(-D_0)^{-1}\mathbf{e} - 1)/c_{var}^2.$$

The system has the finite common buffer space for storing the customers that arrive when the server is busy. The capacity of the buffer is $N$, $N \geq 1$. Therefore, the total number of customers of all types, which can stay in the system simultaneously, is restricted by the number $N + 1$. If a customer of any type arrives when the server is idle, the customer immediately starts processing by the server (service). If the server is busy but the buffer is not full, the customer of any type is placed into the buffer dedicated to this type of customers. There is no specific restriction on the capacity of the dedicated buffers, except that the total number of the customers staying in all these buffers always does not exceed the capacity $N$.

Customers of different types have different priorities. The priority defines the fate of the customer if it arrives when the buffer is full and the order of picking up the customers from the buffer when the server finishes service. We assume that type-$r$, $r = \overline{1,R}$, customers have the non-preemptive priority over type-$l$ customers, $l = \overline{r+1,R}$. This means the following.

(1) If during the arrival of a type-$r$ customer the server is busy and the number of customers in the buffer is $N$ and there are no type-$l$, $l = \overline{r+1,R}$, customers, the arriving customer is lost. If there are type-$l$, $l = \overline{r+1,R}$, customers in the buffer then, with the probability $q$, the arriving customer is accepted to the buffer and one of the customers with the lowest priority among the presenting in the buffer is lost. With the complimentary probability $1-q$, the arriving customer is lost despite the presence in the system of customers with lower priority.
(2) Type-1 customers have the highest priority among all types of customers and if type-1 customers present in the buffer at a service completion epoch, one of these customers starts service, ..., type $R$ customers have the lowest priority. A customer of such a type has a chance to start service only if customers of types $1, 2, \ldots, R-1$ are absent in the buffer. Service of any customer cannot be preempted (interrupted) in the case of an arrival of a customer having a higher priority.

We assume that during the stay in the system, each customer of type-$r$, $r = \overline{2,R}$, can increase its priority. It means that after exponentially distributed time with the parameter $\alpha_r$ a type-$r$ customer becomes a type-$l$ customer with the probability $p_{r,l}$, $l = \overline{1,r-1}$, independently of other customers. Here, $\sum\limits_{l=1}^{r-1} p_{r,l} = 1$, $r = \overline{2,R}$.

It is worth noting that more popular in the existing literature assumption is that only the head-of-the-line customer of each type can make a jump to the end of the queue of higher priority customers. We assume that each customer of any type can jump to higher priority class, independently of other customers. This means that not only the head-of-the-line customer has a clock counting the time till the jump, but each customer (not of the highest priority) has its own clock. Our assumption seems more realistic in some potential applications, e.g., health of any patient, not only the head-of-the-line patient in emergency department modeling example, can suddenly become worse. The same is true in applications where various information units become obsolete independently of the other units or different perishable foods have independent spoiling times. Note also, that, using the slight modification of some matrix blocks defined and constructed in the next section, the presented results can be extended to the models with the head-of-the-line customer priority jumps as well.

Customers staying in the buffer are impatient and can leave the system without service, independently of other customers, if the waiting time is too long. A type-$r$ customer leaves the system without service after an exponentially distributed patience time with the parameter $\gamma_r$, $\gamma_r \geq 0$. Let us denote $\gamma = (\gamma_1, \gamma_2, \ldots, \gamma_R)$. If the customer changes the priority, its patience time starts from the early beginning with the parameter corresponding to the new priority.

We assume that the service time of any type customer has a $PH$ distribution with the underlying Markov process $m_t, t \geq 0$, having a finite state space $\{1, \ldots, M, M+1\}$ and the irreducible representation $(\boldsymbol{\beta}, S)$, see, [28]. We denote $\mathbf{S}_0 = -S\mathbf{e}$. The mean service time is given by $b_1 = \boldsymbol{\beta}(-S)^{-1}\mathbf{e}$. The mean service rate can be compute as $\mu = b_1^{-1}$.

If during the service completion epoch there are customers in the buffer, the first customer among having the highest priority starts service. Otherwise, the server remains idle until the next arrival moment.

## 3. Process of the System States

The behavior of the system under study can be described by the regular irreducible continuous-time Markov chain

$$\xi_t = \{n_t, \nu_t, m_t, \eta_t^{(1)}, \ldots, \eta_t^{(R)}\},\ t \geq 0,$$

where, during the epoch $t$,

- $n_t$ is the number of customers in the system, $n_t = \overline{0, N+1}$;
- $\nu_t$ is the state of the underlying process of the $MMAP$, $\nu_t = \overline{1, W}$;
- $m_t$ is the state of the underlying process of PH service process, $m_t = \overline{1, M}$;
- $\eta_t^{(r)}$ is the number of type-$r$ customers in the buffer, $\eta_t^{(r)} = \overline{0, n_t - 1}$, $r = \overline{1, R}$, $\sum\limits_{r=1}^{R} \eta_t^{(r)} = n_t - 1$, $n_t > 1$.

To investigate the Markov chain $\xi_t,\ t \geq 0$, let us enumerate its states in the direct lexicographic order of the components $\nu_t$ and $m_t$, and in the reverse lexicographic order of the components $\eta_t^{(1)}, \ldots, \eta_t^{(R)}$.

The most technically difficult and important part of the research is the analysis of the transitions of the process of the number of different type customers in the buffer. Let us firstly consider the process $\zeta_t^{(n)} = \{\eta_t^{(1)}, \ldots, \eta_t^{(R)}\},\ t \geq 0,\ \eta_t^{(r)} = \overline{0, n},\ r = \overline{1, R},\ \sum\limits_{r=1}^{R} \eta_t^{(r)} = n$. The process $\zeta_t^{(n)}$ describes the transitions of the number of different types customers in the buffer when the total number of customers in the buffer is $n$. First, we present the algorithms for computing the set of the matrices that define the transition probabilities or transition intensities of the process $\zeta_t^{(n)}$ at the moments of the changes, due to various reasons, of the components of this process when $n$, $n = \overline{1, N}$, customers stay in the buffer.

**Lemma 1.**

(a) *Let $L_n(\gamma)$ be the matrix the entries of which define the intensities of transitions when some customer leaves the buffer due to impatience.*

*The matrices $L_n(\gamma)$, $n = \overline{1, N}$, can be computed by the following way:*

1. *Calculate the matrices $L_n^{(l)}(\gamma)$ using the recursive formulas:*

$$L_n^{(0)}(\gamma) = n\gamma_R,$$

$$L_n^{(l)}(\gamma) = \begin{pmatrix} n\gamma_{R-l}I & O & \cdots & O \\ L_1^{(l-1)}(\gamma) & (n-1)\gamma_{R-l}I & \cdots & O \\ O & L_2^{(l-1)}(\gamma) & \cdots & O \\ \vdots & \vdots & \ddots & \vdots \\ O & O & \cdots & \gamma_{R-l}I \\ O & O & \cdots & L_n^{(l-1)}(\gamma) \end{pmatrix},\ l = \overline{1, R-1}.$$

*Here and after, $I$ is the identity matrix and $O$ is a zero matrix of an appropriate dimension;*

2. *Calculate the matrices $L_n(\gamma)$ as $L_n(\gamma) = L_n^{(R-1)}(\gamma)$, $n = \overline{1, N}$.*

(b) *Let $Y_n = Y_n(H)$ be the matrix the entries of which define the intensities of transitions that occur when some customer increases its priority. Here, the matrix $H$ defines the intensities of priorities increasing and has the following form:*

$$H = \begin{pmatrix} 0 & 0 & 0 & \cdots & 0 & 0 \\ \alpha_2 & 0 & 0 & \cdots & 0 & 0 \\ p_{3,1}\alpha_3 & p_{3,2}\alpha_3 & 0 & \cdots & 0 & 0 \\ \vdots & \vdots & \ddots & \vdots & \vdots & \vdots \\ p_{R-1,1}\alpha_{R-1} & p_{R-1,2}\alpha_{R-1} & p_{R-1,3}\alpha_{R-1} & \cdots & 0 & 0 \\ p_{R,1}\alpha_R & p_{R,2}\alpha_R & p_{R,3}\alpha_R & \cdots & p_{R,R-1}\alpha_R & 0 \end{pmatrix}.$$

*Calculation of the matrices $Y_n(H)$, $n = \overline{1,N}$, can be performed as follows:*

1. *Calculate the matrices $H_j$, $j = \overline{1,R-2}$, which are obtained by deletion of $R-2-j$ first rows and columns from the matrix $H$.*
2. *Calculate the matrices $Z_n^{(l)}(H_j)$ using the recursive formulas:*

$$Z_n^{(0)}(H_j) = nh^j_{r_j,1},\ n = \overline{1,N},\ j = \overline{1,R-2},$$

$$Z_n^{(l)}(H_j) = \begin{pmatrix} nh^j_{r_j-l,1}I & O & \cdots & O \\ Z_1^{(l-1)}(H_j) & (n-1)h^j_{r_j-l,1}I & \cdots & O \\ O & Z_2^{(l-1)}(H_j) & \cdots & O \\ \vdots & \vdots & \ddots & \vdots \\ O & O & \cdots & h^j_{r_j-l,1}I \\ O & O & \cdots & Z_n^{(l-1)}(H_j) \end{pmatrix},$$

$$l = \overline{1,\ldots,r_j-2},\ n = \overline{1,N},\ j = \overline{1,R-2},$$

*where $h^j_{a,b}$ is the $(a,b)$th entry of the matrix $H_j$ and $r_j$ is the number of rows of the matrix $H_j$.*

3. *Calculate the matrices $X_n^{(l)}(H_j)$ using the recursive formulas:*

$$X_n^{(0)}(H_j) = h^j_{1,r_j},\ n = \overline{0,N-1},\ j = \overline{1,R-2},$$

$$X_n^{(l)}(H_j) = \begin{pmatrix} h^j_{1,r_j-l}I & X_0^{(l-1)}(H_j) & O & \cdots & O & O \\ O & h^j_{1,r_j-l}I & X_1^{(l-1)}(H_j) & \cdots & O & O \\ \vdots & \vdots & \vdots & \ddots & \vdots & \vdots \\ O & O & O & \cdots & h^j_{1,r_j-l}I & X_n^{(l-1)}(H_j) \end{pmatrix},$$

$$l = \overline{1,r_j-2},\ n = \overline{0,N-1},\ j = \overline{1,R-2}.$$

4. *Calculate the matrices $Z_n(H_j) = Z_n^{(r_j-2)}(H_j)$, $n = \overline{1,N}$, and $X_n(H_j) = X_n^{(r_j-2)}(H_j)$, $n = \overline{0,N-1}$, $j = \overline{1,R-2}$.*

5. *Calculate the matrices* $Y_n^{(j)}$, $n = \overline{1,N}$, *using the recursive formulas:*

$$Y_n^{(0)} = \begin{pmatrix} 0 & nH_{M-1,M} & 0 & \cdots & 0 & 0 \\ H_{M,M-1} & 0 & (n-1)H_{M-1,M} & \cdots & 0 & 0 \\ 0 & 2H_{M,M-1} & 0 & \cdots & 0 & 0 \\ \vdots & \vdots & \vdots & \ddots & \vdots & \vdots \\ 0 & 0 & 0 & \cdots & 0 & H_{M-1,M} \\ 0 & 0 & 0 & \cdots & nH_{M,M-1} & 0 \end{pmatrix},$$

$$Y_n^{(j)} = \begin{pmatrix} O & nX_0(H_j) & O & \cdots & O & O \\ Z_1(H_j) & Y_1^{(j-1)} & (n-1)X_1(H_j) & \cdots & O & O \\ O & Z_2(H_j) & Y_2^{(j-1)} & \cdots & O & O \\ \vdots & \vdots & \vdots & \ddots & \vdots & \vdots \\ O & O & O & \cdots & Y_{n-1}^{(j-1)} & 1X_{n-1}(H_j) \\ O & O & O & \cdots & Z_n(H_j) & Y_n^{(j-1)} \end{pmatrix},$$

$$j = \overline{1, R-2}.$$

6. *Calculate the matrices* $Y_n(H)$ *as* $Y_n(H) = Y_n^{(R-2)}$, $n = \overline{1,N}$.

(c) *Let* $A_n(\mathbf{h})$, $n = \overline{0, N-1}$, *be the matrix the entries of which define the transition probabilities at the moment when a new customer arrives to the system and the system capacity is not exhausted (there are* $n$, $0 \leq n < N$, *customers in the buffer). Here, the row vector* $\mathbf{h}$ *has the following form* $\mathbf{h} = (h_1, h_2, \dots, h_R)$ *where* $h_r$ *is the probability that the arrived to the system customer has type-*$r$, $r = \overline{1,R}$.

*Computation of the matrices* $A_n(\mathbf{h})$ *can be performed as follows:*

$A_0(\mathbf{h}) = \mathbf{h}$ *and* $A_n(\mathbf{h}) = A_n^{(R-2)}(\mathbf{h})$ *where the matrices* $A_n^{(l)}(\mathbf{h})$ *of block size* $(n+1) \times (n+2)$, $n = \overline{1, N-1}$, *are recursively computed as*

$$A_n^{(0)}(\mathbf{h}) = \begin{pmatrix} h_{R-1} & h_R & 0 & \cdots & 0 & 0 \\ 0 & h_{R-1} & h_R & \cdots & 0 & 0 \\ \vdots & \vdots & \vdots & \ddots & \vdots & \vdots \\ 0 & 0 & 0 & \cdots & h_{R-1} & h_R \end{pmatrix},$$

$$A_n^{(l)}(\mathbf{h}) = \begin{pmatrix} h_{R-l-1} & \bar{\mathbf{h}}^{(l)} & \mathbf{0} & \mathbf{0} & \cdots & \mathbf{0} & \mathbf{0} \\ \mathbf{0}^T & h_{R-l-1}I & A_1^{(l-1)} & O & \cdots & O & O \\ \mathbf{0}^T & O & h_{R-l-1}I & A_2^{(l-1)} & \cdots & O & O \\ \vdots & \vdots & \vdots & \vdots & \ddots & \vdots & \vdots \\ \mathbf{0}^T & O & O & O & \cdots & h_{R-l-1}I & A_n^{(l-1)} \end{pmatrix},$$

$$l = \overline{1, R-2},$$

*where the vectors* $\bar{\mathbf{h}}^{(l)}$ *are defined as* $\bar{\mathbf{h}}^{(l)} = (h_{R-l}, h_{R-l+1}, \dots, h_R)$, $l = \overline{1, R-2}$.

(d) *Let* $E_n^-$, $n = \overline{1,N}$, *be the matrix the entries of which define the transition probabilities at the moment when a customer with the maximal (among currently presenting in the system) priority is chosen for service.*

*The matrices* $E_n^-$ *can be computed as*

$$E_1^- = (\underbrace{1, 1, \dots, 1}_{R})^T,$$

$$E_n^- = \begin{pmatrix} I_{K_R^{(n)}} & \\ O_{K_{R-1}^{(n)} \times (K_R^{(n)} - K_{R-1}^{(n)})} & I_{K_{R-1}^{(n)}} \\ \cdots & \\ O_{K_2^{(n)} \times (K_R^{(n)} - K_2^{(n)})} & I_{K_2^{(n)}} \\ O_{K_1^{(n)} \times (K_R^{(n)} - K_1^{(n)})} & I_{K_1^{(n)}} \end{pmatrix}, \; n = \overline{2, N},$$

*where*

$$K_r^{(n)} = \binom{n+r-2}{r-1}, \; r = \overline{1, R}.$$

*Here,* $\binom{n+r-2}{r-1} = C_{n+r-2}^{r-1}$ *is the binomial coefficient.*

(e) *Let the entries of the square matrix* $\hat{E}_r$, $r = \overline{1, R}$, *of size* $\binom{N+R-1}{R-1}$ *define the transition probabilities at the moment when a type-r customer arrives at the system when there are N customers in the buffer and the arriving customer tries to force out a customer with a lower priority from the buffer. All entries in each row of this matrix are equal to zero except one entry which is equal to 1. We assume that each row and column of the matrix* $\hat{E}_r$ *correspond to some state* $\{\eta_1, \eta_2, \ldots, \eta_R\}$ *of the process* $\zeta_t$, $t \geq 0$. *Note, that all states of the process* $\zeta_t$, $t \geq 0$, *are enumerated in the reverse lexicographical order of components* $\eta_t^{(1)}, \ldots, \eta_t^{(R)}$. *For example, the first row and column of the matrix* $\hat{E}_r$ *correspond to the state* $\{N, 0, 0, \ldots, 0\}$, *the second row and column correspond to the state* $\{N-1, 1, 0, \ldots, 0\}$,*…, the last row and column correspond to the state* $\{0, 0, 0, \ldots, N\}$. *In the row of the matrix* $\hat{E}_r$ *that corresponds to the state* $\{\eta_1, \eta_2, \ldots, \eta_R\}$, *the entry 1 is located in the column that corresponds to the same state* $\{\eta_1, \eta_2, \ldots, \eta_R\}$ *only in the case if* $\eta_l = 0$ *for all* $l$, $R \geq l > r$. *In this case, the arriving type-r customer is lost, because the customers with lower priority are absent in the buffer. If* $\eta_l > 0$ *for some* $l$, $R \geq l > r$ *and* $r^*$ *is a maximum of such values* $l$, *then the entry 1 is located in the column that corresponds to the state* $\{\eta_1, \ldots, \eta_{r-1}, \eta_r + 1, \eta_{r+1}, \ldots, \eta_{r*-1}, \eta_{r*} - 1, 0, \ldots, 0\}$. *In this case, the customer of type-*$r^*$ *has the lowest priority among the customers presenting in the system and an arriving type-r customer forces out one type-*$r^*$ *customer which departs from the system (is lost).*

**Proof.** The derivation of the form of the matrices that describe the transitions of the process $\zeta_t^{(n)}$, $t \geq 0$, is quite complicated and cumbersome. In derivations, we used some ideas of the paper [29]. To explain the scheme of the derivation of the form of the presented matrices, we show here how to compute the matrices $L_n(\gamma)$, $n = \overline{1, R}$, the entries of which define the intensities of transitions of the components of the process $\zeta_t^{(n)}$, $t \geq 0$, when some customer leaves the buffer due to impatience. The rest of the matrices that define the intensities of transition of the components of the process $\zeta_t^{(n)}$, $t \geq 0$, can be obtained by the same way based on the careful account of possible transitions.

Computation of the matrices $L_n(\gamma)$ can be performed as follows. Let us introduce the matrices $L_n^{(l)}(\gamma)$ of the transition intensities of the components $n_t^{(R)}, \ldots, n_t^{(R-l)}$ at the moment when there are $n$ customers in the buffer and one of the customers leaves it due to impatience conditional on the fact that all customers have types $R, R-1, \ldots, R-l$, where $l = \overline{0, R-1}$.

It is clear, that for $l = 0$, the matrices $L_n^{(0)}(\gamma)$ have the scalar form $L_n^{(0)}(\gamma) = n\gamma_R$, because all $n$ customers are of type-$R$ in this situation.

Let us consider the matrix $L_n^{(1)}$. This matrix defines the transition intensities of the components $n_t^{(R)}, n_t^{(R-1)}$ at the moment when there are $n$ customers in the buffer and one of the customers leaves it due to impatience conditional on the fact that all customers have types $R$ or $R-1$. Taking into account the reverse lexicographic order of components, by definition the first row of the matrix $L_n^{(1)}(\gamma)$ corresponds to the state where all $n$ customers are of type-$(R-1)$, the second row corresponds to the state where $n-1$ customers are of type-$(R-1)$ and one customer is of type-$R$, etc., the last row corresponds to the state where all $n$ customers are of type-$R$. After the customer leaves the system, the number of customers in the buffer decreases by 1. Thus, the first column of the matrix $L_n^{(1)}(\gamma)$

corresponds to the state where all $n-1$ customers are of type-$(R-1)$, the second column corresponds to the state where $n-2$ customers are of type-$(R-1)$ and one customer is of type-$R$, etc., the last column corresponds to the state where all $n-1$ customers are of type-$R$. Taking into account these considerations, it is easy to verify that the matrix $L_n^{(1)}(\gamma)$ of size $(n+1)\times n$ has the form

$$L_n^{(1)}(\gamma)=\begin{pmatrix} n\gamma_{R-1} & 0 & \cdots & 0 \\ \gamma_R & (n-1)\gamma_{R-1} & \cdots & 0 \\ 0 & 2\gamma_R & \cdots & 0 \\ \vdots & \vdots & \ddots & \vdots \\ 0 & 0 & \cdots & \gamma_{R-1} \\ 0 & 0 & \cdots & n\gamma_R \end{pmatrix},$$

or

$$L_n^{(1)}(\gamma)=\begin{pmatrix} n\gamma_{R-1} & 0 & \cdots & 0 \\ L_1^{(0)}(\gamma) & (n-1)\gamma_{R-1} & \cdots & 0 \\ 0 & L_2^{(0)}(\gamma) & \cdots & 0 \\ \vdots & \vdots & \ddots & \vdots \\ 0 & 0 & \cdots & \gamma_{R-1} \\ 0 & 0 & \cdots & L_n^{(0)}(\gamma) \end{pmatrix}.$$

Using the same reasonings, it can be shown that the matrix $L_n^{(l)}(\gamma)$ of block size $(n+1)\times n$ has the following form

$$L_n^{(l)}(\gamma)=\begin{pmatrix} n\gamma_{R-l}I & O & \cdots & O \\ L_1^{(l-1)}(\gamma) & (n-1)\gamma_{R-l}I & \cdots & O \\ O & L_2^{(l-1)}(\gamma) & \cdots & O \\ \vdots & \vdots & \ddots & \vdots \\ O & O & \cdots & \gamma_{R-l}I \\ O & O & \cdots & L_n^{(l-1)}(\gamma) \end{pmatrix},\ l=\overline{2,R-1}.$$

It is clear that the required matrices $L_n(\gamma)$ can be computed as $L_n(\gamma)=L_n^{(R-1)}(\gamma)$, $n=\overline{1,N}$. This proves the proposed formulas for computation of the matrices $L_n(\gamma)$. □

**Remark 1.** *Derivation of the form of the matrices defined in Lemma 1 creates an opportunity to analyze not only the system under study in this paper but also many other queueing systems with a finite buffer and many types of customers having different priorities.*

Let us introduce the following notation:

- $\otimes$ and $\oplus$ indicate the symbols of the Kronecker product and sum of matrices, respectively, see [30];
- $\mathbf{h}_r=(\underbrace{0,\ldots,0}_{r-1},1,\underbrace{0,\ldots,0}_{R-r})$, $r=\overline{1,R}$;
- $\hat{I}_n=-\text{diag}\{Y_n\mathbf{e}+L_n\mathbf{e}\}$, $n=\overline{1,N}$, where diag$\{\ldots\}$ denotes the diagonal matrix with the diagonal entries defined by the vector in the brackets;
- $K_n=\binom{n+R-1}{R-1}$, $n=\overline{1,N}$.

By analyzing all possible transitions of the Markov chain $\xi_t, t \geq 0$, during an interval of infinitesimal length and rewriting the intensities of these transitions in the block matrix form, we obtain the following result.

**Theorem 1.** *The infinitesimal generator $Q$ of the Markov chain $\xi_t,\ t \geq 0$, has the following block-tridiagonal structure*

$$Q = \begin{pmatrix} Q_{0,0} & Q_{0,1} & O & O & \dots & O & O \\ Q_{1,0} & Q_{1,1} & Q_{1,2} & O & \dots & O & O \\ O & Q_{2,1} & Q_{2,2} & Q_{2,3} & \dots & O & O \\ \vdots & \vdots & \vdots & \vdots & \ddots & \vdots & \vdots \\ O & O & O & O & \dots & Q_{N+1,N} & Q_{N+1,N+1} \end{pmatrix}.$$

*The non-zero blocks are defined as follows:*

$$Q_{0,0} = D_0,$$

$$Q_{1,1} = D_0 \oplus S,$$

$$Q_{n,n} = D_0 \oplus S \otimes I_{K_{n-1}} + I_{WM} \otimes (Y_{n-1} + \hat{I}_{n-1}),\ n = \overline{2, N},$$

$$Q_{N+1,N+1} = (D_0 \oplus S) \otimes I_{K_N} + I_{WM} \otimes (Y_N + \hat{I}_N) + (1-q) \sum_{r=1}^{R} D_r \otimes I_{MK_N} +$$

$$q \sum_{r=1}^{R} D_r \otimes I_M \otimes \hat{E}_r,$$

$$Q_{0,1} = \sum_{r=1}^{R} D_r \otimes \boldsymbol{\beta},$$

$$Q_{n,n+1} = \sum_{r=1}^{R} D_r \otimes I_M \otimes A_{n-1}(\mathbf{h}_r),\ n = \overline{1, N},$$

$$Q_{1,0} = I_W \otimes \mathbf{S_0},$$

$$Q_{n,n-1} = I_W \otimes \mathbf{S_0}\boldsymbol{\beta} \otimes E^{-}_{n-1} + I_{WM} \otimes L_{n-1}(\gamma),\ n = \overline{1, N+1}.$$

The Markov chain $\xi_t,\ t \geq 0$, is an irreducible and has a finite state space. Therefore, the stationary probabilities of the system states

$$\pi(n, v, m, \eta^{(1)}, \dots, \eta^{(R)}) =$$

$$= \lim_{t \to \infty} P\{n_t = n, v_t = v, m_t = m, \eta_t^{(1)} = \eta^{(1)}, \dots, \eta_t^{(R)} = \eta^{(R)}\}$$

always exist.

Let us form the row vectors $\boldsymbol{\pi}_n,\ n = \overline{0, N+1}$, of these probabilities which are enumerated in the reverse lexicographic order of the components $\eta_t^{(1)}, \dots, \eta_t^{(R)}$ and the direct lexicographic order of the components $v_t$ and $m_t$.

It is well known that the probability vectors $\boldsymbol{\pi}_n,\ n = \overline{0, N+1}$, satisfy the following system of linear algebraic equations:

$$(\boldsymbol{\pi}_0, \boldsymbol{\pi}_1, \dots, \boldsymbol{\pi}_{N+1})Q = \mathbf{0}, \tag{1}$$

$$(\boldsymbol{\pi}_0, \boldsymbol{\pi}_1, \dots, \boldsymbol{\pi}_{N+1})\mathbf{e} = 1$$

where $Q$ is the infinitesimal generator of the Markov chain $\xi_t,\ t \geq 0$.

To compute the steady-state distribution of this Markov chain, it is necessary to solve system (1). The matrix of this system has the block-tridiagonal structure. Markov chains having the structure of the generator similar to the one defined in Theorem 1 are sometimes called in the existing literature as the Level-Dependent Quasi-Birth-and-Death processes; see, e.g., [31]. System (1) is finite and can be directly solved via the use of the variety of the standard computer programs. However, the number of equations of the finite system (1) for queueing model under study can be large especially when the buffer capacity $N$ or the number of priority classes is large. Therefore, to effectively solve this system, it is desirable to apply an algorithm that exploits the sparse block-tridiagonal structure of the generator $Q$. In particular, the algorithm given in [32] can be recommended.

## 4. Performance Measures

The average number of customers in the buffer is

$$N_{buffer} = \sum_{n=2}^{N+1} (n-1)\boldsymbol{\pi}_n \mathbf{e}.$$

The average number $N_{buffer}^{(r)}$ of type-$r$, $r = \overline{1,R}$, customers in the buffer can be computed as

$$N_{buffer}^{(r)} = \sum_{n=2}^{N+1} \boldsymbol{\pi}_n (I_{WM} \otimes L_{n-1}(\mathbf{h}_r))\mathbf{e}.$$

The intensity of the output flow of successfully serviced customers is

$$\lambda_{out} = \sum_{n=1}^{N+1} \boldsymbol{\pi}_n (I_W \otimes \mathbf{S}_0 \otimes I_{K_{n-1}})\mathbf{e}.$$

The intensity of the output flow of customers who leave the buffer due to impatience is

$$\lambda_{imp} = \sum_{n=2}^{N+1} \boldsymbol{\pi}_n (I_{WM} \otimes L_{n-1}(\boldsymbol{\gamma}))\mathbf{e}.$$

The probability $P_{loss}$ of loss of an arbitrary customer is computed

$$P_{loss} = 1 - \frac{\lambda_{out}}{\lambda}.$$

The probability $P_{imp-loss}$ of loss of an arbitrary customer due to impatience is computed

$$P_{imp-loss} = \frac{\lambda_{imp}}{\lambda}.$$

The intensity $\lambda_{imp}^{(r)}$ of the output flow of the type-$r$, $r = \overline{1,R}$, customers who leave the buffer due to impatience is

$$\lambda_{imp}^{(r)} = \sum_{n=2}^{N+1} \boldsymbol{\pi}_n (I_{WM} \otimes L_{n-1}(\boldsymbol{\gamma}_r))\mathbf{e}$$

where $\boldsymbol{\gamma}_r$ is the row vector of size $R$ with all zero entries except the $r$-th entry which is equal to $\gamma_r$.

The average intensity $\tilde{\lambda}^{(r)}$ of the type-$l$, $l = \overline{r+1,R}$, customers transformation to the type-$r$, $r = \overline{1,R-1}$, customers is computed as

$$\tilde{\lambda}^{(r)} = \sum_{l=r+1}^{R} \alpha_l N_{buffer}^{(l)} p_{l,r}.$$

The probability $P^{(r)}_{imp-loss}$, $r = \overline{1,R}$, of loss of an arbitrary type-$r$ customer due to impatience can be computed

$$P^{(r)}_{imp-loss} = \frac{\lambda^{(r)}_{imp}}{\lambda_r + \tilde{\lambda}^{(r)}}.$$

Here, we assume that $\tilde{\lambda}^{(R)} = 0$.

The probability of an arbitrary type-$r$ customer loss upon arrival without trying to force out a customer with lower priority is

$$P^{(r)}_{ent-loss-without-force-out} = (1-q)\lambda_r^{-1}\boldsymbol{\pi}_{N+1}(D_r \otimes I_{MK_N})\mathbf{e},\ r = \overline{1,R}.$$

The probability of an arbitrary type-$r$ customer loss upon arrival despite an attempt to force out a customer with lower priority is

$$P^{(r)}_{ent-loss-with-force-out} = q\lambda_r^{-1}\boldsymbol{\pi}_{N+1}(D_r \otimes I_M \otimes \tilde{E}_r)\mathbf{e},\ r = \overline{1,R},$$

where the matrix $\tilde{E}_r$ has all zero entries except the diagonal entries which are equal to the diagonal entries of the matrix $\hat{E}_r$.

The probability of an arbitrary customer loss upon arrival is

$$P_{ent-loss} = \frac{\sum\limits_{r=1}^{R}((1-q)\boldsymbol{\pi}_{N+1}(D_r \otimes I_{MK_N})\mathbf{e} + q\boldsymbol{\pi}_{N+1}(D_r \otimes I_M \otimes \tilde{E}_r)\mathbf{e})}{\lambda}.$$

The probability of an arbitrary type-$r$ customer loss upon arrival is

$$P^{(r)}_{ent-loss} = P^{(r)}_{ent-loss-with-force-out} + P^{(r)}_{ent-loss-without-force-out},\ r = \overline{1,R}.$$

The probability that an arbitrary type-$r$ customer meets the full buffer upon arrival and forces out a customer with lower priority is

$$P^{(r)}_{force-out} = q\lambda_r^{-1}\boldsymbol{\pi}_{N+1}(D_r \otimes I_M \otimes \bar{E}_r)\mathbf{e},\ r = \overline{1,R},$$

where the matrix $\bar{E}_r = \hat{E}_r - \tilde{E}_r$.

Let the square matrix $\hat{E}_{r,l}$, $r = \overline{1,R-1}$, $l = \overline{r+1,R}$, of size $\binom{N+R-1}{R-1}$ define the transition probabilities of the process $\zeta_t^{(N)}$, $t \geq 0$, at the moment when a type-$r$ customer arrives to the system when there are $N$ customers in the buffer and the arriving customer forces out a type-$l$ customer from the buffer. This matrix is defined by analogy with the matrix $\hat{E}_r$ defined above. All entries in each row of this matrix are equal to zero except one entry which can be equal to 1. We assume that each row and column of the matrix $\hat{E}_{r,l}$ correspond to some state $\{\eta_1, \eta_2, \ldots, \eta_R\}$ of the process $\zeta_t^{(N)}$, $t \geq 0$. In the row of the matrix $\hat{E}_{r,l}$ that corresponds to the state $\{\eta_1, \eta_2, \ldots, \eta_R\}$, the entry 1 is located in the column that corresponds to the state $\{\eta_1, \ldots, \eta_{r-1}, \eta_r + 1, \eta_{r+1}, \ldots, \eta_{l-1}, \eta_l - 1, 0, \ldots, 0\}$ only in the case if $\eta_m = 0$ for all $m$, $R \geq m > l$, and $\eta_l > 0$. If this condition is false, all entries of this row are zero entries.

The intensity $\lambda^{(r)}_{force-out}$ of forcing out from the buffer type-$r$, $r = \overline{2,R}$, customers is

$$\lambda^{(r)}_{force-out} = q\sum_{l=1}^{r-1}\boldsymbol{\pi}_{N+1}(D_l \otimes I_M \otimes \hat{E}_{l,r})\mathbf{e}.$$

The probability $P_{force-loss}$ of the loss of an arbitrary customer due to forcing out is

$$P_{force-loss} = \frac{\sum_{r=2}^{R} \lambda^{(r)}_{force-out}}{\lambda}.$$

The probability $P^{(r)}_{force-loss}$ of the loss of an arbitrary type-$r$, $r = \overline{2,R}$, customer due to forcing out is

$$P^{(r)}_{force-loss} = \frac{\lambda^{(r)}_{force-out}}{\lambda_r + \tilde{\lambda}^{(r)}}.$$

## 5. Numerical Example

In this section, we illustrate the dependencies of some performance measures of the system on the buffer capacity $N$ and show the poor quality of evaluation of the value of the loss probability via the following three simplifications of the model: (i) the arrival flow is assumed to be described not by the $MMAP$, but by the superposition of the stationary Poisson processes; (ii) the service time distribution is assumed to be not of a general phase-type, but exponential; (iii) the arrival flow is assumed to be the superposition of the stationary Poisson processes and the service time distribution is assumed to be exponential.

In this illustrative example, we consider a small information transmission device that is designed for transmission of four types of information. We assume that the distribution of the size of various types information units is the same. The information units of various types have different importance for the system and, correspondingly, have different priority. Let us assume that the arrivals of the units (customers) of different types are modeled by the $MMAP$ arrival process defined by the matrices:

$$D_0 = \begin{pmatrix} -1.8 & 0.0 \\ 0.0 & -0.4458 \end{pmatrix}, \; D_1 = \begin{pmatrix} 0.51 & 0.04 \\ 0.006 & 0.1047 \end{pmatrix},$$

$$D_2 = \begin{pmatrix} 0.31 & 0.01 \\ 0.0 & 0.2641 \end{pmatrix}, \; D_3 = \begin{pmatrix} 0.41 & 0.01 \\ 0.002 & 0.058 \end{pmatrix}, \; D_4 = \begin{pmatrix} 0.5 & 0.01 \\ 0.001 & 0.01 \end{pmatrix}.$$

It has the average arrival intensity $\lambda = 0.600076$, the coefficient of correlation $c_{cor} = 0.148534$, and the coefficient of variation $c^2_{var} = 1.46139$. The intensities of type-$r$ customer arrivals are $\lambda_1 = 0.160747$, $\lambda_2 = 0.270468$, $\lambda_3 = 0.101013$, $\lambda_4 = 0.0678481$, respectively.

The $PH$ service process is defined by the vector $\boldsymbol{\beta} = (0.01, 0.99)$ and the sub-generator

$$S = \begin{pmatrix} -0.1 & 0.1 \\ 0.02 & -2 \end{pmatrix}.$$

The average service time is $b_1 = 0.706060$ and the coefficient of variation is $c^2_{var} = 8.781$.

The rest parameters are as follows: $\gamma_1 = 0.012$, $\gamma_2 = 0.011$, $\gamma_3 = 0.01$, $\gamma_4 = 0.009$, $\alpha_r = 0.1$, $r = \overline{2,4}$, $p_{2,1} = 1$, $p_{3,1} = p_{3,2} = 0.5$, $p_{4,1} = p_{4,2} = p_{4,3} = \frac{1}{3}$, $q = 0.5$.

Let us vary the buffer capacity $N$ over the interval $[1, 25]$ and calculate the main performance measures of the system. It is worth to note that capacity of the buffer not exceeding 25 is realistic in many real-world applications, e.g., in application for modeling emergency departments in a hospital, the number of waiting patients cannot be large because if this number grows, the ambulance cars will deliver new patients to other neighboring hospitals. In modeling the operation of an information transmission device, the capacity of the buffer can also be not very large due to fast obsolescence of the transmitted information.

For computations, we use a PC with an Intel Core i7-8700 CPU and 16 GB RAM, Mathematica 11.0. The computation time for all 25 different buffer capacities is about 15 min.

Figure 2 illustrates the dependence of the average number of customers in the buffer $N_{buffer}$ and the average numbers $N^{(r)}_{buffer}$, $r = \overline{1,R}$, of type-$r$ customers in the buffer on the buffer capacity $N$. As it is expected, the values $N_{buffer}$ and $N^{(r)}_{buffer}$, $r = \overline{1,R}$, increase with the growth of the buffer capacity $N$.

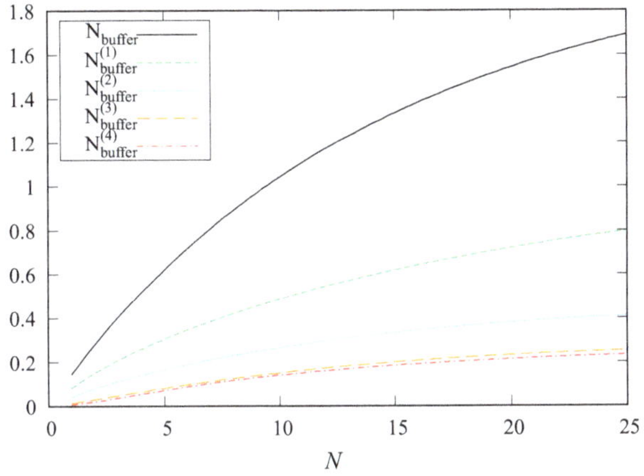


**Figure 2.** The dependence of $N_{buffer}$ and $N^{(r)}_{buffer}$, $r = \overline{1,R}$, on the buffer capacity $N$.

Figure 3 illustrates the dependence of the average intensities $\tilde{\lambda}^{(r)}$ of type-$l$, $l = \overline{r+1,R}$, customers transformation to the type-$r$, $r = \overline{1,R-1}$, customers on the buffer capacity $N$. All these intensities increase with the growth of the buffer capacity $N$ because the larger capacity of the buffer implies the longer stay of a customer in the buffer and, therefore, higher chances to increase the priority. The highest value of the intensity $\tilde{\lambda}^{(1)}$ among the values $\tilde{\lambda}^{(r)}$, $r = \overline{1,R-1}$, is easily explained by the fact that about 45 percent of arriving customers are type-2 customers that can increase their priority only to type-1, a half of type-3 customers may increase the priority directly to type-1 and one third of type-4 customers may also increase the priority directly to type-1.

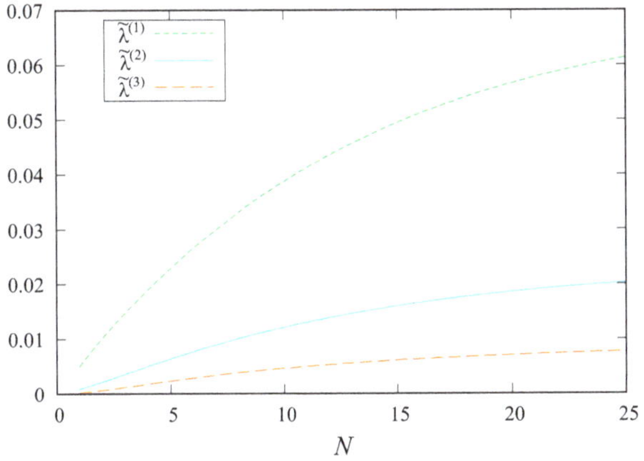


**Figure 3.** The dependence of the average intensities $\tilde{\lambda}^{(r)}$, $r = \overline{1,R-1}$, on the buffer capacity $N$.

Figure 4 illustrates the dependence of the probability of an arbitrary customer loss upon arrival $P_{ent-loss}$ and the probabilities of an arbitrary type-$r$, $r = \overline{1,R}$, customer loss upon arrival $P^{(r)}_{ent-loss}$ on the buffer capacity $N$. This figure confirms the intuitively clear fact that all these loss probabilities decrease with the growth of the buffer capacity.

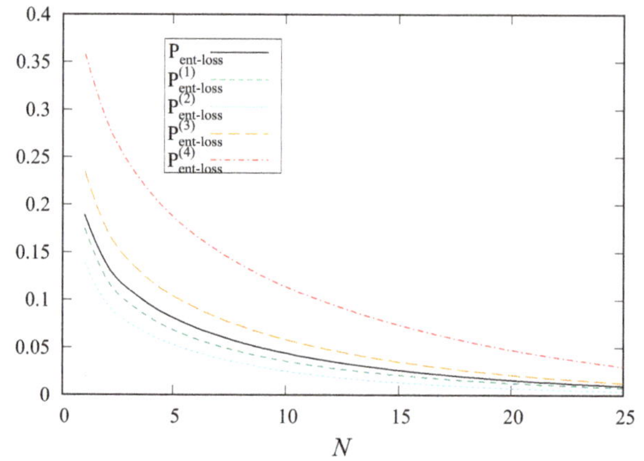


**Figure 4.** The dependence of the probabilities $P_{ent-loss}$ and $P^{(r)}_{ent-loss}$, $r = \overline{1,R}$, on the buffer capacity $N$.

Figure 5 illustrates the dependence of the probability $P_{force-loss}$ of the loss of an arbitrary customer due to forcing out and the probability $P^{(r)}_{force-loss}$ of the loss of an arbitrary type-$r$, $r = \overline{2,R}$, customer on the buffer capacity $N$. The behavior of these probabilities for type-3 and type-4 customers is explained as follows. For small values of $N$, these probabilities are small because there is a high probability that such customers are not admitted to the system at all (are lost at the entrance to the system). Then, when the buffer capacity $N$ increases, fewer customers of these types are lost at the entrance and, therefore, more customers are accepted to the buffer and are forced out by the high priority customers. After the buffer capacity $N$ reaches the values about 2 or 3, the probability that the high priority customers will meet full buffer essentially decreases and these customers have no need to force out type-3 and type-4 customers. Consequently, the probabilities $P^{(r)}_{force-loss}$, $r = 3,4$, decrease when $N$ further increases.

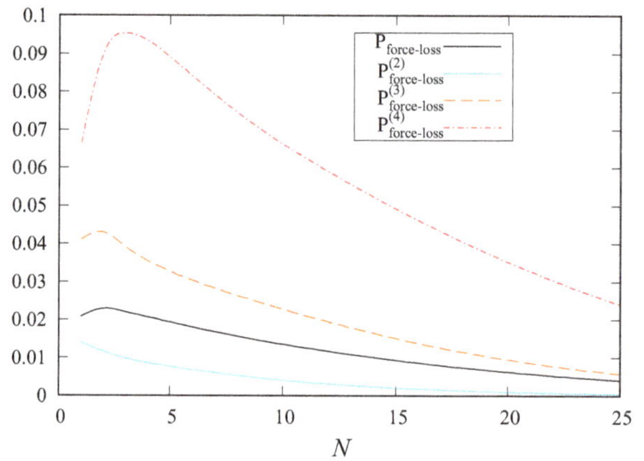


**Figure 5.** The dependence of the probabilities $P_{force-loss}$ and $P^{(r)}_{force-loss}$, $r = \overline{2,R}$, on the buffer capacity $N$.

Figure 6 illustrates the dependence of the probability $P_{imp-loss}$ of the loss of an arbitrary customer due to impatience and the probability $P^{(r)}_{imp-loss}$, $r = \overline{1,R}$, of loss of an arbitrary type-$r$ customer due to impatience on the buffer capacity $N$. When the buffer capacity increases, customers of all types spend more time in the buffer and are lost due to the impatience more frequently.

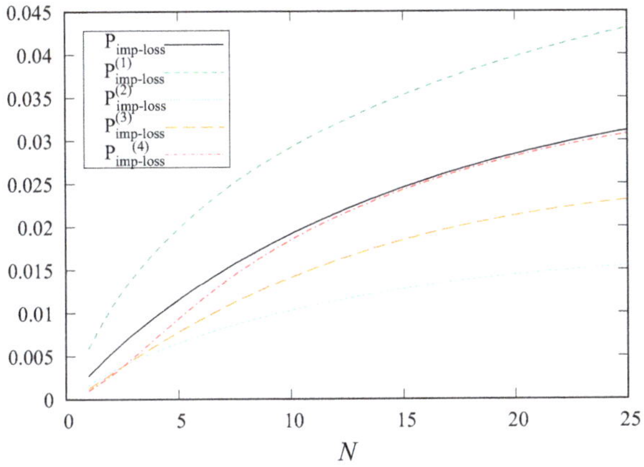


**Figure 6.** The dependence of the probabilities $P_{imp-loss}$ and $P^{(r)}_{imp-loss}$, $r = \overline{1,R}$, on the buffer capacity $N$.

As it was announced above, one of the important goals of our numerical example is to demonstrate the poor quality of approximation of the value of the loss probability in the considered $MMAP/PH/1/N$ model with dynamically variable non-preemptive priorities by the value of the loss probability in more simple models coded below as $MMAP/M/1/N$, $M/PH/1/N$ and $M/M/1/N$ type priority models with the same rates of the arrival of different types of customers and the service rate. Using the $MMAP/M/1/N$ model, one ignores that we assumed that the service time has the coefficient of variation $c^2_{var} = 8.781$, not $c^2_{var} = 1$, as the exponential distribution of the service time suggests. Using the $M/PH/1/N$ model, one ignores that the inter-arrival times have the coefficient of correlation $c_{cor} = 0.148534$, and the coefficient of variation $c^2_{var} = 1.46139$, not $c^2_{var} = 1$, as the exponential distribution of inter-arrival times of different types of customers suggests. Using the $M/M/1/N$ model, one assumes a zero coefficient of inter-arrival times and the coefficient of variation of inter-arrival of all types of customers and the service times equal to 1.

Figure 7 illustrates the dependence of the probability $P_{loss}$ of the loss of an arbitrary customer on the buffer capacity $N$ for the considered $MMAP/PH/1/N$ priority system and its particular cases coded as the $MMAP/M/1/N$, $M/PH/1/N$ and $M/M/1/N$ type systems.

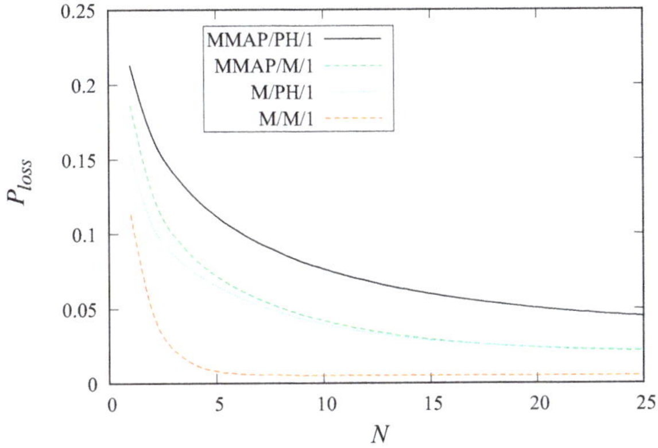


**Figure 7.** The dependence of the probability $P_{loss}$ on the buffer capacity $N$ for the considered set of the system parameters.

One can see that the values of the loss probabilities computed for the approximating models are essentially smaller than the actual value. It is well known that queueing models with a finite buffer can help to solve the important problem of computing the required capacity $N$ of the buffer, e.g., the problem of finding the minimum value of $N$ such as the loss probability $P_{loss}$ is less than 0.05 can be considered. Using the approximate value of this loss probability computed via the $M/M/1/N$ type system, one can compute that the buffer capacity $N = 2$ is enough to guarantee the fulfillment of the inequality $P_{loss} < 0.05$. Using the approximate value of this loss probability computed via the $M/PH/1/N$ type system, one can compute that the required buffer capacity is $N = 8$. Using the approximate value of this loss probability computed via the $MMAP/M/1/N$ type system, one can compute that the required buffer capacity is $N = 9$. Furthermore, finally, if one properly accounts the values of the coefficients of correlation and variation via the use of the $MMAP/PH/1/N$ model, he/she obtains that the required buffer capacity is $N = 21$. For $N = 2, 8$ and 9 the loss probability has values 0.1659179, 0.087093, and 0.081367, correspondingly, and is essentially larger than 0.05. Therefore, the simplified models give a quite poor estimation of the required capacity of the buffer.

## 6. Conclusions

We analyzed a quite general single-server queue with heterogeneous customers and a finite buffer. The arrival flow is defined by the $MMAP$ what allows us to take into account the possible correlation of inter-arrival intervals of customers of different types. The service time distribution is of phase-type which allows to approximate more general distributions. Customers of various types have different impatience. It is assumed that the problem of assigning the non-preemptive priorities to different types of customers is solved in the assumption that during staying in the buffer customers can improve their priority. Presented above results allow computing the steady-state distribution of the system and the key performance measures of the system under any fixed set of the system parameters. This creates an opportunity for further use of the obtained results for the optimal scheduling of the flows (assigning the priorities and permissions to increase the priority) under any fixed cost criterion. The criterion may include, e.g., the profit gained via the service of different types of customers or the coefficient of utilization of the server and loss probabilities (rejection at the entrance of the system, pushing out by a high priority customer, leaving the system due to impatience) of different types of customers.

Results can be applied for optimization of the scheduling of: (i) information flows in communication networks where users are categorized into several groups according to their importance, in particular, possible damage caused by the loss or obsolescence of the corresponding information; (ii) patients with different degree of life threat in emergency departments; (iii) perishable goods and foods in warehouses, etc. As future directions of generalization of the considered model we can mention the account of possibility of different distribution of service time for different types of customers and possibility of unreliable service of customers similar to [33].

**Author Contributions:** Conceptualization, S.L., S.D. and V.K.; methodology, S.D., O.D., and C.K.; software, S.L., S.D. and O.D.; validation, S.L., S.D. and O.D.; formal analysis, S.D., V.K., and C.K.; investigation, C.K.; writing, original draft preparation, S.L. and C.K.; writing, review and editing V.K., and C.K.; supervision S.L. and C.K.; project administration O.D. and V.K. All authors read and agreed to the published version of the manuscript.

**Funding:** This work has been supported by Sangji University Grant 2018. This work was also partially supported by the Basic Science Research Program through the National Research Foundation of Korea (NRF) funded by the Ministry of Education (NRF-2018K2A9A1A06072058) and grant No. F19KOR-001 of the Belarusian Republican Foundation for Fundamental Research.

**Conflicts of Interest:** The authors declare no conflict of interest.

## References

1. Kalashnikov, V.V. *Mathematical Methods in Queuing Theory*; Springer: Berlin/Heidelberg, Germany, 2013.
2. Dudin, S.; Dudina, O.; Samouylov, K.; Dudin, A. Improvement of the fairness of non-preemptive priorities in the transmission of heterogeneous traffic. *Mathematics* **2020**, *8*, 929. [CrossRef]
3. Fratini, S. Analysis of a dynamic priority queue. *Commun. Stat. Stoch. Model.* **1990**, *6*, 415–444. [CrossRef]

4. Kim, C.S.; Klimenok, V.; Dudin, A. Priority tandem queueing system with retrials and reservation of channels as a model of call center. *Comput. Ind. Eng.* **2016**, *96*, 61–71. [CrossRef]
5. Knessl, C.; Tier, C.; Cho, D. A dynamic priority queue model for simultaneous service of two traffic types. *SIAM J. Appl. Math.* **2003**, *63*, 398–422. [CrossRef]
6. Ramaswami, V.; Lucantoni, D.M. Algorithmic analysis of a dynamic priority queue. In *Applied Probability—Computer Science: The Interface*; Birkhäuser: Boston, MA, USA, 1982; pp. 157–206,
7. Xin, J.; Zhu, Q.; Liang, G.; Zhang, T. Performance Analysis of D2D Underlying Cellular Networks Based on Dynamic Priority Queuing Model. *IEEE Access* **2019**, *7*, 27479–27489. [CrossRef]
8. De Clercq, S.; Steyaert, B.; Wittevrongel, S.; Bruneel, H. Analysis of a discrete-time queue with time-limited overtake priority. *Ann. Oper. Res.* **2015**, *238*, 69–97. [CrossRef]
9. De Boeck, K.; Carmen, R.; Vandaele, N. Needy boarding patients in emergency departments: An exploratory case study using discrete-event simulation. *Oper. Res. Health Care* **2019**, *21*, 19–31. [CrossRef]
10. Bilodeau, B.; Stanford, D.A. Average Waiting Times in the Two-Class $M/G/1$ Delayed Accumulating Priority Queue. *arXiv* **2020**, arXiv:2001.06054.
11. Fajardo, V.A.; Drekic, S. Waiting Time Distributions in the Preemptive Accumulating Priority Queue. *Methodol. Comput. Appl. Probab.* **2017**, *19*, 255–284. [CrossRef]
12. Mojalal, M.; Stanford, D.A.; Caron, R.J. The lower-class waiting time distribution in the delayed accumulating priority queue. *INFOR Inf. Syst. Oper. Res.* **2020**, *58*, 60–86. [CrossRef]
13. Sharma, K.C.; Sharma, G.C. A delay dependent queue without preemption with general linearly increasing priority function. *J. Oper. Res. Soc.* **1994**, *45*, 948–953. [CrossRef]
14. Stanford, D.A.; Taylor, P.; Ziedins, I. Waiting time distributions in the accumulating priority queue. *Queueing Syst.* **2014**, *77*, 297–330. [CrossRef]
15. Lim, Y.; Kobza, J.E. Analysis of a delay-dependent priority discipline in an integrated multiclass traffic fast packet switch. *IEEE Trans. Commun.* **1990**, *38*, 659–665. [CrossRef]
16. Maertens, T.; Bruneel, H.; Walraevens, J. On priority queues with priority jumps. *Perform. Eval.* **2006**, *63*, 1235–1252. [CrossRef]
17. Klimenok, V.; Dudin, A.; Dudina, O.; Kochetkova, I. Queuing System with Two Types of Customers and Dynamic Change of a Priority. *Mathematics* **2020**, *8*, 824. [CrossRef]
18. Xie, O.; He, Q.-M.; Zhao, X. On the stationary distribution of queue lengths in a multi-class priority queueing system with customer transfers. *Queueing Syst.* **2009**, *62*, 255–277. [CrossRef]
19. He, Q.M.; Xie, J.G.; Zhao, X.B. Stability conditions of a preemptive repeat priority $MMAP[N]/PH[N]/S$ queue with customer transfers (short version). In Proceedings of the 2009 Conference Proceedings on ASMDA(Advanced Stochastic Models and Data Analysis), Vilnius, Lithuania, 30 June–3 July 2009; pp. 463–467.
20. He, Q.-M.; Xie, J.; Zhao, X. Priority Queue with Customer Upgrades. *Nav. Res. Logist.* **2012**, *59*, 362–375. [CrossRef]
21. Chakravarthy, S.R. The batch Markovian arrival process: A review and future work. In *Advances in Probability Theory and Stochastic Processes*; Krishnamoorthy, A., Raju, N., Ramaswami, V., Eds.; Notable Publications Inc.: Branchburg, NJ, USA, 2001; pp. 21–29.
22. Lucantoni, D. New results on the single server queue with a batch Markovian arrival process. *Commun. Stat. Stoch. Model.* **1991**, *7*, 1–46. [CrossRef]
23. Dudin, A.N.; Klimenok, V.I.; Vishnevsky, V.M. *The Theory of Queuing Systems with Correlated Flows*; Springer: Berlin/Heidelberg, Germany, 2019.
24. He, Q.M. Queues with marked customers. *Adv. Appl. Probab.* **1996**, *28*, 567–587. [CrossRef]
25. Kim, C.S.; Dudin, S.; Dudina, O.; Dudin, A.N. Mathematical Model of a Cell With Bandwidth Sharing and Moving Users. *IEEE Trans. Wirel. Commun.* **2020**, *19*, 744–755. [CrossRef]
26. Sun, B.; Dudin, S.; Dudina, O.; Samouylov, K. Optimization of admission control in tandem queue with heterogeneous customers and pre-service. *Optimization* **2020**, *69*, 165–185. [CrossRef]
27. Dudin, S.; Dudin, A.; Dudina, O.; Samouylov, K. Competitive queueing systems with comparative rating dependent arrivals. *Oper. Res. Perspect.* **2020**, *7*, 100139. [CrossRef]
28. Neuts, M. *Matrix-Geometric Solutions in Stochastic Models*; The Johns Hopkins University Press: Baltimore, MD, USA, 1981.

29. Ramaswami, V.; Lucantoni,D. Algorithms for the multi-server queue with phase-type service. *Commun. Stat. Stoch. Model.* **1985**, *1*, 393–417. [CrossRef]
30. Graham, A. *Kronecker Products and Matrix Calculus with Applications*; Horwood, E., Ed.; Courier Dover Publications: Cichester, UK, 1981.
31. Latouche, G.; Ramaswami, V. *Introduction to Matrix Analytic Methods in Stochastic Modeling*; Society for Industrial and Applied Mathematics: Philadelphia, PA, USA, 1999.
32. Baumann, H.; Sandmann, W. Numerical solution of level dependent quasi-birth-and-death processes. *Procedia Comput. Sci.* **2010**, *1*, 1561–1569. [CrossRef]
33. Dudin, S.; Dudina, O. Retrial multi-server queuing system with PHF service time distribution as a model of a channel with unreliable transmission of information. *Appl. Math. Model.* **2019**, *65*, 676–695. [CrossRef]

mathematics MDPI

Review

# Highly Efficient Robust and Stable *M*-Estimates of Location

**Georgy Shevlyakov** †

Department of Applied Mathematics, Peter the Great St. Petersburg Polytechnic University, 195251 St. Petersburg, Russia; georgy.shevlyakov@phmf.spbstu.ru

† Prof. Georgy Shevlyakov passed away during the revision cycle of the manuscript, comments have been adressed by Dr. Maya Shevlyakova.

Received: 27 August 2020; Accepted: 30 December 2020; Published: 5 January 2021

**Abstract:** This article is partially a review and partially a contribution. The classical two approaches to robustness, Huber's minimax and Hampel's based on influence functions, are reviewed with the accent on distribution classes of a non-neighborhood nature. Mainly, attention is paid to the minimax Huber's *M*-estimates of location designed for the classes with bounded quantiles and Meshalkin-Shurygin's stable *M*-estimates. The contribution is focused on the comparative performance evaluation study of these estimates, together with the classical robust *M*-estimates under the normal, double-exponential (Laplace), Cauchy, and contaminated normal (Tukey gross error) distributions. The obtained results are as follows: (i) under the normal, double-exponential, Cauchy, and heavily-contaminated normal distributions, the proposed robust minimax *M*-estimates outperform the classical Huber's and Hampel's *M*-estimates in asymptotic efficiency; (ii) in the case of heavy-tailed double-exponential and Cauchy distributions, the Meshalkin-Shurygin's radical stable *M*-estimate also outperforms the classical robust *M*-estimates; (iii) for moderately contaminated normal, the classical robust estimates slightly outperform the proposed minimax *M*-estimates. Several directions of future works are enlisted.

**Keywords:** robustness; minimax approach; stable estimation

---

## 1. Introduction

Robust statistics, as a new field of mathematical statistics, originates from the pioneering works of John Tukey (1960) [1], Peter Huber (1964) [2], and Frank Hampel (1968) [3]. The term "robust" (Latin: strong, vigorous, sturdy, tough, powerful) was introduced into statistics by George Box (1953) [4].

The reasons of research in this field of statistics are of a general mathematical nature: the conceptions of "optimality" and "stability" are mutually complementary in performance evaluation for almost all mathematical procedures, and the trade-off between them is often a sought goal.

It is not rare that the performance of optimal solutions is rather sensitive to small violations of the assumed conditions of optimality. In statistics, the classical example of such unstable optimal procedure is given by the least squares estimates, in which performance under small deviations from normality can become disastrous [5].

Since the term "stability" is overloaded in mathematics, the term "robustness" being its synonym is at present conventionally used in statistics and in optimal control theory: in general, it means the stability of statistical inference under uncontrolled violations of accepted distribution models.

In present, there are two main approaches to robustness: historically, the first global minimax approach of Huber (quantitative robustness) [5] and the local approach of Hampel based on influence functions (qualitative robustness) [6]. Within the first approach, the least informative (favorable) distribution minimizing Fisher information over a certain distribution class is obtained with the subsequent use of the asymptotically optimal maximum likelihood parameter estimate for this distribution. In this case, the minimax approach gives the guaranteed accuracy of robust estimates, that is, the asymptotic variance of the optimal parameter estimate is upper-bounded for distributions from the aforementioned class.

Within the second approach, a parameter estimate is defined by its desired influence function, which determines the qualitative robustness properties of an estimate, such as its low sensitivity to the presence of gross outliers in the data, to the data rounding-off, to the missing data, etc.

In what follows, we consider these methodologies in detail focusing on the optimization and variational calculus tools used in both aforementioned approaches. Within Huber's minimax approach, we review the conventional least informative (favorable) distributions obtained for the neighborhoods of a Gaussian [5], as well as those designed for a variety of the non-standard distribution classes of a non-neighborhood nature [7]. Within Hampel's local approach [6], we mostly emphasize its recently developed stable estimation branch with the originally posed variational calculus problems and rather prospective results on their application to robust statistics [8].

While this paper focuses on particular topics in the field of robust statistics, it is worth noting a few comprehensive reviews also covering the present state of art in this field, namely Reference [9–14].

An outline of the remainder of the article is as follows. In Section 2, a general problem setting for the design of minimax variance *M*-estimates of location is recalled. In Section 3, the globally stable Meshalkin-Shurygin's *M*-estimates are described. In Section 4, a comparative performance evaluation of the conventional robust *M*-estimates of location with the several novel proposed *M*-estimates is examined (univariate setting is considered throughout the paper), and several unforeseen and unexpected results have been obtained. In Section 5, some conclusions are given.

## 2. Huber's Minimax Variance Robust M-Estimates of Location

### 2.1. Preliminaries

The minimax principle aims at the worst case suggesting for it the best solution [2]; thus, this approach provides a guaranteed result [5]. However, being applied to the problem of estimation of location, it yields a robust version of the principle of the maximum likelihood [2]. Usually, estimation of location is of a primary interest, and, in this study, we focus on it.

Let $x_1, \ldots, x_n$ be a sample from a distribution with density $p(x-\theta)$ from a convex class $\mathcal{P}$, where $\theta$ is a location parameter. Further, we assume that $p$ is a symmetric distribution density; hence, $\theta$ is the center of symmetry to be estimated.

An *M*-estimate $T_n$ of $\theta$ is a solution to the following minimization problem:

$$T_n = \arg\min_{\theta} \sum_{i=1}^{n} \rho(x_i - \theta), \tag{1}$$

where $\rho(u)$ is called the *function of contrast* [15]: $\rho(x_i - \theta)$ is a measure of difference between the observation $x_i$ and the estimated center of symmetry. The following particular cases of (**??**) are of a particular interest: (i) for $\rho(u) = u^2$, we have the least squares method with the sample mean $\bar{x}$ as the estimate of location; (ii) for $\rho(u) = |u|$, we arrive at the least absolute values method with the sample median estimate *med* $x$; and (iii), mostly importantly, for a given density $p$, the choice $\rho(u) = -\log p(u)$ yields the maximum likelihood (*ML*) estimate of location.

It is more convenient to formulate the properties of *M*-estimates in terms of the derivative of the function of contrast $\psi(u) = \rho'(u)$ called a *score function*. In this case, an *M*-estimate is defined as a solution to the following implicit estimating equation

$$\sum_{i=1}^{n} \psi(x_i - T_n) = 0. \tag{2}$$

Under rather general conditions of regularity imposed on the class $\Psi$ of score functions $\psi$ and on the class $\mathcal{P}$ of densities $p$ (their various forms can be found in Reference [2,5,6]), $M$-estimates are consistent and asymptotically normal with the asymptotic variance $AV$:

$$Var\left(n^{1/2}T_n\right) = AV(\psi,p) = \frac{A}{B^2} = \frac{\int \psi^2(x)p(x)\,dx}{\left(\int \psi'(x)p(x)\,dx\right)^2}, \tag{3}$$

where

$$A(\psi,p) = \int \psi^2(x)p(x)\,dx,$$

$$B(\psi,p) = \int \psi'(x)p(x)\,dx.$$

For $M$-estimates (??), the following result holds [5].

**Theorem 1.** *(Huber, 1964) Under regularity conditions, M-estimates satisfy the minimax property*

$$AV(\psi^*,p) \leq AV(\psi^*,p^*) = \sup_{p\in\mathcal{P}} \inf_{\psi\in\Psi} AV(\psi,p) \leq AV(\psi,p^*), \tag{4}$$

*where $p^*(x)$ is the least informative (favorable) density $p^*$ minimizing Fisher information for location $I(p)$ over the class $\mathcal{P}$:*

$$p^* = \arg\min_{p\in\mathcal{P}} I(p), I(p) = \int \left[\frac{p'(x)}{p(x)}\right]^2 p(x)\,dx. \tag{5}$$

From (??) and (??), it follows that the minimax function of contrast and score function are given by the maximum likelihood method for the least informative density $p^*$:

$$\rho^*(x) = -\log p^*(x), \psi^*(x) = -p^*(x)'/p^*(x). \tag{6}$$

Thus, the pair $(\psi^*,p^*)$ is the saddle-point of the functional $AV(\psi,p)$. The right-hand part of inequality (??) is the Rao–Cramér inequality:

$$AV(\psi,p^*) \geq AV(-p^{*\prime}/p^*,p^*) = 1/\int \left(p^*(x)'\right)^2/p^*(x)\,dx = 1/I(p^*),$$

whereas its left-hand part guarantees the asymptotic accuracy of robust minimax estimation with the following upper bound upon the asymptotic variance of the minimax variance robust $M$-estimate of location: $AV(\psi^*,p) \leq 1/I(p^*)$.

The key point of the minimax approach is the solution of the variational problem (??): further, various classes $\mathcal{P}$ with the corresponding least informative densities $p^*$ and minimax estimates are enlisted.

Now, we recall the Huber's classical solution for $\varepsilon$-contaminated normal distributions (Tukey's gross-error model):

$$\mathcal{P}_\varepsilon = \left\{p : p(x) \geq (1-\varepsilon)\varphi(x),\right. \tag{7}$$

where $\varphi(x) = (2\pi)^{-1/2}\exp\left(-x^2/2\right)$ is the standard normal distribution density.

**Theorem 2.** *(Huber, 1964) In the class $\mathcal{P}_\varepsilon$, the least informative density $p^*$ and the optimal score function has the following form [2]:*

$$p^*(x) = p_{Huber}(x) = \begin{cases} (1-\varepsilon)\varphi(x), & |x| \leq k, \\ C\exp(-D|x|), & |x| > k, \end{cases} \tag{8}$$

$$\psi^*(x) = \psi_{Huber}(x) = \begin{cases} x, & |x| \leq k, \\ CD\,\mathrm{sgn}\,(x), & |x| > k, \end{cases} \tag{9}$$

*where the parameters C, D, and k satisfy the conditions of norming, continuity, and continuous differentiability of the solution at* $x = k$*:*

$$\int p^*(x)\,dx = 1, p^*(k-0) = p^*(k+0), p^{*\prime}(k-0) = p^{*\prime}(k+0).$$

Finally, we get the linear bounded score $\psi_{Huber}(x) = \max\{-k, \min\{x, k\}\}$, where $k$ depends on the value of the contamination parameter $\varepsilon$, as follows:

$$\frac{2\,\varphi\,(k)}{k} - 2\,\Phi(-k) = \frac{\varepsilon}{1-\varepsilon}, \Phi(x) = \int_{-\infty}^{x} \varphi\,(t)\,dt;$$

the values of the parameter $k = k(\varepsilon)$ are tabulated in Reference [5].

The particular cases of this solution for $\varepsilon = 0$ and $\varepsilon \to 1$ are given by $k \to \infty$ (the sample mean) and $k = 0$ (the sample median), respectively.

The Huber's score function $\psi_{Huber}(x)$ is a robust version of the *ML* estimation: in the center $|x_i - \theta| \leq k$, the data are processed by the *ML* method, and they are trimmed within the exponential distribution tails $|x_i - \theta| > k$. In the limiting case of a completely unknown density as $\varepsilon \to 1$, the minimax variance *M*-estimate of location tends to the sample median.

### 2.2. Free Extremals of the Basic Variational Problem

Consider the problem of minimization of Fisher information for location (**??**) under two basic side conditions of non-negativeness and norming: $p(x) \geq 0$, $\int_{-\infty}^{\infty} p(x)\,dx = 1$.

Set $\sqrt{p(x)} = q(x)$, and rewrite this minimization problem as

$$\text{minimize} I(p) = 4\int_{-\infty}^{\infty} (q'(x))^2\,dx \text{under} \int_{-\infty}^{\infty} q^2(x)\,dx = 1.$$

Introducing the Lagrange multiplier $\lambda$ related to the norming condition, we obtain the following differential equation for the function $q(x)$: $4q''(x) + \lambda q(x) = 0$.

The general solutions of this harmonic oscillator equation have the well-known forms depending on the sign of $\lambda = 4k^2$: (i) the exponential $q(x) = C_1 e^{kx} + C_2 e^{-kx}$, the cosine $q(x) = C_1 \sin kx + C_2 \cos kx$, and the linear $q(x) = C_1 + C_2 x$, where $k = \pm\sqrt{\lambda}/2$.

Further, we show that all these forms work in the structures of least informative distribution densities.

### 2.3. Least Informative Distributions

The neighborhoods of normal, generally, are not the only models of interest. In real-life applications, the information about the distribution central part tails, its moments, and/or subranges is rather often available. The empirical distribution and quantile functions, histograms, and kernel estimates, together with their tolerance limits, provide other examples. To enhance the efficiency of robust estimates, this information can be used in minimax settings.

Further, we deal with symmetric distribution densities $p(-x) = p(x)$. Evidently, distribution densities must also satisfy the non-negativeness and norming conditions common for all classes. For brevity, we do not write out all these conditions any time we define a distribution density class.

Now, we enlist several examples of distribution classes convenient for the description of a prior knowledge about data distributions [7].

(1) The class of non-degenerate densities [15]: $\mathcal{P}_1 = \{p : p(0) \geq 1/(2a) > 0\}$. The parameter $a$ of this class characterizes the dispersion of the central part of a distribution. The least informative density in this class is given by the double-exponential or Laplace density [16]: $p_1^*(x) = L(x;0,a) = 2a^{-1}\exp(-|x|/a)$. This result is quite natural if one recalls the exponential form of free extremals for the basic variational problem.

(2) The class $\mathcal{P}_2$ of distribution densities with a bounded variance:

$$\mathcal{P}_2 = \left\{p : \sigma^2(p) = \int_{-\infty}^{\infty} x^2 p(x)\,dx \leq \overline{\sigma}^2\right\}.$$

All distribution densities with bounded variances are the members of this class. Evidently, the Cauchy-type distributions do not belong to it. The least informative density in this class is normal [17]:

$$p_2^*(x) = N(x;0,\overline{\sigma}) = \frac{1}{\sqrt{2\pi}\overline{\sigma}}\exp\left(-\frac{x^2}{2\overline{\sigma}^2}\right).$$

(3) The class $\mathcal{P}_3$ of approximately normal distribution densities is defined by Equation (??): $\{p :\ p(x) \geq (1-\varepsilon)\ \varphi\ (x),$

(4) The class of finite distributions: $\mathcal{P}_4 = \left\{p : \int_{-l}^{l} p(x)\,dx = 1\right\}$. This class defines the boundaries of the data (i.e., the inequality $|X|\leq l$ holds with probability one), and there is no more information about this distribution besides the boundary conditions of smoothness: $p(\pm l) = p'(\pm l) = 0$. The least informative density in this class has the cosine-squared form [15]:

$$p_4^*(x) = \begin{cases} \frac{1}{l}\cos^2\left(\frac{\pi x}{2l}\right) & \text{for} \quad |x|\leq l, \\ 0 & \text{for} \quad |x|> l. \end{cases}$$

(5) The class of distributions with a bounded interquantile distribution mass:

$$\mathcal{P}_5 = \left\{p : \int_{-l}^{l} p(x)\,dx \geq 1-\beta\right\}, 0 \leq \beta < 1.$$

The parameters $l$ and $\beta$ are assumed given. The restriction upon the interquantile mass means that $P(|X|\leq l) \geq 1-\beta$. We can redefine this class in a different way as the class with an upper-bounded interquantile range $IQR_\beta = P^{-1}\left(\frac{1+\beta}{2}\right) - P^{-1}\left(\frac{1-\beta}{2}\right) \leq 2l$, where $P(x)$ is a probability density function. The least informative density in this class has both the cosine and exponential parts working at the center and tail areas, respectively [7],

$$p_5^*(x) = \begin{cases} A_1\cos^2(B_1 x) & \text{for} \quad |x|\leq l, \\ A_2\exp(-B_2|x|) & \text{for} \quad |x|> l, \end{cases}$$

where the constants $A_1$, $A_2$, $B_1$, and $B_2$ are determined from the simultaneous equations (restrictions) of the class $\mathcal{P}_5$, namely the conditions of normalization and upon the distribution interquantile range, and the conditions of continuity and continuous differentiability at $x = l$ (for details, see Reference [7]). It is worth noting that the classes $\mathcal{P}_1$ and $\mathcal{P}_4$ are the particular cases of the class $\mathcal{P}_5$ when $\beta \to 0$ and $\beta = 1$, respectively.

## 3. Hampel's Robust and Shurygin's Stable Estimates of Location

Robust methods have lower sensitivity to possible departures from the accepted distribution models as compared to conventional statistical methods. To analyze the sensitivity of estimation, it is natural to have its specific indicator. In what follows, we introduce these indicators, namely the influence function and related to it measures.

### 3.1. Hampel's Local Approach to Robustness

Let $P$ be a distribution function corresponding to $p \in \mathcal{P}$, the class of distribution densities, and let $T(P)$ be a functional defined in a subset of all distribution functions. The natural estimate defined by $T$ is $T_n = T(P_n)$, i.e., the functional computed at the sample distribution function $P_n$.

The influence function $IF(x; T, p)$ of this functional is defined as

$$IF(x; T, p) = \lim_{t \to 0} \frac{T((1-t)P + t\Delta_x) - T(P)}{t}, \tag{10}$$

where $\Delta_x$ is the degenerate distribution taking mass 1 at $x$ [6].

The influence function measures the impact of an infinitesimal contamination at $x$ on the value of an estimate, formally being the Gâteaux derivative of the functional $T(P)$. For an $M$-estimate with a score function $\psi$, the influence function is proportional to it: $IF(x; \psi, p) = \psi(x)/B(\psi, p)$, where the term $B(\psi, p)$ stands in Equation (**??**) for the asymptotic variance of $M$-estimates.

Based on the influence function, several local measures of robustness are defined [6], including *the gross-error sensitivity* of $T$ at $p$:

$$\gamma^*(T, p) = \sup_x \left| IF(x; T, p) \right|.$$

This indicator of sensitivity gives an upper bound upon the asymptotic estimate bias and measures the worst influence of an infinitesimal contamination on the value of an estimate. Maximizing the efficiency of an $M$-estimate of location under the condition of a bounded gross-error sensitivity at the normal distribution

$$\max_{\psi} eff(\psi, \varphi) \text{under} \gamma^* \leq \overline{\gamma},$$

where $eff(\psi, p) = \frac{1}{AV(\psi,p)I(p)}$ formally leads to the Huber's minimax linear bounded score $\psi_{Huber}(x) = \max\{-k, \min\{x, k\}\}$ in the class of contaminated normal distributions [6]. This particular result confirms the following general observation: the best estimates within both approaches, Huber's minimax and Hampel's local, are rather close in their performances.

### 3.2. Meshalkin-Shurygin's Stable Estimates of Location

This topic is partially reversal to the conventional setting: the maximum of some measure of sensitivity is minimized under the guaranteed value of the estimate variance or efficiency.

The conventional point-wise local measures of sensitivity, such as the influence and change-of-variance functions [6] are not appropriate here—a global indicator of sensitivity is desirable. We show that a novel global indicator of robustness proposed by Shurygin [18], the estimate *stability*, is closely related to the classical variation of the functional of the estimate asymptotic variance. Although the related theory has been developed for stable estimation of an arbitrary parameter of the underlying distribution, here, we focus on stable estimation of location.

A measure of the $M$-estimate sensitivity called *variance sensitivity* ($VS$) is introduced in Reference [19]. Formally, it is defined as the Lagrange functional derivative of the asymptotic variance (**??**):

$$VS(\psi, p) = \frac{\partial AV(\psi, p)}{\partial p} = \frac{\partial}{\partial p}\left( \frac{\int \psi^2(x) p(x)\, dx}{\left(\int \psi'(x) p(x)\, dx\right)^2} \right) = \frac{\int \psi^2(x)\, dx}{\left(\int \psi'(x) p(x)\, dx\right)^2}. \tag{11}$$

Equation (**??**) gives a global measure of the stability of an $M$-estimate in a model, where the outliers occur uniformly anywhere on the real line. The boundness of the Lagrange derivative (**??**)

holds under the condition of square integrability of $\psi$ with the corresponding *redescending* scores when $\psi(x) \to 0$ for $|x| \to \infty$.

In Reference [7], it is shown that the principal part of the variation $\delta AV(\psi, p)$ of the asymptotic variance $AV(\psi, p)$ with respect to $\|\delta p\|$ is proportional to the variance sensitivity (**??**) or to the Lagrange derivative of the asymptotic variance: $\delta AV(\psi, p) \propto \partial AV(\psi, p)/\partial p$.

Further, consider the following optimization problem: what is the minimum variance sensitive score function for a given distribution density $p$? The solution of this optimization problem is given by the minimum variance sensitive ($MVS$) score function:

$$\psi_{MVS}(x) = \arg\min_{\psi} VS(\psi, p) = -p'(x). \tag{12}$$

The estimate with this optimal score function (**??**) is called as the estimate of *minimum variance sensitivity* with $VS_{min} = VS(\psi_{MVS}, p)$. We define a global measure of the *stability* of any $M$-estimate comparing an estimate variance sensitivity with its minimum

$$0 \le stb(\psi, p) = VS_{min}(p)/VS(\psi, p) \le 1.$$

A number of optimization criteria settings with different weights for efficiency and stability of $M$-estimates of location have been proposed and solved; in other examples, efficiency is maximized under guaranteed stability, and vice versa [18]. Practically in all these cases, we deal with the score functions $\psi_{opt}(x)$ with the following limiting forms: the maximum likelihood case $\psi_{opt}(x) \to \psi_{ML}(x) = -p'(x)/p(x)$ when the requirement of high efficiency mostly matters and the opposite redescending case $\psi_{opt}(x) \to \psi_{MVS}(x) = -p'(x)$ when the requirement of high stability is important.

A compromise case is given by a stable estimate called *radical* with equal efficiency and stability, $eff(\psi, p) = stb(\psi, p)$, desirably both highly efficient and stable: its score function is given by

$$\psi_{rad}(x) = \psi_{ML}(x)\sqrt{p(x)} = -p'(x)/\sqrt{p(x)}. \tag{13}$$

Finally, it should be noted that the minimum sensitivity and radical estimates belong to the class of $M$-estimates with the exponentially weighted maximum likelihood score functions previously proposed by Meshalkin [20].

## 4. Asymptotic Efficiency of M-Estimates: A Comparative Study

Here, we compare the asymptotic efficiency performance of various robust and stable estimates of location at the conventional in robustness studies distributions: some particular results can be found in Reference [19,21]. We mainly focus on the minimax variance estimates for distributions with bounded interquantile ranges, as until present, their performance has not been thoroughly studied. It is important that some obtained results are unexpected and surprising.

### 4.1. *Robust and Stable* M-*Estimates of a Location Parameter*

We test the following $M$-estimates of location: (i) the sample mean and median, (ii) the Huber's minimax variance $M$-estimate with the linear bounded score $\psi_{Huber}(x) = \max[-1.14, \min(x, 1.14)]$ optimal for $\varepsilon$-contaminated standard normal distributions with the parameter of contamination $\varepsilon = 0.1$, (iii) the Hampel's $M$-estimate with the redescending three-part score

$$\psi_{Hampel}(x) = \begin{cases} x & \text{for} 0 \le |x| \le a, \\ a\,\text{sgn}(x) & \text{for} a \le |x| \le b, \\ a\,\frac{r-|x|}{r-b}\text{sgn}(x) & \text{for} a \le |x| \le b, \\ 0 & \text{for} r \le |x|, \end{cases}$$

where the parameters $a = 1.31$, $b = 2.039$, and $r = 4$ (see Reference [6], pp. 166–167), (iv) the minimax variance $M$-estimates with the scores $\psi_\beta(x)$ for various values of the parameter $\beta : 0.01, 0.05, 0.1, 0.5, 0.9, 0.95, 0.99$, and (v) the stable Shurygin's minimum variance sensitivity and radical $M$-estimates with the scores $\psi_{MVS}(x) = -p'(x)$ and $\psi_{rad}(x) = -p'(x)/\sqrt{p(x)}$, respectively.

### 4.2. Data Distributions

In our study, the following distribution densities are used:
(i) the standard normal $p(x) = N(x;0,1) = (2\pi)^{-1/2}\exp\left(-x^2/2\right)$,
(ii) the standard Laplace $p(x) = L\left(x;0,1/\sqrt{2}\right) = 2^{-1/2}\exp(-\sqrt{2}|x|)$,
(iii) the standard Cauchy $p(x) = C(x;0,1) = \pi^{-1}\left(1+x^2\right)^{-1}$, and
(iv) the heavy-tailed Tukey gross-error model as the Cauchy contaminated normal density $p(x) = 0.9N(x;0,1) + 0.1C(x;0,1)$.

### 4.3. Asymptotic Efficiency

The asymptotic efficiency of $M$-estimates is numerically computed as follows:

$$eff(T_n) = \frac{Var(T_{ML})}{Var(T_n)} = \frac{\left(\int \psi'(x)p(x)\,dx\right)^2}{I(p)\int \psi^2(x)p(x)\,dx}.$$

## 5. Conclusions

From Table 1, it follows:

(1) As usual, the performance of the sample mean under heavy-tailed Cauchy and contaminated normal distributions is awful. Designed for these models, Huber's and Hampel's $M$-estimates perform well except the Laplace distribution case. This distribution with moderately heavy tails against a sharp peak at the center is a rather tough test for the asymptotic performance of $M$-estimates of location, especially as compared to the Cauchy distribution case. Recall that the Laplace and distributions close to it are the least informative ones in wide classes of distributions, for instance, in the class $\mathcal{P}_\beta$ with the parameter $\beta$ close to unit (the corresponding minimax variance $M$-estimates perform quite well in these cases). For a statistical user, the version with a bounded $IQR$ (interquartile range, $\beta = 1/2$) seems a reasonable choice.

(2) Surprisingly, the proposed minimax variance $M$-estimates with the scores $\psi_\beta$ outperform the classical robust Huber's and Hampel's $M$-estimates at the normal, although the shape of the least informative distribution is not at all normal: note that the Taylor expansion of the *cosine*$^2$-bell shape is close to the exponential one with small values of $\beta$. Moreover, these $M$-estimates are better than the classical robust estimates in heavy-tailed distribution models. We explain this effect by the nature of the distribution class $\mathcal{P}_\beta$—it is one of the most wide possible distribution classes. Finally, these $M$-estimates and their statistical properties can be obtained in a closed analytical form.

(3) The globally stable Meshalkin-Shurygin's radical $M$-estimate also outperforms the classical robust Huber's and Hampel's $M$-estimates at the heavy-tailed Laplace and Cauchy distributions.

**Table 1.** Asymptotic efficiency (two best values are boldfaced).

| *M*-Estimate | Normal | Laplace | Cauchy | Tukey Gross Error |
|---|---|---|---|---|
| mean | **1** | 0.500 | 0 | 0 |
| median | 0.636 | **1** | 0.811 | 0.678 |
| Huber's linear bounded | 0.924 | 0.669 | 0.759 | **0.953** |
| Hampel's three-part | 0.911 | 0.644 | 0.869 | 0.946 |
| $\psi_{0.01}$ | 0.895 | 0.310 | 0.250 | 0.747 |
| $\psi_{0.1}$ | **0.976** | 0.503 | 0.484 | 0.948 |
| $\psi_{0.5}$ | 0.844 | 0.799 | 0.765 | **0.962** |
| $\psi_{0.9}$ | 0.679 | 0.965 | **0.898** | 0.825 |
| $\psi_{0.99}$ | 0.641 | **0.995** | 0.859 | 0.727 |
| *MVS* | 0.650 | 0.750 | 0.800 | 0.710 |
| radical | 0.840 | 0.890 | **0.920** | 0.890 |

Finally, we outline the prospective future works: (i) an extension of the proposed minimax variance *M*-estimates to the multivariate case and the classes with simultaneously bounded subranges of different parameter $\beta$ values—in the latter case, we may expect a uniformly better performance; (ii) a generalization of the Meshalkin-Shurygin's stable estimates to the multivariate case; and (iii) a thorough comparative study of the small sample performance of *M*-estimates of location—Monte Carlo experiments show a slightly better performance of the classical Huber's and Hampel's estimates in this case.

**Funding:** This research received no external funding.

**Institutional Review Board Statement:** Not applicable.

**Informed Consent Statement:** Not applicable.

**Conflicts of Interest:** The author declares no conflict of interest.

## References


1. Tukey, J.W. *A Survey of Sampling from Contaminated Distributions. Contributions to Probability and Statistics*; Olkin, I., Ed.; Stanford Univ. Press: Redwood City, CA, USA, 1960; pp. 448–485.
2. Huber, P.J. Robust estimation of a location parameter. *Ann. Math. Stat.* **1964**, *35*, 73–101. [CrossRef]
3. Hampel, F.R. Contributions to the Theory of Robust Estimation. Ph.D. Thesis, University of California, Berkeley, CA, USA, 1968.
4. Box, G.E.P. Non-normality and tests on variances. *Biometrika* **1953**, *40*, 318–335. [CrossRef]
5. Huber, P.J.; Ronchetti, E. *Robust Statistics*; Wiley: New York, NY, USA, 2009.
6. Hampel, F.R.; Ronchetti, E.; Rousseeuw, P.J.; Stahel, W.A. *Robust Statistics. The Approach Based on Influence Functions*; Wiley: New York, NY, USA, 2005.
7. Shevlyakov, G.L.; Oja, H. *Robust Correlation. Theory and Applications*; Wiley, John Wiley & Sons Ltd.: Chichester, UK, 2016.
8. Shevlyakov, G.L.; Morgenthaler, S.; Shurygin, A.M. Redescending *M*-estimators. *J. Stat. Plann. Inference* **2008**, *138*, 2906–2916. [CrossRef]
9. Daszykowski, M.; Kaczmarek, K.; Vander Heyden, Y.; Walczak, B. Robust statistics in data analysis—A review: Basic concepts. *Chemom. Intell. Lab. Syst.* **2007**, *85*, 203–219. [CrossRef]
10. Augustin, T.; Hable, R. On the impact of robust statistics on imprecise probability models: A review. *Struct. Saf.* **2010**, *32*, 358–365. [CrossRef]
11. Rousseeuw, P.J.; Hubert, M. Robust statistics for outlier detection. *Wiley Interdiscip. Rev. Data Min. Knowl. Discov.* **2011**, *1*, 73–79. [CrossRef]
12. Maronna, R.A.; Martin, R.D.; Yohai, V.J. *Robust Statistics: Theory and Methods (with R)*; Wiley: John Wiley & Sons Ltd.: Chichester, UK, 2019.

13. Rousseeuw, P.J.; Hubert, M. Anomaly detection by robust statistics. *Wiley Interdiscip. Rev. Data Min. Knowl. Discov.* **2018**, *8*, e1236. [CrossRef]
14. Ronchetti, E. The main contributions of robust statistics to statistical science and a new challenge. *METRON* **2020**, 1–9. [CrossRef]
15. Polyak, B.T.; Tsypkin, Y.Z. Robust identification. In *Identification of Systems and Parameter Estimation, Part 1, Proceedings of the 4th IFAC Symposium, Tbilisi, GA, USA, 21–27 September 1976*; North-Holland Pub. Co.: Amsterdam, The Netherlands, 1976; pp. 203–224.
16. Shevlyakov, G.L.; Vilchevski, N.O. *Robustness in Data Analysis: Criteria and Methods*; VSP BV: Zeist, The Netherlands, 2002.
17. Kagan, A.M.; Linnik, Y.V.; Rao, S.R. *Characterization Problems in Mathematical Statistics*; Wiley: New York, NY, USA, 1972.
18. Shurygin, A.M. New approach to optimization of stable estimation. In *Proceedings of the 1 US/Japan Conference on Frontiers of Statist. Modeling*; Bozdogan, H., Ed.; Kluwer Academic Publishers: Dordrecht, The Netherlands, 1994; pp. 315–340.
19. Shevlyakov, G.L.; Shagal, A.A.; Shin, V.I. A comparative study of robust and stable estimates of multivariate location. *J. Math. Sci.* **2019**, *237*, 831–845. [CrossRef]
20. Meshalkin, L.D. Some mathematical methods for the study of non-communicable diseases. In Proceedings of the 6th International Meeting of Uses of Epidemiology in Planning Health Services, Primosten, Yugoslavia, 29 August–3 September 1971; Volume 1, pp. 250–256.
21. Shevlyakov, G.L.; Tkhakushinova, R.V.; Snin, V.I. Robust minimax variance estimation of location under bounded interquantile ranges. *J. Math. Sci.* **2020**, *248*, 25–32. [CrossRef]

mathematics

MDPI

Article

# Local Limit Theorem for the Multiple Power Series Distributions

**Arsen L. Yakymiv**

Steklov Mathematical Institute of Russian Academy of Sciences, 8 Gubkina St., Moscow 119991, Russia; arsen@mi-ras.ru

Received: 30 September 2020; Accepted: 13 November 2020; Published: 19 November 2020

**Abstract:** We study the behavior of multiple power series distributions at the boundary points of their existence. In previous papers, the necessary and sufficient conditions for the integral limit theorem were obtained. Here, the necessary and sufficient conditions for the corresponding local limit theorem are established. This article is dedicated to the memory of my teacher, professor V.M. Zolotarev.

**Keywords:** multiple power series distribution; integral limit theorem; local limit theorem; Tauberian lemma; $R$-weakly one-sided oscillation of the multiple sequence at infinity along the given multiple sequence

## 1. Introduction

Let $(a(i) \geq 0,\ i = 0, 1, 2, \dots)$ be a sequence with

$$B(x) = \sum_{i=0}^{\infty} a(i)x^i < \infty$$

for $x \in (0,1)$. The trivial case $a(i) \equiv 0,\ i = 0, 1, 2, \dots$ is excluded. It is said that a random variable $\xi_x$ has a power series distribution iff

$$\mathsf{P}\{\xi_x = i\} = \frac{a(i)x^i}{B(x)},$$

for some $B(x)$ and for any $i \in Z_+$.

Power series distributions were introduced in the fundamental paper of Noack [1] (1950).

Systematic studies of their properties (moments, generating functions, convolutions, limit properties, statistical applications, etc.) began immediately. References may be found in the encyclopedias of Johnson, Kotz, and Kemp [2] (for the one-dimensional case) and Johnson, Kotz, and Balakrishnan [3] (for the multidimensional case). For example, the binomial, Poisson, negative binomial, and logarithmic distributions, as well as their multidimensional analogues are among the important distributions in this class.

Note that power series distributions are widely useful in a generalized allocation scheme (in the one-dimensional case). This scheme was introduced by V. Kolchin [4]. His results and, in particular, those obtained with the use of this scheme, play an important role in probabilistic combinatorics (see, for example, his books [5,6]). So, one can express distributions of various characteristics of random permutations (a(i) = $1/i$), random mappings (($a(i) = i^{-1}\sum_{k=0}^{i-1} i^k/k!$)) [5]), and random mappings with various constraints (on cycle length, height, component sizes, etc.; see, for example, the books of Timashev [7,8]), random trees, and random forests (i.e., random mappings with cycles of only unit length (see the book of Yu. Pavlov [9])) in terms of power series distributions. An analogue of Kolchin's generalized allocation scheme [4] with a bounded number of particles was introduced in the work of A.N. Chuprunov and I. Fezekash [10]. A corresponding multivariate scheme was recently introduced by A.N. Chuprunov, G. Alsaied, and M. Alkhuzani [11]. For another investigation of A.N. Chuprunov and his students, see the paper [11] and the references therein. We also note the successful work of the representatives of the Karelian Scientific Center in the study of the asymptotic properties of configuration graphs under the leadership and participation of Yu.L. Pavlov by I.A. Cheplyukova, M.M. Leri, and E.V. Khvorostyanskaya [12–17].

Suppose that $B(x)$ regularly varies as $x \uparrow 1$ with index $\varrho > 0$ [18,19]. It is known, in this case, that

$$\mathsf{P}\{\xi_x(1-x) \leq y\} \to \int_0^y e^{-u} u^{\varrho-1}\, du, \ \forall y \geq 0$$

as $x \uparrow 1$. In addition, the corresponding local limit theorem is true when $a(i)$ is regularly varying at infinity with index $\varrho - 1 > -1$. See, for instance, Timashev [8].

The multidimensional integral limit theorem was obtained in [20]. It is supposed in [20] that the corresponding multiple power series regularly varies at the boundary point of its convergence (see Definition 2). In [21], it was shown that this condition is necessary and sufficient.

In this paper, we prove the corresponding local limit theorem. For this aim, we introduce in Section 2 some generalizations of multivariate regularly varying sequences in the orthant. Namely, the notion of R-weakly one-sided oscillatory sequences at infinity along some sequence (see Definition 3). This concept allows us to give adequate conditions for the validity of both the local limit theorem and the corresponding statement of Tauberian type (Lemma 2). The definition of multiple power series distribution and the main result are given in the next section (see Definition 1 and Theorem 1, respectively). Here, we also formulate the corresponding integral limit result from [21] as Lemma 1. The statement of this lemma also gives the necessary and sufficient conditions but describes them in terms of regular variation of the power series $B(x)$ at the boundary point of their existence. Proofs of Lemma 2 and the main result (Theorem 1) are given in the Sections 3 and 4, respectively. In Section 5, we describe some previous results in this direction.

## 2. Main Result

### 2.1. Some Notations

We introduce the following notations. Let the vectors $x = (x_1, \dots, x_n)$ and $y = (y_1, \dots, y_n)$ belong to $R^n$. Denote $xy = (x_1y_1, \dots, x_ny_n)$ and $x/y = (x_1/y_1, \dots, x_n/y_n)$ (the last in the case, when $y_k \neq 0\ \forall k = 1, \dots, n$). Put $\exp(x) = (\exp(x_1), \dots, \exp(x_n))$, $\ln x = (\ln x_1, \dots, \ln x_n)$. The notation $x \uparrow \mathbf{1}$ means that $x \to \mathbf{1}$, $x \in (0,1)^n$. Here $\mathbf{1} = (1, \dots, 1)$. Set $R^n_+ = \{x : x = (x_1, \dots, x_n) \in R^n,\ x_k \geq 0\ \forall k = 1, \dots, n\}$, $Z^n_+ = \{x : x = (x_1, \dots, x_n) \in R^n_+,\ x_k \in Z_+ = N \cup 0\ \forall k = 1, \dots, n\}$. For $\alpha = (\alpha_1, \dots, \alpha_n) \in R^n_+$, $x = (x_1, \dots, x_n) \in R^n_+$ we use an abbreviation

$$x^\alpha = \prod_{k=1}^n x_k^{\alpha_k},$$

assuming that $0^0 = 1$. Let $(\eta_k,\ k \in N)$ be a sequence of random vectors (r.v.) from $R^n$. Further, the notation $\eta_k \xrightarrow{d} \eta$ means the weak convergence of the corresponding distributions with $\mathsf{P}\{\eta \in R^n\} = 1$.

### 2.2. Multiple Power Series Distributions

First we give the necessary definitions. Let $(a(i) \geq 0,\ i \in Z^n_+)$ be a multiple sequence with

$$B(x) = \sum_{i \in Z^n_+} a(i) x^i < \infty$$

for $x = (x_1, \dots, x_n) \in [0,1)^n$. The trivial case $a(i) \equiv 0$, $i \in Z^n_+$ is excluded.

**Definition 1.** *For $x \in [0,1)^n$ a random vector (r.v.) $\xi_x$ has a multiple power series distribution iff*

$$\mathsf{P}\{\xi_x = i\} = \frac{a(i)x^i}{B(x)}, \tag{1}$$

*for some $B(x)$ and for any $i \in Z^n_+$.*

It is clear that $\mathsf{P}\{\xi_x \in R^n\} = 1$. The history of this notion and some bibliographic references are given in encyclopedias [2,3], also see the articles [20,22]. Let the sequence of vectors $b = b(k) \in (0,\infty)^n$, $k \in N$ be given with $b_j = b_j(k) \to \infty$, $\forall j = 1,\ldots,n$ as $k \to \infty$.

**Definition 2** ([23])**.** *We say that $B(x)$ regularly varies as $x \uparrow \mathbf{1}$ along the sequence $b = b(k)$, iff*

$$\frac{B(\exp(-\lambda/b))}{B(\exp(-\mathbf{1}/b))} \to \Psi(\lambda) \in (0,\infty), \tag{2}$$

*for an arbitrary fixed $\lambda = (\lambda_1,\ldots,\lambda_n) > 0$ as $k \to \infty$.*

(Notations $\lambda/b$ and $\exp(-\lambda/b)$ are defined in the Section 2.1).

The following statement has been proved in [21] (we formulate it as a lemma).

**Lemma 1.** *A series $B(x)$ regularly varies as $x \uparrow \mathbf{1}$ along the sequence $b = b(k)$ iff for any (some) fixed vector $u \in G$ and $x = \exp(-u/b)$*

$$\xi_x(\mathbf{1} - x) \xrightarrow{d} \eta = \eta(u), \quad (k \to \infty) \tag{3}$$

*In both cases, the function $\Psi(\lambda)$ from Equation (2) is the Laplace transform of some $\sigma$-finite measure $\Phi(\cdot)$ and r.v. $\eta(u)$ has Laplace transform $\Psi((\lambda+\mathbf{1})u)/\Psi(u)$.*

Let $R(k)$ be some positive sequence. To formulate the resulting limit theorem, we need to give the following definition.

**Definition 3.** *We say that the sequence $a(i)$ is R-weakly one-sided oscillatory at infinity along the sequence $b = b(k)$ if for every $j = 1,\ldots,n$ and for any sequence $z_j = z_j(k) > 1, z_j = 1 + o(1)$ one of the following inequalities*

$$\liminf_{k\to\infty}(a(r_1,\ldots,r_{j-1},z_jr_j,r_{j+1},\ldots,r_n) - a(r))/R(k) \geq 0; \tag{4}$$

$$\limsup_{k\to\infty}(a(r_1,\ldots,r_{j-1},z_jr_j,r_{j+1},\ldots,r_n) - a(r))/R(k) \leq 0. \tag{5}$$

*holds for every fixed $y = (y_1,\ldots,y_n) \in G$. Here $r = r(k) = (r_1(k),\ldots,r_n(k))$ is an arbitrary function of $k$ with*

$$r_1 \sim y_1b_1,\ldots,r_n \sim y_nb_n.$$

Hereinafter, we define $a(x) = a([x])$ for $x \notin Z^n_+$. The simplest examples of such sequences are monotone in each variable sequence $(a(i) \geq 0,\ i \in Z^n_+)$.

**Theorem 1.** *Suppose that $B(x)$ regularly varies as $x \uparrow \mathbf{1}$ along the sequence $b = b(k)$ (i.e., the the assumption of integral limit Lemma 1 is true). Then, for any compact $K \subset G$ and for any (some) fixed vector $u \in G$ and $x = \exp(-u/b)$*

$$\frac{\mathsf{P}\{\xi_x = [y/(\mathbf{1}-x)]\}}{\prod_{j=1}^n(1-x_j)} \overset{y\in K}{\rightrightarrows} \psi_u(y) < \infty \quad (k \to \infty) \tag{6}$$

*where function $\psi_u(\cdot)$ is continuous in G, iff the sequence $a(i)$ is R-weakly one-sided oscillatory at infinity along the sequence $b = b(k)$ with*

$$R(k) = B(\exp(-\mathbf{1}/b(k)))/\prod_{i=1} b_i(k). \tag{7}$$

*In both cases, the measure $\Phi(\cdot)$ from Lemma 1 has the continuous density $\varphi(\cdot)$ in G and the following equality holds:*

$$\psi_u(y) = \frac{\varphi(y/u)e^{-(y,\mathbf{1})}}{\prod_{j=1}^{n} u_j \Psi(u)}, \quad \forall y \in G. \tag{8}$$

Note that, in Theorem 1, the case when $\Phi(\partial G) > 0 (\Leftrightarrow \mathsf{P}\{\eta(u) \in \partial G\} > 0)$ is not excluded. In addition, we admit that $\psi_u(y) = 0,\ y \in V$, for some nonempty set $V \subseteq G$ in this theorem.

## 3. Tauberian Lemma

The next lemma gives some generalization of the Tauberian Theorem 2 from [23].

**Lemma 2.** *Assume that $B(x)$ regularly varies as $x \uparrow \mathbf{1}$ along the sequence $b = b(k)$ (i.e., (2) holds). Then, for some continuous function $\varphi(\cdot)$ in G the relation*

$$\frac{a(bv)}{R(k)} \overset{v \in K}{\rightrightarrows} \varphi(u) < \infty \tag{9}$$

*holds for any compact $K \subset G$ iff the sequence $a(i)$ is R-weakly one-sided oscillatory at infinity along the sequence $b = b(k)$ with $R(k)$ from Equation (7). In both cases, the measure $\Phi(\cdot)$ from Lemma 1 is absolutely continuous in G with density $\varphi(\cdot)$.*

**Proof.** For an arbitrary bounded set $A \subset R^n_+$, put

$$\Phi_k(A) = \sum_{i \in Z^n_+,\ i/b \in A} \frac{a(i)}{\prod_{j=1}^{n} m_j(k) R(k)} \tag{10}$$

It follows from Equations (2) and (7) that

$$\widetilde{\Phi}_k(y) \equiv \int_{R^n_+} e^{-(x,y)} \Phi_k(dx) = \frac{B(exp(-y/b))}{\prod_{j=1}^{n} b_j(k) R(k)}$$

$$\to \psi(y) = \widetilde{\Phi}(y) \equiv \int_{R^n_+} e^{-(x,y)} \Phi(dx)$$

for any fixed $y \in G$. The last equality follows from the statement of Lemma 1. Thus, according to the continuity theorem for Laplace transforms of measures, it follows from Equation (10) that

$$\Phi_k(\cdot) \Rightarrow \Phi(\cdot). \tag{11}$$

(see, for example the theorem 1.3.2 from [24]). Suppose that the sequence $a(i)$ is $R$-weakly one-sided oscillatory at infinity along the sequence $b = b(k)$. Set $m(j) = 1$ if Equation (4) holds and $m(j) = -1$ if Equation (5) is valid. Fix $v \in G$. For an arbitrary $\delta \in (0,1)$, put

$$A(\delta) = \left\{ y = (y_1, \ldots, y_n),\ y_j \in \left(v_j, v_j(1+\delta)^{b(j)}\right),\ \forall j = 1, \ldots, n \right\} \tag{12}$$

(for $c > d$, we put $(c,d) = (d,c)$). Further, for an arbitrary $\varepsilon \in (0,1)$, there exists such $\delta \in (0,\varepsilon)$ that

$$\frac{a(i) - a(bv)}{R(k)} \geq -\varepsilon \tag{13}$$

for any $i \in mA(\delta)$. The proof of this fact repeats the proof of Lemma 5 from [23]. Without loss of generality, we assume that $\Phi(\partial A(\delta)) = 0$. It follows from Equations (9) and (13) that

$$\Phi_k(A(\delta)) = \sum_{i \in Z_+^n,\, i/b \in A(\delta)} \frac{a(i)}{\prod_{j=1}^n b_j(k) R(k)}$$

$$\geq -\varepsilon + \frac{a(bv)}{R(k)} \frac{1}{\prod_{j=1}^n b_j(k)} \sum_{i \in Z_+^n,\, i/b \in A(\delta)} 1.$$

$$\geq -\varepsilon + \frac{a(bv)}{R(k)} (1 + \eta_k) |A(\delta)|$$

where $\eta_k \to 0$ as $k \to \infty$. By $|A(\delta)|$, we denote here the Lebesque measure of the set $A(\delta)$. Therefore,

$$\frac{a(bv)}{R(k)} \leq \left( \frac{\Phi_k(A(\delta))}{|A(\delta)|} + \frac{\varepsilon}{|A(\delta)|} \right) \frac{1}{1 + \eta_k}. \tag{14}$$

Since $\Phi(\partial A(\delta)) = 0$, we have from Equations (9) and (11) that

$$\Phi_k(A(\delta)) = \sum_{i \in Z_+^n,\, i/b \in A(\delta)} \frac{a(i)}{\prod_{j=1}^n b_j(k) R(k)} \to \Phi(A(\delta)). \tag{15}$$

Tending in Equation (14) $k$ to $\infty$ and using Equation (15), we have

$$\limsup_{k \to \infty} \frac{a(bv)}{R(k)} \leq \frac{\Phi(A(\delta))}{|A(\delta)|} + \frac{\varepsilon}{|A(\delta)|}. \tag{16}$$

Since the left side of Equation (16) does not depend on $\varepsilon$, we have

$$\limsup_{k \to \infty} \frac{a(bv)}{R(k)} \leq \frac{\Phi(A(\delta))}{|A(\delta)|}. \tag{17}$$

Put $\Delta = \{ \delta \in (0,1) : \Phi(\partial A(\delta)) = 0 \}$. Since the left side of Equation (17) does not depend on $\delta$, we have

$$\limsup_{k \to \infty} \frac{a(bv)}{R(k)} \leq \liminf_{\delta \to 0,\, \delta \in \Delta} \frac{\Phi(A(\delta))}{|A(\delta)|}. \tag{18}$$

Similarly, we obtain the inequality

$$\liminf_{k \to \infty} \frac{a(bv)}{R(k)} \geq \limsup_{\delta \to 0,\, \delta \in \Delta} \frac{\Phi(A(\delta))}{|A(\delta)|}. \tag{19}$$

It follows from Equations (18) and (19) that there exist the next two limits:

$$\lim_{k \to \infty} \frac{a(bv)}{R(k)} = \lim_{\delta \to 0,\, \delta \in \Delta} \frac{\Phi(A(\delta))}{|A(\delta)|} \left( \stackrel{def}{=} \varphi(v) \right).$$

□

The next proof repeats the proof of Theorem 2 from [23]. The inverse assertion of Lemma 2 follows immediately from Equation (9). Lemma 2 is proved.

## 4. Proof of Theorem 1

Suppose that Equation (2) holds. Put for $z \in N^n$ and $x \in [0,1]^n$

$$p(z,x) = \mathsf{P}\{\xi_x = z\} = \frac{a(z)}{B(x)} \exp(z, \ln x).$$

We have

$$a(z) = p(z,x)B(x)\exp(z, -\ln x). \tag{20}$$

Suppose that Equation (6) takes place for some $u \in G$ and continuous in $G$ function $\psi_u(\cdot)$. For fixed $y \in G$, put in Equation (20) $x = \exp(-u/b(k))$ and $z = [y/(\mathbf{1} - x)]$. We have $x = \mathbf{1} - (u + \varepsilon(k))/b(k)$ and $z = [b(k)y/(u + \varepsilon(k))] = b(k)(y/u + \delta(k))$. Here $\varepsilon(k)$ and $\delta(k)$ are some functions tending to zero as $k \to \infty$. Thus $(z, -\ln x) = (y, \mathbf{1}) + o(1)$ as $k \to \infty$. So, it follows from Equations (20), (2), and (6) that

$$\begin{aligned}
a(z) &= p(z,x)B(x)\exp(z, -\ln x) = (1 + o(1))p(z,x)B(x)\exp(y, \mathbf{1}) \\
&= \prod_{j=1}^{n}(1 - x_j)B(x)(\psi_u(y) + o(1))\exp(y, \mathbf{1}) \\
&= \prod_{j=1}^{n}(1 - x_j)B(\exp(-\mathbf{1}/b(k))\Psi(u)(\psi_u(y) + o(1))\exp(y, \mathbf{1}) \\
&= \prod_{j=1}^{n}(u_j/b_j(k))B(\exp(-\mathbf{1}/b(k))\Psi(u)(\psi_u(y) + o(1))\exp(y, \mathbf{1}) \\
&= R(k)\prod_{j=1}^{n} u_j \Psi(u)(\psi_u(y) + o(1))\exp(y, \mathbf{1})
\end{aligned} \tag{21}$$

according to Equation (7). Since Equation (6) holds locally uniformly on $y$ then it follows from Equation (21) that

$$\frac{a(by/u)}{R(k)} \to \prod_{j=1}^{n} u_j \Psi(u)\psi_u(y)\exp(y, \mathbf{1}) = \varphi(y/u) \tag{22}$$

and the last relation also holds locally uniformly on $y$. The equality Equation (8) follows directly from Equation (22). Replacing in Equation (22) $y/u$ by $v$, we obtain Equation (9). One-sided $R$-oscillation of $a(\cdot)$ along $b(k)$ follows immediately from Equation (9). The proof of inverse assertion repeats the proof of Theorem 2 from [20].

## 5. On Some Previous Results

The definition of regularly varying functions of one variable was given in Karamata's well-known work [25]. The notion of regularly varying functions at infinity along some sequence in an orthant was introduced in Omey [26]. The definition of regularly varying multiple power series is given in [23]. A brief overview of various definitions of multivariate regularly varying functions is available in [27]. The history of different class functions having slow (one-sided or ordinary) oscillation can be seen in the book [24]. In [22], we give the integral representation and Abelian statements (Theorems 3.1 and 3.2). With the help of these theorems, it is easy to set such sequences $a(i)$ explicitly.

As the source, for $n = 1$ the sufficient condition for Equation (3) was given in Timashev [28], see also [8]. In [20], we show that conditions from [8,28] are equivalent to Equation (2). Timashev used the method of moments in his aforementioned result. In the papers [20,22,27] and in this article, we use the corresponding Tauberian statements. All these statements go back to Karamata's well-known Tauberian theorems [29,30].

**Funding:** This research received no external funding.

**Acknowledgments:** The author expresses his deep gratitude to reviewers for valuable comments.

**Conflicts of Interest:** The author declare no conflict of interest.

## References


1. Noack, A. A class of random variables with discrete distributions. *Ann. Math. Statist.* **1950**, *21*, 127–132. [CrossRef]
2. Johnson, N.L.; Kotz, S.; Kemp, A.W. *Univariate Discrete Distributions*, 2nd ed.; Wiley Series in Probability and Statistics; John Wiley and Sons: New York, NY, USA, 1992.
3. Johnson, N.L.; Kotz, S.; Balakrishnan, N. *Discrete Multivariate Distributions*; Wiley Series in Probability and Statistics; Wiley: New York, NY, USA, 1997; Volume xxii, 299p.
4. Kolchin, V.F. A certain Class of Limit Theorems for Conditional Distributions. In *Selected Translations in Mathematical Statistics and Probability*; American Mathematical Society: Providence, RI, USA, 1973; Voume 11, pp. 185–197.
5. Kolchin, V.F. *Random Mappings*; Translations Series in Mathematics and Engineering; Optimization Software, Inc., Publications Division: New York, NY, USA, 1986.
6. Kolchin, V.F. *Random Graphs, Encyclopedia Math. Appl. 53*; Cambridge Univ. Press: Cambridge, UK, 1999.
7. Timashev, A.N. *Random Components in Generalized Allocation Scheme*; Akademiya: Moscow, Russia, 2017.
8. Timashev, A.N. Limit theorems for power-series distributions with finite radius of convergence. *Theory Probab. Appl.* **2018**, *63*, 45–56. [CrossRef]
9. Pavlov, Y.L. *Random Forests*; VSP: Utrecht, The Netherlands, 2000.
10. Chuprunov, A.N.; Fazekas, I. An analogue of the generalised allocation scheme: limit theorems for the number of cells containing a given number of particles. *Discrete Math. Appl.* **2012**, *22*, 101–122 [CrossRef]
11. Chuprunov, A.N.G.; Alsaied, M. Alkhuzani, On maximal quantity of particles of one color in analogs of multicolor urn schemes. *Russ. Math. (Iz. VUZ)* **2017**, *61*, 83–88. [CrossRef]
12. Pavlov, Y.L. Conditional configuration graphs with discrete power-law distribution of vertex degrees. *Sb. Math.* **2018**, *209*, 258–275. [CrossRef]
13. Pavlov, Y.L. On the connectedness of configuration graphs. *Diskr. Mat.* **2019**, *31*, 114–122.
14. Pavlov, Y.L. On the asymptotics of the cluster coefficient of a configuration graph with an unknown distribution of vertex degrees. *Inform. Appl.* **2019**, *13*, 9–13.
15. Pavlov, Y.L.; Cheplyukova, I.A. On the asymptotics of degree structure of configuration graphs with bounded number of edges. *Discret. Math. Appl.* **2019**, *29*, 219–232. [CrossRef]
16. Pavlov, Y.L.; Khvorostyanskaya, E.V. On the limit distributions of the degrees of vertices in configuration graphs with a bounded number of edges. *Sb. Math.* **2016**, *207*, 400–417. [CrossRef]
17. Leri, M.M.; Pavlov, Y.L. On the stability of configuration graphs in a random environment. *Inform. Appl.* **2018**, *12*, 2–10.
18. Bingham, N.H.; Goldie, C.M.; Teugels, J. *Regular Variation*; Cambridge University Press: Cambridge, UK, 1987.
19. Seneta, E. *Regularly Varying Functions*; Springer-Verlag: Berlin/Heidelberg, Germany; New York, NY, USA, 1976.
20. Yakymiv, A.L. On the multiple power series distribution regularly varying at the boundary point. *Discret. Math. Appl.* **2019**, *29*, 409–421. [CrossRef]
21. Yakymiv, A.L. Some Properties of Regularly Varying Functions and Series in the Orthant, Probability-Analytical models. In *Methods and Applications, Springer Proceedings in Mathematics and Statistics*; Shiryaev, A.N., Pavlov, I.V., Eds.; Springer: Berlin/Heidelberg, Germany, 2020. in press.
22. Yakymiv, A.L. Abelian theorem for the regularly varying measure and its density in orthant. *Theory Probab. Appl.* **2019**, *64*, 385–400. [CrossRef]
23. Yakymiv, A.L. A Tauberian theorem for multiple power series. *Sb. Math.* **2016**, *207*, 286–313. [CrossRef]
24. Yakymiv, A.L. *Probabilistic Applications of Tauberian Theorems.-Modern Probability and Statistics*; VSP: Leiden, The Netherlands; Boston, MA, USA, 2005; Volume viii, 225p.
25. Karamata, J. Sur un mode croissanse régulière des fonctions. *Mathematica* **1930**, *4*, 38–53.

26. Omey, E. *Multivariate Regular Variation and Application in Probability Theory, Eclectica*; EHSAL: Brussel, Belgium, 1989; Volume 74.
27. Yakymiv A.L. Multivariate Regular Variation in Probability Theory. *J. Math. Sci.* **2020**, *246*, 580–586. [CrossRef]
28. Timashev, A.N. *Power Series Distributions and Generalized Allocation Scheme*; Akademiya: Moscow, Russia, 2016.
29. Karamata, J. Über die Hardy–Littelwoodsche Umkehrungen des Abelschen Steitigkeitssatzes. *Math. Z.* **1930**, *32*, 319–320. [CrossRef]
30. Karamata, J. Neuer Beweis und Verallgemeinerung einiger Tauberian-Sätze. *Math. Z.* **1931**, *33*, 294–299. [CrossRef]

mathematics

MDPI

Article

# Multivariate Scale-Mixed Stable Distributions and Related Limit Theorems

Yury Khokhlov [1,2,*], Victor Korolev [1,2,3,4] and Alexander Zeifman [1,4,5,6]

1 Moscow Center for Fundamental and Applied Mathematics, Moscow State University, 119991 Moscow, Russia; vkorolev@cs.msu.ru (V.K.); a_zeifman@mail.ru (A.Z.)
2 Faculty of Computational Mathematics and Cybernetics, Moscow State University, 119991 Moscow, Russia
3 Department of Mathematics, School of Science, Hangzhou Dianzi University, Hangzhou 310018, China
4 Institute of Informatics Problems, Federal Research Center «Computer Science and Control» of the Russian Academy of Sciences, 119993 Moscow, Russia
5 Department of Applied Mathematics, Vologda State University, 160000 Vologda, Russia
6 Vologda Research Center of the Russian Academy of Sciences, 160014 Vologda, Russia
* Correspondence: yskhokhlov@yandex.ru

Received: 13 April 2020; Accepted: 3 May 2020; Published: 8 May 2020

**Abstract:** In the paper, multivariate probability distributions are considered that are representable as scale mixtures of multivariate stable distributions. Multivariate analogs of the Mittag–Leffler distribution are introduced. Some properties of these distributions are discussed. The main focus is on the representations of the corresponding random vectors as products of independent random variables and vectors. In these products, relations are traced of the distributions of the involved terms with popular probability distributions. As examples of distributions of the class of scale mixtures of multivariate stable distributions, multivariate generalized Linnik distributions and multivariate generalized Mittag–Leffler distributions are considered in detail. Their relations with multivariate 'ordinary' Linnik distributions, multivariate normal, stable and Laplace laws as well as with univariate Mittag–Leffler and generalized Mittag–Leffler distributions are discussed. Limit theorems are proved presenting necessary and sufficient conditions for the convergence of the distributions of random sequences with independent random indices (including sums of a random number of random vectors and multivariate statistics constructed from samples with random sizes) to scale mixtures of multivariate elliptically contoured stable distributions. The property of scale-mixed multivariate elliptically contoured stable distributions to be both scale mixtures of a non-trivial multivariate stable distribution and a normal scale mixture is used to obtain necessary and sufficient conditions for the convergence of the distributions of random sums of random vectors with covariance matrices to the multivariate generalized Linnik distribution.

**Keywords:** geometrically stable distribution; generalized Linnik distribution; random sum; transfer theorem; multivariate normal scale mixtures; heavy-tailed distributions; multivariate stable distribution; multivariate Linnik distribution; generalized Mittag–Leffler distribution; multivariate generalized Mittag–Leffler distribution

**MSC:** 60E07; 60F05; 62E10; 62E20; 62H05

## 1. Introduction

Actually, this paper can be regarded as variations on the theme of 'multiplication theorem' 3.3.1 in the famous book of V. M. Zolotarev [1]. Here, multivariate probability distributions are considered that are representable as scale mixtures of multivariate stable distributions. Some properties of these distributions are discussed. Attention is paid to the representations of the corresponding random vectors as products of independent random variables and vectors. In these products, relations of the distributions of the involved terms with popular probability distributions are traced.

As examples of distributions of the class of scale mixtures of multivariate stable distributions, multivariate generalized Linnik distributions and multivariate generalized Mittag–Leffler distributions are considered in detail. Limit theorems are proved presenting necessary and sufficient conditions for the convergence of the distributions of random sequences with independent random indices (including sums of a random number of random vectors and multivariate statistics constructed from samples with random sizes) to scale mixtures of multivariate elliptically contoured stable distributions. As particular cases, conditions are obtained for the convergence of the distributions of random sums of random vectors with covariance matrices to the multivariate generalized Linnik distribution.

Along with general multiplicative properties of the class of scale mixtures of multivariate stable distributions, some important and popular special cases are considered in detail. Multivariate analogs of the Mittag–Leffler distribution are proposed. We study the multivariate (generalized) Linnik and related (generalized) Mittag–Leffler distributions, their interrelation and their relations with multivariate 'ordinary' Linnik distributions, multivariate normal, stable and Laplace laws as well as with univariate 'ordinary' Mittag–Leffler distributions. Namely, we consider mixture representations for the multivariate generalized Mittag–Leffler and multivariate generalized Linnik distributions. We continue the research we started in [2–5]. In most papers (see, e.g., [6–18]), the properties of the (multivariate) generalized Mittag–Leffler and Linnik distributions were deduced by analytical methods from the properties of the corresponding probability densities and/or characteristic functions. Instead, here we use the approach which can be regarded as arithmetical in the space of random variables or vectors. Within this approach, instead of the operation of scale mixing in the space of distributions, we consider the operation of multiplication in the space of random vectors/variables provided the multipliers are independent. This approach considerably simplifies the reasoning and makes it possible to notice some general features of the distributions under consideration. We prove mixture representations for general scale mixtures of multivariate stable distributions and their particular cases in terms of normal, Laplace, generalized gamma (including exponential, gamma and Weibull) and stable laws and establish the relationship between the mixing distributions in these representations. In particular, we prove that the multivariate generalized Linnik distribution is a multivariate normal scale mixture with the univariate generalized Mittag–Leffler mixing distribution and, moreover, show that this representation can be used as the definition of the multivariate generalized Linnik distribution. Based on these representations, we prove some limit theorems for random sums of independent random vectors with covariance matrices. As a particular case, we prove some theorems in which the multivariate generalized Linnik distribution plays the role of the limit law. By doing so, we demonstrate that the scheme of geometric (or, in general, negative binomial) summation is not the only asymptotic setting (even for sums of independent random variables) in which the multivariate generalized Linnik law appears as the limit distribution.

In [2], we showed that along with the traditional and well-known representation of the univariate Linnik distribution as the scale mixture of a strictly stable law with exponential mixing distribution, there exists another representation of the Linnik law as the normal scale mixture with the Mittag–Leffler mixing distribution. The former representation makes it possible to treat the Linnik law as the limit distribution for geometric random sums of independent identically distributed random variables (random variables) in which summands have very large variances. The latter normal scale mixture representation opens the way to treating the Linnik distribution as the limit distribution in the central limit theorem for random sums of independent random variables in which summands have *finite* variances. Moreover, being scale mixtures of normal laws, the Linnik distributions can serve as the one-dimensional distributions of a special subordinated Wiener process. Subordinated Wiener processes with various types of subordinators are often used as models of the evolution of stock prices and financial indexes, e.g., [19]. Strange as it may seem, the results concerning the possibility of representation of the Linnik distribution as a scale mixture of normals were never explicitly presented in the literature in full detail before [2], although the property of the Linnik distribution to be a normal scale mixture is something almost obvious. Perhaps, the paper [10] was the closest to this conclusion and exposed the representability of the Linnik law as a scale mixture of Laplace distributions with the mixing distribution written out explicitly. These results became the base for our efforts to extend them from the Linnik distribution to the multivariate generalized Linnik law and more general scale mixtures of multivariate stable distributions. Methodically, the present paper is very close to the work of L. Devroye [20] where many examples of mixture representations of popular probability distributions were discussed from the simulation point of view. The presented material substantially relies on the results of [2,5,15].

In many situations related to experimental data analysis, one often comes across the following phenomenon: although conventional reasoning based on the central limit theorem of probability theory concludes that the expected distribution of observations should be normal, instead, the statistical procedures expose the noticeable non-normality of real distributions. Moreover, as a rule, the observed non-normal distributions are more leptokurtic than the normal law, having sharper vertices and heavier tails. These situations are typical in financial data analysis (see, e.g., Chapter 4 in [19] or Chapter 8 in [21] and the references therein), in experimental physics (see, e.g., [22]) and other fields dealing with statistical analysis of experimental data. Many attempts were undertaken to explain this heavy-tailedness. Most significant theoretical breakthrough is usually associated with the results of B. Mandelbrot and others [23–25] who proposed, instead of the standard central limit theorem, to use reasoning based on limit theorems for sums of random summands with very large variances (also see [26,27]) resulting in non-normal stable laws as heavy-tailed models of the distributions of experimental data. However, in most cases, the key assumption within this approach, the lareg size of the variances of elementary summands, can hardly be believed to hold in practice. To overcome this contradiction, in [28], we considered an extended limit setting where it may be assumed that the intensity of the flow of informative events is random resulting in that the number of jumps up to a certain time in a random-walk-type model or the sample size is random. We show that in this extended setting, actually, heavy-tailed scale mixtures of stable laws can also be limit distributions for sums of a random number of random vectors with *finite* covariance matrices.

The paper is organized as follows. Section 2 contains basic notations and definitions. Some properties of univariate scale distributions are recalled in Section 3. In Section 4, we introduce multivariate stable distributions and prove a multivariate analog of the univariate 'multiplication theorem' (see Theorem 3.3.1 in [1]). In Section 5 we discuss some properties of scale-mixed multivariate elliptically contoured stable laws. In particular, we prove that these mixtures are identifiable. Section 6 contains the description of the properties of uni- and multi-variate generalized Mittag–Leffler distributions. In Section 7, we consider the multivariate generalized Linnik distribution. Here, we discuss different approaches to the definition of this distribution and prove some new mixture representations for the multivariate generalized Linnik distribution. General properties of scale-mixed multivariate stable distributions are discussed in Section 8. In Section 9, we first prove a general transfer theorem presenting necessary and sufficient conditions for the convergence of the distributions of random sequences with independent random indices (including sums of a random number of random vectors and multivariate statistics constructed from samples with random sizes) to scale mixtures of multivariate elliptically contoured stable distributions. As particular cases, conditions are obtained for the convergence of the distributions of scalar normalized random sums of random vectors with covariance matrices to scale mixtures of multivariate stable distributions and their special cases: 'pure' multivariate stable distributions and the multivariate generalized Linnik distributions. The results of this section extend and refine those proved in [29].

## 2. Basic Notation and Definitions

Let $r \in \mathbb{N}$. We will consider random elements taking values in the $r$-dimensional Euclidean space $\mathbb{R}^r$. The Euclidean norm of a vector $\boldsymbol{x} \in \mathbb{R}^r$ will be denoted $\|\boldsymbol{x}\|$. Assume that all the random variables and random vectors are defined on one and the same probability space $(\Omega, \mathcal{A}, \mathsf{P})$. The distribution of a random variable $Y$ or an $r$-variate random vector $\boldsymbol{Y}$ with respect to the measure $\mathsf{P}$ will be denoted $\mathcal{L}(Y)$ and $\mathcal{L}(\boldsymbol{Y})$, respectively. The weak convergence, the coincidence of distributions and the convergence in probability with respect to a specified probability measure will be denoted by the symbols $\Longrightarrow$, $\stackrel{d}{=}$ and $\stackrel{P}{\longrightarrow}$, respectively. The product of *independent* random elements will be denoted by the symbol $\circ$. The vector with all zero coordinates will be denoted $\mathbf{0}$.

A univariate random variable with the standard normal distribution function $\Phi(x)$ will be denoted $X$,

$$\mathsf{P}(X < x) = \Phi(x) = \frac{1}{\sqrt{2\pi}} \int_{-\infty}^{x} e^{-z^2/2} dz, \quad x \in \mathbb{R}.$$

Let $\Sigma$ be a positive definite $(r \times r)$-matrix. The normal distribution in $\mathbb{R}^r$ with zero vector of expectations and covariance matrix $\Sigma$ will be denoted $\mathfrak{N}_{\Sigma}$. This distribution is defined by its density

$$\phi(\boldsymbol{x}) = \frac{\exp\{-\frac{1}{2}\boldsymbol{x}^{\top}\Sigma^{-1}\boldsymbol{x}\}}{(2\pi)^{r/2}|\Sigma|^{1/2}}, \quad \boldsymbol{x} \in \mathbb{R}^r.$$

The characteristic function $\mathfrak{f}^{(\boldsymbol{X})}(\boldsymbol{t})$ of a random vector $\boldsymbol{X}$ such that $\mathcal{L}(\boldsymbol{X}) = \mathfrak{N}_{\Sigma}$ has the form

$$\mathfrak{f}^{(\boldsymbol{X})}(\boldsymbol{t}) = \mathsf{E}\exp\{i\boldsymbol{t}^{\top}\boldsymbol{X}\} = \exp\big\{-\tfrac{1}{2}\boldsymbol{t}^{\top}\Sigma\boldsymbol{t}\big\}, \quad \boldsymbol{t} \in \mathbb{R}^r.$$

A random variable having the gamma distribution with shape parameter $r > 0$ and scale parameter $\lambda > 0$ will be denoted $G_{r,\lambda}$,

$$\mathsf{P}(G_{r,\lambda} < x) = \int_0^x g(z; r, \lambda) dz, \ \text{with} \ g(x; r, \lambda) = \frac{\lambda^r}{\Gamma(r)} x^{r-1} e^{-\lambda x}, \ x \geqslant 0,$$

where $\Gamma(r)$ is Euler's gamma-function,

$$\Gamma(r) = \int_0^{\infty} x^{r-1} e^{-x} dx, \quad r > 0.$$

In this notation, obviously, $G_{1,1}$ is a random variable with the standard exponential distribution: $\mathsf{P}(G_{1,1} < x) = \big[1 - e^{-x}\big]\mathbf{1}(x \geqslant 0)$ (here and in what follows $\mathbf{1}(A)$ is the indicator function of a set $A$).

The gamma distribution is a particular representative of the class of generalized gamma distributions (GG distributions), that was first described in [30] as a special family of lifetime distributions containing both gamma and Weibull distributions. A *generalized gamma* (*GG*) *distribution* is the absolutely continuous distribution defined by the density

$$\overline{g}(x; r, \alpha, \lambda) = \frac{|\alpha|\lambda^r}{\Gamma(r)} x^{\alpha r - 1} e^{-\lambda x^{\alpha}}, \quad x \geqslant 0,$$

with $\alpha \in \mathbb{R}$, $\lambda > 0$, $r > 0$. A random variable with the density $\overline{g}(x; r, \alpha, \lambda)$ will be denoted $\overline{G}_{r,\alpha,\lambda}$. It is easy to see that

$$\overline{G}_{r,\alpha,\mu} \stackrel{d}{=} G_{r,\mu}^{1/\alpha} \stackrel{d}{=} \mu^{-1/\alpha} G_{r,1}^{1/\alpha} \stackrel{d}{=} \mu^{-1/\alpha} \overline{G}_{r,\alpha,1}. \tag{1}$$

Let $\gamma > 0$. The distribution of the random variable $W_\gamma$:

$$\mathsf{P}(W_\gamma < x) = \big[1 - e^{-x^{\gamma}}\big]\mathbf{1}(x \geqslant 0),$$

is called the *Weibull distribution* with shape parameter $\gamma$. It is obvious that $W_1$ is the random variable with the standard exponential distribution: $\mathsf{P}(W_1 < x) = \big[1 - e^{-x}\big]\mathbf{1}(x \geqslant 0)$. The Weibull distribution is a particular case of GG distributions corresponding to the density $\overline{g}(x; 1, \gamma, 1)$. It is easy to see that

$W_1^{1/\gamma} \stackrel{d}{=} W_\gamma$. Moreover, if $\gamma > 0$ and $\gamma' > 0$, then $\mathsf{P}(W_{\gamma'}^{1/\gamma} \geqslant x) = \mathsf{P}(W_{\gamma'} \geqslant x^\gamma) = e^{-x^{\gamma\gamma'}} = \mathsf{P}(W_{\gamma\gamma'} \geqslant x)$, $x \geqslant 0$, that is, for any $\gamma > 0$ and $\gamma' > 0$

$$W_{\gamma\gamma'} \stackrel{d}{=} W_{\gamma'}^{1/\gamma}.$$

In the paper [31], it was shown that any gamma distribution with shape parameter no greater than one is mixed exponential. Namely, the density $g(x;r,\mu)$ of a gamma distribution with $0 < r < 1$ can be represented as

$$g(x;r,\mu) = \int_0^\infty ze^{-zx}p(z;r,\mu)dz,$$

where

$$p(z;r,\mu) = \frac{\mu^r}{\Gamma(1-r)\Gamma(r)} \cdot \frac{\mathbf{1}(z \geqslant \mu)}{(z-\mu)^r z}. \tag{2}$$

Moreover, a gamma distribution with shape parameter $r > 1$ cannot be represented as a mixed exponential distribution.

In [32] it was proved that if $r \in (0,1)$, $\mu > 0$ and $G_{r,1}$ and $G_{1-r,1}$ are independent gamma-distributed random variables, then the density $p(z;r,\mu)$ defined by (2) corresponds to the random variable

$$Z_{r,\mu} = \frac{\mu(G_{r,1} + G_{1-r,1})}{G_{r,1}} \stackrel{d}{=} \mu Z_{r,1} \stackrel{d}{=} \mu(1 + \tfrac{1-r}{r}V_{1-r,r}), \tag{3}$$

where $V_{1-r,r}$ is the random variable with the Snedecor–Fisher distribution defined by the probability density

$$q(x;1-r,r) = \frac{(1-r)^{1-r}r^r}{\Gamma(1-r)\Gamma(r)} \cdot \frac{1}{x^r[r+(1-r)x]}, \quad x \geqslant 0.$$

In other words, if $r \in (0,1)$, then

$$G_{r,\mu} \stackrel{d}{=} W_1 \circ Z_{r,\mu}^{-1}. \tag{4}$$

## 3. Univariate Stable Distributions

Let $r \in \mathbb{N}$. Recall that the distribution of an $r$-variate random vector $\mathbf{S}$ is called *stable*, if for any $a, b \in \mathbb{R}$ there exist $c \in \mathbb{R}$ and $\mathbf{d} \in \mathbb{R}^r$ such that $a\mathbf{S}_1 + b\mathbf{S}_2 \stackrel{d}{=} c\mathbf{S} + \mathbf{d}$, where $\mathbf{S}_1$ and $\mathbf{S}_2$ are independent and $\mathbf{S}_1 \stackrel{d}{=} \mathbf{S}_2 \stackrel{d}{=} \mathbf{S}$. In what follows we will concentrate our attention on a special sub-class of stable distributions called *strictly stable*. This sub-class is characterized by that in the definition given above $\mathbf{d} = \mathbf{0}$.

In the univariate case, the characteristic function $\mathfrak{f}(t)$ of a strictly stable random variable can be represented in several equivalent forms (see, e.g., [1]). For our further constructions the most convenient form is

$$\mathfrak{f}_{\alpha,\theta}(t) = \exp\{-|t|^\alpha + i\theta w(t,\alpha)\}, \quad t \in \mathbb{R}, \tag{5}$$

where

$$w(t,\alpha) = \begin{cases} \tan\frac{\pi\alpha}{2} \cdot |t|^\alpha \mathrm{sign}\, t, & \alpha \neq 1, \\ -\frac{2}{\pi} \cdot t\log|t|, & \alpha = 1. \end{cases} \tag{6}$$

Here $\alpha \in (0,2]$ is the *characteristic exponent*, $\theta \in [-1, 1]$ is the *skewness* parameter (for simplicity we consider the "standard" case with unit scale coefficient at $t$). Any random variable with characteristic function (5) will be denoted $S(\alpha,\theta)$ and the characteristic function (5) itself will be written as $\mathfrak{f}_{\alpha,\theta}(t)$. For definiteness, $S(1, 1) = 1$.

From (5) it follows that the characteristic function of a symmetric ($\theta = 0$) strictly stable distribution has the form

$$\mathfrak{f}_{\alpha,0}(t) = e^{-|t|^{\alpha}}, \quad t \in \mathbb{R}. \tag{7}$$

From (7) it is easy to see that $S(2,0) \stackrel{d}{=} \sqrt{2}X$.

Univariate stable distributions are popular examples of heavy-tailed distributions. Their moments of orders $\delta \geqslant \alpha$ do not exist (the only exception is the normal law corresponding to $\alpha = 2$), and if $0 < \delta < \alpha$, then

$$\mathsf{E}|S(\alpha,0)|^{\delta} = \frac{2^{\delta}}{\sqrt{\pi}} \cdot \frac{\Gamma(\frac{\delta+1}{2})\Gamma(1-\frac{\delta}{\alpha})}{\Gamma(\frac{2}{\delta}-1)} \tag{8}$$

(see, e.g., [33]). Stable laws and only they can be limit distributions for sums of a non-random number of independent identically distributed random variables with very large variance under linear normalization.

Let $0 < \alpha \leqslant 1$. By $S(\alpha,1)$ we will denote a positive random variable with the one-sided stable distribution corresponding to the characteristic function $\mathfrak{f}_{\alpha,1}(t)$, $t \in \mathbb{R}$. The Laplace–Stieltjes transform $\psi^{(S)}_{\alpha,1}(s)$ of the random variable $S(\alpha,1)$ has the form

$$\psi^{(S)}_{\alpha,1}(s) = \mathsf{E}\exp\{-sS(\alpha,1)\} = e^{-s^{\alpha}}, \quad s > 0.$$

The moments of orders $\delta \geqslant \alpha$ of the random variable $S(\alpha,1)$ are very large and for $0 < \delta < \alpha$ we have

$$\mathsf{E}S^{\delta}(\alpha,1) = \frac{2^{\delta}\Gamma(1-\frac{\delta}{\alpha})}{\Gamma(1-\delta)}$$

(see, e.g., [33]). For more details see [27] or [1].

The following product representations hold for strictly stable random variables. Let $\alpha \in (0,2]$, $|\theta| \leqslant \min\{1, \frac{2}{\alpha}-1\}$, $\alpha' \in (0,1]$. Then

$$S(\alpha\alpha',\theta) \stackrel{d}{=} S^{1/\alpha}(\alpha',1) \circ S(\alpha,\theta), \tag{9}$$

see Theorem 3.3.1 in [1]. In particular,

$$S(\alpha,0) \stackrel{d}{=} \sqrt{2S(\alpha/2,1)} \circ X. \tag{10}$$

Another particular case of (9) concerns one-sided strictly stable random variables: if $0 < \alpha \leqslant 1$ and $0 < \alpha' \leqslant 1$, then

$$S(\alpha\alpha',1) \stackrel{d}{=} S^{1/\alpha}(\alpha',1) \circ S(\alpha,1), \tag{11}$$

see Corollary 1 to Theorem 3.3.1 in [1]. Finally, if $0 < \alpha \leqslant 1$, then

$$S(\alpha,\theta) \stackrel{d}{=} S(\alpha,1) \circ S(1,\theta), \tag{12}$$

see Corollary to Theorem 3.3.2 (relation (3.3.10)) in [1].

## 4. Multivariate Stable Distributions

Now turn to the multivariate case. By $\mathbb{Q}^r$ we denote the unit sphere: $\mathbb{Q}^r = \{\boldsymbol{u} \in \mathbb{R}^r : \|\boldsymbol{u}\| = 1\}$. Let $\mu$ be a finite ('spectral') measure on $\mathbb{Q}^r$. It is known that the characteristic function of a strictly stable random vector $\boldsymbol{S}$ has the form

$$\mathsf{E}\exp\{i\boldsymbol{t}^{\top}\boldsymbol{S}\} = \exp\Big\{-\int_{\mathbb{Q}^r}\big(|\boldsymbol{t}^{\top}\boldsymbol{s}|^{\alpha} + iw(\boldsymbol{t}^{\top}\boldsymbol{s},\,\alpha)\big)\mu(d\boldsymbol{s})\Big\}, \quad \boldsymbol{t} \in \mathbb{R}^r, \tag{13}$$

with $w(\cdot, \alpha)$ defined in (6), see [34–37]. An $r$-variate random vector with the characteristic function (13) will be denoted $\boldsymbol{S}(\alpha, \mu)$. We will sometimes use the notation $\mathfrak{S}_{\alpha,\mu}$ for $\mathcal{L}\left(\boldsymbol{S}(\alpha, \mu)\right)$.

As is known, a random vector $\boldsymbol{S}$ has a strictly stable distribution with some characteristic exponent $\alpha$ if and only if for any $\boldsymbol{u} \in \mathbb{R}^r$ the random variable $\boldsymbol{u}^\top \boldsymbol{S}$ (the projection of $\boldsymbol{S}$) has the univariate strictly stable distribution with the same characteristic exponent $\alpha$ and some skewness parameter $\theta(\boldsymbol{u})$ up to a scale coefficient $\gamma(\boldsymbol{u})$:

$$\boldsymbol{u}^\top \boldsymbol{S}(\alpha, \mu) \stackrel{d}{=} \gamma(\boldsymbol{u}) S(\alpha, \theta(\boldsymbol{u})), \tag{14}$$

see [38]. Moreover, the projection parameter functions are related with the spectral measure $\mu$ as

$$(\gamma(\boldsymbol{u}))^\alpha = \int_{\mathbb{Q}^r} |\boldsymbol{u}^\top \boldsymbol{s}|^\alpha \mu(d\boldsymbol{s}), \tag{15}$$

$$\theta(\boldsymbol{u})(\gamma(\boldsymbol{u}))^\alpha = \int_{\mathbb{Q}^r} |\boldsymbol{u}^\top \boldsymbol{s}|^\alpha \operatorname{sign}(\boldsymbol{u}^\top \boldsymbol{s}) \mu(d\boldsymbol{s}), \quad \boldsymbol{u} \in \mathbb{R}^r, \tag{16}$$

see [36–38]. Conversely, the spectral measure $\mu$ is uniquely determined by the projection parameter functions $\gamma(\boldsymbol{u})$ and $\theta(\boldsymbol{u})$. However, there is no simple formula for this [37].

An $r$-variate analog of a one-sided univariate strictly stable random variable $S(\alpha, 1)$ is the random vector $\boldsymbol{S}(\alpha, \mu^+)$ where $0 < \alpha \leqslant 1$ and $\mu^+$ is a finite measure concentrated on the set $\mathbb{Q}^r_+ = \{\boldsymbol{u} = (u_1, \ldots, u_r)^\top : u_i \geqslant 0,\ i = 1, \ldots, r\}$.

Consider multivariate analogs of product representations (9) and (11).

**Theorem 1.** *Let $0 < \alpha \leqslant 2$, $0 < \alpha' \leqslant 1$, $\mu$ be a finite measure on $\mathbb{Q}^r$, $\boldsymbol{S}(\alpha, \mu)$ be an $r$-variate random vector having the strictly stable distribution with characteristic exponent $\alpha$ and spectral measure $\mu$. Then*

$$\boldsymbol{S}(\alpha\alpha', \mu) \stackrel{d}{=} S^{1/\alpha}(\alpha', 1) \circ \boldsymbol{S}(\alpha, \mu). \tag{17}$$

*If, in addition, $0 < \alpha < 1$, and $\mu^+$ is a finite measure on $\mathbb{Q}^r_+$, then*

$$\boldsymbol{S}(\alpha\alpha', \mu^+) \stackrel{d}{=} S^{1/\alpha}(\alpha', 1) \circ \boldsymbol{S}(\alpha, \mu^+). \tag{18}$$

**Proof.** Let $\gamma(\boldsymbol{u})$ and $\theta(\boldsymbol{u})$, $\boldsymbol{u} \in \mathbb{R}^r$, be the projection parameter functions corresponding to the measure $\mu$ (see (15) and (16)). Then, in accordance with (9) and (14), for any $\boldsymbol{u} \in \mathbb{R}^r$ we have

$$\boldsymbol{u}^\top (S^{1/\alpha}(\alpha', 1) \circ \boldsymbol{S}(\alpha, \mu)) = S^{1/\alpha}(\alpha', 1) \circ \boldsymbol{u}^\top \boldsymbol{S}(\alpha, \mu) \stackrel{d}{=} S^{1/\alpha}(\alpha', 1) \circ (\gamma(\boldsymbol{u}) S(\alpha, \theta(\boldsymbol{u}))) \stackrel{d}{=}$$

$$\stackrel{d}{=} \gamma(\boldsymbol{u}) \cdot S^{1/\alpha}(\alpha', 1) \circ S(\alpha, \theta(\boldsymbol{u})) \stackrel{d}{=} \gamma(\boldsymbol{u}) S(\alpha\alpha', \theta(\boldsymbol{u})).$$

The remark that $\gamma(\boldsymbol{u})$ and $\theta(\boldsymbol{u})$ uniquely determine $\mu$ concludes the proof of (17). Representation (18) is a particular case of (17). □

**Remark 1.** *Actually, the essence of Theorem 1 is that all multivariate strictly stable distributions with $\alpha < 2$ are scale mixtures of multivariate scale laws with no less characteristic exponent, the mixing distribution being univariate one-sided strictly stable law. The case $\alpha = 2$ is not an exception: in this case the mixing distribution is degenerate concentrated in the unit point. This degenerate law formally satisfies the definition of a stable distribution being the only stable law that is not absolutely continuous.*

Let $\Sigma$ be a symmetric positive definite $(r \times r)$-matrix, $\alpha \in (0, 2]$. If the characteristic function $\mathfrak{f}_{\alpha,\mu}(\boldsymbol{t})$ of a strictly stable random vector $\boldsymbol{S}(\alpha, \mu)$ has the form

$$\mathfrak{f}_{\alpha,\mu}(\boldsymbol{t}) = \mathsf{E}\exp\{i\boldsymbol{t}^\top \boldsymbol{S}_{\alpha,\mu}\} = \exp\{-(\boldsymbol{t}^\top \Sigma \boldsymbol{t})^{\alpha/2}\}, \quad \boldsymbol{t} \in \mathbb{R}^r, \tag{19}$$

then the random vector $S(\alpha, \mu)$ is said to have the (centered) *elliptically contoured* stable distribution with characteristic exponent $\alpha$. In this case for better vividness we will use the special notation $S(\alpha, \mu) = S(\alpha, \Sigma)$. The corresponding characteristic function (19) will be denoted $\mathfrak{f}_{\alpha,\Sigma}(t)$ and the elliptically contoured stable distribution with characteristic function (19) will be denoted $\mathfrak{S}_{\alpha,\Sigma}$. It is easy to see that $\mathfrak{S}_{2,\Sigma} = \mathfrak{N}_{2\Sigma}$.

Let $\alpha \in (0, 2]$. If $X$ is a random vector such that $\mathcal{L}(X) = \mathfrak{N}_\Sigma$ independent of the random variable $S(\alpha/2, 1)$, then from (17) it follows that

$$S(\alpha, \Sigma) \stackrel{d}{=} S^{1/2}(\alpha/2, 1) \circ S(2, \Sigma) \stackrel{d}{=} \sqrt{2S(\alpha/2, 1)} \circ X \tag{20}$$

(also see Proposition 2.5.2 in [27]). More general, If $0 < \alpha \leqslant 2$ and $0 < \alpha' \leqslant 1$, then

$$S(\alpha\alpha', \Sigma) \stackrel{d}{=} S^{1/\alpha}(\alpha', 1) \circ S(\alpha, \Sigma). \tag{21}$$

If $\alpha = 2$, then (21) turns into (20).

## 5. Scale Mixtures of Multivariate Elliptically Contoured Stable Distributions

Let $U$ be a nonnegative random variable. The symbol $\mathsf{E}\mathfrak{N}_{U\Sigma}(\cdot)$ will denote the distribution which for each Borel set $A$ in $\mathbb{R}^r$ is defined as

$$\mathsf{E}\mathfrak{N}_{U\Sigma}(A) = \int_0^\infty \mathfrak{N}_{u\Sigma}(A) d\mathsf{P}(U < u).$$

It is easy to see that if $X$ is a random vector such that $\mathcal{L}(X) = \mathfrak{N}_\Sigma$, then $\mathsf{E}\mathfrak{N}_{U\Sigma} = \mathcal{L}(\sqrt{U} \circ X)$. In this notation, relation (20) can be written as

$$\mathfrak{S}_{\alpha,\Sigma} = \mathsf{E}\mathfrak{N}_{2S(\alpha/2,1)\Sigma}. \tag{22}$$

By analogy, the symbol $\mathsf{E}\mathfrak{S}_{\alpha,U^{2/\alpha}\Sigma}$ will denote the distribution that for each Borel set $A$ in $\mathbb{R}^r$ is defined as

$$\mathsf{E}\mathfrak{S}_{\alpha,U^{2/\alpha}\Sigma}(A) = \int_0^\infty \mathfrak{S}_{\alpha,u^{2/\alpha}\Sigma}(A) d\mathsf{P}(U < u).$$

The characteristic function corresponding to the distribution $\mathsf{E}\mathfrak{S}_{\alpha,U^{2/\alpha}\Sigma}$ has the form

$$\int_0^\infty \exp\big\{-\big(t^\top(u^{2/\alpha}\Sigma)t\big)^{\alpha/2}\big\} d\mathsf{P}(U < u) = \int_0^\infty \exp\big\{-\big((u^{1/\alpha}t)^\top \Sigma (u^{1/\alpha}t)\big)^{\alpha/2}\big\} d\mathsf{P}(U < u) =$$

$$= \mathsf{E}\exp\big\{it^\top U^{1/\alpha} \circ S(\alpha, \Sigma)\big\}, \quad t \in \mathbb{R}^r, \tag{23}$$

where the random variable $U$ is independent of the random vector $S(\alpha, \Sigma)$, that is, the distribution $\mathsf{E}\mathfrak{S}_{\alpha,U^{2/\alpha}\Sigma}$ corresponds to the product $U^{1/\alpha} \circ S(\alpha, \Sigma)$.

Let $\mathcal{U}$ be the set of all nonnegative random variables. Now consider an auxiliary statement dealing with the identifiability of the family of distributions $\{\mathsf{E}\mathfrak{S}_{\alpha,U^{2/\alpha}\Sigma} : U \in \mathcal{U}\}$.

**Lemma 1.** *Whatever a nonsingular positive definite matrix $\Sigma$ is, the family $\{\mathsf{E}\mathfrak{S}_{\alpha,U^{2/\alpha}\Sigma} : U \in \mathcal{U}\}$ is identifiable in the sense that if $U_1 \in \mathcal{U}$, $U_2 \in \mathcal{U}$ and*

$$\mathsf{E}\mathfrak{S}_{\alpha,U_1^{2/\alpha}\Sigma}(A) = \mathsf{E}\mathfrak{S}_{\alpha,U_2^{2/\alpha}\Sigma}(A) \tag{24}$$

*for any set $A \in \mathcal{B}(\mathbb{R}^r)$, then $U_1 \stackrel{d}{=} U_2$.*

**Proof.** The proof of this lemma is very simple. If $U \in \mathcal{U}$, then it follows from (13) that the characteristic function $\mathfrak{v}_{\alpha,\Sigma}^{(U)}(t)$ corresponding to the distribution $\mathsf{E}\mathfrak{S}_{\alpha,U^{2/\alpha}\Sigma}$ has the form

$$\mathfrak{v}_{\alpha,\Sigma}^{(U)}(t) = \int_0^\infty \exp\big\{-\big(t^\top(u^{2/\alpha}\Sigma)t\big)^{\alpha/2}\big\}d\mathsf{P}(U<u) =$$

$$= \int_0^\infty \exp\{-us\}d\mathsf{P}(U<u),\ \ s=(t^\top\Sigma t)^{\alpha/2},\quad t\in\mathbb{R}^r, \tag{25}$$

But on the right-hand side of (25) there is the Laplace–Stieltjes transform of the random variable $U$. From (24) it follows that $\mathfrak{v}_{\alpha,\Sigma}^{(U_1)}(t) = \mathfrak{v}_{\alpha,\Sigma}^{(U_2)}(t)$ whence by virtue of (25) the Laplace–Stieltjes transforms of the random variables $U_1$ and $U_2$ coincide, whence, in turn, it follows that $U_1 \stackrel{d}{=} U_2$. The lemma is proved. □

**Remark 2.** *When proving Lemma 1 we established a simple but useful by-product result: if $\psi^{(U)}(s)$ is the Laplace–Stieltjes transform of the random variable U, then the characteristic function $\mathfrak{v}_{\alpha,\Sigma}^{(U)}(t)$ corresponding to the distribution $\mathsf{E}\mathfrak{S}_{\alpha,U^{2/\alpha}\Sigma}$ has the form*

$$\mathfrak{v}_{\alpha,\Sigma}^{(U)}(t) = \psi^{(U)}\big((t^\top\Sigma t)^{\alpha/2}\big),\ \ t\in\mathbb{R}^r. \tag{26}$$

Let $X$ be a random vector such that $\mathcal{L}(X) = \mathfrak{N}_\Sigma$ with some positive definite $(r\times r)$-matrix $\Sigma$. Define the multivariate Laplace distribution as $\mathcal{L}(\sqrt{2W_1}\circ X) = \mathsf{E}\mathfrak{N}_{2W_1\Sigma}$. The random vector with this multivariate Laplace distribution will be denoted $\Lambda_\Sigma$. It is well known that the Laplace—Stieltjes transform $\psi^{(W_1)}(s)$ of the random variable $W_1$ with the exponential distribution has the form

$$\psi^{(W_1)}(s) = (1+s)^{-1},\quad s>0. \tag{27}$$

Hence, in accordance with (27) and Remark 2, the characteristic function $\mathfrak{f}_\Sigma^{(\Lambda)}(t)$ of the random variable $\Lambda_\Sigma$ has the form

$$\mathfrak{f}_\Sigma^{(\Lambda)}(t) = \psi^{(W_1)}(t^\top\Sigma t) = \big(1+t^\top\Sigma t\big)^{-1},\quad t\in\mathbb{R}^r.$$

## 6. Generalized Mittag–Leffler Distributions

We begin with the univariate case. The probability distribution of a nonnegative random variable $M_\delta$ whose Laplace transform is

$$\psi_\delta^{(M)}(s) = \mathsf{E}e^{-sM_\delta} = \big(1+\lambda s^\delta\big)^{-1},\quad s\geqslant 0, \tag{28}$$

where $\lambda>0$, $0<\delta\leqslant 1$, is called *the Mittag–Leffler distribution*. For simplicity, in what follows we will consider the standard scale case and assume that $\lambda=1$.

The origin of the term *Mittag–Leffler distribution* is due to that the probability density corresponding to Laplace transform (28) has the form

$$f_\delta^{(M)}(x) = \frac{1}{x^{1-\delta}}\sum\nolimits_{n=0}^\infty \frac{(-1)^n x^{\delta n}}{\Gamma(\delta n+1)} = -\frac{d}{dx}E_\delta(-x^\delta),\quad x\geqslant 0,$$

where $E_\delta(z)$ is the Mittag–Leffler function with index $\delta$ that is defined as the power series

$$E_\delta(z) = \sum\nolimits_{n=0}^\infty \frac{z^n}{\Gamma(\delta n+1)},\quad \delta>0,\ z\in\mathbb{Z}.$$

With $\delta=1$, the Mittag–Leffler distribution turns into the standard exponential distribution, that is, $F_1^M(x) = [1-e^{-x}]\mathbf{1}(x\geqslant 0)$, $x\in\mathbb{R}$. But with $\delta<1$ the Mittag–Leffler distribution density has the heavy power-type tail: from the well-known asymptotic properties of the Mittag–Leffler function it can be deduced that if $0 < \delta < 1$, then

$$f_{\delta}^{(M)}(x) \sim \frac{\sin(\delta\pi)\Gamma(\delta+1)}{\pi x^{\delta+1}}$$

as $x \to \infty$, see, e.g., [39].

It is well-known that the Mittag–Leffler distribution is geometrically stable. This means that if $X_1, X_2, \ldots$ are independent random variables whose distributions belong to the domain of attraction of a one-sided $\alpha$-strictly stable law $\mathcal{L}(S(\alpha, 1))$ and $NB_{1,p}$ is the random variable independent of $X_1, X_2, \ldots$ and having the geometric distribution

$$\mathsf{P}(NB_{1,p} = n) = p(1-p)^{n-1}, \quad n = 1, 2, \ldots, \quad p \in (0, 1), \tag{29}$$

then for each $p \in (0, 1)$ there exists a constant $a_p > 0$ such that $a_p(X_1 + \ldots + X_{NB_{1,p}}) \Longrightarrow M_\delta$ as $p \to 0$, see, e.g., [40].

The history of the Mittag–Leffler distribution was discussed in [2]. For more details see e.g., [2,3] and the references therein. The Mittag–Leffler distributions are of serious theoretical interest in the problems related to thinned (or rarefied) homogeneous flows of events such as renewal processes or anomalous diffusion or relaxation phenomena, see [41,42] and the references therein.

Let $\nu > 0$, $\delta \in (0, 1]$. It can be easily seen that the Laplace transform $\psi_{\delta}^{(M)}(s)$ (see (28)) is greatly divisible. Therefore, any its positive power is a Laplace transform, and, moreover, is greatly divisible as well. The distribution of a nonnegative random variable $M_{\delta,\nu}$ defined by the Laplace–Stieltjes transform

$$\psi_{\delta,\nu}^{(M)}(s) = \mathsf{E}e^{-sM_{\delta,\nu}} = \left(1 + s^{\delta}\right)^{-\nu}, \quad s \geqslant 0, \tag{30}$$

is called the *generalized Mittag–Leffler distribution*, see [43,44] and the references therein. Sometimes this distribution is called the *Pillai distribution* [20], although in the original paper [18] R. Pillai called it *semi-Laplace*. In the present paper we will keep to the first term *generalized Mittag–Leffler distribution*.

The properties of univariate generalized Mittag–Leffler distribution are discussed in [4,43–45]. In particular, if $\delta \in (0, 1]$ and $\nu > 0$, then

$$M_{\delta,\nu} \stackrel{d}{=} S(\delta,\, 1) \circ \overline{G}_{\nu,\delta,1} \stackrel{d}{=} S(\delta,\, 1) \circ G_{\nu,1}^{1/\delta} \tag{31}$$

(see [43,44]). If $\nu = 1$, then (31) turns into

$$M_{\delta} \stackrel{d}{=} S(\delta,\, 1) \circ W_1^{1/\delta}. \tag{32}$$

If $\beta \geqslant \delta$, then the moments of order $\beta$ of the random variable $M_{\delta,\nu}$ are very large, and if $0 < \beta < \delta < 1$, then

$$\mathsf{E}M_{\delta,\nu}^{\beta} = \frac{\Gamma(1-\frac{\beta}{\delta})\Gamma(\nu+\frac{\beta}{\delta})}{\Gamma(1-\beta)\Gamma(\nu)},$$

see [4].

In [4] it was demonstrated that the generalized Mittag–Leffler distribution can be represented as a scale mixture of 'ordinary' Mittag–Leffler distributions: if $\nu \in (0, 1]$ and $\delta \in (0, 1]$, then

$$M_{\delta,\nu} \stackrel{d}{=} Z_{\nu,1}^{-1/\delta} \circ M_{\delta}. \tag{33}$$

In [4] it was also shown that any generalized Mittag–Leffler distribution is a scale mixture a one-sided stable law with any greater characteristic parameter, the mixing distribution being the generalized Mittag–Leffler law: if $\delta \in (0,1]$, $\delta' \in (0,1)$ and $\nu > 0$, then

$$M_{\delta\delta',\nu} \stackrel{d}{=} S(\delta,1) \circ M_{\delta',\nu}^{1/\delta}. \tag{34}$$

Now turn to the multivariate case. As the starting point for our consideration we take representation (31). The nearest aim is to obtain its multivariate generalization. Let $\boldsymbol{S}(\alpha,\mu)$ be a strictly stable random vector with $\alpha \neq 1$. Consider the characteristic function $\mathfrak{h}_{\alpha,\nu,\mu}(\boldsymbol{t})$ of the random vector $G_{\nu,1}^{1/\alpha} \circ \boldsymbol{S}(\alpha,\mu)$. From (13) and (6) we have

$$\mathfrak{h}_{\alpha,\nu,\mu}(\boldsymbol{t}) = \mathsf{E}\exp\big\{i\boldsymbol{t}^{\top}(G_{\nu,1}^{1/\alpha} \circ \boldsymbol{S}(\alpha,\mu))\big\} = \frac{1}{\Gamma(\nu)}\int_0^\infty \mathfrak{f}_{\alpha,\mu}(s^{1/\alpha}\boldsymbol{t})s^{\nu-1}e^{-s}ds =$$

$$= \frac{1}{\Gamma(\nu)}\int_0^\infty \exp\big\{-s\big(1-\log \mathfrak{f}_{\alpha,\mu}(\boldsymbol{t})\big)\big\}s^{\nu-1}ds =$$

$$= \frac{1}{\Gamma(\nu)\big(1-\log \mathfrak{f}_{\alpha,\mu}(\boldsymbol{t})\big)^{\nu}}\int_0^\infty e^{-s}s^{\nu-1}ds = \big(1-\log \mathfrak{f}_{\alpha,\mu}(\boldsymbol{t})\big)^{-\nu}. \tag{35}$$

That is, from (1) and (35) we obtain the following result.

**Lemma 2.** *The characteristic function $\mathfrak{h}_{\alpha,\nu,\mu}(\mathbf{t})$ of the product of the random variable $\overline{G}_{\nu,\alpha,1}$ with the generalized gamma distribution with parameters $\nu > 0$, $0 < \alpha \leqslant 2$, $\alpha \neq 1$, $\lambda = 1$ and the random vector $\boldsymbol{S}(\alpha,\mu)$ with the multivariate strictly stable distribution with the characteristic exponent $\alpha$ and spectral measure $\mu$, independent of $\overline{G}_{\nu,\alpha,1}$, has the form (35).*

Now, comparing the right-hand side of (35) with (30) we can conclude that, if $\overline{G}_{\nu,\alpha,1}$ is the random variable with the generalized gamma distribution with parameters $\nu > 0$, $0 < \alpha \leqslant 2$, $\alpha \neq 1$, $\lambda = 1$, and $\boldsymbol{S}(\alpha,\mu^+)$ is a random vector with the one-sided strictly stable distribution with characteristic exponent $\alpha \in (0,1)$ and spectral measure $\mu^+$ concentrated on $\mathbb{Q}_+^r$, then we have all grounds to call the distribution of the random vector $\overline{G}_{\nu,\alpha,1} \circ \boldsymbol{S}(\alpha,\mu^+)$ *multivariate generalized Mittag–Leffler distribution* with parameters $\alpha$, $\nu$ and $\mu^+$. To provide the possibility to consider the univariate generalized Mittag–Leffler distribution as a particular case of a more general multivariate definition, here we use the measure $\mu^+$ and $\alpha \in (0,1)$ characterising the "one-sided" stable law, although from the formal viewpoint this is not obligatory. Moreover, as we will see below, in the multivariate case the (generalized) Mittag–Leffler distribution can be regarded as a special case of the (generalized) Linnik law defined in the same way but with $\mu$ and $\alpha \in (0,2]$.

By $\boldsymbol{M}_{\alpha,\nu,\mu^+}$ we will denote the random vector with the multivariate generalized Mittag–Leffler distribution, $\boldsymbol{M}_{\alpha,\nu,\mu^+} \stackrel{d}{=} \overline{G}_{\nu,\alpha,1} \circ \boldsymbol{S}(\alpha,\mu^+)$.

Setting $\nu = 1$ we obtain the definition of the 'ordinary' multivariate Mittag–Leffler distribution as the distribution of the random vector $\boldsymbol{M}_{\alpha,\mu^+} \stackrel{d}{=} W_1^{1/\alpha} \circ \boldsymbol{S}(\alpha,\mu^+)$ given by the characteristic function $\mathfrak{h}_{\alpha,\mu^+}(\boldsymbol{t}) = \big(1-\log \mathfrak{f}_{\alpha,\mu^+}(\boldsymbol{t})\big)^{-1}$.

Some properties of the multivariate generalized Mittag–Leffler distributions generalizing (33) and (34) to the multivariate case are presented in the following theorem.

**Theorem 2.** *Let $\delta \in (0,1)$, $\delta' \in (0,1)$ and $\nu > 0$. Then*

$$\boldsymbol{M}_{\delta\delta',\nu,\mu^+} \stackrel{d}{=} M_{\delta',\nu}^{1/\delta} \circ \boldsymbol{S}(\delta,\mu^+), \tag{36}$$

$$\boldsymbol{M}_{\delta,\nu,\mu^+} \stackrel{d}{=} Z_{\nu,1}^{-1/\delta} \circ \boldsymbol{M}_{\delta,\mu^+} \tag{37}$$

*with the random variable $Z_{\nu,1}$ defined in* (3).

**Proof.** To prove (36) use (1) together with representation (18) in Theorem 1 and obtain

$$M_{\delta\delta',\nu,\mu^+} \stackrel{d}{=} \overline{G}_{\nu,\delta\delta',1} \circ S(\delta\delta',\mu^+) \stackrel{d}{=} \overline{G}^{1/\delta}_{\nu,\delta',1} \circ S^{1/\delta}(\delta',1) \circ S(\delta,\mu^+) \stackrel{d}{=} M^{1/\delta}_{\delta',\nu} \circ S(\delta,\mu^+).$$

To prove (37) use (1) together with (4) and obtain

$$M_{\delta,\nu,\mu^+} \stackrel{d}{=} \overline{G}_{\nu,\delta,1} \circ S(\delta,\mu^+) \stackrel{d}{=} Z^{-1/\delta}_{\nu,1} \circ W^{1/\delta}_1 \circ S(\delta,\mu^+) \stackrel{d}{=} Z^{1/\delta}_{\nu,1} \circ M_{\delta,\mu^+}.$$

The theorem is proved. □

## 7. Generalized Linnik Distributions

In 1953 Yu. V. Linnik [46] introduced a class of symmetric distributions whose characteristic functions have the form

$$\mathfrak{f}^{(L)}_\alpha(t) = \big(1+|t|^\alpha\big)^{-1}, \quad t \in \mathbb{R}, \tag{38}$$

where $\alpha \in (0,2]$. The distributions with the characteristic function (38) are traditionally called the *Linnik distributions*. Although sometimes the term *$\alpha$-Laplace distributions* [18] is used, we will use the first term which has already become conventional. If $\alpha = 2$, then the Linnik distribution turns into the Laplace distribution corresponding to the density

$$f^{(\Lambda)}(x) = \tfrac{1}{2}e^{-|x|}, \quad x \in \mathbb{R}. \tag{39}$$

A random variable with density (39) will be denoted $\Lambda$. A random variable with the Linnik distribution with parameter $\alpha$ will be denoted $L_\alpha$.

Perhaps, most often Linnik distributions are recalled as examples of symmetric geometric stable distributions. This means that if $X_1, X_2, \ldots$ are independent random variables whose distributions belong to the domain of attraction of an $\alpha$-strictly stable symmetric law and $N\!B_{1,p}$ is the random variable independent of $X_1, X_2, \ldots$ and having the geometric distribution (29), then for each $p \in (0,1)$ there exists a constant $a_p > 0$ such that $a_p(X_1 + \ldots + X_{N\!B_{1,p}}) \Longrightarrow L_\alpha$ as $p \to 0$, see, e.g., [47] or [40].

The properties of the Linnik distributions were studied in many papers. We should mention [7–9,48] and other papers, see the survey in [2].

In [2,7] it was demonstrated that

$$L_\alpha \stackrel{d}{=} W_1^{1/\alpha} \circ S(\alpha,0) \stackrel{d}{=} \sqrt{2M_{\alpha/2}} \circ X, \tag{40}$$

where the random variable $M_{\alpha/2}$ has the Mittag–Leffler distribution with parameter $\alpha/2$.

The multivariate Linnik distribution was introduced by D. N. Anderson in [49] where it was proved that the function

$$\mathfrak{f}^{(L)}_{\alpha,\Sigma}(\boldsymbol{t}) = \big[1 + (\boldsymbol{t}^\top\Sigma\boldsymbol{t})^{\alpha/2}\big]^{-1}, \quad \boldsymbol{t} \in \mathbb{R}^r, \ \alpha \in (0,2), \tag{41}$$

is the characteristic function of an $r$-variate probability distribution, where $\Sigma$ is a positive definite $(r \times r)$-matrix. In [49] the distribution corresponding to the characteristic function (41) was called the *$r$-variate Linnik distribution*. For the properties of these distributions see [16,49]. To distinguish from the general case, in what follows, the distribution corresponding to characteristic function (41) will be called *multivariate (centered) elliptically contoured Linnik distribution*.

The $r$-variate elliptically contoured Linnik distribution can also be defined in another way. Let $\boldsymbol{X}$ be a random vector such that $\mathcal{L}(\boldsymbol{X}) = \mathfrak{N}_{\Sigma}$, where $\Sigma$ is a positive definite $(r \times r)$-matrix, independent of the random variable $M_{\alpha/2}$. By analogy with (40) introduce the random vector $\boldsymbol{L}_{\alpha,\Sigma}$ as

$$\boldsymbol{L}_{\alpha,\Sigma} = \sqrt{2M_{\alpha/2}} \circ \boldsymbol{X}.$$

Then, in accordance with what has been said in Section 5,

$$\mathcal{L}(\boldsymbol{L}_{\alpha,\Sigma}) = \mathsf{E}\mathfrak{N}_{2M_{\alpha/2}\Sigma}. \tag{42}$$

Using Remark 1 we can easily make sure that the two definitions of the multivariate elliptically contoured Linnik distribution coincide. Indeed, with the account of (28), according to Remark 2, the characteristic function of the random vector $\boldsymbol{L}_{\alpha,\Sigma}$ defined by (42) has the form

$$\mathsf{E}\exp\{i\boldsymbol{t}^{\top}\boldsymbol{L}_{\alpha,\Sigma}\} = \psi^{(M)}_{\alpha/2}(\boldsymbol{t}^{\top}\Sigma\boldsymbol{t}) = \left[1 + (\boldsymbol{t}^{\top}\Sigma\boldsymbol{t})^{\alpha/2}\right]^{-1} = \mathfrak{f}^{(L)}_{\alpha,\Sigma}(\boldsymbol{t}), \;\; \boldsymbol{t} \in \mathbb{R}^r,$$

that coincides with Anderson's definition (41).

Based on (40), one more equivalent definition of the multivariate elliptically contoured Linnik distribution can be proposed. Namely, let $\boldsymbol{L}_{\alpha,\Sigma}$ be an $r$-variate random vector such that

$$\boldsymbol{L}_{\alpha,\Sigma} = W_1^{1/\alpha} \circ \boldsymbol{S}(\alpha,\Sigma). \tag{43}$$

In accordance with (27) and Remark 2 the characteristic function of the random vector $\boldsymbol{L}_{\alpha,\Sigma}$ defined by (43) again has the form

$$\mathsf{E}\exp\{i\boldsymbol{t}^{\top}\boldsymbol{L}_{\alpha,\Sigma}\} = \psi^{(W_1)}\big((\boldsymbol{t}^{\top}\Sigma\boldsymbol{t})^{\alpha/2}\big) = \left[1 + (\boldsymbol{t}^{\top}\Sigma\boldsymbol{t})^{\alpha/2}\right]^{-1} = \mathfrak{f}^{(L)}_{\alpha,\Sigma}(\boldsymbol{t}), \;\; \boldsymbol{t} \in \mathbb{R}^r.$$

The definitions (42) and (43) open the way to formulate limit theorems stating that the multivariate elliptically contoured Linnik distribution can not only be limiting for geometric random sums of independent identically distributed random vectors with very large second moments [50], but it also can be limiting for random sums of independent random vectors with finite covariance matrices.

It can be easily seen that the characteristic function $\mathfrak{f}^{(L)}_{\alpha}(t)$ (see (38)) is very largely divisible. Therefore, any its positive power is a characteristic function and, moreover, is also very largely divisible. In [17], Pakes showed that the probability distributions known as *generalized Linnik distributions* which have characteristic functions

$$\mathfrak{f}^{(L)}_{\alpha,\nu}(t) = \big(1 + |t|^{\alpha}\big)^{-\nu}, \quad t \in \mathbb{R},\; 0 < \alpha \leqslant 2,\; \nu > 0, \tag{44}$$

play an important role in some characterization problems of mathematical statistics. The class of probability distributions corresponding to characteristic function (44) have found some interesting properties and applications, see [6,7,10–12,14,51,52] and related papers. In particular, they are good candidates to model financial data which exhibits high kurtosis and heavy tails [53].

Any random variable with the characteristic function (44) will be denoted $L_{\alpha,\nu}$.

Recall some results containing mixture representations for the generalized Linnik distribution. The following well-known result is due to Devroye [7] and Pakes [17] who showed that

$$L_{\alpha,\nu} \stackrel{d}{=} S(\alpha,0) \circ G_{\nu,1}^{1/\alpha} \stackrel{d}{=} S(\alpha,0) \circ \overline{G}_{\nu,\alpha,1} \tag{45}$$

for any $\alpha \in (0,2]$ and $\nu > 0$.

It is well known that

$$\mathsf{E}G_{\nu,1}^{\gamma} = \frac{\Gamma(\nu+\gamma)}{\Gamma(\nu)}$$

for $\gamma > -\nu$. Hence, for $0 \leqslant \beta < \alpha$ from (8) and (45) we obtain

$$\mathsf{E}|L_{\alpha,\nu}|^{\beta} = \mathsf{E}|S(\alpha,0)|^{\beta} \cdot \mathsf{E}G_{\nu,1}^{\beta/\alpha} = \frac{2^{\beta}}{\sqrt{\pi}} \cdot \frac{\Gamma(\frac{\beta+1}{2})\Gamma(1-\frac{\beta}{\alpha})\Gamma(\nu+\frac{\beta}{\alpha})}{\Gamma(\frac{2}{\beta}-1)\Gamma(\nu)}.$$

Generalizing and improving some results of [15,17], with the account of (31) in [5] it was demonstrated that for $\nu > 0$ and $\alpha \in (0,2]$

$$L_{\alpha,\nu} \stackrel{d}{=} X \circ \sqrt{2S(\alpha/2,1)} \circ G_{\nu,1}^{1/\alpha} \stackrel{d}{=} X \circ \sqrt{2S(\alpha/2,1) \circ \overline{G}_{\nu,\alpha/2,1}} \stackrel{d}{=} X \circ \sqrt{2M_{\alpha/2,\nu}}. \tag{46}$$

that is, the generalized Linnik distribution is a normal scale mixture with the generalized Mittag–Leffler mixing distribution.

It is easy to see that for any $\alpha > 0$ and $\alpha' > 0$

$$\overline{G}_{\nu,\alpha\alpha',1} \stackrel{d}{=} G_{\nu,1}^{1/\alpha\alpha'} \stackrel{d}{=} (G_{\nu,1}^{1/\alpha'})^{1/\alpha} \stackrel{d}{=} \overline{G}_{\nu,\alpha',1}^{1/\alpha}.$$

Therefore, for $\alpha \in (0,2]$, $\alpha' \in (0,1)$ and $\nu > 0$ using (45) and the univariate version of (14) we obtain the following chain of relations:

$$L_{\alpha\alpha',\nu} \stackrel{d}{=} S(\alpha\alpha',0) \circ G_{\nu,1}^{1/\alpha\alpha'} \stackrel{d}{=} S(\alpha,0) \circ S^{1/\alpha}(\alpha',1) \circ G_{\nu,1}^{1/\alpha\alpha'} \stackrel{d}{=}$$

$$\stackrel{d}{=} S(\alpha,0) \circ \big(S(\alpha',1)\overline{G}_{\nu,\alpha',1}\big)^{1/\alpha} \stackrel{d}{=} S(\alpha,0) \circ M_{\alpha',\nu}^{1/\alpha}.$$

Hence, the following statement, more general than (46), holds representing the generalized Linnik distribution as a scale mixture of a symmetric stable law with any greater characteristic parameter, the mixing distribution being the generalized Mittag–Leffler law: if $\alpha \in (0,2]$, $\alpha' \in (0,1)$ and $\nu > 0$, then

$$L_{\alpha\alpha',\nu} \stackrel{d}{=} S(\alpha,0) \circ M_{\alpha',\nu}^{1/\alpha}.$$

Now let $\nu \in (0,1]$. From (45) and (4) it follows that

$$L_{\alpha,\nu} \stackrel{d}{=} S(\alpha,0) \circ G_{\nu,1}^{1/\alpha} \stackrel{d}{=} S(\alpha,0) \circ W_1^{1/\alpha} \circ Z_{\nu,1}^{-1/\alpha} \stackrel{d}{=} L_{\alpha} \circ Z_{\nu,1}^{-1/\alpha}$$

yielding the following relation proved in [5]: if $\nu \in (0,1]$ and $\alpha \in (0,2]$, then

$$L_{\alpha,\nu} \stackrel{d}{=} L_{\alpha} \cdot Z_{\nu,1}^{-1/\alpha}.$$

In other words, with $\nu \in (0,1]$ and $\alpha \in (0,2]$, the generalized Linnik distribution is a scale mixture of 'ordinary' Linnik distributions. In the same paper the representation of the generalized Linnik distribution via the Laplace and 'ordinary' Mittag–Leffler distributions was obtained.

For $\delta \in (0,1]$ denote

$$R_{\delta} = \frac{S(\delta,1)}{S'(\delta,1)},$$

where $S(\delta,1)$ and $S'(\delta,1)$ are independent random variables with one and the same one-sided stable distribution with the characteristic exponent $\delta$. In [2] it was shown that the probability density $f_{\delta}^{(R)}(x)$ of the ratio $R_{\delta}$ of two independent random variables with one and the same one-sided strictly stable distribution with parameter $\delta$ has the form

$$f_{\delta}^{(R)}(x) = \frac{\sin(\pi\delta)x^{\delta-1}}{\pi[1+x^{2\delta}+2x^{\delta}\cos(\pi\delta)]}, \quad x > 0,$$

also see [1], Section 3.3, where it was hidden among other calculations, but was not stated explicitly. In [5] it was proved that if $\nu \in (0,1]$ and $\alpha \in (0,2]$, then

$$L_{\alpha,\nu} \stackrel{d}{=} X \circ Z_{\nu,1}^{-1/\alpha} \circ \sqrt{2M_{\alpha/2}} \stackrel{d}{=} \Lambda \circ Z_{\nu,1}^{-1/\alpha} \circ \sqrt{R_{\alpha/2}}.$$

So, the density of the univariate generalized Linnik distribution admits a simple integral representation via known elementary densities (2), (39) and (45).

As concerns the property of geometric stability, the following statement holds.

**Lemma 3.** *Any univariate symmetric random variable $Y_\alpha$ is geometrically stable if and only if it is representable as*

$$Y_\alpha = W_1^{1/\alpha} \circ S(\alpha, 0), \quad 0 < \alpha \leqslant 2.$$

*Any univariate positive random variable $Y_\alpha$ is geometrically stable if and only if it is representable as*

$$Y_\alpha = W_1^{1/\alpha} \circ S(\alpha, 1), \quad 0 < \alpha \leqslant 1.$$

**Proof.** These representations immediately follow from the definition of geometrically stable distributions and the transfer theorem for cumulative geometric random sums, see, e.g., [54]. □

**Corollary 1.** *If $\nu \neq 1$, then from the identifiability of scale mixtures of stable laws (see Lemma 1) it follows that the generalized Linnik distribution and the generalized Mittag–Leffler distributions are not geometrically stable.*

Let $\Sigma$ be a positive definite $(r \times r)$-matrix, $\alpha \in (0,2]$, $\nu > 0$. As the 'ordinary' multivariate Linnik distribution, the multivariate elliptically contoured generalized Linnik distribution can be defined in at least two equivalent ways. First, it can be defined by its characteristic function. Namely, a multivariate distribution is called (centered) elliptically contoured generalized Linnik law, if the corresponding characteristic function has the form

$$\mathfrak{f}_{\alpha,\nu,\Sigma}^{(L)}(\boldsymbol{t}) = \left[1 + (\boldsymbol{t}^\top \Sigma \boldsymbol{t})^{\alpha/2}\right]^{-\nu}, \quad \boldsymbol{t} \in \mathbb{R}^r. \tag{47}$$

Second, let $\boldsymbol{X}$ be a random vector such that $\mathcal{L}(\boldsymbol{X}) = \mathfrak{N}_\Sigma$, independent of the random variable $M_{\alpha/2,\nu}$ with the generalized Mittag–Leffler distribution. By analogy with (46), introduce the random vector $\boldsymbol{L}_{\alpha,\nu,\Sigma}$ as

$$\boldsymbol{L}_{\alpha,\nu,\Sigma} = \sqrt{2M_{\alpha/2,\nu}} \circ \boldsymbol{X}.$$

Then, in accordance with what has been said in Section 5,

$$\mathcal{L}(\boldsymbol{L}_{\alpha,\nu,\Sigma}) = \mathsf{E}\mathfrak{N}_{2M_{\alpha/2,\nu}\Sigma}. \tag{48}$$

The distribution (42) will be called the multivariate (centered) elliptically contoured generalized Linnik distribution.

Using Remark 2 we can easily make sure that the two definitions of the multivariate elliptically contoured generalized Linnik distribution coincide. Indeed, with the account of (30), according to Remark 2, the characteristic function of the random vector $\boldsymbol{L}_{\alpha,\nu,\Sigma}$ defined by (48) has the form

$$\mathsf{E}\exp\{i\boldsymbol{t}^\top \boldsymbol{L}_{\alpha,\nu,\Sigma}\} = \psi_{\alpha/2,\nu}^{(M)}(\boldsymbol{t}^\top \Sigma \boldsymbol{t}) = \left[1 + (\boldsymbol{t}^\top \Sigma \boldsymbol{t})^{\alpha/2}\right]^{-\nu} = \mathfrak{f}_{\alpha,\nu,\Sigma}^{(L)}(\boldsymbol{t}), \quad \boldsymbol{t} \in \mathbb{R}^r,$$

that coincides with (47).

Based on (45), one more equivalent definition of the multivariate elliptically contoured generalized Linnik distribution can be proposed. Namely, let $\boldsymbol{L}_{\alpha,\nu,\Sigma}$ be an $r$-variate random vector such that

$$\boldsymbol{L}_{\alpha,\nu,\Sigma} = G_{\nu,1}^{1/\alpha} \circ \boldsymbol{S}(\alpha, \Sigma). \tag{49}$$

If $\nu = 1$, then, by definition, we obtain the random vector $W_1^{1/\alpha} \circ \boldsymbol{S}(\alpha, \Sigma) = \boldsymbol{L}_{\alpha,\Sigma}$ with the 'ordinary' multivariate elliptically contoured Linnik distribution.

It is well known that the Laplace—Stieltjes transform $\psi_{\nu,1}^{(G)}(s)$ of the random variable $G_{\nu,1}$ having the gamma distribution with the shape parameter $\nu$ has the form

$$\psi_{\nu,1}^{(G)}(s) = (1+s)^{-\nu}, \quad s > 0.$$

Then in accordance with Remark 1 the characteristic function of the random vector $\boldsymbol{L}_{\alpha,\nu,\Sigma}$ defined by (49) again has the form

$$\mathsf{E}\exp\{i\boldsymbol{t}^\top \boldsymbol{L}_{\alpha,\nu,\Sigma}\} = \psi_{\nu,1}^{(G)}\big((\boldsymbol{t}^\top\Sigma\boldsymbol{t})^{\alpha/2}\big) = \big[1 + (\boldsymbol{t}^\top\Sigma\boldsymbol{t})^{\alpha/2}\big]^{-\nu} = \mathfrak{f}_{\alpha,\nu,\Sigma}^{(L)}(\boldsymbol{t}), \quad \boldsymbol{t} \in \mathbb{R}^r.$$

Definitions (48) and (49) open the way to formulate limit theorems stating that the multivariate elliptically contoured generalized Linnik distribution can be limiting both for random sums of independent identically distributed random vectors with very large second moments, and for random sums of independent random vectors with finite covariance matrices.

There are some different ways of generalization of the univariate symmetric Linnik and generalized Linnik laws to the asymmetric case. The traditional (and formal) approach to the asymmetric generalization of the Linnik distribution (see, e.g., [15,55,56]) consists in the consideration of geometric sums of random summands whose distributions are attracted to an asymmetric strictly stable distribution. The variances of such summands are very large. Since in modeling real phenomena, as a rule, there are no solid reasons to reject the assumption of the finiteness of the variances of elementary summands, in [57], two alternative asymmetric generalizations were proposed based on the representability of the Linnik distribution as a scale mixture of normal laws or a scale mixture of Laplace laws.

Nevertheless, for our purposes it is convenient to deal with the traditional asymmetric generalization of the generalized Linnik distribution. Let $S(\alpha,\theta)$ be a random variable with the strictly stable distribution defined by the characteristic exponent $\alpha \in (0,2]$ and asymmetry parameter $\theta \in [-1,1]$, $\overline{G}_{\nu,\alpha,1}$ be a random variable having the GG distribution with shape parameter $\nu > 0$ and exponent power parameter $\alpha$ independent of $S(\alpha,\theta)$. Based on representation (45), we define the *asymmetric generalized Linnik distribution* as $\mathcal{L}\big(\overline{G}_{\nu,\alpha,1} \circ S(\alpha,\theta)\big)$. A random variable with this distribution will be denoted $L_{\alpha,\nu,\theta}$.

In the multivariate case a natural way of construction of the asymmetric Linnik laws consists in the application of Lemma 2 with not necessarily elliptically contoured strictly stable distribution. Namely, let the random variable $\overline{G}_{\nu,\alpha,1}$ have the generalized gamma distribution and be independent of the random vector $\boldsymbol{S}(\alpha,\mu)$ with the strictly stable distribution with characteristic exponent $\alpha \in (0,2]$ and spectral measure $\mu$. Extending the definitions of multivariate elliptically contoured generalized Linnik distribution given above, we will say that the distribution of the random vector $\overline{G}_{\nu,\alpha,1} \circ \boldsymbol{S}(\alpha,\mu)$ is the *multivariate generalized Linnik distribution*. Formally, this definition embraces both multivariate elliptically contoured generalized Linnik laws and, moreover, multivariate generalized Mittag–Leffler laws (if $\mu = \mu^+$). A random vector with the multivariate generalized Linnik distribution will be denoted $\boldsymbol{L}_{\alpha,\nu,\mu}$.

If $\nu = 1$, then we have the 'ordinary' multivariate Linnik distribution. By definition, $\boldsymbol{L}_{\alpha,1,\mu} = \boldsymbol{L}_{\alpha,\mu}$.

Mixture representations for the generalized Mittag–Leffler distribution were considered in [5] and discussed in Section 6 together with their extensions to the multivariate case. Here, we will focus on the mixture representations for the multivariate generalized Linnik distribution. Our reasoning is based on the definition of the multivariate generalized Linnik distribution given above and Theorem 1.

For $\alpha \in (0,2]$, $\alpha' \in (0,1)$ and $\nu > 0$ using (1), (17) and (31) we obtain the following chain of relations:

$$\boldsymbol{L}_{\alpha\alpha',\nu,\mu} \stackrel{d}{=} \overline{G}_{\nu,\alpha\alpha',1} \circ \boldsymbol{S}(\alpha\alpha',\mu) \stackrel{d}{=} G_{\nu,1}^{1/\alpha\alpha'} \circ \boldsymbol{S}(\alpha\alpha',\mu) \stackrel{d}{=} S^{1/\alpha}(\alpha',1) \circ G_{\nu,1}^{1/\alpha\alpha'} \circ \boldsymbol{S}(\alpha,\mu) \stackrel{d}{=}$$

$$\stackrel{d}{=} \left(S(\alpha',1) \circ \overline{G}_{\nu,\alpha',1}\right)^{1/\alpha} \circ \mathbf{S}(\alpha,\mu) \stackrel{d}{=} M_{\alpha',\nu}^{1/\alpha} \circ \mathbf{S}(\alpha,\mu).$$

Hence, the following statement holds representing the multivariate generalized Linnik distribution as a scale mixture of a multivariate stable law with any greater characteristic parameter, the mixing distribution being the univariate generalized Mittag–Leffler law.

**Theorem 3.** *If $\alpha \in (0,2]$, $\alpha' \in (0,1)$ and $\nu > 0$, then*

$$\mathbf{L}_{\alpha\alpha',\nu,\mu} \stackrel{d}{=} M_{\alpha',\nu}^{1/\alpha} \circ \mathbf{S}(\alpha,\mu). \tag{50}$$

Now let $\nu \in (0,1]$. From (45) and (4) it follows that

$$\mathbf{L}_{\alpha,\nu,\mu} \stackrel{d}{=} G_{\nu,1}^{1/\alpha} \circ \mathbf{S}(\alpha,\mu) \stackrel{d}{=} Z_{\nu,1}^{-1/\alpha} \circ W_1^{1/\alpha} \circ \mathbf{S}(\alpha,\mu) \stackrel{d}{=} Z_{\nu,1}^{-1/\alpha} \stackrel{d}{=} Z_{\nu,1}^{-1/\alpha} \circ \mathbf{L}_{\alpha,\mu}$$

yielding the following statement.

**Theorem 4.** *If $\nu \in (0,1]$ and $\alpha \in (0,2]$, then*

$$\mathbf{L}_{\alpha,\nu,\mu} \stackrel{d}{=} Z_{\nu,1}^{-1/\alpha} \circ \mathbf{L}_{\alpha,\mu}.$$

*In other words, with $\nu \in (0,1]$ and $\alpha \in (0,2]$, the multivariate generalized Linnik distribution is a scale mixture of 'ordinary' multivariate Linnik distributions.*

Consider projections of a random vector with the multivariate generalized Linnik distribution. For an arbitrary $\boldsymbol{u} \in \mathbb{R}^r$ we have

$$\boldsymbol{u}^\top \mathbf{L}_{\alpha,\nu,\mu} \stackrel{d}{=} \boldsymbol{u}^\top\left(\overline{G}_{\nu,\alpha,1} \circ \mathbf{S}(\alpha,\mu)\right) = \overline{G}_{\nu,\alpha,1} \circ \boldsymbol{u}^\top \mathbf{S}(\alpha,\mu) \stackrel{d}{=} \overline{G}_{\nu,\alpha,1} \circ \left(\gamma(\boldsymbol{u}) S(\alpha,\theta(\boldsymbol{u}))\right) =$$

$$= \gamma(\boldsymbol{u}) \cdot \overline{G}_{\nu,\alpha,1} \circ S(\alpha,\theta(\boldsymbol{u})) \stackrel{d}{=} \gamma(\boldsymbol{u}) L_{\alpha,\nu,\theta(\boldsymbol{u})}.$$

This means that the following statement holds.

**Theorem 5.** *Let the random vector $\mathbf{L}_{\alpha,\nu,\mu}$ have the multivariate generalized Linnik distribution with $\alpha \in (0,2]$, $\nu > 0$ and spectral measure $\mu$. Let to the spectral measure $\mu$ there correspond the projection scale parameter function $\gamma(\boldsymbol{u})$ and projection asymmetry parameter function $\theta(\boldsymbol{u})$, $\boldsymbol{u} \in \mathbb{R}^r$. Then any projection of the random vector $\mathbf{L}_{\alpha,\nu,\mu}$ has the univariate asymmetric Linnik distribution with the asymmetry parameter $\theta(\boldsymbol{u})$ scaled by $\gamma(\boldsymbol{u})$:*

$$\boldsymbol{u}^\top \mathbf{L}_{\alpha,\nu,\mu} \stackrel{d}{=} \gamma(\boldsymbol{u}) L_{\alpha,\nu,\theta(\boldsymbol{u})}.$$

Now consider the elliptically contoured case. Let $\alpha \in (0,2]$ and the random vector $\Lambda_\Sigma$ have the multivariate Laplace distribution with some positive definite $(r \times r)$-matrix $\Sigma$. In [33] it was shown that if $\delta \in (0,1]$, then

$$W_\delta \stackrel{d}{=} W_1 \circ S^{-1}(\delta,1). \tag{51}$$

Hence, it can be easily seen that

$$\mathbf{L}_{\alpha,\Sigma} \stackrel{d}{=} W_1^{1/\alpha} \circ \mathbf{S}(\alpha,\Sigma) \stackrel{d}{=} \sqrt{2W_{\alpha/2} \circ S(\alpha/2,1)} \circ \mathbf{X} \stackrel{d}{=} \sqrt{2W_1 \circ R_{\alpha/2}} \circ \mathbf{X} \stackrel{d}{=} \sqrt{R_{\alpha/2}} \circ \Lambda_\Sigma. \tag{52}$$

So, from Theorem 4 and (52) we obtain the following statement.

**Corollary 2.** *If $\nu \in (0,1]$ and $\alpha \in (0,2]$, then the multivariate elliptically contoured generalized Linnik distribution is a scale mixture of multivariate Laplace distributions:*

$$\mathbf{L}_{\alpha,\nu,\Sigma} \stackrel{d}{=} Z_{\nu,1}^{-1/\alpha} \circ \sqrt{R_{\alpha/2}} \circ \Lambda_{\Sigma}.$$

From (31) with $\nu = 1$ and (51) it can be seen that

$$\mathbf{L}_{\alpha,\Sigma} \stackrel{d}{=} \sqrt{2M_{\alpha/2}} \circ \mathbf{X}.$$

Therefore we obtain one more corollary of Theorem 4 representing the multivariate generalized Linnik distribution via 'ordinary' Mittag–Leffler distributions.

**Corollary 3.** *If $\nu \in (0,1]$ and $\alpha \in (0,2]$, then*

$$\mathbf{L}_{\alpha,\nu,\Sigma} \stackrel{d}{=} Z_{\nu,1}^{-1/\alpha} \circ \sqrt{2M_{\alpha/2}} \circ \mathbf{X}.$$

## 8. General Scale-Mixed Stable Distributions

In the preceding sections we considered special scale-mixed stable distributions in which the mixing distribution was generalized gamma leading to popular Mittag–Leffler and Linnik laws. Now turn to the case where the mixing distribution can be arbitrary.

Let $\alpha \in (0,2]$, let $U$ be a positive random variable and $\mathbf{S}(\alpha,\mu)$ be a random vector with the strictly stable distribution defined by the characteristic exponent $\alpha$ and spectral measure $\mu$. An $r$-variate random vector $\mathbf{Y}_{\alpha,\mu}$ is said to have the *U-scale-mixed stable distribution*, if

$$\mathbf{Y}_{\alpha,\mu} \stackrel{d}{=} U^{1/\alpha} \circ \mathbf{S}(\alpha,\mu)$$

Correspondingly, for $0 < \alpha \leqslant 1$, a univariate positive random variable $Y_{\alpha,1}$ is said to have the $U$-scale-mixed one-sided stable distribution, if $Y_{\alpha,1}$ is representable as

$$Y_{\alpha,1} \stackrel{d}{=} U^{1/\alpha} \circ S(\alpha,1).$$

As above, in the elliptically contoured case, where to the spectral measure $\mu$ there corresponds a positive definite $(r \times r)$-matrix $\Sigma$, instead of $\mathbf{Y}_{\alpha,\mu}$ we will write $\mathbf{Y}_{\alpha,\Sigma}$.

The following statement generalizes Theorem 2 (Equation (36)) and Theorem 3.

**Theorem 6.** *Let $U$ be a positive random variable, $\alpha \in (0,2]$, $\alpha' \in (0,1]$. Let $\mathbf{S}(\alpha,\mu)$ be a random vector with the strictly stable distribution defined by the characteristic exponent $\alpha$ and spectral measure $\mu$. Let an $r$-variate random vector $\mathbf{Y}_{\alpha\alpha',\Sigma}$ have the U-scale-mixed stable distribution and a random variable $Y_{\alpha',1}$ have the U-scale-mixed one-sided stable distribution. Assume that $\mathbf{S}_{\alpha,\mu}$ and $Y_{\alpha',1}$ are independent. Then*

$$\mathbf{Y}_{\alpha\alpha',\mu} \stackrel{d}{=} Y_{\alpha',1}^{1/\alpha} \circ \mathbf{S}(\alpha,\mu).$$

**Proof.** From the definition of a $U$-scale-mixed stable distribution and (17) we have

$$\mathbf{Y}_{\alpha\alpha',\mu} \stackrel{d}{=} U^{1/\alpha\alpha'} \circ \mathbf{S}(\alpha\alpha',\mu) \stackrel{d}{=} U^{1/\alpha\alpha'} \circ S^{1/\alpha}(\alpha',1) \circ \mathbf{S}(\alpha,\mu) \stackrel{d}{=}$$

$$\stackrel{d}{=} \big(U^{1/\alpha'} \circ S(\alpha',1)\big)^{1/\alpha} \circ \mathbf{S}(\alpha,\mu) \stackrel{d}{=} Y_{\alpha',1}^{1/\alpha} \circ \mathbf{S}(\alpha,\mu).$$

□

In the elliptically contoured case with $\alpha = 2$, from Theorem 6 we obtain the following statement.

**Corollary 4.** *Let $\alpha \in (0,2)$, $U$ be a positive random variable, $\Sigma$ be a positive definite $(r \times r)$-matrix, $\mathbf{X}$ be a random vector such that $\mathcal{L}(X) = \mathfrak{N}_\Sigma$. Then*

$$\mathbf{Y}_{\alpha,\Sigma} \stackrel{d}{=} \sqrt{2Y_{\alpha/2,1}} \circ \mathbf{X}.$$

*In other words, any multivariate scale-mixed symmetric stable distribution is a scale mixture of multivariate normal laws. On the other hand, since the normal distribution is stable with $\alpha = 2$, any multivariate normal scale mixture is a 'trivial' multivariate scale-mixed stable distribution.*

To give particular examples of 'non-trivial' scale-mixed stable distributions, note that

- if $U \stackrel{d}{=} W_1$, then $Y_{\alpha,1} \stackrel{d}{=} M_\alpha$ and $\mathbf{Y}_{\alpha,\mu} \stackrel{d}{=} \mathbf{L}_{\alpha,\mu}$;
- if $U \stackrel{d}{=} G_{\nu,1}$, then $Y_{\alpha,1} \stackrel{d}{=} M_{\alpha,\nu}$ and $\mathbf{Y}_{\alpha,\mu} \stackrel{d}{=} \mathbf{L}_{\alpha,\nu,\mu}$;
- if $U \stackrel{d}{=} S(\alpha',1)$ with $0 < \alpha' \leqslant 1$, then $Y_{\alpha,1} \stackrel{d}{=} S(\alpha\alpha',1)$ and $\mathbf{Y}_{\alpha,\mu} \stackrel{d}{=} \mathbf{S}(\alpha\alpha',\mu)$.

Among possible mixing distributions of the random variable $U$, we will distinguish a special class that can play important role in modeling observed regularities by heavy-tailed distributions. Namely, assume that $V$ is a positive random variable and let

$$U \stackrel{d}{=} V \circ G_{\nu,1},$$

that is, the distribution of $U$ is a scale mixture of gamma distributions. We will denote the class of these distributions as $\mathcal{G}^{(V)}$. This class is rather wide and besides the gamma distribution and its particular cases (exponential, Erlang, chi-square, etc.) with exponentially fast decreasing tail, contains, for example, Pareto and Snedecor–Fisher laws with power-type decreasing tail. In the last two cases the random variable $V$ is assumed to have the corresponding gamma and inverse gamma distributions, respectively.

For $\mathcal{L}(U) \in \mathcal{G}^{(V)}$ we have

$$Y_{\alpha,1} \stackrel{d}{=} (V \circ G_{\nu,1})^{1/\alpha} \circ S(\alpha,1) \stackrel{d}{=} V^{1/\alpha} \circ \left(G_{\nu,1}^{1/\alpha} \circ S(\alpha,1)\right) \stackrel{d}{=} V^{1/\alpha} \circ M_{\alpha,\nu}$$

and

$$\mathbf{Y}_{\alpha,\mu} \stackrel{d}{=} (V \circ G_{\nu,1})^{1/\alpha} \circ \mathbf{S}(\alpha,\mu) \stackrel{d}{=} V^{1/\alpha} \circ \left(G_{\nu,1}^{1/\alpha} \circ \mathbf{S}(\alpha,\mu)\right) \stackrel{d}{=} V^{1/\alpha} \circ \mathbf{L}_{\alpha,\nu,\mu}.$$

This means that with $\mathcal{L}(U) \in \mathcal{G}^{(V)}$, the $U$-scale-mixed stable distributions are scale mixtures of the generalized Mittag–Leffler and multivariate generalized Linnik laws.

Therefore, we pay a special attention to mixture representations of the generalized Mittag–Leffler and multivariate generalized Linnik distributions. These representations can be easily extended to any $U$-scale-mixed stable distributions with $\mathcal{L}(U) \in \mathcal{G}^{(V)}$.

## 9. Convergence of the Distributions of Random Sequences with Independent Indices to Multivariate Scale-Mixed Stable Distributions

In applied probability it is a convention that a model distribution can be regarded as well-justified or adequate, if it is an *asymptotic approximation*, that is, if there exists a rather simple limit setting (say, schemes of maximum or summation of random variables) and the corresponding limit theorem in which the model under consideration manifests itself as a limit distribution. The existence of such limit setting can provide a better understanding of real mechanisms that generate observed statistical regularities, see e.g., [54].

In this section we will prove some limit theorems presenting necessary and sufficient conditions for the convergence of the distributions of random sequences with independent random indices (including sums of a random number of random vectors and multivariate statistics constructed from samples with random sizes) to scale mixtures of multivariate elliptically contoured stable distributions. As particular cases, conditions will be obtained for the convergence of the distributions of random sums of random vectors with both very large and finite covariance matrices to the multivariate generalized Linnik distribution.

Consider a sequence $\{S_n\}_{n\geqslant 1}$ of random elements taking values in $\mathbb{R}^r$. Let $\Xi(\mathbb{R}^r)$ be the set of all nonsingular linear operators acting from $\mathbb{R}^r$ to $\mathbb{R}^r$. The identity operator acting from $\mathbb{R}^r$ to $\mathbb{R}^r$ will be denoted $I_r$. Assume that there exist sequences $\{B_n\}_{n\geqslant 1}$ of operators from $\Xi(\mathbb{R}^r)$ and $\{\boldsymbol{a}_n\}_{n\geqslant 1}$ of elements from $\mathbb{R}^r$ such that

$$Q_n = B_n^{-1}(S_n - \boldsymbol{a}_n) \Longrightarrow Q \quad (n \to \infty) \tag{53}$$

where $Q$ is a random element whose distribution with respect to P will be denoted $H$, $H = \mathcal{L}(Q)$.

Along with $\{S_n\}_{n\geqslant 1}$, consider a sequence of integer-valued positive random variables $\{N_n\}_{n\geqslant 1}$ such that for each $n \geqslant 1$ the random variable $N_n$ is independent of the sequence $\{S_k\}_{k\geqslant 1}$. Let $\boldsymbol{c}_n \in \mathbb{R}^r$, $D_n \in \Xi(\mathbb{R}^r)$, $n \geqslant 1$. Now we will formulate sufficient conditions for the weak convergence of the distributions of the random elements $\boldsymbol{Z}_n = D_n^{-1}(S_{N_n} - \boldsymbol{c}_n)$ as $n \to \infty$.

For $\boldsymbol{g} \in \mathbb{R}^r$ denote $\boldsymbol{W}_n(\boldsymbol{g}) = D_n^{-1}(B_{N_n}\boldsymbol{g} + \boldsymbol{a}_{N_n} - \boldsymbol{c}_n)$. By measurability of a random field we will mean its measurability as a function of two variates, an elementary outcome and a parameter, with respect to the Cartesian product of the $\sigma$-algebra $\mathcal{A}$ and the Borel $\sigma$-algebra $\mathcal{B}(\mathbb{R}^r)$ of subsets of $\mathbb{R}^r$.

In [58,59] the following theorem was proved which establishes sufficient conditions of the weak convergence of multivariate random sequences with independent random indices under operator normalization.

**Theorem 7.** *[58,59]. Let $\|D_n^{-1}\| \to \infty$ as $n \to \infty$ and let the sequence of random variables $\{\|D_n^{-1}B_{N_n}\|\}_{n\geqslant 1}$ be tight. Assume that there exist a random element $Q$ with distribution $H$ and an r-dimensional random field $\boldsymbol{W}(\boldsymbol{g})$, $\boldsymbol{g} \in \mathbb{R}^r$, such that (53) holds and*

$$\boldsymbol{W}_n(\boldsymbol{g}) \Longrightarrow \boldsymbol{W}(\boldsymbol{g}) \quad (n \to \infty)$$

*for H-almost all $\boldsymbol{g} \in \mathbb{R}^r$. Then the random field $\boldsymbol{W}(\boldsymbol{g})$ is measurable, linearly depends on $\boldsymbol{g}$ and*

$$\boldsymbol{Z}_n \Longrightarrow \boldsymbol{W}(Q) \quad (n \to \infty),$$

*where the random field $\boldsymbol{W}(\,\cdot\,)$ and the random element $Q$ are independent.*

Now consider a special case of the general limit setting and assume that the normalization is scalar and the limit random vector $Q$ in (53) has an elliptically contoured stable distribution. Namely, let $\{b_n\}_{n\geqslant 1}$ be an very largely increasing sequence of positive numbers and, instead of the general condition (53) assume that

$$\mathcal{L}(b_n^{-1/\alpha}S_n) \Longrightarrow \mathfrak{S}_{\alpha,\Sigma} \tag{54}$$

as $n \to \infty$, where $\alpha \in (0,2]$ and $\Sigma$ is some positive definite matrix. In other words, let

$$b_n^{-1/\alpha}S_n \Longrightarrow S(\alpha,\Sigma) \quad (n \to \infty).$$

Let $\{d_n\}_{n\geqslant 1}$ be an very largely increasing sequence of positive numbers. As $\boldsymbol{Z}_n$ take the scalar normalized random vector

$$\boldsymbol{Z}_n = d_n^{-1/\alpha}S_{N_n}.$$

The following result can be considered as a multivariate generalization of the main theorem of [29].

**Theorem 8.** *Let $N_n \to \infty$ in probability as $n \to \infty$. Assume that the random vectors $\mathbf{S}_1, \mathbf{S}_2, \ldots$ satisfy condition (54) with $\alpha \in (0,2]$ and a positive definite matrix $\Sigma$. Then a distribution $F$ such that*

$$\mathcal{L}(\mathbf{Z}_n) \Longrightarrow F \quad (n \to \infty), \tag{55}$$

*exists if and only if there exists a distribution function $V(x)$ satisfying the conditions*

*(i)* $V(x) = 0$ *for* $x < 0$;
*(ii)* *for any* $A \in \mathcal{B}(\mathbb{R}^r)$

$$F(A) = \mathsf{E}\mathfrak{S}_{\alpha,U^{2/\alpha}\Sigma}(A) = \int_0^\infty \mathfrak{S}_{\alpha,u^{2/\alpha}\Sigma}(A)dV(u), \quad x \in \mathbb{R}^1;$$

*(iii)* $\mathsf{P}(b_{N_n} < d_n x) \Longrightarrow V(x),\ n \to \infty$.

**Proof.** *The 'if' part.* We will essentially exploit Theorem 7. For each $n \geqslant 1$ let $\boldsymbol{a}_n = \boldsymbol{c}_n = 0$, $B_n = b_n^{1/\alpha} I_r$, $D_n = d_n^{1/\alpha} I_r$. Let $U$ be a random variable with the distribution function $V(x)$. Note that the conditions of the theorem guarantee the tightness of the sequence of random variables

$$\|D_n^{-1} B_{N_n}\| = (b_{N_n}/d_n)^{1/\alpha}, \quad n = 1, 2, \ldots$$

implied by its weak convergence to the random variable $U^{1/\alpha}$. Further, in the case under consideration we have $\mathbf{W}_n(\boldsymbol{g}) = (b_{N_n}/d_n)^{1/\alpha} \cdot \boldsymbol{g}$, $\boldsymbol{g} \in \mathbb{R}^r$. Therefore, the condition $b_{N_n}/d_n \Longrightarrow U$ implies $\mathbf{W}_n(\boldsymbol{g}) \Longrightarrow U^{1/\alpha}\boldsymbol{g}$ for all $\boldsymbol{g} \in \mathbb{R}^r$.

Condition (54) means that in the case under consideration $H = \mathfrak{S}_{\alpha,\Sigma}$. Hence, by Theorem 7 $\mathbf{Z}_n \Longrightarrow U^{1/\alpha} \circ \mathbf{S}(\alpha, \Sigma)$ (recall that the symbol $\circ$ stands for the product of *independent* random elements). The distribution of the random element $U^{1/\alpha} \circ \mathbf{S}(\alpha, \Sigma)$ coincides with $\mathsf{E}\mathfrak{S}_{\alpha,U^{2/\alpha}\Sigma}$, see Section 5.

*The 'only if' part.* Let condition (55) hold. Make sure that the sequence $\{\|D_n^{-1} B_{N_n}\|\}_{n \geqslant 1}$ is tight. Let $\mathbf{Q} \stackrel{d}{=} \mathbf{S}(\alpha, \Sigma)$. There exist $\delta > 0$ and $\rho > 0$ such that

$$\mathsf{P}(\|\mathbf{Q}\| > \rho) > \delta. \tag{56}$$

For $\rho$ specified above and an arbitrary $x > 0$ we have

$$\mathsf{P}(\|\mathbf{Z}_n\| > x) \geqslant \mathsf{P}\big(\|d_n^{-1/\alpha}\mathbf{S}_{N_n}\| > x;\ \|b_{N_n}^{-1/\alpha}\mathbf{S}_{N_n}\| > \rho\big) =$$

$$= \mathsf{P}\big((b_{N_n}/d_n)^{1/\alpha} > x \cdot \big\|b_{N_n}^{-1/\alpha}\mathbf{S}_{N_n}\big\|^{-1};\ \|b_{N_n}^{-1/\alpha}\mathbf{S}_{N_n}\| > \rho\big) \geqslant$$

$$\geqslant \mathsf{P}\big((b_{N_n}/d_n)^{1/\alpha} > x/\rho;\ \|b_{N_n}^{-1/\alpha}\mathbf{S}_{N_n}\| > \rho\big) =$$

$$= \sum\nolimits_{k=1}^\infty \mathsf{P}(N_n = k)\mathsf{P}\big((b_k/d_n)^{1/\alpha} > x/\rho;\ \|b_k^{-1/\alpha}\mathbf{S}_k\| > \rho\big) =$$

$$= \sum\nolimits_{k=1}^\infty \mathsf{P}(N_n = k)\mathsf{P}\big((b_k/d_n)^{1/\alpha} > x/\rho\big)\mathsf{P}\big(\|b_k^{-1/\alpha}\mathbf{S}_k\| > \rho\big) \tag{57}$$

(the last equality holds since any constant is independent of any random variable). Since by (54) the convergence $b_k^{-1/\alpha}\mathbf{S}_k \Longrightarrow Y$ takes place as $k \to \infty$, from (56) it follows that there exists a number $k_0 = k_0(\rho, \delta)$ such that

$$\mathsf{P}\big(\|b_k^{-1/\alpha}\mathbf{S}_k\| > \rho\big) > \delta/2$$

for all $k > k_0$. Therefore, continuing (57) we obtain

$$\mathsf{P}(\|\mathbf{Z}_n\| > x) \geqslant \frac{\delta}{2} \sum\nolimits_{k=k_0+1}^\infty \mathsf{P}(N_n = k)\mathsf{P}\big((b_k/d_n)^{1/\alpha} > x/\rho\big) =$$

$$= \frac{\delta}{2}\big[\mathsf{P}\big((b_{N_n}/d_n)^{1/\alpha} > x/\rho\big) - \sum\nolimits_{k=1}^{k_0} \mathsf{P}(N_n = k)\mathsf{P}\big((b_k/d_n)^{1/\alpha} > x/\rho\big)\big] \geqslant$$

$$\geqslant \frac{\delta}{2}\big[\mathsf{P}\big((b_{N_n}/d_n)^{1/\alpha} > x/\rho\big) - \mathsf{P}(N_n \leqslant k_0)\big].$$

Hence,

$$\mathsf{P}\big((b_{N_n}/d_n)^{1/\alpha} > x/R\big) \leqslant \frac{2}{\delta}\mathsf{P}(\|\mathbf{Z}_n\| > x) + \mathsf{P}(N_n \leqslant k_0). \tag{58}$$

From the condition $N_n \xrightarrow{P} \infty$ as $n \to \infty$ it follows that for any $\epsilon > 0$ there exists an $n_0 = n_0(\epsilon)$ such that $\mathsf{P}(N_n \leqslant n_0) < \epsilon$ for all $n \geqslant n_0$. Therefore, with the account of the tightness of the sequence $\{\mathbf{Z}_n\}_{n\geqslant 1}$ that follows from its weak convergence to the random element $\mathbf{Z}$ with $\mathcal{L}(\mathbf{Z}) = F$ implied by (55), relation (58) implies

$$\lim_{x\to\infty} \sup_{n\geqslant n_0(\epsilon)} \mathsf{P}\big((b_{N_n}/d_n)^{1/\alpha} > x/\rho\big) \leqslant \epsilon, \tag{59}$$

whatever $\epsilon > 0$ is. Now assume that the sequence

$$\|D_n^{-1}B_{N_n}\| = (b_{N_n}/d_n)^{1/\alpha}, \quad n = 1, 2, \ldots$$

is not tight. In that case there exists an $\gamma > 0$ and sequences $\mathcal{N}$ of natural and $\{x_n\}_{n\in\mathcal{N}}$ of real numbers satisfying the conditions $x_n \uparrow \infty$ $(n \to \infty,\ n \in \mathcal{N})$ and

$$\mathsf{P}\big((b_{N_n}/d_n)^{1/\alpha} > x_n\big) > \gamma, \quad n \in \mathcal{N}. \tag{60}$$

But, according to (59), for any $\epsilon > 0$ there exist $M = M(\epsilon)$ and $n_0 = n_0(\epsilon)$ such that

$$\sup_{n\geqslant n_0(\epsilon)} \mathsf{P}\big((b_{N_n}/d_n)^{1/\alpha} > M(\epsilon)\big) \leqslant 2\epsilon. \tag{61}$$

Choose $\epsilon < \gamma/2$ where $\gamma$ is the number from (60). Then for all $n \in \mathcal{N}$ large enough, in accordance with (60), the inequality opposite to (61) must hold. The obtained contradiction by the Prokhorov theorem proves the tightness of the sequence $\{\|D_n^{-1}B_{N_n}\|\}_{n\geqslant 1}$ or, which in this case is the same, of the sequence $\{b_{N_n}/d_n\}_{n\geqslant 1}$.

Introduce the set $\mathcal{W}(\mathbf{Z})$ containing all nonnegative random variables $U$ such that $\mathsf{P}(\mathbf{Z} \in A) = \mathsf{E}\mathfrak{S}_{\alpha,U^{2/\alpha}\Sigma}(A)$ for any $A \in \mathcal{B}(\mathbb{R}^r)$. Let $\lambda(\cdot,\cdot)$ be any probability metric that metrizes weak convergence in the space of $r$-variate random vectors, or, which is the same in this context, in the space of distributions, say, the Lévy–Prokhorov metric. If $\mathbf{X}_1$ and $\mathbf{X}_2$ are random variables with the distributions $F_1$ and $F_2$ respectively, then we identify $\lambda(\mathbf{X}_1, \mathbf{X}_2)$ and $\lambda(F_1, F_2)$). Show that there exists a sequence of random variables $\{U_n\}_{n\geqslant 1}$, $U_n \in \mathcal{W}(\mathbf{Z})$, such that

$$\lambda\big(b_{N_n}/d_n, U_n\big) \longrightarrow 0 \ \ (n \to \infty). \tag{62}$$

Denote

$$\beta_n = \inf\big\{\lambda\big(b_{N_n}/d_n, U\big) : U \in \mathcal{W}(\mathbf{Z})\big\}.$$

Prove that $\beta_n \to 0$ as $n \to \infty$. Assume the contrary. In that case $\beta_n \geqslant \delta$ for some $\delta > 0$ and all $n$ from some subsequence $\mathcal{N}$ of natural numbers. Choose a subsequence $\mathcal{N}_1 \subseteq \mathcal{N}$ so that the sequence $\{b_{N_n}/d_n\}_{n\in\mathcal{N}_1}$ weakly converges to a random variable $U$ (this is possible due to the tightness of the family $\{b_{N_n}/d_n\}_{n\geqslant 1}$ established above). But then $W_n(\mathbf{g}) \Longrightarrow U^{1/\alpha}\mathbf{g}$ as $n \to \infty$, $n \in \mathcal{N}_1$ for any $\mathbf{g} \in \mathbb{R}^r$. Applying Theorem 7 to $n \in \mathcal{N}_1$ with condition (54) playing the role of condition (53), we make sure that $U \in \mathcal{W}(\mathbf{Z})$, since condition (55) provides the coincidence of the limits of all weakly convergent subsequences. So, we arrive at the contradiction to the assumption that $\beta_n \geqslant \delta$ for all $n \in \mathcal{N}_1$. Hence, $\beta_n \to 0$ as $n \to \infty$.

For any $n = 1, 2, \ldots$ choose a random variable $U_n$ from $\mathcal{W}(\mathbf{Z})$ satisfying the condition

$$\lambda\big(b_{N_n}/d_n,\, U_n\big) \leqslant \beta_n + \tfrac{1}{n}.$$

This sequence obviously satisfies condition (62). Now consider the structure of the set $\mathcal{W}(\mathbf{Z})$. This set contains all the random variables defining the family of special mixtures of multivariate centered elliptically contoured stable laws considered in Lemma 1, according to which this family is identifiable. So, whatever a random element $\mathbf{Z}$ is, the set $\mathcal{W}(\mathbf{Z})$ contains at most one element. Therefore, actually condition (62) is equivalent to

$$b_{N_n}/d_n \Longrightarrow U \quad (n \to \infty),$$

that is, to condition (iii) of the theorem. The theorem is proved. □

**Corollary 5.** *Under the conditions of Theorem 7, non-randomly normalized random sequences with independent random indices $d_n^{-1/\alpha}\mathbf{S}_{N_n}$ have the limit stable distribution $\mathfrak{S}_{\alpha,\Sigma'}$ with some positive definite matrix $\Sigma'$ if and only if there exists a number $c > 0$ such that*

$$b_{N_n}/d_n \Longrightarrow c \quad (n \to \infty).$$

*Moreover, in this case $\Sigma' = c^{2/\alpha}\Sigma$.*

This statement immediately follows from Theorem 8 with the account of Lemma 1.

Now consider convergence of the distributions of random sums of random vectors to special scale-mixed multivariate elliptically contoured stable laws.

In Section 4 (see (20)) we made sure that all scale-mixed centered elliptically contoured stable distributions are representable as multivariate normal scale mixtures. Together with Theorem 8 this observation allows to suspect at least two principally different limit schemes in which each of these distributions can appear as limiting for random sums of independent random vectors. We will illustrate these two cases by the example of the multivariate generalized Linnik distribution.

As we have already mentioned, 'ordinary' Linnik distributions are geometrically stable. Geometrically stable distributions are only possible limits for the distributions of geometric random sums of independent identically distributed random vectors. As this is so, the distributions of the summands belong to the domain of attraction of the multivariate strictly stable law with some characteristic exponent $\alpha \in (0, 2]$ and hence, for $0 < \alpha < 2$ the univariate marginals have very large moments of orders greater or equal to $\alpha$. As concerns the case $\alpha = 2$, where the variances of marginals are finite, within the framework of the scheme of geometric summation in this case the only possible limit law is the multivariate Laplace distribution.

Correspondinly, as we will demonstrate below, the multivariate generalized Linnik distributions can be limiting for negative binomial sums of independent identically distributed random vectors. Negative binomial random sums turn out to be important and adequate models of characteristics of precipitation (total precipitation volume, etc.) during wet (rainy) periods in meteorology [60–62]. However, in this case the summands (daily rainfall volumes) also must have distributions from the domain of attraction of a strictly stable law with some characteristic exponent $\alpha \in (0, 2]$ and hence, with $\alpha \in (0, 2)$, have very large variances, that seems doubtful, since to have an very large variance, the random variable *must* be allowed to take *arbitrarily large* values with *positive* probabilities. If $\alpha = 2$, then the only possible limit distribution for negative binomial random sums is the so-called variance gamma distribution which is well known in financial mathematics [54].

However, when the (generalized) Linnik distributions are used as models of statistical regularities observed in real practice and an additive structure model is used of type of a (stopped) random walk for the observed process, the researcher cannot avoid thinking over the following question: which of the two combinations of conditions can be encountered more often:

- the distribution of the number of summands (the number of jumps of a random walk) is asymptotically gamma (say, negative binomial), but the distributions of summands (jumps) have so heavy tails that, at least, their variances are very large, or
- the second moments (variances) of the summands (jumps) are finite, but the number of summands exposes an irregular behavior so that its very large values are possible?

Since, as a rule, when real processes are modeled, there are no serious reasons to reject the assumption that the variances of jumps are finite, the second combination at least deserves a thorough analysis.

As it was demonstrated in the preceding sections, the scale-mixed multivariate elliptically contoured stable distributions (including multivariate (generalized) Linnik laws) even with $\alpha < 2$ can be represented as multivariate normal scale mixtures. This means that they can be limit distributions in analogs of the central limit theorem for random sums of independent random vectors *with finite covariance matrices*. Such analogs with univariate 'ordinary' Linnik limit distributions were presented in [2] and extended to generalized Linnik distributions in [5]. In what follows we will present some examples of limit settings for random sums of independent random vectors with principally different tail behavior. In particular, it will de demonstrated that the scheme of negative binomial summation is far not the only asymptotic setting (even for sums of independent random variables!) in which the multivariate generalized Linnik law appears as the limit distribution.

**Remark 3.** *Based on the results of [63], by an approach that slightly differs from the one used here by the starting point, in the paper [64] it was demonstrated that if the random vectors $\{\mathbf{S}_n\}_{n\geqslant 1}$ are formed as cumulative sums of independent random vectors:*

$$\mathbf{S}_n = \mathbf{X}_1 + \ldots + \mathbf{X}_n \tag{63}$$

*for $n \in \mathbb{N}$, where $\mathbf{X}_1, \mathbf{X}_2, \ldots$ are independent r-valued random vectors, then the condition $N_n \xrightarrow{P} \infty$ in the formulations of Theorem 8 and Corollary 4 can be omitted.*

Throughout this section we assume that the random vectors $\mathbf{S}_n$ have the form (63).

Let $U \in \mathcal{U}$ (see Section 5), $\alpha \in (0,2]$, $\Sigma$ be a positive definite matrix. In Section 8 the $r$-variate random vector $\mathbf{Y}_{\alpha,\Sigma}$ with the the multivariate $U$-scale-mixed elliptically contoured stable distribution was introduced as $\mathbf{Y}_{\alpha,\Sigma} = U^{1/\alpha} \circ \mathbf{S}(\alpha,\Sigma)$. In this section we will consider the conditions under which multivariate $U$-scale-mixed stable distributions can be limiting for sums of independent random vectors.

Consider a sequence of integer-valued positive random variables $\{N_n\}_{n\geqslant 1}$ such that for each $n \geqslant 1$ the random variable $N_n$ is independent of the sequence $\{\mathbf{S}_k\}_{k\geqslant 1}$. First, let $\{b_n\}_{n\geqslant 1}$ be an very largely increasing sequence of positive numbers such that convergence (54) takes place. Let $\{d_n\}_{n\geqslant 1}$ be an very largely increasing sequence of positive numbers. The following statement presents necessary and sufficient conditions for the convergence

$$d_n^{-1/\alpha}\mathbf{S}_{N_n} \Longrightarrow \mathbf{Y}_{\alpha,\Sigma} \quad (n \to \infty). \tag{64}$$

**Theorem 9.** *Under condition (54), convergence (64) takes place if and only if*

$$b_{N_n}/d_n \Longrightarrow U \quad (n \to \infty).$$

**Proof.** This theorem is a direct consequence of Theorem 8 and the definition of $\mathbf{Y}_{\alpha,\Sigma}$ with the account of Remark 3. □

**Corollary 6.** *Assume that $\nu > 0$. Under condition (54), the convergence*

$$d_n^{-1/\alpha}\mathbf{S}_{N_n} \Longrightarrow \mathbf{L}_{\alpha,\nu,\Sigma} \quad (n \to \infty)$$

*takes place if and only if*

$$b_{N_n}/d_n \Longrightarrow G_{\nu,1} \quad (n \to \infty). \tag{65}$$

**Proof.** To prove this statement it suffices to notice that the multivariate generalized Linnik distribution is a $U$-scale-mixed stable distribution with $U \stackrel{d}{=} G_{\nu,1}$ (see representation (49)) and refer to Theorem 9 with the account of Remark 3.

Condition (65) holds, for example, if $b_n = d_n = n$, $n \in \mathbb{N}$, and the random variable $N_n$ has the negative binomial distribution with shape parameter $\nu > 0$, that is, $N_n = N\!B_{\nu,p_n}$,

$$\mathsf{P}(N\!B_{\nu,p_n} = k) = \frac{\Gamma(\nu + k - 1)}{(k-1)!\Gamma(r)} \cdot p_n^{\nu}(1-p_n)^{k-1}, \quad k = 1, 2, ...,$$

with $p_n = n^{-1}$ (see, e.g., [65,66]). In this case $\mathsf{E}N\!B_{\nu,p_n} = n\nu$. □

Now consider the conditions providing the convergence in distribution of scalar normalized random sums of independent random vectors satisfying condition (54) with some $\alpha \in (0,2]$ and $\Sigma$ to a random vector $\mathbf{Y}_{\beta,\Sigma}$ with the $U$-scale-mixed stable distribution $\mathsf{E}\mathfrak{S}_{\beta,U^{2/\beta}\Sigma}$ with some $\beta \in (0,\alpha)$. For convenience, let $\beta = \alpha\alpha'$ where $\alpha' \in (0,1)$.

Recall that in Section 8, for $\alpha' \in (0,1]$ the positive random variable $Y_{\alpha',1}$ with the univariate one-sided $U$-scale-mixed stable distribution was introduced as $Y_{\alpha',1} \stackrel{d}{=} U^{1/\alpha} \circ S(\alpha',1)$.

**Theorem 10.** *Let $\alpha' \in (0,1]$. Under condition (54), the convergence*

$$d_n^{-1/\alpha}\mathbf{S}_{N_n} \Longrightarrow \mathbf{Y}_{\alpha\alpha',\Sigma} \quad (n \to \infty)$$

*takes place if and only if*

$$b_{N_n}/d_n \Longrightarrow Y_{\alpha',1} \quad (n \to \infty).$$

**Proof.** This statement directly follows from Theorems 5 and 7 with the account of Remark 3. □

**Corollary 7.** *Let $\alpha' \in (0,1]$, $\nu > 0$. Under condition (54), the convergence*

$$d_n^{-1/\alpha}\mathbf{S}_{N_n} \Longrightarrow \mathbf{L}_{\alpha\alpha',\Sigma,\nu} \quad (n \to \infty)$$

*takes place if and only if*

$$b_{N_n}/d_n \Longrightarrow M_{\alpha',\nu} \quad (n \to \infty).$$

**Proof.** This statement directly follows from Theorems 3 (see representation (50)) and 9 with the account of Remark 3. □

From the case of heavy tails turn to the 'light-tails' case where in (54) $\alpha = 2$. In other words, assume that the properties of the summands $\mathbf{X}_j$ provide the asymptotic normality of the sums $\mathbf{S}_n$. More precisely, instead of (54), assume that

$$b_n^{-1/2}\mathbf{S}_n \Longrightarrow \mathbf{X} \quad (n \to \infty). \tag{66}$$

The following results show that even under condition (66), heavy-tailed $U$-scale-mixed multivariate stable distributions can be limiting for random sums.

**Theorem 11.** *Under condition (66), convergence (64) takes place if and only if*

$$b_{N_n}/d_n \Longrightarrow Y_{\alpha/2,1} \quad (n \to \infty).$$

**Proof.** This theorem is a direct consequence of Theorem 8 and Corollary 4, according to which $\mathbf{Y}_{\alpha,\Sigma} \stackrel{d}{=} \sqrt{2Y_{\alpha/2,1}} \circ \mathbf{X}$ with the account of Remark 3. □

**Corollary 8.** *Assume that $N_n \to \infty$ in probability as $n \to \infty$. Under condition (66), non-randomly normalized random sums $d_n^{-1/2}\mathbf{S}_{N_n}$ have the limit stable distribution $\mathfrak{S}_{\alpha,\Sigma}$ if and only if*

$$b_{N_n}/d_n \Longrightarrow 2S(\alpha/2,1) \quad (n \to \infty).$$

**Proof.** This statement follows from Theorem 11 with the account of (13) and Remark 3. □

**Corollary 9.** *Assume that $N_n \to \infty$ in probability as $n \to \infty$, $\nu > 0$. Under condition (66), the convergence*

$$d_n^{-1/2}\mathbf{S}_{N_n} \Longrightarrow \mathbf{L}_{\alpha,\nu,\Sigma} \quad (n \to \infty)$$

*takes place if and only if*

$$b_{N_n}/d_n \Longrightarrow 2M_{\alpha/2,\nu} \quad (n \to \infty).$$

**Proof.** To prove this statement it suffices to notice that the multivariate generalized Linnik distribution is a multivariate normal scale mixture with the generalized Mittag–Leffler mixing distribution (see definition (48)) and refer to Theorem 11 with the account of Remark 3.

Another way to prove Corollary 9 is to deduce it from Corollary 7. □

Product representations for limit distributions in these theorems proved in the preceding sections allow to use other forms of the conditions for the convergence of random sums of random vectors to particular scale mixtures of multivariate stable laws.

## 10. Conclusions

In this paper, multivariate probability distributions were considered that are representable as scale mixtures of multivariate stable distributions. Multivariate analogs of the Mittag–Leffler distribution were introduced. Some properties of these distributions were discussed. Attention was paid to the representations of the corresponding random vectors as products of independent random variables and vectors. In these products, relations were traced of the distributions of the involved terms with popular probability distributions. As examples of distributions of the class of scale mixtures of multivariate stable distributions, multivariate generalized Linnik distributions and multivariate generalized Mittag–Leffler distributions were considered in detail. Their relations with multivariate 'ordinary' Linnik distributions, multivariate normal, stable and Laplace laws as well as with univariate Mittag–Leffler and generalized Mittag–Leffler distributions were discussed. Limit theorems were proved presenting necessary and sufficient conditions for the convergence of the distributions of random sequences with independent random indices (including sums of a random number of random vectors and multivariate statistics constructed from samples with random sizes) to scale mixtures of multivariate elliptically contoured stable distributions. The property of scale-mixed multivariate elliptically contoured stable distributions to be both scale mixtures of a non-trivial multivariate stable distribution and a normal scale mixture was used to obtain necessary and sufficient conditions for the convergence of the distributions of random sums of random vectors with covariance matrices to the multivariate generalized Linnik distribution.

The key points of the paper are:

- analogs of the multiplication theorem for stable laws were proved for scale-mixed multivariate stable distributions relating these laws with different parameters;
- some alternative but equivalent definitions are proposed for the generalized multivariate Linnik distributions based on their property to be scale-mixed multivariate stable distributions;
- The multivariate analog of the (generalized) Mittag–Leffler distribution was introduced and it was noticed that the multivariate (generalized) Mittag–Leffler distribution can be regarded as a special case of the multivariate (generalized) Linnik distribution;
- new mixture representations were presented for the multivariate generalized Mittag–Leffler and Linnik distributions;
- a general transfer theorem was proved establishing necessary and sufficient conditions for the convergence of the distributions of sequences of multivariate random vectors with independent random indices (including sums of a random number of random vectors and multivariate statistics constructed from samples with random sizes) to multivariate elliptically contoured scale-mixed stable distributions;
- the property of scale-mixed multivariate elliptically contoured stable distributions to be both scale mixtures of a non-trivial multivariate stable distribution and a normal scale mixture was used to obtain necessary and sufficient conditions for the convergence of the distributions of random sums of random vectors to the multivariate elliptically contoured generalized Linnik distribution in covariance matrices.

**Author Contributions:** Conceptualization, Y.K., V.K. and A.Z.; methodology, Y.K. and V.K.; validation, Y.K. and V.K.; investigation, Y.K., V.K. and A.Z.; supervision, V.K.; project administration, V.K. and A.Z. All authors have read and agreed to the published version of the manuscript.

**Funding:** This research was supported by Russian Science Foundation, project 18-11-00155.

**Conflicts of Interest:** The authors declare no conflict of interest.

## References

1. Zolotarev, V.M. One-Dimensional Stable Distributions. In *Translation of Mathematical Monographs*; American Mathematical Society: Providence, RI, USA, 1986; Volume 65.
2. Korolev, V.Y.; Zeifman, A.I. Convergence of statistics constructed from samples with random sizes to the Linnik and Mittag–Leffler distributions and their generalizations. *J. Korean Stat. Soc.* **2017**, *46*, 161–181.
3. Korolev, V.Y.; Zeifman, A.I. A note on mixture representations for the Linnik and Mittag–Leffler distributions and their applications. *J. Math. Sci.* **2017**, *218*, 314–327. [CrossRef]
4. Korolev, V.Y.; Gorshenin, A.K.; Zeifman, A.I. New mixture representations of the generalized Mittag–Leffler distributions and their applications. *Inform. Appl.* **2018**, *12*, 75–85.
5. Korolev, V.Y.; Gorshenin, A.K.; Zeifman, A.I. On mixture representations for the generalized Linnik distribution and their applications in limit theorems. *J. Math. Sci.* **2020**, *246*, 503–518.
6. Anderson, D.N.; Arnold, B.C. Linnik distributions and processes. *J. Appl. Prob.* **1993**, *30*, 330–340. [CrossRef]
7. Devroye, L. A note on Linnik's distribution. *Stat. Probab. Lett.* **1990**, *9*, 305–306. [CrossRef]
8. Kotz, S.; Ostrovskii, I.V.; Hayfavi, A. Analytic and asymptotic properties of Linnik's probability densities, I. *J. Math. Anal. Appl.* **1995**, *193*, 353–371. [CrossRef]
9. Kotz, S.; Ostrovskii, I.V.; Hayfavi, A. Analytic and asymptotic properties of Linnik's probability densities, II. *J. Math. Anal. Appl.* **1995**, *193*, 497–521. [CrossRef]
10. Kotz, S.; Ostrovskii, I.V. A mixture representation of the Linnik distribution. *Stat. Probab. Lett.* **1996**, *26*, 61–64. [CrossRef]
11. Kotz, S.; Kozubowski, T.J.; Podgorski, K. *The Laplace Distribution and Generalizations: A Revisit with Applications to Communications, Economics, Engineering, and Finance*; Birkhauser: Boston, MA, USA, 2001.
12. Kozubowski, T.J. Mixture representation of Linnik distribution revisited. *Stat. Probab. Lett.* **1998**, *38*, 157–160. [CrossRef]
13. Kozubowski, T.J. Exponential mixture representation of geometric stable distributions. *Ann. Inst. Stat. Math.* **1999**, *52*, 231–238. [CrossRef]

14. Lin, G.D. A note on the Linnik distributions. *J. Math. Anal. Appl.* **1998**, *217*, 701–706. [CrossRef]
15. Lim, S.C.; Teo, L.P. Analytic and asymptotic properties of multivariate generalized Linnik's probability densities. *J. Fourier Anal. Appl.* **2010**, *16*, 715–747.
16. Ostrovskii, I.V. Analytic and asymptotic properties of multivariate Linniks distribution. *Math. Phys. Anal. Geom. [Mat. Fiz. Anal. Geom.]* **1995**, *2*, 436–455.
17. Pakes, A.G. Mixture representations for symmetric generalized Linnik laws. *Stat. Probab. Lett.* **1998**, *37*, 213–221. [CrossRef]
18. Pillai, R.N. Semi-$\alpha$-Laplace distributions. *Commun. Stat. Theory Methods* **1985**, *14*, 991–1000. [CrossRef]
19. Shiryaev, A.N. Foundations of Financial Mathematics. In *Facts, Models*; World Scientific: Singapore, 1998; Volume 1.
20. Devroye, L. Random variate generation in one line of code. In Proceedings of the 1996 Winter Simulation Conference, Coronado, CA, USA, 8–11 December 1996; Charnes, J.M., Morrice, D.J., Brunner, D.T., Swain, J.J., Eds.; IEEE Press: Piscataway, NJ, USA, 1996; pp. 265–272.
21. Bening, V.; Korolev, V. *Generalized Poisson Models and Their Applications in Insurance and Finance*; VSP: Utrecht, The Netherlands, 2002.
22. Meerschaert, M.M.; Scheffler, H.-P. Limit theorems for continuous-time random walks with very large mean waiting times. *J. Appl. Probab.* **2004**, *41*, 623–638. [CrossRef]
23. Fama, E. The behavior of stock market prices. *J. Bus.* **1965**, *38*, 34–105. [CrossRef]
24. Mandelbrot, B.B. The variation of certain speculative prices. *J. Bus.* **1963**, *36*, 394–419. [CrossRef]
25. Mandelbrot, B.B. The variation of some other speculative prices. *J. Bus.* **1967**, *40*, 393–413. [CrossRef]
26. McCulloch, J.H. Financial applications of stable distributions. In *Handbook of Statistics*; Maddala, G.S., Rao, C.R., Eds.; Elsevier Science: Amsterdam, The Netherlands, 1996; Volume 14, pp. 393–425.
27. Samorodnitsky, G.; Taqqu, M.S. Stable Non-Gaussian Random Processes. In *Stochastic Models with Infinite Variance*; Chapman and Hall: New York, NY, USA, 1994.
28. Korolev, V.Y.; Zeifman, A.I. From Asymptotic Normality to Heavy-Tailedness via Limit Theorems for Random Sums and Statistics with Random Sample Sizes. In *Probability, Combinatorics and Control*; Korolev, V., Kostogryzov, A., Eds.; IntechOpen: London, UK, 2019; pp. 1–23.
29. Korolev, V. On the convergence of distributions of random sums of independent random variables to stable laws. *Theory Probab. Appl.* **1998**, *42*, 695–696. [CrossRef]
30. Stacy, E.W. A generalization of the gamma distribution. *Ann. Math. Stat.* **1962**, *33*, 1187–1192. [CrossRef]
31. Gleser, L.J. The gamma distribution as a mixture of exponential distributions. *Am. Stat.* **1989**, *43*, 115–117.
32. Korolev, V.Y. Analogs of Gleser's theorem for negative binomial and generalized gamma distributions and some their applications. *Inform. Appl.* **2017**, *11*, 2–17.
33. Korolev, V.Y. Product representations for random variables with Weibull distributions and their applications. *J. Math. Sci.* **2016**, *218*, 298–313. [CrossRef]
34. Feldheim, M.E. Étude de la Stabilité des Lois de Probabilité. Ph.D. Thesis, Faculté des Sciences de Paris, Paris, France, 1937.
35. Lévy, P. *Théorie de Laddition des Variables Aléatoires*, 2nd ed.; Gauthier-Villars: Paris, France, 1937.
36. Nolan, J.P. Modeling financial data with stable distributions. In *Handbook of Heavy Tailed Distributions in Finance*; Rachev, S.T., Ed.; Elsevier: Boston, MA, USA, 2003; Chapter 3, pp. 105–130.
37. Nolan, J.P. Multivariate stable densities and distribution functions: General and elliptical case. In Proceedings of the Deutsche Bundesbank's 2005 Annual Autumn Conference, Eltville, Germany, 11 November 2005; pp. 1–20.
38. Press, S.J. Multivariate stable distributions. *J. Multivar. Anal.* **1972**, *2*, 444–462. [CrossRef]
39. Gorenflo, R.; Kilbas, A.A.; Mainardi, F.; Rogosin, S.V. *Mittag–Leffler Functions, Related Topics and Applications*; Springer: Berlin, Germany; New York, NY, USA, 2014.
40. Klebanov, L.B.; Rachev, S.T. Sums of a random number of random variables and their approximations with $\varepsilon$-accompanying very largely divisible laws. *Serdica* **1996**, *22*, 471–498.
41. Gorenflo, R.; Mainardi, F. Continuous time random walk, Mittag–Leffler waiting time and fractional diffusion: Mathematical aspects. In *Anomalous Transport: Foundations and Applications*; Klages, R., Radons, G., Sokolov, I.M., Eds.; Wiley-VCH: Weinheim, Germany, 2008; Chapter 4, pp. 93–127. Available online: http://arxiv.org/abs/0705.0797 (accessed on 6 May 2007).

42. Weron, K.; Kotulski, M. On the Cole-Cole relaxation function and related Mittag–Leffler distributions. *Physica A* **1996**, *232*, 180–188. [CrossRef]
43. Jose, K.K.; Uma, P.; Lekshmi, V.S.; Haubold, H.J. Generalized Mittag–Leffler Distributions and Processes for Applications in Astrophysics and Time Series Modeling. *Astrophys. Space Sci. Proc.* **2010**, *202559*, 79–92.
44. Mathai, A.M.; Haubold, H.J. Matrix-variate statistical distributions and fractional calculus. *Fract. Calc. Appl. Anal. Int. J. Theory Appl.* **2011**, *14*, 138–155. [CrossRef]
45. Mathai, A.M. Some properties of Mittag–Leffler functions and matrix-variate analogues: A statistical perspective. *Fract. Calc. Appl. Anal. Int. J. Theory Appl.* **2010**, *13*, 113–132.
46. Linnik, Y.V. Linear forms and statistical criteria, I, II. *Sel. Transl. Math. Stat. Probab.* **1963**, *3*, 41–90. (Original Paper Appeared in Ukrainskii Matematicheskii Zhournal **1953**, *5*, 207–243, 247–290).
47. Bunge, J. Compositions semigroups and random stability. *Ann. Probab.* **1996**, *24*, 1476–1489. [CrossRef]
48. Laha, R.G. On a class of unimodal distributions. *Proc. Am. Math. Soc.* **1961**, *12*, 181–184. [CrossRef]
49. Anderson, D.N. A multivariate Linnik distribution. *Stat. Probab. Lett.* **1992**, *14*, 333–336. [CrossRef]
50. Kozubowski, T.J.; Rachev, S.T. Multivariate geometric stable laws. *J. Comput. Anal. Appl.* **1999**, *1*, 349–385.
51. Baringhaus, L.; Grubel, R. On a class of characterization problems for random convex combinations. *Ann. Inst. Statist. Math.* **1997**, *49*, 555–567. [CrossRef]
52. Jayakumar, K.; Kalyanaraman, K.; Pillai, R.N. $\alpha$-Laplace processes. *Math. Comput. Model.* **1995**, *22*, 109–116. [CrossRef]
53. Mittnik, S.; Rachev, S.T. Modeling asset returns with alternative stable distributions. *Econom. Rev.* **1993**, *12*, 261–330. [CrossRef]
54. Gnedenko, B.V.; Korolev, V.Y. *Random Summation: Limit Theorems and Applications*; CRC Press: Boca Raton, FL, USA, 1996.
55. Kozubowski, T.J.; Podgórski, K.; Rychlik, I. Multivariate generalized Laplace distribution and related random fields. *J. Multivar. Anal.* **2013**, *113*, 59–72. [CrossRef]
56. Pakes, A.G. A characterization of gamma mixtures of stable laws motivated by limit theorems. *Stat. Neerl.* **1992**, *46*, 209–218. [CrossRef]
57. Korolev, V.Y.; Zeifman, A.I.; Korchagin, A.Y. Asymmetric Linnik distributions as limit laws for random sums of independent random variables with finite variances. *Inform. Primen.* **2016**. *10*, 21–33.
58. Korolev, V.Y.; Kossova, E.V. On limit distributions of randomly indexed multidimensional random sequences with an operator normalization. *J. Math. Sci.* **1992**, *72*, 2915–2929. [CrossRef]
59. Korolev, V.Y.; Kossova, E.V. Convergence of multidimensional random sequences with independent random indices. *J. Math. Sci.* **1995**, *76*, 2259–2268. [CrossRef]
60. Korolev, V.Y.; Gorshenin, A.K.; Gulev, S.K.; Belyaev, K.P.; Grusho, A.A. Statistical Analysis of Precipitation Events. *AIP Conf. Proc.* **2017**, *1863*, 090011.
61. Korolev, V.Y.; Gorshenin, A.K. The probability distribution of extreme precipitation. *Dokl. Earth Sci.* **2017**, *477*, 1461–1466. [CrossRef]
62. Korolev, V.Y.; Gorshenin, A.K. Probability models and statistical tests for extreme precipitation based on generalized negative binomial distributions. *Mathematics* **2020**, *8*, 604. [CrossRef]
63. Korolev, V.; Zeifman, A. On normal variancemean mixtures as limit laws for statistics with random sample sizes. *J. Stat. Plan. Inference* **2016**, *169*, 34–42. [CrossRef]
64. Korchagin, A.Y. On convergence of random sums of independent random vectors to multivariate generalized variance-gamma distributions. *Syst. Means. Inform.* **2015**, *25*, 127–141.
65. Bening, V.E.; Korolev, V.Y. On an application of the Student distribution in the theory of probability and mathematical statistics. *Theory Probab. Appl.* **2005**, *49*, 377–391. [CrossRef]
66. Korolev, V.Y. On the relationship between the generalized Student t-distribution and the variance gamma distribution in statistical analysis of random-size samples. *Dokl. Math.* **2012**, *86*, 566–570. [CrossRef]

mathematics MDPI

*Article*

# On Convergence Rates of Some Limits

**Edward Omey [1,*] and Meitner Cadena [2]**

1 Research Centre for Mathematics, Education, Econometrics and Statistics (MEES), Catholic University Leuven at Campus Brussels, Warmoesberg 26, 1000 Brussels, Belgium
2 Departamento de Ciencias Exactas, Universidad de las Fuerzas Armadas, Sangolqui 171103, Ecuador; meitner.cadena@gmail.com
* Correspondence: edward.omey@kuleuven.be

Received: 9 March 2020; Accepted: 17 April 2020; Published: 21 April 2020

**Abstract:** In 2019 Seneta has provided a characterization of slowly varying functions $L$ in the Zygmund sense by using the condition, for each $y > 0$, $x\left(\frac{L(x+y)}{L(x)} - 1\right) \to 0$ as $x \to \infty$. Very recently, we have extended this result by considering a wider class of functions $U$ related to the following more general condition. For each $y > 0$, $r(x)\left(\frac{U(x+yg(x))}{U(x)} - 1\right) \to 0$ as $x \to \infty$, for some functions $r$ and $g$. In this paper, we examine this last result by considering a much more general convergence condition. A wider class related to this new condition is presented. Further, a representation theorem for this wider class is provided.

**Keywords:** slowly varying; monotony in the Zygmund sense; class $\Gamma_a(g)$; self-neglecting function; convergence rates

---

## 1. Introduction

The notion of ultimately monotony introduced by Zygmund says that a function $U \geq 0$ is slowly varying if for each $\epsilon > 0$ the function $x^{\epsilon}U(x)$ is ultimately increasing and $x^{-\epsilon}U(x)$ is ultimately decreasing ([1], p. 186). A different kind of slowly varying functions was defined by Karamata [2] known as simply the class of slowly varying functions (KSV). It is known that any ZSV function is a KSV function (see [1], p. 186 and, e.g., [3], p. 49).

Recently, Seneta [4] found that the slowly varying functions $L$ in the sense of Zygmund are characterized by the relation:

$$\lim_{x\to\infty} x\left(\frac{L(x+y)}{L(x)} - 1\right) = 0, \forall y.$$

More recently, Omey and Cadena's [5] functions extended the results of Seneta, and they considered functions for which the following relation holds:

$$\lim_{x\to\infty} r(x)\left(\frac{L(x+yg(x))}{L(x)} - 1\right) = 0, \quad \forall y.$$

Here, the function $g(x)$ is self-neglecting (notation: $g \in SN$) and $r$ is in the class $\Gamma_0(g)$ with $r(x) \to \infty$. The class $\Gamma_0(g)$ is deeply studied in [6]. Recall that $g \in SN$ if it satisfies

$$\lim_{x\to\infty} \frac{g(x+yg(x))}{g(x)} = 1,$$

locally uniformly in $y$. In addition, recall that, for $g \in SN$, we have $f \in \Gamma_{\alpha}(g)$ if $f$ satisfies

$$\lim_{x\to\infty} \frac{f(x+yg(x))}{f(x)} = e^{\alpha y}, \forall y.$$

Now, we study more general relations of the form

$$\lim_{x\to\infty} r(x)\left(\frac{U(x+yg(x))}{U(x)} - e^{\alpha y}\right) = \theta(y), \quad \forall y,$$

where we assume that the convergence is l.u. in $y$. As before, we assume that $r \in \Gamma_0(g)$, $r(x) \to \infty$ and that $g \in SN$.

Throughout this paper, we use the notation $f(x) \sim g(x)$ for representing $f(x)/g(x) \to 1$ as $x \to \infty$.

We study in detail the two cases: $\alpha = 0$ and $\alpha \neq 0$. The case $\alpha = 0$ can be considered as the class $SN$ with a rate of convergence in the definition. This case is presented in the following section. The case where $\alpha \neq 0$ can be considered as the class $\Gamma_\alpha(g)$ with a rate of convergence in the definition. This case is presented in Section 3. For each case, characterizations of the involved functions are provided. Concluding remarks are presented in the last section.

## 2. The Case $\alpha = 0$

### 2.1. The Limit Function

Suppose that $U, g, r > 0$ are measurable functions and suppose that the following relation holds:

$$\lim_{x\to\infty} r(x)\left(\frac{U(x+yg(x))}{U(x)} - 1\right) = \theta(y), \tag{1}$$

and we assume that Equation (1) holds locally uniformly in $y$. As before, we assume that $r(x) \to \infty$, $r \in \Gamma_0(g)$ and that $g \in SN$.

Clearly, Equation (1) holds if and only if

$$\lim_{x\to\infty} r(x)(W(x+yg(x)) - W(x)) = \theta(y), \tag{2}$$

where $W(x) = \log U(x)$.

Now, we replace $x$ by $x = t + zg(t)$. Note that $g(t)/t \to 0$ so that $x/t \to 1$ l.u. in $z$. We find

$$\lim_{t\to\infty} r(t+zg(t))(W(t+zg(t)+yg(t+zg(t))) - W(t+zg(t))) = \theta(y).$$

Using $r \in \Gamma_0(g)$, we have

$$\lim_{t\to\infty} r(t)(W(t+zg(t)+yg(t+zg(t))) - W(t+zg(t))) = \theta(y),$$

and then it follows that

$$\lim_{t\to\infty} r(t)(W(t+zg(t)+yg(t+zg(t))) - W(t)) = \theta(y) + \theta(z).$$

Now, we have

$$W(t+zg(t)+yg(t+zg(t))) - W(t) = W\left(t + \left(z + y\frac{g(t+zg(t))}{g(t)}\right)g(t)\right) - W(t).$$

Using l.u. convergence, we obtain that

$$\lim_{t\to\infty} r(t)(W(t+zg(t)+yg(t+zg(t))) - W(t)) = \theta(y+z).$$

We conclude that

$$\theta(z+y) = \theta(z) + \theta(y),$$

and (since $\theta$ is measurable) hence also that $\theta(y) = \theta y$ for some constant $\theta$.

Conversely, we have the following (cf. [6]): if

$$\lim_{x\to\infty} r(x)\left(\frac{U(x+yg(x))}{U(x)} - 1\right) = \theta y,$$

then this relation holds l.u. in $y$.

To conclude, we have the following theorem.

**Theorem 1.** *Assume that $g \in SN$ and that $r \in \Gamma_0(g)$ with $r(x) \to \infty$.*

(a) *If Equation (1) or Equation (2) holds l.u. in $y$, then $\theta(x) = \theta x$ for some constant $\theta$.*
(b) *If Equation (1) or Equation (2) holds with $\theta(x) = \theta x$ for some constant $\theta$, then Equation (2) holds l.u. in $y$.*

### 2.2. *Representation*

Three different ways to represent the functions satisfying Equation (1) follow.

#### 2.2.1. First Form

For further use, let $A(x) = \int_a^x 1/g(t)dt$. Clearly, we have

$$A(x+yg(x)) - A(x) = \int_0^y \frac{g(x)}{g(x+zg(x))}dz \to y$$

l.u. in $y$. Note that $A(x)$ is an increasing function so that $f_x(y) = A(x+yg(x)) - A(x)$ is an increasing function of $y$ for which $f_x(y) \to y$ as $x \to \infty$. As a consequence, the inverse function also satisfies $f_x^{-1}(y) \to y$. To calculate the inverse, we set

$$f_x(y) = A(x+yg(x)) - A(x) = t$$

so that $x + yg(x) = A^{-1}(t + A(x))$ and

$$y = f_x^{-1}(t) = \frac{A^{-1}(t+A(x)) - A^{-1}(A(x))}{g(x)}.$$

We conclude that

$$\frac{A^{-1}(t+A(x)) - A^{-1}(A(x))}{g(x)} \to t,$$

so that (replacing $A(x)$ by $x$ and $t$ by $y$)

$$\frac{A^{-1}(x+y) - A^{-1}(x)}{g(A^{-1}(x))} \to y,$$

l.u. in $y$.

Now, let $K(x) := W(A^{-1}(x))$. We have (using l.u. convergence in the last step):

$$\begin{aligned} r(x)(K(A(x)+y) - K(A(x))) &= r(x)(W(A^{-1}(A(x)+y)) - W(x)) \\ &= r(x)\left(W\left(x + g(x)\frac{A^{-1}(A(x)+y) - x}{g(x)}\right) - W(x)\right) \\ &\to \theta y. \end{aligned}$$

It follows that

$$r(A^{-1}(x))(K(x+y) - K(x)) \to \theta y \tag{3}$$

l.u. in $y$. Taking the integral $\int_{y=0}^{1}(.)dy$ in Equation (3) we have

$$\int_{y=0}^{1} r(A^{-1}(x))(K(x+y)-K(x))dy \to \int_{0}^{1} \theta y dy$$

or

$$r(A^{-1}(x))\int_{x}^{x+1} K(z)dz - r(A^{-1}(x))K(x) \to \frac{\theta}{2}.$$

We see that $K(x)$ is of the form

$$\begin{aligned} K(x) &= C + \int_{x}^{x+1} K(z)dz + \frac{C(x)}{r(A^{-1}(x))} \\ &= L(x) + \frac{C(x)}{r(A^{-1}(x))}, \end{aligned}$$

where $C(x) \to C(=\theta/2)$ and $L(x) = \int_{x}^{x+1} K(z)dz$. Note that

$$r(A^{-1}(x)L'(x) = r(A^{-1}(x)(K(x+1)-K(x)) \to \theta.$$

Using $W(x) = K(A(x))$, we find that

$$W(x) = T(x) + \frac{C^{\circ}(x)}{r(x)},$$

where $C^{\circ}(x) = C(A(x)) \to C$ and $T(x) = L(A(x))$. Note that

$$r(x)g(x)T'(x) = r(x)L'(A(x))g(x)A'(x) = r(x)L'(A(x)) \to \theta.$$

We prove the following result:

**Theorem 2.** *Assume that $g \in SN$ and that $r \in \Gamma_0(g)$, $r(x) \to \infty$.*

*(a) If Equation (1) holds with $\theta(x) = \theta x$, then $W(x) = \log U(x)$ is of the form*

$$W(x) = T(x) + \frac{C(x)}{r(x)},$$

*where $C(x) \to C$ and $r(x)g(x)T'(x) \to \theta$.*

*(b) If $W(x) = T(x) + C(x)/r(x)$, where $C(x) \to 0$ and $r(x)g(x)T'(x) \to \theta$, then Equation (1) holds with $\theta(y) = \theta y$.*

**Proof.** The proof of (a) is given above. To prove (b), we have

$$\begin{aligned} &W(x+yg(x)) - W(x) \\ =\ & T(x+yg(x)) - T(x) + \frac{C(x+yg(x))}{r(x+yg(x))} - \frac{C(x)}{r(x)}. \end{aligned}$$

Clearly, we have

$$r(x)(T(x+yg(x)) - T(x)) = yr(x)g(x)T'(x+\beta g(x))$$

for some $\beta \in (0, y)$. It follows that

$$\begin{aligned} r(x)(T(x+yg(x)) - T(x)) &= y(\theta + o(1))\frac{r(x)g(x)}{r(x+\beta g(x))r(x+\beta g(x))} \\ &\to y\theta. \end{aligned}$$

For the second term, we have

$$r(x)\left(\frac{C(x+yg(x))}{r(x+yg(x))} - \frac{C(x)}{r(x)}\right) = \frac{r(x)}{r(x+yg(x))}C(x+yg(x)) - C(x) \to 0.$$

The result follows. □

**Remark 1.**

1. *In the special case where $g(x) = 1$, we have*

$$\lim_{x\to\infty} r(x)(W(x+y) - W(x)) = \theta y$$

*iff W is of the form $W(x) = C + T(x) + \epsilon(x)/r(x)$ where $\epsilon(x) \to 0$ and $r(x)T'(x) \to \theta$.*

2. *From Equation (1), it follows that*

$$\frac{r(x)}{U(x)}(U(x+yg(x)) - U(x)) \to \theta y.$$

*The previous representation result shows that*

$$U(x) = T(x) + C(x)\frac{U(x)}{r(x)}$$

*where $r(x)g(x)T'(x) \sim \theta U(x)$.*

3. *Using $U(x) = e^{W(x)}$, we also have that $U(x) = R(x)e^{C(x)/r(x)}$ where $R(x) = e^{T(x)}$. Note that*

$$r(x)g(x)\frac{R'(x)}{R(x)} = r(x)g(x)T'(x) \to \theta.$$

2.2.2. Second Form

In Equation (3), we find that $r(A^{-1}(x))(K(x+y) - K(x)) \to \theta y$, where $K(x) = W(A^{-1}(x))$. Using logarithms, we get that

$$\frac{K(\log xy) - K(x)}{L(x)} \to \theta \log y$$

where $L(x) = r(A^{-1}(\log x))$. From de Haan's theorem ([7], Theorem 3.7.3), we find that $K(\log x)$ can be written as

$$K(\log x) = C + \theta L_1(x) + \int_a^x \theta L_1(t)t^{-1}dt,$$

where $L_1(x) \sim L(x)$. It follows that

$$K(x) = C + \theta L_2(x) + \theta \int_{a^\circ}^x L_2(t)dt,$$

where $L_2(x) = L_1(\exp x) \sim r(A^{-1}(x))$.

2.2.3. Third Form

In [5], we found that relations of the form in Equation (1) hold with limit function $\theta(x) = 0$. In that case, we have

$$r(x)\left(\frac{U(x+yg(x))}{U(x)} - 1\right) \to 0.$$

As usual, we assume that $g \in SN$, $r \in \Gamma_0(g)$ and $r(x) \to \infty$. From Theorem 3 in [5], we get the following representation:

$$U(x) = \exp\left(c + \int_0^x f(t)dt\right)$$

where $f$ satisfies $r(x)g(x)f(x) \to 0$.

*2.3. Sufficient Conditions*

In the next result, we assume that the $k$th derivative of $U$ exists and we assume that

$$h_k(x) = g(x)\frac{U^{(k)}(x)}{U^{(k-1)}(x)} \to 0,$$

where $U^{(0)}(x) = U(x)$.

(a) If $k = 1$, we have $U'(x)/U(x) = \epsilon(x)/g(x)$ with $\epsilon(x) \to 0$ and

$$\int_x^{x+yg(x)} \frac{U'(z)}{U(z)}dz = \int_x^{x+yg(x)} \frac{\epsilon(z)}{g(z)}dz,$$

so that

$$\log\frac{U(x+yg(x))}{U(x)} = \int_0^y \frac{\epsilon(x+zg(x))g(x)}{g(x+zg(x))}dz \to 0,$$

and hence

$$\frac{U(x+yg(x))}{U(x)} \to 1.$$

(b) If $k = 2$, then we have

$$\frac{U''(x)}{U'(x)} = \frac{\epsilon(x)}{g(x)}$$

and

$$\int_x^{x+yg(x)} \frac{U''(z)}{U'(z)}dz = \int_x^{x+yg(x)} \frac{\epsilon(z)}{g(z)}dz,$$

so that

$$\log\frac{U'(x+yg(x))}{U'(x)} = \int_0^y \frac{\epsilon(x+zg(x))}{g(x+zg(x))}g(x)dz \to 0.$$

We find that

$$\frac{U'(x+yg(x))}{U'(x)} \to 1.$$

Now, consider

$$\begin{aligned} U(x+yg(x)) - U(x) &= \int_x^{x+yg(x)} U'(z)dz \\ &= g(x)\int_0^y U'(x+zg(x))dz \\ &= g(x)U'(x)\int_0^y \frac{U'(x+zg(x))}{U'(x)}dz \\ &\sim g(x)U'(x)y, \end{aligned}$$

and then

$$\frac{U(x+yg(x))}{U(x)} - 1 \sim h_1(x)y,$$

and thus Equation (1) holds with $r(x) = 1/h_1(x)$.

(c) If $k = 3$, as before, we have

$$\frac{U''(x+yg(x))}{U''(x)} \to 1$$

and

$$\begin{aligned} U'(x+yg(x)) - U'(x) &= g(x)U''(x)\int_0^y \frac{U''(x+zg(x))}{U''(x)}dz \\ &\sim g(x)U''(x)y. \end{aligned}$$

Further, we have

$$U(x+yg(x)) - U(x) = g(x)\int_0^y U'(x+zg(x))dz$$

and

$$\begin{aligned} &U(x+yg(x)) - U(x) - g(x)U'(x)y \\ &= g(x)\int_0^y (U'(x+zg(x)) - U'(x))dz \\ &\sim g^2(x)U''(x)\frac{y^2}{2}. \end{aligned}$$

We conclude that

$$\frac{U(x+yg(x))}{U(x)} - 1 - h_1(x)y \sim h_1(x)h_2(x)\frac{y^2}{2}.$$

(d) In general, we get a result of the type

$$\frac{U(x+yg(x))}{U(x)} - 1 - \sum_{i=1}^{k-1}\prod_{j=1}^{i} h_j(x)\frac{y^i}{i!} \sim \prod_{j=1}^{k} h_j(x)\frac{y^k}{k!}.$$

As a special case, we can take $g(x) = 1$: if $U'''(x)/U''(x) \to 0$, then

$$\frac{U(x+y)}{U(x)} - 1 - \frac{U'(x)}{U(x)}y \sim \frac{U''(x)}{U(x)}\frac{y^2}{2}.$$

## 2.4. More Results

**Proposition 1.** *Suppose that $\overline{F}(x) = x^{-\alpha}L(x)$ where $L(\cdot)$ is a normalized slowly varying (SV) function (that is, $xL'(x)/L(x) \to 0$). Assume that $g(x)$ and $r(x)$ satisfy $g(x)/x \to 0$ and $r(x)g(x)/x \to \delta > 0$. Then,*

$$r(x)\left(\frac{\overline{F}(x+yg(x))}{\overline{F}(x)} - 1\right) \to -\alpha\delta y.$$

**Proof.** We have $\overline{F}(x) = L(x)x^{-\alpha}$ and then

$$\frac{\overline{F}(x+yg(x))}{\overline{F}(x)} = \frac{L(x+yg(x))}{L(x)} \times \left(1+y\frac{g(x)}{x}\right)^{-\alpha}.$$

It follows that

$$\begin{aligned}\frac{\overline{F}(x+yg(x))}{\overline{F}(x)} - 1 &= \frac{L(x+yg(x))}{L(x)} \times \left(\left(1+y\frac{g(x)}{x}\right)^{-\alpha} - 1\right) + \frac{L(x+yg(x))}{L(x)} - 1 \\ &= I(a) + I(b).\end{aligned}$$

For $I(a)$, we have

$$\frac{L(x+yg(x))}{L(x)} = \frac{L\left(x\left(1+y\frac{g(x)}{x}\right)\right)}{L(x)} \to 1,$$

because $L$ is SV and $g(x)/x \to 0$. We also have

$$\left(1+y\frac{g(x)}{x}\right)^{-\alpha} - 1 \sim -\alpha y\frac{g(x)}{x},$$

so that

$$r(x)\left(\left(1+y\frac{g(x)}{x}\right)^{-\alpha} - 1\right) \sim -\alpha y\frac{r(x)g(x)}{x} \to -\alpha\delta y.$$

For the second term, we have

$$\begin{aligned}r(x)(\frac{L(x+yg(x))}{L(x)} - 1) &= \frac{r(x)}{L(x)}\int_x^{x+yg(x)} L'(t)dt \\ &= \frac{r(x)}{L(x)}\int_x^{x+yg(x)} \frac{tL'(t)}{L(t)}\frac{L(t)}{t}dt \\ &= o(1)\frac{r(x)g(x)}{L(x)}\int_0^y \frac{L(x+\theta g(x))}{x+\theta g(x)}d\theta \\ &= o(1)\frac{r(x)g(x)}{x}.\end{aligned}$$

We conclude that

$$r(x)\left(\frac{L(x+yg(x))}{L(x)} - 1\right) \to 0.$$

Combining these results, we obtain the desired result. □

**Remark 2.** *The condition on $L(x)$ in the previous theorem is equivalent to the requirement that*

$$\frac{xf(x)}{\overline{F}(x)} \to \alpha,$$

*where $f(x) = F'(x)$ is the density of $F$.*

### 2.5. Examples

#### 2.5.1. Example 1

Assume that $U(x) = \exp x^{\beta}$ with $\beta > 1$. We have

$$\frac{U'(x)}{U(x)} = \beta x^{\beta-1}$$

and

$$\frac{U''(x)}{U'(x)} = \beta x^{\beta-1} + (\beta-1)x^{-1}.$$

Using $g(x) = x^{-\gamma}$, we find

$$h_1(x) = g(x)\frac{U'(x)}{U(x)} = \beta x^{\beta-\gamma-1}$$

and

$$h_2(x) = g(x)\frac{U''(x)}{U'(x)} = \beta x^{\beta-\gamma-1} + (\beta-1)x^{-\gamma-1}.$$

If $0 < \beta - 1 < \gamma$, we find that $h_1(x) \to 0$ and $h_2(x) \to 0$. The results of this section show that

$$\frac{U(x+yg(x))}{U(x)} - 1 \sim h_1(x)y,$$

and Equation (1) holds with $r(x) = 1/h_1(x) \sim x^{\gamma+1-\beta}/\beta$.

#### 2.5.2. Example 2

Assume that $U(x) = \exp x^{-\beta}$ with $\beta > 0$. Clearly, we have

$$\begin{aligned}\frac{U'(x)}{U(x)} &= -\beta x^{-\beta-1}\\ \frac{U''(x)}{U'(x)} &= -\beta x^{-\beta-1} - (\beta+1)x^{-1}.\end{aligned}$$

We use $g(x) = x^{\gamma}$ and find

$$\begin{aligned}h_1(x) &= -\beta x^{\gamma-\beta-1}\\ h_2(x) &= h_1(x) - (\beta+1)x^{\gamma-1}.\end{aligned}$$

If $\gamma < \beta + 1$, we have $h_1(x) \to 0$. If $\gamma < 1$, we have $h_1(x) \to 0$ and $h_2(x) \to 0$. The results of the previous section show that

$$\frac{U(x+yg(x))}{U(x)} - 1 \sim h_1(x)y,$$

and Equation (1) holds with $r(x) = 1/h_1(x) \sim -x^{\beta+1-\gamma}/\beta$.

#### 2.5.3. Example 3

Assume that $U(x) = x^{\beta}$ where $\beta \neq 0$. We have

$$h_1(x) = g(x)\frac{U'(x)}{U(x)} = \beta\frac{g(x)}{x}$$

and

$$h_2(x) = g(x)\frac{U''(x)}{U'(x)} = (\beta - 1)\frac{g(x)}{x}.$$

Taking $g \in SN$ and $r(x) = x/g(x)$ ($\to \infty$) we find

$$r(x)\left(\frac{U(x+yg(x))}{U(x)} - 1\right) \to \beta y.$$

2.5.4. Example 4

Proposition 1 can be extended for some stable distributions. For instance, consider the density of an asymmetric stable distribution. The representation of such a stable density in the form of a convergent series is, for $0 < \alpha < 1$ and for any $x > 0$ (see, e.g., [8]),

$$q(x,\alpha,\rho) = \frac{1}{\pi}\sum_{n=1}^{\infty}\frac{(-1)^{n-1}\Gamma(\alpha n+1)}{n!}\sin(n\rho\pi)x^{-\alpha n-1}.$$

Additionally, assume $xq'(x,\alpha,\rho)/q(x,\alpha,\rho) \to \tau$ ($\neq 0$) as $x \to \infty$.
Let $g(x)$ and $r(x)$ be positive functions satisfying $g(x)/x \to 0$ and $r(x)g(x)/x \to \delta > 0$.
Note that, for each $n > 1$ and for $x$ large enough, we have, making use of $z - 1 \sim \log z$ as $z \to 1$,

$$\left(1 + y\frac{g(x)}{x}\right)^{-\alpha n-1} - 1 \sim -(\alpha n+1)\log\left(1 + y\frac{g(x)}{x}\right) \sim -(\alpha n+1)y\frac{g(x)}{x}.$$

Then, we have for $x$ large enough

$$\begin{aligned}
&\frac{q(x+yg(x),\alpha,\rho)}{q(x,\alpha,\rho)} - 1 \\
= \;& \frac{1}{\pi q(x,\alpha,\rho)}\sum_{n=1}^{\infty}\frac{(-1)^{n-1}\Gamma(\alpha n+1)}{n!}\sin(n\rho\pi)x^{-\alpha n-1}\left(\left(1+y\frac{g(x)}{x}\right)^{-\alpha n-1} - 1\right) \\
\sim \;& yg(x)\frac{-1}{\pi q(x,\alpha,\rho)}\sum_{n=1}^{\infty}(\alpha n+1)\frac{(-1)^{n-1}\Gamma(\alpha n+1)}{n!}\sin(n\rho\pi)x^{-\alpha n-2} \\
= \;& yg(x)\frac{q'(x,\alpha,\rho)}{q(x,\alpha,\rho)}.
\end{aligned}$$

Hence, we have

$$\lim_{x\to\infty} r(x)\left(\frac{q(x+yg(x),\alpha,\rho)}{q(x,\alpha,\rho)} - 1\right) = y\delta\tau.$$

## 3. The Case $\alpha \neq 0$

Now, suppose that $\alpha \neq 0$ and that

$$\lim_{x\to\infty} r(x)\left(\frac{U(x+yg(x))}{U(x)} - e^{\alpha y}\right) = \theta(y),$$

holds l.u. in $y$.

Equivalently, we have

$$\lim_{x\to\infty} r(x)\left(\frac{e^{-\alpha y}U(x+yg(x))}{U(x)} - 1\right) = e^{-\alpha y}\theta(y),$$

and then (using $\log z \sim z - 1$)

$$\lim_{x\to\infty} r(x)(W(x + yg(x)) - W(x) - \alpha y) = \Omega(y), \tag{4}$$

where $W(x) = \log U(x)$ and $\Omega(y) = e^{-\alpha y}\theta(y)$.

### 3.1. The Limit

In Equation (4), we replace $x$ by $x = t + zg(t)$ to find

$$\lim_{x\to\infty} r(x)\big(W(t + zg(t) + yg(t + zg(t))) - W(t + zg(t)) - \alpha y\big) = \Omega(y),$$

and

$$\begin{aligned}&\lim_{x\to\infty} r(x)\\ &\quad ((W(t + zg(t) + yg(t + zg(t))) - W(t) - \alpha(y + z)) - (W(t + zg(t)) - W(t) - \alpha z))\\ &\quad = \Omega(y).\end{aligned}$$

The second term converges to $\Omega(z)$ and thus we have

$$\begin{aligned}&\lim_{x\to\infty} r(x)\\ &\quad \left(W\left(t + \left(z + y\frac{g(t + zg(t))}{g(t)}\right)g(t)\right) - W(t) - \alpha(y + z)\right)\\ &\quad = \Omega(z) + \Omega(y)\end{aligned}$$

or

$$\begin{aligned} r(x)\Bigg(W\left(t + \left(z + y\frac{g(t + zg(t))}{g(t)}\right)g(t)\right) - W(t)\\ -\alpha\left(z + y\frac{g(t + zg(t))}{g(t)}\right)\Bigg) + \alpha y\left(\frac{g(t + zg(t))}{g(t)} - 1\right)\\ \to \quad \Omega(z) + \Omega(y).\end{aligned}$$

By l.u. convergence, the first part converges to $\Omega(z + y)$ and then we have

$$r(x)\alpha y\left(\frac{g(t + zg(t))}{g(t)} - 1\right) \to \Omega(z) + \Omega(y) - \Omega(y + z).$$

Using the result of the previous subsection, we find that

$$\alpha y\beta z = \Omega(z) + \Omega(y) - \Omega(y + z).$$

We propose a solution of the form $\Omega(z) = dx + cx^2$. The previous equation gives

$$\alpha\beta yz = cz^2 + cy^2 - c(y^2 + z^2 + 2yz),$$

and hence $\alpha\beta yz = 2cyz$ so that $c = \alpha\beta/2$. We conclude that $\Omega(x) = dx + \alpha\beta x^2/2$ and that $\theta(x) = (dx + \alpha\beta x^2/2)e^{\alpha x}$.

We conclude:

**Theorem 3.** *Suppose that $\alpha \neq 0$. If*

$$\lim_{x\to\infty} r(x)\left(\frac{U(x+yg(x))}{U(x)} - e^{\alpha y}\right) = \theta(y),$$

*holds l.u. in $y$, or equivalently if*

$$\lim_{x\to\infty} r(x)\big(W(x+yg(x)) - W(x) - \alpha y\big) = \Omega(y),$$

*holds l.u. in $y$, then $g(x)$ satisfies Equation (2), $\Omega(x) = dx + \alpha\beta x^2/2$ and $\theta(x) = (dx + \alpha\beta x^2/2)e^{\alpha x}$.*

### 3.2. Special Case

We assume that $W$ is differentiable and that $g(x)W'(x) \to \alpha$.

In this case, we have

$$\begin{aligned} W(x+yg(x)) - W(x) &= g(x)\int_0^y W'(x+zg(x))dz \\ &= \int_0^y \frac{g(x)}{g(x+zg(x))} g(x+zg(x))W'(x+zg(x))dz \\ &\to \alpha y. \end{aligned}$$

Now, suppose in addition that $r(x)\big(g(x)W'(x) - \alpha\big) \to \delta$ and that

$$r(t)\left(\frac{g(x+tg(x))}{g(x)} - 1\right) \to \beta t.$$

We have

$$\begin{aligned} &W(x+yg(x)) - W(x) - \alpha y \\ &= g(x)\int_0^y W'(x+tg(x))dt - \alpha y \\ &= \int_0^y g(x+tg(x))W'(x+tg(x))\frac{g(x)}{g(x+tg(x))}dt - \alpha y \\ &= \int_0^y g(x+tg(x))W'(x+tg(x))\left(\frac{g(x)}{g(x+tg(x))} - 1\right)dt \\ &+ \int_0^y (g(x+tg(x))W'(x+tg(x)) - \alpha)dt. \end{aligned}$$

For the first integral, by assumption, we have

$$r(t)\left(\frac{g(x+tg(x))}{g(x)} - 1\right) \to \beta t,$$

or

$$r(t)\frac{g(x+tg(x))}{g(x)}\left(1 - \frac{g(x)}{g(x+tg(x))}\right) \to \beta t,$$

or

$$r(t)\left(\frac{g(x)}{g(x+tg(x))} - 1\right) \to -\beta t.$$

Since $r(x)\big(g(x)W'(x)-\alpha\big) \to \delta$, we obtain

$$\begin{aligned} & r(x)\big(W(x+yg(x))-W(x)-\alpha y\big) \\ = & \int_0^y g(x+tg(x))W'(x+tg(x))r(x)\left(\frac{g(x)}{g(x+tg(x))}-1\right)dt \\ & +\int_0^y r(x)(g(x+tg(x))W'(x+tg(x))-\alpha)dt \\ \to & \ \alpha(-\beta)\frac{y^2}{2}+\delta y. \end{aligned}$$

### 3.3. Representation Theorem

Now, consider $Q(x)=W(x)+\alpha A(x)$, where $A(x)=\int_a^x 1/g(t)dt$ as before. We prove above that

$$A(x+yg(x))-A(x) \to y$$

l.u. in $y$. If $g \in SN$ satisfies

$$\lim_{x\to\infty} x\left(\frac{g(x+yg(x))}{g(x)}-1\right)=\beta y,$$

then we also have

$$\begin{aligned} A(x+yg(x))-A(x)-y &= \int_0^y\left(\frac{g(x)}{g(x+zg(x))}-1\right)dz \\ &= -\int_0^y \frac{g(x)}{g(x+zg(x))}\left(\frac{g(x+zg(x))}{g(x)}-1\right)dz, \end{aligned}$$

so that

$$r(x)(A(x+yg(x))-A(x)-y) \to -\beta\frac{y^2}{2}.$$

Using $Q(x)=W(x)-\alpha A(x)$, we see that

$$\begin{aligned} & Q(x+yg(x))-Q(x) \\ = & \ W(x+yg(x))-W(x)-\alpha y \\ & -\alpha\big(A(x+yg(x))-A(x)-y\big). \end{aligned}$$

Hence, using Equation (3),

$$\begin{aligned} & r(x)\big(Q(x+yg(x))-Q(x)\big) \\ = & \ r(x)\big(W(x+yg(x))-W(x)-\alpha y\big) \\ & -\alpha r(x)\big(A(x+yg(x))-A(x)-y\big) \\ \to & \ \Omega(y)+\alpha\beta\frac{y^2}{2}=\Psi(y) \end{aligned}$$

l.u. in $y$. As in the previous subsection, we conclude that

$$r(x)\big(Q(x+yg(x))-Q(x)\big) \to \Psi(y)=\lambda y$$

for some real number $\lambda$. The first representation of the previous subsection gives

$$Q(x)=T(x)+\frac{C(x)}{r(x)},$$

or

$$W(x) = \alpha A(x) + T(x) + \frac{C(x)}{r(x)},$$

where $C(x) \to C$ and $r(x)g(x)T'(x) \to \lambda$.

**Theorem 4.** *We have Equation (3) if and only if $W(x)$ is of the form*

$$W(x) = \alpha A(x) + T(x) + \frac{C(x)}{r(x)},$$

*where $C(x) \to C$ and $r(x)g(x)T'(x) \to \lambda$.*

### 3.4. More Results

In our next result, we consider the function $h(x) = f(x)/\overline{F}(x)$, where $f$ is the density of $F$. We make the following assumptions about $h$:

(a) $h \in SN$.
(b) $r(x)h'(x)/h^2(x) \to -\beta > 0$, where $r(x) \to \infty$, $r(x) \in \Gamma_0(g)$ with $g(x) = 1/h(x)$.
Recall that $r \in \Gamma_0(g)$ means that $r(x + yg(x))/r(x) \to 1$ as $x \to \infty$.

**Lemma 1.** *If (a) and (b) hold, then*

$$r(x)\left(\frac{h(x+yg(x))}{h(x)} - 1\right) \to -\beta y.$$

**Proof.** We have

$$h(x+yg(x)) - h(x) = g(x)\int_0^y h'(x+zg(x))dz.$$

Since $r(x)h'(x)/h^2(x) \to -\beta > 0$, we have that $h'(x) \in \Gamma_0(g)$ and, using $g(x) = 1/h(x)$, we obtain that

$$r(x)\left(\frac{h(x+yg(x))}{h(x)} - 1\right) = \frac{r(x)}{h^2(x)}\int_0^y h'(x+zg(x))dz \to -\beta y.$$

□

Now, we study the tail $\overline{F}(x)$.

**Lemma 2.** *If (a) and (b) hold, then*

$$r(x)\left(\log\frac{\overline{F}(x+yg(x))}{\overline{F}(x)} + y\right) \to \beta\frac{y^2}{2}.$$

**Proof.** Using $h(x) = f(x)/\overline{F}(x)$, we obtain that

$$\int_x^{x+yg(x)} h(z)dz = \int_x^{x+yg(x)} \frac{f(z)}{\overline{F}(z)}dz,$$

so that

$$g(x)\int_0^y h(x+zg(x))dz = -\log\frac{\overline{F}(x+yg(x))}{\overline{F}(x)}.$$

It follows that (recall $g(x) = 1/h(x)$)

$$r(x)\int_0^y \left(\frac{h(x+zg(x))}{h(x)} - 1\right)dz = -r(x)\left(\log\frac{\overline{F}(x+yg(x))}{\overline{F}(x)} + y\right),$$

and using Lemma 1, it follows that

$$r(x)\left(\log\frac{\overline{F}(x+yg(x))}{\overline{F}(x)}+y\right)\to\beta\frac{y^2}{2}.$$

This proves the result. □

Now, we arrive at the main result here.

**Theorem 5.** *If (a) and (b) hold, then*

$$r(x)\left(\frac{\overline{F}(x+yg(x))}{\overline{F}(x)}-e^{-y}\right)\to\beta\frac{y^2}{2}e^{-y}.$$

**Proof.** Using Lemma 2, we have

$$r(x)\log e^{y}\frac{\overline{F}(x+yg(x))}{\overline{F}(x)}\to\beta\frac{y^2}{2}.$$

Using $\log z\sim z-1$, it follows that

$$r(x)\left(e^{y}\frac{\overline{F}(x+yg(x))}{\overline{F}(x)}-1\right)\to\beta\frac{y^2}{2},$$

or

$$r(x)\left(\frac{\overline{F}(x+yg(x))}{\overline{F}(x)}-e^{-y}\right)\to\beta\frac{y^2}{2}e^{-y}.$$

□

The previous theorem can be useful in extreme value theory as follows.

We assume that (a) and (b) hold and that $F$ is strictly increasing. We define $a_n$ by the equality $n\overline{F}(a_n)=1$. It is clear that $a_n\uparrow\infty$. In the result of Theorem 5, we replace $x$ by $a_n$ to see that

$$r(a_n)(n\overline{F}(a_n+yg(a_n))-e^{-y})\to\beta\frac{y^2}{2}e^{-y}.$$

Now, we use $\log(z)+(1-z)=O(1)(1-z)^2$ and write

$$\begin{aligned} n\overline{F}(a_n+yg(a_n)) &= n\overline{F}(a_n+yg(a_n))+n\log F(a_n+yg(a_n))-n\log F(a_n+yg(a_n)) \\ &= O(1)n\overline{F}^2(a_n+yg(a_n))-\log F^n(a_n+yg(a_n)). \end{aligned}$$

Now, notice that

$$r(a_n)n\overline{F}^2(a_n+yg(a_n))=O(1)r(a_n)n\overline{F}^2(a_n)=O(1)\frac{r(a_n)}{n}.$$

If $r(a_n)/n\to 0$, we obtain that

$$r(a_n)\left(\log\overline{F}^n(a_n+yg(a_n))+e^{-y}\right)\to-\beta\frac{y^2}{2}e^{-y},$$

and hence also that

$$r(a_n)\log e^{\exp -y}\overline{F}^n(a_n+yg(a_n))\to-\beta\frac{y^2}{2}e^{-y},$$

and

$$r(a_n)\left(e^{\exp -y}\overline{F}^n(a_n+yg(a_n))-1\right)\to -\beta\frac{y^2}{2}e^{-y},$$

or

$$r(a_n)\left(\overline{F}^n(a_n+yg(a_n))-\exp -e^{-y}\right)\to -\beta\frac{y^2}{2}e^{-y}\exp -e^{-y}.$$

It means that, if $X_i$ are independent and identically distributed random variables with distribution function $F$, then

$$r(a_n)\left(P\left(\frac{M_n-a_n}{g(a_n)}\leq y\right)-\Lambda(y)\right)\to \Phi(y),$$

where $M_n=\max(X_1,X_2,...,X_n)$, $\Lambda(y)=\exp -e^{-y}$ and $\Phi(y)=-\beta\frac{y^2}{2}e^{-y}\exp -e^{-y}$.

It means that $F$ is in the max-domain of attraction of the double exponential and the convergence rate is determined by $r(a_n)$.

### 3.5. Examples

#### 3.5.1. Example 1

The following example is related to Theorem 5.

Let $U(x)=\exp -x^2$ for $x>0$. Using $g(x)=1/(2x)$, we have $U(x)\in\Gamma_{-1}(g)$. Now, we consider the difference

$$\frac{U(x+yg(x))}{U(x)}-e^{-y}.$$

We have

$$\begin{aligned}\frac{U(x+yg(x))}{U(x)}-e^{-y} &= e^{-y-y^2g^2(x)}-e^{-y}\\ &= e^{-y}(e^{-y^2g^2(x)}-1)\\ &\sim -e^{-y}y^2g^2(x)\end{aligned}$$

and

$$x^2\left(\frac{U(x+yg(x))}{U(x)}-e^{-y}\right)\to -\frac{1}{4}y^2e^{-y}.$$

#### 3.5.2. Example 2

Let $U(x)=\exp x^\beta, \beta>1$. We have $W(x)=\log U(x)=x^\beta$ and $W'(x)=\beta x^{\beta-1}$. Taking $g(x)=x^{1-\beta}$, we have

$$g(x)W'(x)=\beta.$$

As for $g(x)$, we have $g(x)/x\to 0$ and

$$\frac{g(x+yg(x))}{g(x)}-1=\left(1+y\frac{g(x)}{x}\right)^{1-\beta}-1\sim(1-\beta)y\frac{g(x)}{x}.$$

Taking $r(x)=x/g(x)=x^\beta$, we have

$$r(x)\left(\frac{g(x+yg(x))}{g(x)}-1\right)\to(1-\beta)y.$$

The result of Section 3.2 shows that

$$r(x)(W(x+yg(x))-W(x)-\beta y)\to\beta(\beta-1)\frac{y^2}{2},$$

and then

$$r(x)\left(\frac{U(x+yg(x))}{U(x)}-e^{\beta y}\right)\to\beta(\beta-1)\frac{y^2}{2}e^{\beta y}.$$

## 4. Concluding Remarks

In this paper, new results on the condition, for some functions $r$ and $g$,

$$\lim_{x\to\infty} r(x)\left(\frac{U(x+yg(x))}{U(x)}-e^{\alpha y}\right)=\theta(y),\quad \forall y,$$

where we assume that the convergence is l.u. in $y$, are presented. This limit generalizes the ones analyzed by Seneta [4] and Omey and Cadena [5], both of them being related to the monotony of functions in the Zygmund sense. Under this analysis, properties of $\theta(y)$ are described. Representations of the functions $U$ involved in this limit are provided.

**Author Contributions:** The authors have equally contributed to the writing, editing and style of the paper. All authors have read and agreed to the published version of the manuscript.

**Funding:** This research received no external funding.

**Conflicts of Interest:** The authors declare no conflict of interest.

## References


1. Zygmund, A. *Trigonometric Series*, 2nd ed.; Cambridge University Press: Cambridge, UK, 1959; Volume 1.
2. Karamata, J. Sur un mode de croissance régulière des fonctions. *Mathematica* **1930**, *4*, 38–53.
3. Seneta, E. *Regularly Varying Functions, Lecture Notes in Mathematics*; Springer: Berlin/Heidelberg, Germany, 1976; Volume 508.
4. Seneta, E. Slowly varying functions in the Zygmund sense and generalized regular variation. *J. Math. Anal. Appl.* **2019**, *475*, 1647–1657. [CrossRef]
5. Omey, E.; Cadena, M. New results on slowly varying functions in the Zygmund sense. *Proc. Jap. Acad. Ser. A* **2019**, accepted.
6. Omey, E. On the class gamma and related classes of functions. *Publ. Inst. Math.* **2013**, *93*, 1–18. [CrossRef]
7. Bingham, N.H.; Goldie, C.M.; Teugels, J.L. *Regular Variation*; Cambridge University Press: Cambridge, UK, 1987.
8. Uchaikin, V.V.; Zolotarev, V.M. *Chance and Stability: Stable Distributions and Their Applications*; De Gruyter: Utrecht, The Netherlands, 1999.

mathematics

MDPI

Article

# Optimal Filtering of Markov Jump Processes Given Observations with State-Dependent Noises: Exact Solution and Stable Numerical Schemes

**Andrey Borisov [1,*] and Igor Sokolov [2]**

1 Institute of Informatics Problems of Federal Research Center "Computer Science and Control" RAS, 44/2 Vavilova str., 119333 Moscow, Russia
2 Faculty of Computational Mathematics and Cybernetics, Lomonosov Moscow State University, GSP-1, 1-52 Leninskiye Gory, 119991 Moscow, Russia; ISokolov@cs.msu.ru
* Correspondence: ABorisov@frccsc.ru

Received: 14 March 2020; Accepted: 30 March 2020; Published: 2 April 2020

**Abstract:** The paper is devoted to the optimal state filtering of the finite-state Markov jump processes, given indirect continuous-time observations corrupted by Wiener noise. The crucial feature is that the observation noise intensity is a function of the estimated state, which breaks forthright filtering approaches based on the passage to the innovation process and Girsanov's measure change. We propose an equivalent observation transform, which allows usage of the classical nonlinear filtering framework. We obtain the optimal estimate as a solution to the discrete–continuous stochastic differential system with both continuous and counting processes on the right-hand side. For effective computer realization, we present a new class of numerical algorithms based on the exact solution to the optimal filtering given the time-discretized observation. The proposed estimate approximations are stable, i.e., have non-negative components and satisfy the normalization condition. We prove the assertions characterizing the approximation accuracy depending on the observation system parameters, time discretization step, the maximal number of allowed state transitions, and the applied scheme of numerical integration.

**Keywords:** stochastic differential observation system; nonlinear filtering problem; state-dependent observation noise; numerical filtering algorithm; filtering given time-discretized observations; stable approximation; approximation accuracy

## 1. Introduction

The Wonham filter [1], as well as the Kalman–Bucy filter [2], is one of the most practically used filtering algorithms for the states of the stochastic differential observation systems. It is applied extensively for signal processing in technics, communications, finance and economy, biology, medicine, etc. [3–6]. The filter provides the optimal in *the Mean Square* (MS) sense on-line estimate of the finite-state *Markov Jump Process*. (MJP) given indirect continuous-time observations, corrupted by the Wiener noise. The elegant algorithm represents the desired estimate as a solution to *a Stochastic Differential System* (SDS) with continuous random processes on *the Right-Hand Side* (RHS).

The fundamental condition for the solution to the filtering problem is the independence of the observation noise intensity of the estimated state. It provides the continuity from the right for the natural flow of $\sigma$-algebras induced by the observations, with subsequent utilization of the innovation process framework. The condition violation breaks these advantages. In the case of the state-dependent observation noise, the author of [7] presents the optimal estimate within the class of the linear estimates. Further, the authors of [8,9] use filters of a linear structure for the solution to the $\mathcal{H}_2$-optimal state filtering problem. To find the absolute optimal filtering estimate, one has to make extra efforts. First, for proper utilization of the stochastic analysis framework, one needs to reformulate the optimal filtering problem, "smoothing forward" the flow of $\sigma$-algebras induced by the observations. Second, in the case of state-dependent noise, the innovation process contains less information than the original observations. One has to supplement the innovation by the observation quadratic characteristic, which represents a continuous-time noiseless function of the estimated MJP state. In general, the optimal filtering given partially noiseless observations is a challenging problem. Its solution can be expressed either as a sequence of some regularized estimates [10] or by the additional differentiation of the smooth observation components or their quadratic characteristics [11–14]. In both cases, one needs to realize a limit passage, which is difficult in computers.

Even in the traditional settings, the numerical realization of the MJP state filtering is a complicated problem. For example, the explicit numerical methods based on the Itô–Taylor expansion applied to the Wonham filter equation, diverge: the produced approximations do not meet component-wise non-negativity condition. Over time the approximation components reach arbitrary large absolute values. Further, in the presentation, we refer to the approximations, preserving both the component non-negativity and normalization condition as *the stable ones*.

The Wonham filtering equation is a particular case of the nonlinear Kushner–Stratonovich equation. To solve it, one can use various numerical algorithms

- the procedures based on the weak approximation of the original processes by Markov chains [15,16],
- some variants of the splitting methods [17],
- the robust procedures based on the Clark transform [18,19],
- the schemes, which represent the conditional probability distributions through the logarithm [20], etc.

All the algorithms are developed for the case of additive observation noise and based on the Girsanov's measure transform. Hence, they are useless for the estimation of the MJP given the observations with state-dependent noise.

The goal of the paper is two-fold. First, it presents a theoretical solution to the MS-optimal filtering problem, given the observations with state-dependent noise. Second, it introduces a new class of stable numerical algorithms for filter realization and investigates its accuracy. We organize the paper as follows. Section 2 contains a description of the studying observation system with state-dependent observation noise along with the MS-optimal filtering problem statement. To solve the problem, one needs to transform the available observations both to preserve the information equivalence and suit for application of the known results of the optimal nonlinear filtering. Section 3 describes both the observation transformation and the SDS defining the optimal filtering estimate. The SDS is discrete–continuous and contains both continuous and counting random processes on the RHS. Previously, the author of the note [21] presents a sketch of the observation transform, but it cannot guarantee the uniqueness of that SDS solution.

Section 4 presents a new class of the stable numerical algorithms of the nonlinear filtering. The main idea is to discretize original continuous-time observations and then find the MS-optimal filtering estimate given the sampled observations. The authors of [22] use this idea to solve a particular case of the estimation problem, namely the classification problem of a finite-state random vector given continuous-time observations with multiplicative noise. Section 4.1 contains a general solution to the problem. The corresponding estimate represents a ratio, which numerator and denominator are the infinite sums of integrals. They are shift-scale mixtures of the Gaussians. The mixing distributions, in turn, describe the occupation time of the system state in each admissible value during the time discretization interval. In Section 4.2, we suggest approximating the estimates by a convergent sequence bounding number $s$ of possible state transitions, which occurred over the discretization interval. We replace the infinite sums in the formula of the optimal estimate by their finite analogs and also investigate the accuracy of the approximations. We refer these approximations as *the analytical ones of the s-th order*. One cannot calculate the integrals analytically and have to replace them with some integral sums, and this brings an extra error. Section 4.3 analyzes the value of this error and the total distance between the optimal filtering estimate given the discretized observations and its numerical realization. Section 4.4 presents a numerical example that illustrates the conformity of theoretical estimates and their numerical realization. Section 5 contains discussion and concluding remarks.

## 2. Continuous-Time Filtering Problem Statement

On the probability triplet with filtration $(\Omega, \mathcal{F}, \mathcal{P}, \{\mathcal{F}\}_{t\geqslant 0})$ we consider the observation system

$$X_t = X_0 + \int_0^t \Lambda^\top(s) X_s ds + M_t^X, \tag{1}$$

$$Y_t = \int_0^t f(s) X_s ds + \int_0^t \sum_{n=1}^N X_s^n G_n^{1/2}(s) dW_s. \tag{2}$$

Here

- $X_t = \mathrm{col}(X_t^1, \ldots, X_t^N) \in \mathbb{S}^N$ is an unobservable state which is a finite-state *Markov jump process* (MJP) with the state space $\mathbb{S}^N \triangleq \{e_1, \ldots, e_N\}$ ($\mathbb{S}^N$ stands for the set of all unit coordinate vectors of the Euclidean space $\mathbb{R}^N$) with the transition matrix $\Lambda(t)$ and the initial distribution $\pi = \mathrm{col}(\pi^1, \ldots, \pi^N)$; the process $M_t^X$ is an $\mathcal{F}_t$-adapted martingale,
- $Y_t = \mathrm{col}(Y_t^1, \ldots, Y_t^M) \in \mathbb{R}^M$ is an observation process: $W_t = \mathrm{col}(W_t^1, \ldots, W_t^M) \in \mathbb{R}^M$ is an $\mathcal{F}_t$-adapted standard Wiener process characterizing the observation noise, $f(t)$ is an $M \times N$-dimensional observation matrix and the collection of $M \times M$-dimensional matrices $\{G_n(t)\}_{n=\overline{1,N}}$ defines the conditional observation noise intensities given $X_t = e_n$.

The natural flow of $\sigma$-algebras generated by the observations $Y$ up to the moment $t$ is denoted by $\mathcal{Y}_t \triangleq \sigma\{Y_s : s \in [0,t]\}$, $\mathcal{Y}_0 \triangleq \{\emptyset, \Omega\}$.

*The optimal state filtering given the observations* $Y$ is to find *the Conditional Mathematical Expectation* (CME)

$$\widehat{X}_t \triangleq \mathbf{E}\{X_t | \mathcal{Y}_{t+}\}. \tag{3}$$

## 3. Observation Transform and Optimal Filtering Equation

Before derivation of the optimal filtering equation we specify the properties of the observation system (1) and (2).

1. All trajectories of $\{X_t\}_{t\geqslant 0}$ are continuous from the left and have finite limits from the right, i.e., are *cádlág-processes*.
2. Nonrandom matrix-valued functions $\Lambda(t)$, $f(t)$ and $\{G_n(t)\}_{n=\overline{1,N}}$ consist of the *cádlág*-components.
3. The noises in $Y$ are uniformly nondegenerate [10], i.e., $\min_{\substack{1\leqslant n\leqslant N,\\ t\geqslant 0}} G_n(t) > \alpha I$ for some $\alpha > 0$; here and after, $I$ is a unit matrix of appropriate dimensionality.
4. The processes
$$K_{ij}(t) \triangleq \mathbf{I}_{\{\mathbf{0}\}}(G_i(t) - G_j(t)), \quad i,j = \overline{1,N} \tag{4}$$
have a finite variation; here and after, $\mathbf{I}_{\mathcal{A}}(x)$ is an indicator function of the set $\mathcal{A}$, and $\mathbf{0}$ is a zero matrix of appropriate dimensionality.

Conditions 1–3 are standard for the filtering problems [10]. They guarantee the proper description of MJP distribution $\pi(t) \triangleq \mathbf{E}\{X_t\}$ by the Kolmogorov system $\pi(t) = \pi + \int_0^t \Lambda^\top(s)\pi(s)ds$. Condition 4 relates to the quadratic characteristic of the observation process as a key information source itself. Below we show that collection of $G_n(\cdot)$, distinguished for different $n$, allows to restore the state $X_t$ *precisely* given the available *noisy* observations. Condition 4 guarantees the local regularity of the time subsets, where $G_n(\cdot)$ coincide and/or differ each other: one can express them as finite unions of the intervals. The condition is not too restrictive: for instance, they are valid when $G_n(\cdot)$ are piece-wise continuous with bounded derivatives.

Both the system state and observation are special square-integrable semimartingales [6,23] with the predictable characteristics

$$\langle X, X\rangle_t \triangleq X_t X_t^\top - \int_0^t X_{s-} dX_s^\top - \int_0^t dX_s X_{s-}^\top =$$
$$= \int_0^t \left(\operatorname{diag}\left(\Lambda^\top(s) X_s\right) - \Lambda^\top(s) \operatorname{diag} X_s - \operatorname{diag}(X_s)\Lambda(s)\right) ds \quad (5)$$

and

$$\langle Y, Y\rangle_t \triangleq Y_t Y_t^\top - \int_0^t Y_{s-} dY_s^\top - \int_0^t dY_s Y_{s-}^\top = \sum_{n=1}^N \int_0^t X_s^n G_n(s) ds. \quad (6)$$

Conditions 1–3 and the properties of $X_t$ guarantee $\mathcal{P}$-a.s. fulfilment of the following equalities for the one-sided derivatives of $\langle Y, Y\rangle_t$:

$$\begin{aligned} \tfrac{d\langle Y,Y\rangle_s}{ds}\Big|_{s=t-} &= \textstyle\sum_{n=1}^N X_{t-}^n G_n(t-) = \sum_{n=1}^N X_t^n G_n(t-), \\ \tfrac{d\langle Y,Y\rangle_s}{ds}\Big|_{s=t+} &= \textstyle\sum_{n=1}^N X_{t-}^n \left(G_n(t-) + \Delta G_n(t)\right) = \sum_{n=1}^N X_t^n G_n(t), \end{aligned} \quad (7)$$

where $\Delta G_n(t) \triangleq G_n(t) - G_n(t-)$ is a jump function of $G_n(t)$. So, if there exists a nonrandom instant $t^* > 0$ such that $\sum_{n=1}^N \pi^n(t^*)\Delta G_n(t^*) \neq 0$, then $\mathcal{Y}_{t^*} \subset \mathcal{Y}_{t^*+} = \mathcal{Y}_{t^*} \vee \sigma\{\sum_{n=1}^N X_{t^*}^n \Delta G_n(t^*)\}$. The inclusion presumes the flow of $\sigma$-subalgebras $\{\mathcal{Y}_t\}_{t\geqslant 0}$ is not necessarily continuous from the right for the considered observations [24]. This is a reason to define a filtering estimate as a CME of $X_t$ with respect to the "smoothed" flow $\mathcal{Y}_{t+}$ for subsequent correct usage of the stochastic analysis framework.

Let us transform the available observations in such a way to derive the optimal filtering estimate by the standard methods [6,23]. Initially, the idea of this transform is suggested in [11]. As the result, the authors introduce the pair

$$U_t \triangleq \int_0^t \left(\tfrac{d\langle Y,Y\rangle_u}{du}\Big|_{u=s+}\right)^{-1/2} dY_s, \quad (8)$$

$$\langle Y, Y\rangle_t = \sum_{n=1}^N \int_0^t X_s^n G_n(s) ds. \quad (9)$$

The authors of [11] prove coincidence of the $\sigma$-algebras $\mathcal{Y}_t = \sigma\{U_s,\ 0 \leqslant s \leqslant t\} \vee \sigma\{\langle Y, Y\rangle_s,\ 0 \leqslant s \leqslant t\}$ for the general diffusion observation systems. However, they do not pay attention to the continuity of $\{\mathcal{Y}_t\}$ from the right. The authors of [12,14] suggest to replace the observations $\langle Y, Y\rangle_t$ by their derivative

$$Q(t) \triangleq \tfrac{d\langle Y,Y\rangle_s}{ds}\Big|_{s=t-} = \sum_{n=1}^N X_{t-}^n G_n(t-). \quad (10)$$

Then, one can construct the optimal estimate either to use $Q_t$ as a linear constraint or to differentiate (10) for extraction of the dynamic noises. The papers [12,14] contain a rather pessimistic conclusion: the number of differentiations is unbounded in the general case of diffusion observation system. In contrast, we estimate a finite-state MJP and can construct the optimal filtering estimate using $Q$ without additional differentiation.

So, the transformed observations will contain

- diffusion processes with the unit diffusion,
- counting stochastic processes,
- indirect state observations obtained at the nonrandom discrete moments.

The first transformed observation part is the process $U_t$ (8), and in view of (2) and (7) it can be rewritten as

$$U_t = \int_0^t \overline{f}(s)X_s ds + \overline{W}_t, \qquad (11)$$

where $\overline{f}(s) \triangleq \sum_{n=1}^N G_n^{-1/2}(s)f(s)\,\mathrm{diag}(e_n)$ and $\overline{W}_t$ is an $\mathcal{F}_t$-adapted standard Wiener [10].

The process $Q_t$ could play the role of the second part of the transformed observations since $\mathcal{Y}_t = \sigma\{U_s, Q_s,\ s \in [0,t]\}$ [11], however the natural flow of $\sigma$-algebras generated by the couple $(U,Q)$ is not continuous from the right yet. Moreover, the process $Q_t$ is matrix-valued and looks overabundant for the filter derivation. The point is, $Q_t = Q(t, X_{t-})$ (10) is a function of the finite-set argument $X_t$, and it affects the estimate performance through its complete preimage

$$Q_t = Q(t, X_{t-}) \xrightarrow{Q^{-1}} \{e_n \in \mathbb{S}^N :\ G_n(t-)e_n = Q_t\}.$$

To go to the preimage we introduce the following transformation of $Q_t$:

$$H_t \triangleq \sum_{n=1}^N \mathbf{I}_{\{0\}}\left(Q_t - G_n(t)\right)e_n.$$

$H_t$ is a $\mathcal{Y}_t$-adapted vector process with components 0 or 1, but the trajectories $H_t$ are not *cádlág* processes. Due to the fact $X_{t-} = X_t$ $\mathcal{P}$-a.s. for $\forall\, t \geq 0$ the equalities below are valid

$$H_t = \sum_{n,k=1}^N \mathbf{I}_{\{0\}}\left(G_k(t) - G_n(t)\right)X_t^k e_n = K(t)X_t = K(t)X_{t-} \quad \mathcal{P}-\text{a.s.}, \qquad (12)$$

where $K(t)$ is the $N \times N$-dimensional matrix with the components (4).

The function $K(t)$ has the following properties.

1. $K(t) \equiv K^\top(t)$ for any $t \geqslant 0$.
2. The number of $K(\cdot)$ jumps occurred in any finite time interval is finite due to condition 4.
3. $K(t)$ is not a *cádlág*-function [25].
4. $\mathcal{P}\left\{\|\Delta K(t)\|\|\Delta X_t\| > 0\right\} = 0$ for any $t \geqslant 0$.
5. For any $t \geqslant 0$ there exists a transformation $T(t)$ such that the matrix $T(t)K(t)$ is trapezoid with orthogonal strings and 0 and 1 as the components.
6. $\mathcal{P}\left\{T(t)H_t \in \mathbb{S}^N\right\} = 1$ for any $t \geqslant 0$.

Let us define a $\mathcal{Y}_{t+}$-adapted process $V_t = \mathrm{col}(V_t^1, \ldots, V_t^N)$ with the *cádlág*-trajectories:

$$V_t \triangleq T(t+)H_{t+}. \qquad (13)$$

From (12) and (13) it follows that $V_t = J(t)X_t$ $\mathcal{P}$-a.s., where $J(t) \triangleq T(t+)K(t+)$.

We denote the set of the process $V$ discontinuity by $\mathcal{V}$, $\mathcal{X}$ stands for the set of $X$ discontinuity and $\mathcal{J}$ for the analogous set of the process $J$. The sets $\mathcal{V}$ and $\mathcal{X}$ are random, in contrast $\mathcal{J}$ is nonrandom. The process $V_t$ is purely discontinuous, and due to property 4 it can be rewritten in the form

$$V_t = J(0)X_0 + \sum_{\kappa \in \mathcal{V}:\ \kappa \leqslant t} \Delta V_\kappa = J(0)X_0 + \sum_{\kappa \in \mathcal{J}:\ \kappa \leqslant t} \Delta J(\kappa)X_\kappa + \sum_{\kappa \in \mathcal{V}\setminus\mathcal{J}:\ \kappa \leqslant t} J(\kappa)\Delta X_\kappa =$$

$$= J(0)X_0 + \sum_{\kappa \in \mathcal{J}:\ \kappa \leqslant t} \Delta J(\kappa)X_\kappa + \sum_{\kappa \in \mathcal{X}:\ \kappa \leqslant t} J(\kappa)\Delta X_\kappa = \underbrace{J(0)X_0 + \sum_{\kappa \in \mathcal{J}:\ \kappa \leqslant t} \Delta J(\kappa)X_\kappa}_{\triangleq D_t} + \underbrace{\int_0^t J(s)dX_s}_{\triangleq R_t}. \qquad (14)$$

Due to the definition $V_t \in \mathbb{S}^N$ for $\forall\, t \geqslant 0$. The process $D_t$ characterizes the observable jumps at the nonrandom moments caused by $J(t)$ changes, and $R_t$ is an observable part of the state $X_t$ jumps, occurred, at some random instants.

As a second part of the transformed observations, we choose the $N$-dimensional random process $C_t \triangleq \text{col}(C_t^1, \dots, C_t^N)$: the components $C_t^n$ count the jumps of the process $V_t$ into the state $e_n$, occurred *at the random instants* over the interval $[0, t]$:

$$C_t^n = \int_0^t (1 - e_n^\top V_{s-}) e_n^\top dR_s. \tag{15}$$

The third part of the transformed observations is the $N$-dimensional process $D_t$ with the jumps at the nonrandom moments.

**Lemma 1.** *If $\overline{\mathcal{Y}}_t \triangleq \sigma\{(U_s, C_s, D_s),\ s \in [0, t]\}$, then the coincidence $\overline{\mathcal{Y}}_t = \mathcal{Y}_{t+}$ is true for any $t \geqslant 0$.*

Correctness of the Lemma assertion follows immediately from the fact the composite process $(U_t, C_t, D_t)$ is constructed to be $\mathcal{Y}_{t+}$-adapted, and one-to-one correspondence of the $(U, C, D)$ and $Y$ paths:

$$\begin{cases} U_t = \int_0^t \left( \frac{d\langle Y,Y \rangle_u}{du} |_{u=s+} \right)^{-1/2} dY_s, \\ C_t = \int_0^t (I - \text{diag}\, V_{s-}) dV_s - \sum\limits_{\kappa \in \mathcal{J}:\, \kappa \leqslant t} (I - \text{diag}\, V_{\kappa-}) \Delta V_\kappa, \\ D_t = \sum\limits_{\kappa \in \mathcal{J}:\, \kappa \leqslant t} (I - \text{diag}\, V_{\kappa-}) \Delta V_\kappa, \\ V_t = T(t+) H_{t+}, \\ H_t \triangleq \sum\limits_{n=1}^N \mathbf{I}_{\{0\}} \left( \frac{d\langle Y,Y \rangle_s}{ds} |_{s=t-} - G_n(t) \right) e_n, \end{cases} \tag{16}$$

$$\begin{cases} V_t = D_t + \int_0^t \sum\limits_{(i,j):\, i \neq j}^N V_{s-}^i (e_j - e_i) dC_s^j. \\ Y_t = \int_0^t \sum\limits_{n=1}^N V_s^n G_n^{1/2}(s) dU_s, \end{cases} \tag{17}$$

Below we use the following notations: $\mathbf{1}$ is a row vector of the appropriate dimensionality formed by units, $J_n(s) \triangleq e_n^\top J(s)$ is the $n$-th row of the matrix $J(s)$,

$$\Gamma_n(s) \triangleq \text{diag}(J_n(s)) \Lambda^\top(s) (I - \text{diag}\, J_n(s)). \tag{18}$$

**Lemma 2.** *The process $C_t = \text{col}(C_t^1, \dots, C_t^N)$ has the following properties.*

1. *$n$-th component $C_t^n$ allows the martingale representation*

$$C_t^n = \int_0^t \mathbf{1}\Gamma_n(s) X_s ds + \int_0^t (1 - J_n(s) X_{s-}) J_n(s) dM_s^X.$$

2. $[C^n, C^m]_t \equiv 0$ *for any* $n \neq m$;

$$\langle C^n, C^n \rangle_t = \int_0^t \mathbf{1}\Gamma_n(s) X_s ds. \tag{19}$$

3. *The innovation processes*

$$\nu_t^n \triangleq \int_0^t \left( dC_s^n - \mathbf{1}\Gamma_n(s) \widehat{X}_s ds \right), \quad n = \overline{1, N} \tag{20}$$

*are $\overline{\mathcal{Y}}_t$-adapted martingales with the quadratic characteristics*

$$\langle \nu^n, \nu^n \rangle_t = \int_0^t \mathbf{1}\Gamma_n(s) \widehat{X}_s ds. \tag{21}$$

Proof of Lemma 2 is given in Appendix A.

Finally, the transformed observations $(U, C, D)$ take the form

$$\begin{cases} U_t = \int_0^t \overline{f}(s) X_s ds + \overline{W}_t, \\ C_t^n = \int_0^t \mathbf{1}\Gamma_n(s) X_s ds + \int_0^t (1 - J_n(s) X_{s-}) J_n(s) dM_s^X, \ n = \overline{1, N}, \\ D_t = J(0) X_0 + \sum_{\kappa \in \mathcal{J}:\, \kappa \leqslant t} \Delta J(\kappa) X_\kappa. \end{cases} \quad (22)$$

**Theorem 1.** *The optimal filtering estimate $\widehat{X}_t$ is a strong solution to the SDS*

$$\widehat{X}_t = \left( (D_0)^\top J(0) \pi_0 \right)^+ \mathrm{diag}(D_0) J(0) \pi_0 + \int_0^t \Lambda^\top(s) \widehat{X}_s ds + \int_0^t \left( \mathrm{diag}\, \widehat{X}_s - \widehat{X}_s \widehat{X}_s^\top \right) \overline{f}^\top(s) d\omega_s +$$
$$+ \sum_{n=1}^N \int_0^t \left( \Gamma_n(s) - \mathbf{1}\Gamma_n(s) \widehat{X}_{s-} I \right) \widehat{X}_{s-} \left( \mathbf{1}\Gamma_n(s) \widehat{X}_{s-} \right)^+ d\nu_s^n +$$
$$+ \sum_{\kappa \in \mathcal{J}:\, \kappa \leqslant t} \left( \left( \Delta D_\kappa^\top \Delta J(\kappa) \widehat{X}_{\kappa-} \right)^+ \mathrm{diag}(\Delta D_\kappa) \Delta J(\kappa) - I \right) \widehat{X}_{\kappa-}, \quad (23)$$

*where*

$$\omega_t \triangleq U_t - \int_0^t \overline{f}(s) \widehat{X}_s ds \quad (24)$$

*and $A^+$ is a Moore–Penrose pseudoinverse. The solution is unique within the class of nonnegative piecewise-continuous $\mathcal{Y}_{t+}$-adapted processes with discontinuity set lying in $\mathcal{V}$.*

Proof of Theorem 1 is given in Appendix B.

The transformed observations (22) along with Theorem 1 prompt a condition of the *exact* identifiability of the state $X_t$ given indirect noisy observations $Y_t$ (2).

**Corollary 1.** *If for any $n \neq m$ ($n, m = \overline{1, N}$) the inequalities $G_n(s) \neq G_m(s)$ are true almost everywhere on $[0, t]$, then $\widehat{X}_t = X_t$ $\mathcal{P}$-a.s., and $X_t$ is the solution to SDS (23).*

The proof of Corollary 1 is given in Appendix C.

## 4. Numerical Algorithms of Optimal Filtering

### 4.1. Optimal Filtering Given Discretized Observations

The latter section contains the stochastic system (23) defining the optimal filtering estimate $\widehat{X}_t$. The problem of its numerical realization seems routine: we should apply the corresponding methods of numerical integration of SDS with jumps on the RHS [26]. However, this simplicity is illusory. The problem is that the "new" countable observation $C_t$ and discrete-time one $D_t$ are results of certain transform of the available observation $Y$, and this transform includes a limit passage operation. In fact, to obtain $C_t$ we have to estimate/restore the current value of the derivative $\frac{d\langle Y,Y\rangle_{t+}}{dt}$. First, this leads to some time delay to accumulate observations $Y_t$. Second, any pre-limit variant of $C_t$ either has a.s. continuous trajectories or represents their sampling, which demonstrates oscillating nature. Third, the considered filtering estimate is the CME of the state $X_t$ given the observations $Y$ up to the moment $t$. The CME has natural properties: its components are a.s. non-negative and satisfy the normalization condition. The estimates and approximations having these properties are referred in the paper as the stable ones. Mostly, the conventional numerical algorithms do not provide these properties for the calculated approximations. They can preserve the normalization condition only, but the components can have the arbitrary signs and absolute values.

In the paper we present another approach to the numerical realization of the filtering algorithm above. We discretize the available observations $Y$ by time with the increment $h$ and then solve the optimal state filtering problem given discretized observations. The estimate can be considered as approximation of the one given the initial continuous-time observations. Properties of the CME guarantee the stability of the proposed approximation.

To simplify derivation of the numerical algorithm and its accuracy analysis we investigate the time-invariant subset of the observation system (1), (2), i.e., $\Lambda(t) \equiv \Lambda$, $A(t) \equiv A$, $G_n(t) \equiv G_n$, $n = \overline{1,N}$. The observations are discretized with the time increment $h$:

$$\mathsf{Y}_r \triangleq \int_{t_{r-1}}^{t_r} f X_s ds + \int_{t_{r-1}}^{t_r} \sum_{n=1}^{N} X_s^n G_n^{1/2} dW_s, \quad r \in \mathbb{N}, \tag{25}$$

where $t_r \triangleq rh$ are equidistant time instants. We denote $\mathfrak{Y}_r \triangleq \sigma\{\mathsf{Y}_s : 1 \leqslant s \leqslant r\}$ non-decreasing collection of $\sigma$-algebras generated by the time-discretized observations; $\mathfrak{Y}_0 \triangleq \{\emptyset, \Omega\}$.

*The optimal state filtering problem given discretized observations* is to find $\widehat{\mathsf{X}}_r \triangleq \mathbf{E}\{X_{t_r}|\mathfrak{Y}_r\}$.

Let us consider asymptotics of $\widehat{\mathsf{X}}$. We fix some $T > 0$ and consider a condensed sequence of binary meshes $\{\frac{rT}{2^n}\}_{r=\overline{1,2^n}}$ with time increments $h_n \triangleq \frac{T}{2^n}$ and corresponding increasing sequence of $\sigma$-subalgebras $\{\mathfrak{Y}_{2^n}^n\}$: $\mathfrak{Y}_{2^n}^n \triangleq \sigma\{\mathsf{Y}_r,\ 1 \leqslant r \leqslant 2^n\}$. The observation process $\{Y_t\}$ is separable, hence $\sigma\{\bigcup_{n=1}^{\infty}\mathfrak{Y}_n\} = \mathcal{Y}_T$. Then, by Levy theorem $\widehat{\mathsf{X}}_{2^n} \triangleq \mathbf{E}\{X_T|\mathfrak{Y}_n\} \xrightarrow{n\to\infty} \mathbf{E}\{X_T|\mathcal{Y}_T\} = \mathbf{E}\{X_T|\mathcal{Y}_{T+}\} \triangleq \widehat{X}_T$ $\mathcal{P}$-a.s. Moreover, since $\mathbf{E}\left\{\widehat{X}_T\right\} \equiv \mathbf{E}\left\{\widehat{\mathsf{X}}_{2^n}\right\} = \pi(T)$, the $\mathcal{L}_1$-convergence is also true: $\lim_{n\to\infty} \mathbf{E}\left\{|\widehat{X}_T - \widehat{\mathsf{X}}_{2^n}|\right\} = 0$. The convergence also holds, if we replace the sequence of the binary meshes by any condensed sequence with vanishing step. So, we can conclude that the optimal filtering given the discretized observation is a way to design the stable convergent approximations without observation transform $Y \to (U, C, D)$ introduced in the previous section.

To derive the filtering formula we use the approach of [27] and the mathematical induction.

In the case $r = 0$ we have

$$\widehat{\mathsf{X}}_0 = \mathbf{E}\{X_0|\mathfrak{Y}_0\} = \mathbf{E}\{X_0\} = \pi. \tag{26}$$

Let for some $r \in \mathbb{N}$ the estimate $\widehat{\mathsf{X}}_{r-1} = \mathbf{E}\{X_{t_{r-1}}|\mathfrak{Y}_{r-1}\}$ be known. Now we calculate $\widehat{\mathsf{X}}_r$ at the next time instant. To do this we have to specify the mutual conditional distribution $(X_{t_r}, \mathsf{Y}_r)$ with respect to $\mathfrak{Y}_{r-1}$. From the observation model and ([10] Lemma 7.5) it follows that the conditional distribution of $\mathsf{Y}_r$ given $\sigma$-algebra $\mathcal{F}_{t_r}^X \vee \mathfrak{Y}_{r-1}$ is Gaussian with the parameters

$$\mathbf{E}\left\{\mathsf{Y}_r|\mathcal{F}_{t_r}^X\right\} = f v_r, \qquad \operatorname{cov}(\mathsf{Y}_r, \mathsf{Y}_r|\mathcal{F}_{t_r}^X) = \sum_{n=1}^{N} v_r^n G_n. \tag{27}$$

Here, $v_r = \operatorname{col}(v_r^1, \ldots, v_r^N) \triangleq \int_{t_{r-1}}^{t_r} X_s ds$ is a random vector composed of the occupation times of the process $X$ in each state $e_n$ during the interval $[t_{r-1}, t_r]$.

Below in the presentation we use the following notations:

- $\mathcal{D} \triangleq \{u = \operatorname{col}(u^1, \ldots, u^N) : \ u^n \geqslant 0,\ \sum_{n=1}^{N} u^n = h\}$ is an $(N-1)$-dimensional simplex in the space $\mathbb{R}^M$; $\mathcal{D}$ is a distribution support of the vector $v_r$;
- $\Pi \triangleq \{\pi = \operatorname{col}(\pi^1, \ldots, \pi^N) : \ \pi^n \geqslant 0,\ \sum_{n=1}^{N} \pi^n = 1\}$ is a "probabilistic simplex" formed by the possible values of $\pi$;
- $N_r^X$ is a random number of the state $X_t$ transitions, occurred on the interval $[t_{r-1}, t_r]$,
- $a_r^s \triangleq \{\omega \in \Omega : \ N_r^X(\omega) \leqslant s\}$, $A_r^s \triangleq \prod_{q=1}^{r} a_q^s$;
- $\rho^{k,\ell,q}(du)$ is a conditional distribution of the vector $X_{t_r}^\ell \mathbf{I}_{\{q\}}(N_r^X) v_r$ given $X_{t_{r-1}} = e_k$, i.e., for any $\mathcal{G} \in \mathcal{B}(\mathbb{R}^M)$ the following equality is true:

$$\mathbf{E}\left\{\mathbf{I}_{\mathcal{G}}(v_r)\mathbf{I}_{\{q\}}(N_r^X) X_{t_r}^\ell | X_{t_{r-1}} = e_k\right\} = \int_{\mathcal{G}} \rho^{k,\ell,q}(du);$$

- $\mathcal{N}(y,m,K) \triangleq (2\pi)^{-M/2}\det^{-1/2}K\exp\left\{-\frac{1}{2}\|y-m)\|^2_{K^{-1}}\right\}$ is an $M$-dimensional Gaussian *probability density function* (pdf) with the expectation $m$ and nondegenerate covariance matrix $K$;
- $\|\alpha\|^2_K \triangleq \alpha^\top K\alpha, \quad \langle\alpha,\beta\rangle_K \triangleq \alpha^\top K\beta$.

Markovianity of $\{(X_{t_r},Y_r)\}_{r\geqslant 0}$, formula of the total probability and Fubini theorem provide the equalities below for any set $\mathcal{A}\in\mathcal{B}(\mathbb{R}^M)$

$$\begin{aligned}
\mathbf{E}\{X_{t_r}\mathbf{I}_{\mathcal{A}}(Y_r)|\mathfrak{Y}_{r-1}\} &= \mathbf{E}\left\{\mathbf{E}\left\{X_{t_r}\mathbf{I}_{\mathcal{A}}(Y_r)|\mathcal{F}^X_{t_r}\vee\mathfrak{Y}_{r-1}\right\}|\mathfrak{Y}_{r-1}\right\} = \\
&= \mathbf{E}\left\{X_{t_r}\int_{\mathcal{A}}\mathcal{N}\Big(y,fv_r,\sum_{p=1}^N v_r^p G_p\Big)dy\Big|\mathfrak{Y}_{r-1}\right\} = \\
&= \mathbf{E}\left\{\mathbf{E}\left\{X_{t_r}\int_{\mathcal{A}}\mathcal{N}\Big(y,fv_r,\sum_{p=1}^N v_r^p G_p\Big)dy\Big|X_{t_{r-1}}\vee\mathfrak{Y}_{r-1}\right\}\Big|\mathfrak{Y}_{r-1}\right\} = \\
&= \mathbf{E}\left\{\sum_{\ell=1}^N e_\ell\sum_{q=0}^\infty\sum_{k=1}^N e_k^\top X_{t_{r-1}}\int_{\mathcal{D}}\int_{\mathcal{A}}\mathcal{N}\Big(y,fu,\sum_{p=1}^N u^p G_p\Big)dy\rho^{k,\ell,q}(du)\Big|\mathfrak{Y}_{r-1}\right\} = \\
&= \sum_{\ell=1}^N e_\ell\int_{\mathcal{A}}\left[\sum_{k=1}^N\widehat{X}^k_{r-1}\sum_{q=0}^\infty\int_{\mathcal{D}}\mathcal{N}\Big(y,fu,\sum_{p=1}^N u^p G_p\Big)\rho^{k,\ell,q}(du)\right]dy.
\end{aligned}$$

This means that the integrand in the square brackets defines the conditional distribution $(X_{t_r},Y_r)$ given $\mathfrak{Y}_{r-1}$. Further, the conditional distribution $\widehat{X}_r$ is defined component-wisely by the generalized Bayes rule [10]

$$\widehat{X}^j_r = \frac{\sum_{k=1}^N\widehat{X}^k_{r-1}\sum_{q=0}^\infty\int_{\mathcal{D}}\mathcal{N}\Big(Y_r,fu,\sum_{p=1}^N u^p G_p\Big)\rho^{k,j,q}(du)}{\sum_{i,\ell=1}^N\widehat{X}^i_{r-1}\sum_{c=0}^\infty\int_{\mathcal{D}}\mathcal{N}\Big(Y_r,fv,\sum_{n=1}^N v^n G_n\Big)\rho^{i,\ell,c}(dv)}, \quad j=\overline{1,N}. \tag{28}$$

So, we have proved the following

**Lemma 3.** *If for the observation system (1), (2) conditions 1–3 are valid, then the filtering estimate $\widehat{X}_r$ given the discretized observations is defined by (26) at $r=0$, and by recursion (28) at the instant $t_r$ of the discretized observation $Y_r$ reception.*

### 4.2. Stable Analytic Approximations

Recursion (23) cannot be realized directly because of infinite summation both in the numerator and denominator. We replace them by the finite sums, and the corresponding vector sequence $\overline{X}_r(s)$, calculated by the formula

$$\overline{X}^j_r(s) = \frac{\sum_{k=1}^N\overline{X}^k_{r-1}(s)\sum_{q=0}^s\int_{\mathcal{D}}\mathcal{N}\Big(Y_r,fu,\sum_{p=1}^N u^p G_p\Big)\rho^{k,j,q}(du)}{\sum_{i,\ell=1}^N\overline{X}^i_{r-1}(s)\sum_{c=0}^s\int_{\mathcal{D}}\mathcal{N}\Big(Y_r,fv,\sum_{n=1}^N v^n G_n\Big)\rho^{i,\ell,c}(dv)}, \quad j=\overline{1,N} \tag{29}$$

is called *the analytic approximation of the s-th order* of $\widehat{X}_r$. Obviously, that $\overline{X}_r(s)$ is stable.

Let us introduce the following positive random numbers and matrices:

$$\begin{aligned}
\xi^{kj}_q &\triangleq \sum_{m=0}^s\int_{\mathcal{D}}\mathcal{N}\Big(Y_q,fu,\sum_{p=1}^N u^p G_p\Big)\rho^{k,j,m}(du), \\
\theta^{kj}_q &\triangleq \sum_{m=s+1}^\infty\int_{\mathcal{D}}\mathcal{N}\Big(Y_q,fu,\sum_{p=1}^N u^p G_p\Big)\rho^{k,j,m}(du), \\
\xi_q &\triangleq \|\xi^{kj}_q\|_{k,j=\overline{1,N}}, \qquad \theta_q \triangleq \|\theta^{kj}_q\|_{k,j=\overline{1,N}}.
\end{aligned} \tag{30}$$

The estimates $\widehat{X}_r$ (28) and $\overline{X}_r(s)$ (29) can be rewritten in the recurrent form:

$$\widehat{X}_r = (\mathbf{1}(\xi_r + \theta_r)^\top \widehat{X}_{r-1})^{-1}(\xi_r + \theta_r)^\top \widehat{X}_{r-1}, \tag{31}$$

$$\overline{X}_r(s) = (\mathbf{1}\xi_r^\top \overline{X}_{r-1}(s))^{-1}\xi_r^\top \overline{X}_{r-1}(s). \tag{32}$$

Let us define the global distance [28] between the estimates $\{\overline{X}_r(s)\}$ and $\{\widehat{X}_r\}$ as

$$\Sigma_r(s) \triangleq \sup_{\pi\in\Pi} \mathbf{E}\left\{\|\widehat{X}_r - \overline{X}_r(s)\|_1\right\} = \sup_{\pi\in\Pi} \sum_{j=1}^{N} \mathbf{E}\left\{|\widehat{X}_r^j - \overline{X}_r^j(s)|\right\}. \tag{33}$$

The pretty natural characteristic shows the maximal expected divergence of the recursions (28) and (29) at the $r$-th step.

The assertion below defines an upper bound of the characteristic $\Sigma_r(s)$.

**Lemma 4.** *If the conditions of Lemma 3 are valid, then*

$$\Sigma_r(s) \leqslant 2 - 2\left(1 - C_1 \tfrac{(\overline{\lambda}h)^{s+1}}{(s+1)!}\right)^r, \tag{34}$$

*where* $\overline{\lambda} \triangleq \max_{1\leqslant n\leqslant N} |\lambda_{nn}|$, *and* $C_1 = C_1(h, \overline{\lambda}) \in (0,1)$ *is the following parameter:*

$$C_1 \triangleq e^{-\overline{\lambda}h} \tfrac{(s+1)!}{(\overline{\lambda}h)^{s+1}} \sum_{k=s+1}^{\infty} \tfrac{(\overline{\lambda}h)^k}{k!}, \tag{35}$$

*which is bounded from above:* $C_1 \frac{(\overline{\lambda}h)^{s+1}}{(s+1)!} < 1$.

The proof of Lemma 4 is given in Appendix D.

Assertion of Lemma brings the practical benefit. The Lemma does not contain any asymptotic requirements neither to the approximation order $s$ nor to the discretization step $h$: inequality (34) is universal. Mostly, in the digital control systems the data acquisition rate is fixed or bounded from above. There are some extra algorithmic limitations of the rate: the "raw" data should be preprocessed, smoothed, averaged, refined from outliers, etc. For example, utilization of the central limit theorem [29] and diffusion approximation framework [30] for the the renewal processes is legitimate with significant averaging intervals, and their length depends on the process moments.

Now we fix the time instant $T$ and consider an asymptotic $h \to 0$. In this case $r = \frac{T}{h} \to \infty$ and

$$\Sigma_{\frac{T}{h}}(s) \leqslant 2 - 2\left(1 - C_1 \tfrac{(\overline{\lambda}h)^{s+1}}{(s+1)!}\right)^{\frac{T}{h}} \sim 2\overline{\lambda}T \tfrac{(\overline{\lambda}h)^s}{(s+1)!}.$$

### 4.3. Stable Numerical Approximations

In the recursion (32) we use the integrals $\xi_r^{ij}$, which cannot be calculated analytically. The numerical integration brings some extra approximation error. Let us investigate its affect to the total accuracy of the filter numerical realization.

The integrals $\xi^{ij}(y)$ are usually approximated by the sums

$$\xi^{ij}(y) \approx \psi^{ij}(y) \triangleq \textstyle\sum_{\ell=1}^{L} \mathcal{N}\left(y, fw_\ell, \sum_{p=1}^{N} w_\ell^p g_p\right) \varrho_\ell^{ij}, \qquad \psi(y) \triangleq \|\psi^{ij}(y)\|_{i,j=\overline{1,N}}, \tag{36}$$

which are defined by the collection of the pairs $\{(w_\ell, \varrho_\ell^{ij})\}_{\ell=\overline{1,L}}$. Here, $w_\ell \triangleq \mathrm{col}(w_\ell^1, \ldots, w_\ell^N) \in \mathcal{D}$ are the points, and $\varrho_\ell^{ij} \geqslant 0$ $(\ell = \overline{1,L})$ are the weights: $\sum_{j=1}^{N} \sum_{\ell=1}^{L} \varrho_\ell^{ij} \leqslant Q \leqslant 1$.

In complete analogy with $\xi_q$ we define the approximations $\psi_q \triangleq \|\psi^{ij}(Y_q)\|_{i,j=\overline{1,N}}$. By construction, the elements of $\psi_q$ are positive random values, hence the approximation $\widetilde{X}_r$

$$\widetilde{X}_r \triangleq (\mathbf{1}\psi_r^{\top}\widetilde{X}_{r-1})^{-1}\psi_r^{\top}\widetilde{X}_{r-1}, \quad \widetilde{X}_0 = \pi \tag{37}$$

is stable. Below we denote the numerical integration errors and their absolute values as follows

$$\gamma^{kj} \triangleq \psi^{kj} - \xi^{kj}, \qquad \gamma_r \triangleq \|\gamma^{kj}(Y_r)\|_{k,j=\overline{1,N}} \tag{38}$$

$$\overline{\gamma}^{kj} \triangleq |\gamma^{kj}|, \qquad \overline{\gamma}_r \triangleq \left\||\gamma^{kj}(Y_r)|\right\|_{k,j=\overline{1,N}}. \tag{39}$$

So, the recursion (32) is replaced by the scheme (37), holding the common initial condition $\pi$.

Both (32) and (37) are constructed in light of the event $A_r^s$: the state transition numbers do not exceed the threshold $s$ over any subintervals $[t_{q-1}, t_q]$ belonging to $[0, t_r]$. So, the distance between $\widetilde{X}_r$ and $\overline{X}_r(s)$ should be determined taking into account $A_r^s$. In view of this fact, we propose the pseudo-metrics

$$\mathcal{E}_r(s) \triangleq \sup_{\pi\in\Pi} \mathbf{E}\left\{\mathbf{I}_{A_r^s}(\omega)\|\widetilde{X}_r - \overline{X}_r(s)\|_1\right\} = \sup_{\pi\in\Pi}\sum_{n=1}^{N} \mathbf{E}\left\{\mathbf{I}_{A_r^s}(\omega)|\widetilde{X}_r^n - \overline{X}_r^n(s)|\right\}. \tag{40}$$

This index reflects maximal divergence of the algorithms (32) and (37) after $r$ steps, being started from the arbitrary but common initial condition.

**Theorem 2.** *If the inequality*

$$\max_{i=\overline{1,N}} \sum_{j=1}^{N} \int_{\mathbb{R}^M} |\psi^{ij}(y) - \xi^{ij}(y)|dy < \delta \tag{41}$$

*is true for the numerical integration scheme (36), then the distance $\mathcal{E}_r(s)$ is bounded from above:*

$$\mathcal{E}_r(s) \leqslant 2rQ^{r-1}\delta. \tag{42}$$

The proof of Theorem 2 is given in Appendix E.

The chance to describe the accuracy of the numerical algorithm for the stochastic filtering using only the condition (41), related to the calculus, looks remarkable. Furthermore, if the total weight $Q = \sum_{\ell,j}\varrho_\ell^{ij}$ separates from the unity, i.e., $Q < 1$, then the index $\mathcal{E}_r(s)$ is a *sublinear* function of $r$, so as the index $\Sigma_r(s)$ of the analytic accuracy is. Notably, that in the classic numerical algorithms of the SDS solution the global error grows *linearly* with respect to the number of steps $r$ [26].

The precision characteristics of both the analytical approximation and its numerical realization should be aggregated into the one. If the conditions of Lemma 4 and Theorem 2 are valid, then the local distance (i.e., the distance after one iteration) between the optimal filtering estimate and its numerical approximation can be bounded from above:

$$\begin{aligned}\tau(s) \triangleq \sup_{\pi\in\Pi} \mathbf{E}\left\{\|\widehat{X}_1 - \widetilde{X}_1\|_1\right\} \leqslant \sup_{\pi\in\Pi} \mathbf{E}\left\{\mathbf{I}_{a_1^s}(\omega)\|\widetilde{X}_1 - \overline{X}_1(s) + \overline{X}_1(s) - \widehat{X}_1\|_1 + \mathbf{I}_{\overline{a}_1^s}(\omega)\|\widetilde{X}_1 - \overline{X}_1(s)\|_1\right\} \leqslant \\ \leqslant 2\mathcal{P}\{\overline{a}_1^s\} + \sup_{\pi\in\Pi} \mathbf{E}\left\{\|\overline{X}_1(s) - \widehat{X}_1\|_1\right\} + \sup_{\pi\in\Pi} \mathbf{E}\left\{\mathbf{I}_{a_1^s}(\omega)\|\widetilde{X}_1 - \overline{X}_1(s)\|_1\right\} = \\ = 2\mathcal{P}\{\overline{a}_1^s\} + \sigma(s) + \mathcal{E}_1(s) \leqslant 4\tfrac{(\overline{\lambda}h)^{s+1}}{(s+1)!} + 2\delta.\end{aligned} \tag{43}$$

The global distance between $\widehat{X}_r \triangleq \mathbf{E}\{X_r|\mathfrak{Y}_r\}$ and $\widetilde{X}_r$ can be bounded in the similar way:

$$\mathcal{T}(s) \triangleq \sup_{\pi\in\Pi} \mathbf{E}\left\{\|\widehat{X}_r - \widetilde{X}_r\|_1\right\} \leqslant 4\left[1 - \left(1 - \tfrac{(\overline{\lambda}h)^{s+1}}{(s+1)!}\right)^r\right] + 2rQ^{r-1}\delta. \tag{44}$$

We could choose the parameters $(h, s)$ of the analytical approximation and $\delta$ of the numerical integration independently each other. However, both the limitation of the computational resources and the accuracy requirements lead to the necessity of the mutual optimization of $(h, s, \delta)$.

Let us fix some time horizon $T$ along with the order $s$ of analytical approximation, and consider the asymptotic $r \to \infty$, or, equivalently, $h = \frac{T}{r} \to 0$. Due to the Bernoulli inequality, and condition $0 < Q \leqslant 1$ we have that

$$\sup_{\pi \in \Pi} \mathbf{E}\left\{\|\widetilde{X}_{T/h} - \widehat{X}_{T/h}\|_1\right\} \leqslant 4\left[1 - \left(1 - \frac{(\overline{\lambda}h)^{s+1}}{(s+1)!}\right)^r\right] + 2rQ^{r-1}\delta \leqslant 4r\frac{(\overline{\lambda}h)^{s+1}}{(s+1)!} + 2rQ^{r-1}\delta =$$
$$= 4\overline{\lambda}T\frac{(\overline{\lambda}h)^{s}}{(s+1)!} + 2rQ^{r-1}\delta \leqslant 2T\left(2\overline{\lambda}\frac{(\overline{\lambda}h)^{s}}{(s+1)!} + \frac{\delta}{h}\right). \quad (45)$$

The first summand in the brackets represents the contribution of the analytical approximation error, the second one reflects the error of the specified numerical integration scheme. Obviously, the optimal choice of the parameters provides an equal infinitesimal order for both the summands, and it is possible when $\delta \sim \frac{(\overline{\lambda}h)^{s+1}}{\overline{\lambda}}$.

### 4.4. Numerical Example

To illustrate the correspondence between the theoretical estimate and its realization along with the performance of the numerical algorithm, we consider the filtering problem for the observation system (1) and (2) with the following parameters: $t \in [0, 1]$, $N = 3$,

$$\Lambda = \begin{bmatrix} -1.0 & 0.2 & 0.8 \\ 0.8 & -1.0 & 0.2 \\ 0.2 & 0.8 & -1.0 \end{bmatrix}, \quad \pi = \begin{bmatrix} 0.333 \\ 0.333 \\ 0.334 \end{bmatrix}, \quad f = \begin{bmatrix} 0.0 \\ 0.0 \\ 0.0 \end{bmatrix}, \quad \begin{array}{l} G_1 = 1.0, \\ G_2 = 4.0, \\ G_3 = 9.0. \end{array}$$

The specified observation system is the one with state-dependent noise, and the conditions of Corollary 1 hold, so the optimal filter (23) restores the MJP state precisely under available noisy observations. Let us verify this theoretical fact, using the recursive algorithm (37). We choose the analytical approximation of the order $s = 1$ with numerical integration by the simple midpoint rectangle scheme and calculate estimate approximations with decreasing time-discretization step: $h = 0.01$; $0.001$; $0.0001$; $0.00001$. We expect the descent of the estimation error characterized by the MS-criterion $\mathcal{S}_t(h) = \sqrt{\mathbf{E}\left\{\|X_t - \widetilde{X}_{\frac{t}{h}}\|_2^2\right\}}$. To calculate the criterion, we use the Monte–Carlo method over the test sample of the size 1000. Figure 1 presents the corresponding plots of the quality index $\mathcal{S}_t(h)$ for various values of $h$.

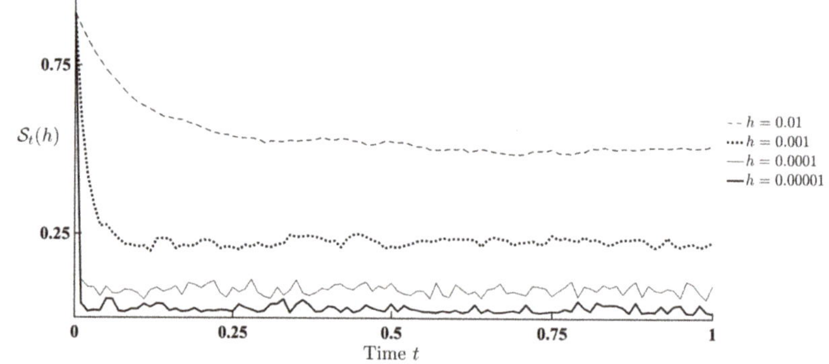


**Figure 1.** Estimation quality index $\mathcal{S}_t(h)$ depending on the time-discretization step $h$.

The determination of the precision order provided by the chosen numerical integration method is out of the scope of this investigation. Nevertheless, one can see the expected decrease of the estimation error when the time-discretization step descends. We appraise this result as a practical confirmation of both the theoretical assertions and numerical algorithm.

## 5. Conclusions

In this paper, we investigated the optimal filtering problem of the MJP states, given the indirect noisy continuous-time observations. The observation noise intensity was a function of the estimated state, so it was impossible to apply the classic Wonham filter to this observation system. To overcome this obstacle, we suggested an observation transform. On the one hand, the transformed observations remained to be equivalent to the original one from the informational point of view. On the other hand, the "new" observations allowed to apply the effective stochastic analysis framework to process them. We derived the optimal filtering estimate theoretically as a unique strong solution to some discrete–continuous stochastic differential system. The transformed observations included derivative of the quadratic characteristics, i.e., the result of some limit passage in the stochastic settings. Hence, the subsequent numerical realization of the filtering became challenging. We proposed to approximate the initial continuous-time filtering problem by a sequence of the optimal ones given the time-discretized observations. We also involved numerical integration schemes to calculate the integrals included in the estimation formula. We prove assertions, characterizing the accuracy of the numerical approximation of the filtering estimate, i.e., the distance between the calculated approximation and optimal discrete-time filtering estimate. The accuracy depended on the observation system parameters, time discretization step, a threshold of state transition number during the time step, and the chosen scheme of the numerical integration. We suggested the whole class of numerical filtering algorithms. In each case, one could choose any specific algorithm individually, taking into account characteristics of the concrete observation system, accuracy requirements, and available computing resources.

We do not consider the presented investigations as completed. First, the characterization of the distance between the initial optimal continuous-time filtering estimate and its proposed approximation is still an open problem. Second, we can use the theoretical solution to the MJP filtering problem as a base of numerical schemes for the diffusion process filtering, given the observations with state-dependent noise. Third, the obtained optimal filtering estimate looks a springboard for a solution to the optimal stochastic control of the Markov jump processes, given both the counting and diffusion observations with state-dependent noise. All of this research is in progress.

**Author Contributions:** Conceptualization, A.B., I.S.; methodology, A.B.; formal analysis and investigation, A.B., I.S.; writing—original draft preparation, A.B.; writing—review and editing, I.S.; supervision, I.S. All authors have read and agreed to the published version of the manuscript.

**Funding:** This research received no external funding.

**Conflicts of Interest:** The authors declare no conflict of interest.

## Abbreviations

The following abbreviations are used in this manuscript:

| | |
|---|---|
| CME | Conditional mathematical expectation |
| MJP | Markov jump process |
| pdf | Probability density function |
| RHS | Right-hand side |
| SDS | Stochastic differential system |

## Appendix A. Proof of Lemma 2

From (14), (15), the identity $\mathrm{diag}(a)b \equiv \mathrm{diag}(b)a$, the fact that $J_n(t) \neq J_n(t-)$ at most at finite points of any finite interval and property 4 of the function $K(t)$, the following equalities are true

$$C_t^n = \int_0^t (1 - e_n^\top V_{s-}) e_n^\top dR_s = \int_0^t (1 - e_n^\top V_{s-}) e_n^\top J(s)(\Lambda^\top(s) X_{s-} ds + dM_s^X) =$$
$$= \int_0^t (1 - J_n(s-) X_{s-}) J_n(s-) \Lambda^\top(s) X_{s-} ds + \int_0^t (1 - e_n^\top V_{s-}) J_n(s) dM_s^X =$$
$$= \int_0^t J_n(s) \Lambda^\top(s)(I - \mathrm{diag}\, J_n(s)) X_s ds + \int_0^t (1 - e_n^\top V_{s-}) J_n(s) dM_s^X =$$
$$= \int_0^t \mathbf{1}\Gamma_n(s) X_s ds + \int_0^t (1 - e_n^\top V_{s-}) J_n(s) dM_s^X. \quad \text{(A1)}$$

Assertion 1 of Lemma is proved.

The definition of the processes $C_t^n$ ($n = \overline{1,N}$) guarantees their strong orthogonality, i.e., $\mathcal{P}\left\{\Delta C_t^i \Delta C_t^j = 0\right\} \equiv 0$ for any $i \neq j$ and $t \geqslant 0$, so $[C^i, C^j]_t \equiv 0$.

Let us use (5), (19) and properties of $X$ and $J_n$ to derive the quadratic characteristics of $C^n$:

$$\langle C^n, C^n \rangle_t = \int_0^t (1 - J_n(s) X_{s-})^2 J_n(s) d\langle X, X \rangle_s J_n^\top(s) =$$
$$= \int_0^t (1 - J_n(s) X_{s-}) J_n(s) \left(\mathrm{diag}(\Lambda^\top(s) X_{s-} - \Lambda^\top(s)\, \mathrm{diag}\, X_{s-} - \mathrm{diag}(X_{s-})\Lambda(s)\right) J_n^\top(s) ds =$$
$$= \int_0^t (1 - J_n(s) X_{s-}) J_n(s)\, \mathrm{diag}(J_n(s)) \Lambda^\top(s) X_{s-} ds = \int_0^t J_n(s) \Lambda^\top(s)(I - \mathrm{diag}\, J_n(s)) X_s ds =$$
$$= \int_0^t \mathbf{1}\Gamma_n(s) X_s ds. \quad \text{(A2)}$$

Assertion 2 of Lemma is proved.

If $s$ and $t$ are two arbitrary moments, such that $s \leqslant t$, then

$$\mathbf{E}\left\{\nu_t^n - \nu_s^n | \overline{\mathcal{Y}}_s\right\} = \mathbf{E}\left\{\int_s^t J_n(u) \Lambda^\top(u)(I - \mathrm{diag}\, J_n(u)) \mathbf{E}\left\{(X_u - \widehat{X}_u) | \overline{\mathcal{Y}}_u\right\} du | \overline{\mathcal{Y}}_s\right\} +$$
$$+ \mathbf{E}\left\{\mathbf{E}\left\{\int_s^t (1 - J_n(s) X_{s-}) J_n(u) dM_u^X | \mathcal{F}_s\right\} | \overline{\mathcal{Y}}_s\right\} = 0,$$

i.e., $\nu_t^n$ is a $\overline{\mathcal{Y}}_t$-adapted martingale. Note, that $\nu_t^n$ is purely discontinuous with unit jumps, hence

$$[\nu^n, \nu^n]_t = \sum_{\tau \leqslant t} (\Delta \nu_\tau^n)^2 = [C^n, C^n]_t = \sum_{\tau \leqslant t} (\Delta C_\tau^n)^2 = C_t^n =$$
$$= \int_0^t J_n(s) \Lambda^\top(s)(I - \mathrm{diag}\, J_n(s)) X_s ds + \int_0^t (1 - J_n(s) X_{s-}) J_n(s) dM_s^X = \int_0^t \mathbf{1}\Gamma_n(s) \widehat{X}_s ds + \mu_t^0,$$

where $\mu_t^0$ is some $\overline{\mathcal{Y}}_t$-adapted martingale. From the uniqueness of the special martingale representation $[\nu^n, \nu^n]_t$ it follows that $\langle \nu^n, \nu^n \rangle_t = \int_0^t \mathbf{1}\Gamma_n(s) \widehat{X}_s ds$. Lemma 2 is proved. □

## Appendix B. Proof of Theorem 1

We use the same approach as in ([6], Part III, Sect. 8.7) to derive the MJP filtering equations. The idea exploits the uniqueness of the representation for a special semimartingale along with the integral representation of a martingale [23].

From the Bayes rule it follows that $\widehat{X}_0 = \mathbf{E}\{X_0|D_0\} = \left(D_0^\top J(0)\pi\right)^+ \mathrm{diag}(D_0)J(0)\pi$. Let $\varkappa_{n-1}$ be a random instant of the $n-1$-th discrete observation $\Delta D_{\varkappa_{n-1}}$. We investigate evolution of $X_t$ over the interval $[\varkappa_{n-1}, \varkappa_n)$:

$$X_t = X_{\varkappa_{n-1}} + \int_{\varkappa_{n-1}}^t \Lambda^\top(s)X_s ds + M_t^X - M_{\varkappa_{n-1}}^X, \quad t \in [\varkappa_{n-1}, \varkappa_n).$$

Conditioning the left and right parts of the latter equality over $\overline{\mathcal{Y}}_t$, one can show that

$$\widehat{X}_t = \widehat{X}_{\varkappa_{n-1}} + \int_{\varkappa_{n-1}}^t \Lambda^\top(s)\widehat{X}_s ds + \mu_t^1,$$

where $\{\mu_t^1\}_{t\in[\varkappa_{n-1},\varkappa_n)}$ is an $\overline{\mathcal{Y}}_t$adapted martingale. For any $t \in [\varkappa_{n-1}, \varkappa_n)$ the equality $\overline{\mathcal{Y}}_t = \overline{\mathcal{Y}}_{\varkappa_{n-1}} \vee \sigma\{U_s,\ s \in (\varkappa_{n-1}, t]\} \vee \sigma\{C_s^j,\ s \in (\varkappa_{n-1}, t],\ j = \overline{1,N}\}$ holds. The process $\{\omega_t\}$ (24) is a $\overline{\mathcal{Y}}_t$ -adapted standard Wiener process [10].

The process $U_t$ is a $\overline{\mathcal{Y}}_t$-adapted semimartingale with $\mathcal{F}^X$-conditionally-independent increments, meanwhile $\{C_t^j\}_{j=\overline{1,N}}$ are $\overline{\mathcal{Y}}_t$-adapted point processes. Hence, the martingale $\mu_t^1$ *admits an integral representation* ([23], Chap. 4, §8, Problem 1), i.e.,

$$\widehat{X}_t = \widehat{X}_{\varkappa_{n-1}} + \int_{\varkappa_{n-1}}^t \Lambda^\top(s)\widehat{X}_s ds + \int_{\varkappa_{n-1}}^t \alpha_s d\omega_s + \int_{\varkappa_{n-1}}^t \sum_{j=1}^N \beta_s^j d\nu_s^j, \tag{A3}$$

where $\alpha_t$ and $\{\beta_t^j\}_{j=\overline{1,N}}$ are $\overline{\mathcal{Y}}_t$-predictable processes of appropriate dimensionality, which should be determined.

Due to the generalized Itô rule

$$X_t U_t^\top = X_{\varkappa_{n-1}} U_{\varkappa_{n-1}}^\top + \int_{\varkappa_{n-1}}^t \left(\Lambda^\top(s)X_s U_s^\top + \mathrm{diag}(X_s)\overline{f}^\top(s)\right) ds + \mu_t^2,$$

where $\mu_t^2$ is an $\mathcal{F}_t$-adapted matringale. Conditioning both sides of the latter equality over $\overline{\mathcal{Y}}_t$, we can show that

$$\widehat{X}_t U_t^\top = \widehat{X}_{\varkappa_{n-1}} U_{\varkappa_{n-1}}^\top + \int_{\varkappa_{n-1}}^t \left(\Lambda^\top(s)\widehat{X}_s U_s^\top + \mathrm{diag}(\widehat{X}_s)\overline{f}^\top(s)\right) ds + \mu_t^3, \tag{A4}$$

where $\mu_t^3$ is a $\overline{\mathcal{Y}}_t$-adapted martingale. On the other hand, using the Itô rule, representation (A3) and the fact that $\omega_t$ is the Wiener process, we can obtain

$$\widehat{X}_t U_t^\top = \widehat{X}_{\varkappa_{n-1}} U_{\varkappa_{n-1}}^\top + \int_{\varkappa_{n-1}}^t \left(\Lambda^\top(s)\widehat{X}_s U_s^\top + \widehat{X}_s \widehat{X}_s^\top \overline{f}^\top(s) + \alpha_s\right) ds + \mu_t^4, \tag{A5}$$

where $\mu_t^4$ is a $\overline{\mathcal{Y}}_t$-adapted martingale. One can see that (A4) and (A5) are two representations of the same special semimartingale $\widehat{X}_t U_t^\top$, hence due to the representation uniqueness the $\overline{\mathcal{Y}}_t$-predictable process $\alpha_t$ should satisfy the equality

$$\int_{\varkappa_{n-1}}^t \mathrm{diag}(\widehat{X}_s)\overline{f}^\top(s) ds = \int_{\varkappa_{n-1}}^t \left(\widehat{X}_s \widehat{X}_s^\top \overline{f}^\top(s) + \alpha_s\right) ds,$$

and $\alpha_t$ may be chosen in the form

$$\alpha_t = \left(\mathrm{diag}\, \widehat{X}_{t-} - \widehat{X}_{t-}\widehat{X}_{t-}^\top\right) \overline{f}^\top(t). \tag{A6}$$

Due to the generalized Itô rule, formulae (5), (18) and the properties of $X$ and $J_j$ we can obtain, that

$$X_t C_t^j = X_{\varkappa_{n-1}} C_{\varkappa_{n-1}}^j + \int_{\varkappa_{n-1}}^t \left(\Lambda^\top(s) X_s C_s^j + \Gamma_j(s) X_s\right) ds + \mu_t^5,$$

where $\mu_t^5$ is an $\mathcal{F}_t$-adapted martingale. Conditioning both sides of this equality over $\overline{\mathcal{Y}}_t$, we get

$$\widehat{X}_t C_t^j = \widehat{X}_{\varkappa_{n-1}} C_{\varkappa_{n-1}}^j + \int_{\varkappa_{n-1}}^t \left(\Lambda^\top(s) \widehat{X}_s C_s^j + \Gamma_j(s) \widehat{X}_s\right) ds + \mu_t^6, \tag{A7}$$

where $\mu_t^6$ is a $\overline{\mathcal{Y}}_t$-adapted martingale. On the other hand, using the Itô rule, representation (A3) and quadratic characteristic (21) we deduce, that

$$\widehat{X}_t C_t^j = \widehat{X}_{\varkappa_{n-1}} C_{\varkappa_{n-1}}^j + \int_{\varkappa_{n-1}}^t \left(\Lambda^\top(s) \widehat{X}_s C_s^j + \widehat{X}_s \mathbf{1}\Gamma_j(s) \widehat{X}_s + \beta_s^j \mathbf{1}\Gamma_j(s) \widehat{X}_s\right) ds + \mu_t^7, \tag{A8}$$

where $\mu_t^7$ is a $\overline{\mathcal{Y}}_t$-adapted martingale. Since the representations (A7) and (A8) correspond to the same special semimartingale $\widehat{X}_t C_t^j$ we conclude that the process $\beta_s^j$ should satisfy the equality

$$\int_{\varkappa_{n-1}}^t \Gamma_j(s) \widehat{X}_s ds = \int_{\varkappa_{n-1}}^t \left[\widehat{X}_s \mathbf{1}\Gamma_j(s) \widehat{X}_s + \beta_s^j \mathbf{1}\Gamma_j(s) \widehat{X}_s\right] ds.$$

Acting as with the coefficient $\alpha_t$, we choose the predictable processes $\beta_t^j$ in the form

$$\beta_t^j = \left(\Gamma_j(t) - \mathbf{1}\Gamma_j(t)\widehat{X}_{t-} I\right) \widehat{X}_{t-} \left(\mathbf{1}\Gamma_j(t)\widehat{X}_{t-}\right)^+, \quad j = \overline{1, N}. \tag{A9}$$

So, on the interval $[\varkappa_{n-1}, \varkappa_n)$ the optimal filtering estimate $\widehat{X}_t$ is described by the SDS

$$\begin{aligned} \widehat{X}_t = \widehat{X}_{\varkappa_{n-1}} &+ \int_{\varkappa_{n-1}}^t \Lambda^\top(s) \widehat{X}_{s-} ds + \int_{\varkappa_{n-1}}^t (\operatorname{diag} \widehat{X}_{s-} - \widehat{X}_{s-}\widehat{X}_{s-}^\top) \overline{f}^\top(s) d\omega_s + \\ &+ \sum_{j=1}^N \int_{\varkappa_{n-1}}^t \left(\Gamma_j(s) - \mathbf{1}\Gamma_j(s)\widehat{X}_{s-} I\right) \widehat{X}_{s-} \left(\mathbf{1}\Gamma_j(s)\widehat{X}_{s-}\right)^+ dv_s^j. \end{aligned} \tag{A10}$$

Since $\mathcal{P}\{\Delta X_{\varkappa_n} = 0\} = 1$, equation (A10) presumes $\mathcal{P}$-a.s. fulfilment of the equality

$$\begin{aligned} &\mathbf{E}\left\{X_{\varkappa_n} | \overline{\mathcal{Y}}_{\varkappa_{n-1}} \vee \sigma\{U_s,\ s \in (\varkappa_{n-1}, \varkappa_n]\} \vee \sigma\{C_s^j,\ s \in (\varkappa_{n-1}, \varkappa_n],\ j = \overline{1, N}\}\right\} = \\ &\quad = \widehat{X}_{\varkappa_{n-1}} + \int_{\varkappa_{n-1}}^{\varkappa_n} \Lambda^\top(s) \widehat{X}_{s-} ds + \int_{\varkappa_{n-1}}^{\varkappa_n} (\operatorname{diag} \widehat{X}_{s-} - \widehat{X}_{s-}\widehat{X}_{s-}^\top) \overline{f}^\top(s) d\omega_s + \\ &\quad\quad + \sum_{j=1}^N \int_{\varkappa_{n-1}}^{\varkappa_n} \left(\Gamma_j(s) - \mathbf{1}\Gamma_j(s)\widehat{X}_{s-} I\right) \widehat{X}_{s-} \left(\mathbf{1}\Gamma_j(s)\widehat{X}_{s-}\right)^+ dv_s^j = \widehat{X}_{\tau_n -}. \end{aligned}$$

Finally,

$$\overline{\mathcal{Y}}_{\varkappa_n} = \overline{\mathcal{Y}}_{\varkappa_{n-1}} \vee \sigma\{U_s,\ s \in (\varkappa_{n-1}, \varkappa_n]\} \vee \sigma\{C_s^j,\ s \in (\varkappa_{n-1}, \varkappa_n],\ j = \overline{1, N}\} \vee \sigma\{\Delta D_{\varkappa_n}\},$$

so, by the Bayes rule we get that

$$\widehat{X}_{\tau_n} = \left(\Delta D_{\tau_n}^\top \Delta J(\tau_n) \widehat{X}_{\tau_n -}\right)^+ \operatorname{diag}(\Delta D_{\tau_n}) \Delta J(\tau_n) \widehat{X}_{\tau_n -}. \tag{A11}$$

Equation (23) can be obtained as "gluing" of local equations (A10), which describe the evolution of $\widehat{X}_t$ on the intervals $[\varkappa_{n-1}, \varkappa_n)$, and formula (A11), which describes the estimate correction given the observations available at the moments $\varkappa_n$.

Uniqueness of the strong solution within the class of nonnegative piecewise-continuous $\mathcal{Y}_{t+}$-adapted processes with discontinuity set lying in $\mathcal{V}$ can be proved in complete analogy with ([31] Chap. 9, Theorem 9.2). Theorem 1 is proved. □

## Appendix C. Proof of Corollary 1

The conditions of Corollary guarantee, that the elements of $K(t)$ (4) satisfy the equality $K_{nm}(t) = \delta_{nm}$ almost everywhere, hence $J(t) \equiv I$. This means that in (23) $D_0 = X_0$, $\mathcal{P}$-a.s., i.e., $\widehat{X}_0 = X_0$. Further, from the properties of transition intensity matrix $\Lambda(\cdot)$ and the identity $J_n(t) \equiv e_n^\top$ it follows that $\Gamma_n(t) = \mathrm{diag}(e_n)\overline{\Lambda}^\top(t)$, where $\overline{\Lambda}(t) \triangleq \Lambda(t) - \lambda(t)$, $\lambda(t) \triangleq \mathrm{diag}(\Lambda_{11}(t), \ldots, \Lambda_{NN})$. In this case

$$C_t = \int_0^t \overline{\Lambda}^\top(s) X_s ds + \int_0^t (I - \mathrm{diag}\, X_{s-}) dM_s^X,$$

and the $n$-th component counts the jumps of $X_t$ into the state $e_n$, occurred on the interval $(0, t]$. This means $X_t$ is the unique solution to the "purely discontinuous" equation

$$X_t = D_0 + \int_0^t (I - X_{s-}\mathbf{1}) dC_s, \tag{A12}$$

i.e., the state $X_t$ is measurable with respect to $\sigma\{D_0,\ C_s,\ 0 \leqslant s \leqslant t\}$, so $\widehat{X}_t = X_t$ $\mathcal{P}$-a.s.

Further, we substitute $X_t$ into (23) and verify its validity. To do this we simplify the RHS of the equality using the explicit form of $J_n(t)$, $\Gamma_n(t)$ and $C_t$, along with the identities $\mathrm{diag}\, X_t - X_t X_t^\top \equiv 0$ and $\Delta J(t) \equiv 0$:

$$\begin{aligned} X_t = D_0 &+ \int_0^t \Lambda^\top(s) X_s ds + \\ &+ \sum_{n=1}^N \int_0^t \left[\mathrm{diag}(e_n)\overline{\Lambda}^\top(s) - e_n^\top \overline{\Lambda}^\top(s) X_{s-} I\right] X_{s-} \left(e_n^\top \overline{\Lambda}^\top(s) X_{s-}\right)^+ \left[dC_s^n - e_n^\top \overline{\Lambda}^\top(s) X_{s-} ds\right] = \\ &= D_0 + \sum_{n=1}^N \int_0^t \left[\mathrm{diag}(e_n)\overline{\Lambda}^\top(s) - e_n^\top \overline{\Lambda}^\top(s) X_{s-} I\right] X_{s-} \left(e_n^\top \overline{\Lambda}^\top(s) X_{s-}\right)^+ dC_s^n. \end{aligned}$$

The properties of counting processes also provides the following implication: if for some $\mathfrak{T} \subseteq [0, T]$ the equality $\int_{\mathfrak{T}} e_n^\top \overline{\Lambda}^\top(s) X_s ds = 0$ holds, then $\int_{\mathfrak{T}} dC_s^n = 0$. Hence, the latter transformation can be continued:

$$X_t = D_0 + \sum_{n=1}^N \int_0^t [e_n - X_{s-}] e_n^\top dC_s = D_0 + \int_0^t (I - X_{s-}\mathbf{1}) dC_s,$$

which leads to (A12). So, we have verified that under conditions of Corollary 1 the state $X_t$ is a solution to the filtering equation (23). Corollary 1 is proved. □

## Appendix D. Proof of Lemma 4

Using notations $\Xi_r \triangleq \xi_1\xi_2 \ldots \xi_r$ and $\Theta_r \triangleq \theta_1\theta_2 \ldots \theta_r$ we can rewrite the estimates $\widehat{X}_r$ and $\overline{X}_r(s)$ in the explicit form

$$\widehat{X}_r = \left(\mathbf{1} (\Xi_r + \Theta_r)^\top \pi\right)^{-1} (\Xi_r + \Theta_r)^\top \pi, \qquad \overline{X}_r(s) = \left(\mathbf{1}\Xi_r^\top \pi\right)^{-1} \Xi_r^\top \pi.$$

To simplify inferences we will omit the index $r$ in $\Xi_r$ and $\Theta_r$. The following relations are valid

$$\mathbf{E}\left\{\left\|\widehat{X}_r - \overline{X}_r(s)\right\|_1\right\} = \mathbf{E}\left\{\left\|\frac{1}{\mathbf{1}(\Xi+\Theta)^\top\pi}(\Xi+\Theta)^\top\pi - \frac{1}{\mathbf{1}\Xi^\top\pi}\Xi^\top\pi\right\|_1\right\} =$$
$$= \mathbf{E}\left\{\frac{1}{\mathbf{1}(\Xi+\Theta)^\top\pi\mathbf{1}\Xi^\top\pi}\left\|\mathbf{1}\Xi^\top\pi\Theta^\top\pi - \mathbf{1}\Theta^\top\pi\Xi^\top\pi\right\|_1\right\} \leqslant$$
$$\leqslant \mathbf{E}\left\{\frac{1}{\mathbf{1}(\Xi+\Theta)^\top\pi\mathbf{1}\Xi^\top\pi}\left(\mathbf{1}\Xi^\top\pi\|\Theta^\top\pi\|_1 + \mathbf{1}\Theta^\top\pi\|\Xi^\top\pi\|_1\right)\right\} = 2\mathbf{E}\left\{\frac{1}{\mathbf{1}(\Xi+\Theta)^\top\pi}\mathbf{1}\Theta^\top\pi\right\}. \quad \text{(A13)}$$

Let us consider an auxiliary estimate $\check{X}_r \triangleq \mathbf{E}\left\{X_{t_r}\mathbf{I}_{A_r^s}(\omega)|\mathfrak{Y}_r\right\}$. From the Bayes rule it follows that $\check{X}_r = \frac{1}{\mathbf{1}(\Xi+\Theta)^\top\pi}\Xi^\top\pi$ and

$$\widehat{X}_r - \check{X}_r = \mathbf{E}\left\{X_{t_r}\mathbf{I}_{\overline{A}_r^s}(\omega)|\mathfrak{Y}_r\right\} = \frac{1}{\mathbf{1}(\Xi+\Theta)^\top\pi}\Theta^\top\pi. \quad \text{(A14)}$$

From (A13) and (A14) we deduce, that for $r = 1$ and $\forall\ \pi \in \Pi$

$$\mathbf{E}\left\{\|\widehat{X}_1 - \overline{X}_1(s)\|_1\right\} \leqslant 2\mathbf{E}\left\{\left\|\mathbf{E}\left\{X_{t_1}\mathbf{I}_{\overline{a}_1^s}(\omega)|\mathfrak{Y}_1\right\}\right\|_1\right\} =$$
$$= 2\mathbf{E}\left\{\sum_{n=1}^{N}\mathbf{E}\left\{X_{t_1}^n\mathbf{I}_{\overline{a}_1^s}(\omega)|\mathfrak{Y}_1\right\}\right\} = 2\mathbf{E}\left\{\mathbf{E}\left\{\mathbf{I}_{\overline{a}_1^s}(\omega)|\mathfrak{Y}_1\right\}\right\} = 2\mathcal{P}\left\{\overline{a}_1^s\right\}. \quad \text{(A15)}$$

The counting process $N_t^X$ has the quadratic characteristic $\langle N^X, N^X\rangle_t = -\int_0^t\sum_{n=1}^N\lambda_{nn}X_s^n ds$, hence the probability $\mathcal{P}\left\{\overline{a}_1^s\right\}$ can be bounded from above as

$$\mathcal{P}\left\{\overline{a}_1^s\right\} \leqslant e^{-\overline{\lambda}h}\sum_{k=s+1}^{\infty}\frac{(\overline{\lambda}h)^k}{k!} = C_1\frac{(\overline{\lambda}h)^{s+1}}{(s+1)!}. \quad \text{(A16)}$$

Formulae (A15) and (A16) lead to the fact, that $\sup_{\pi\in\Pi}\mathbf{E}\left\{\|\widehat{X}_1 - \overline{X}_1(s)\|_1\right\} \leqslant 2C_1\frac{(\overline{\lambda}h)^{s+1}}{(s+1)!}$.

Markovianity of the pair $(X_t, N_t^X)$ and inequality (A16) also allow to bound the probability $\mathcal{P}\left\{\overline{A}_r^s\right\}$ from above: $\mathcal{P}\left\{\overline{A}_r^s\right\} \leqslant 1 - \left(1 - C_1\frac{(\overline{\lambda}h)^{s+1}}{(s+1)!}\right)^r$, that leads to (34). Lemma 4 is proved. □

## Appendix E. Proof of Theorem 2

We have $\widetilde{X}_1 = (\mathbf{1}\psi_1^\top\pi)^{-1}\psi_1^\top\pi$, $\overline{X}_1 = (\mathbf{1}\xi_1^\top\pi)^{-1}\xi_1^\top\pi$ and $\Delta_1 = \widetilde{X}_1 - \overline{X}_1(s)$. Using the matrix algebra it is easy to verify that $[\gamma^\top\pi\mathbf{1} - \mathbf{1}\gamma^\top\pi I]\gamma^\top\pi \equiv 0$. Both the estimates are stable, hence $\|\widetilde{X}_1\|_1 = \|\overline{X}_1(s)\|_1 = 1$. The following relations are valid:

$$\|\Delta_1\|_1 = \frac{1}{\mathbf{1}\psi_1^\top\pi\mathbf{1}\xi_1^\top\pi}\|\mathbf{1}\xi_1^\top\pi\psi_1^\top\pi - \mathbf{1}\psi_1^\top\pi\xi_1^\top\pi\|_1 = \frac{1}{\mathbf{1}\psi_1^\top\pi\mathbf{1}\xi_1^\top\pi}\|\mathbf{1}\xi_1^\top\pi\gamma_1^\top\pi - \mathbf{1}\gamma_1^\top\pi\xi_1^\top\pi\|_1 =$$
$$= \frac{1}{\mathbf{1}\psi_1^\top\pi\mathbf{1}\xi_1^\top\pi}\|[\gamma_1^\top\pi\mathbf{1} - \mathbf{1}\gamma_1^\top\pi I]\xi_1^\top\pi\|_1 =$$
$$= \frac{1}{\mathbf{1}\psi_1^\top\pi\mathbf{1}\xi_1^\top\pi}\|[\gamma_1^\top\pi\mathbf{1} - \mathbf{1}\gamma_1^\top\pi I][\xi_1^\top\pi + \gamma_1^\top\pi]\|_1 = \frac{1}{\mathbf{1}\xi_1^\top\pi}\|[\gamma_1^\top\pi\mathbf{1} - \mathbf{1}\gamma_1^\top\pi I]\widetilde{X}_1\|_1 \leqslant$$
$$\leqslant \frac{1}{\mathbf{1}\xi_1^\top\pi}\|[\gamma_1^\top\pi\mathbf{1} - \mathbf{1}\gamma_1^\top\pi I]\|_1\|\widetilde{X}_1\|_1 \leqslant 2\frac{\mathbf{1}\overline{\gamma}_1^\top\pi}{\mathbf{1}\xi_1^\top\pi} = \sum_{i=1}^N\pi_i\frac{\sum_{j=1}^N\overline{\gamma}_1^{ij}}{\sum_{k,\ell=1}^N\xi_1^{k\ell}\pi_k}.$$

Using the last inequality, (41) and (A20), it can be shown that

$$\mathbf{E}\left\{\mathbf{I}_{a_1^s}(\omega)\|\Delta_1\|_1\right\} \leqslant 2\sum_{i=1}^N\pi_i\int_{\mathbb{R}^M}\sum_{i=1}^N\overline{\gamma}^{ij}(y)dy \leqslant 2\delta.$$

Since the latter inequality is valid for any $\pi \in \Pi$, we have an upper bound for the local distance characteristic:

$$\sup_{\pi\in\Pi} \mathbf{E}\left\{\mathbf{I}_{a_1^s}(\omega)\|\widetilde{\mathsf{X}}_1 - \overline{\mathsf{X}}_1(s)\|_1\right\} \leqslant 2\delta. \tag{A17}$$

Let us define the following products of the random matrices $\xi_r$ and $\psi_r$:

$$\Xi_{q,r} \triangleq \begin{cases} \xi_q\xi_{q+1}\dots\xi_r, & \text{if } q \leqslant r, \\ I & \text{otherwise,} \end{cases}$$

$$\Psi_{q,r} \triangleq \begin{cases} \psi_q\xi_{q+1}\dots\psi_r, & \text{if } q \leqslant r, \\ I & \text{otherwise,} \end{cases}$$

$$\Gamma_{q,r} \triangleq \Psi_{q,r} - \Xi_{q,r}.$$

To proceed the proof of Theorem 2 we need the following auxiliary

**Lemma A1.** *If $\phi_r \triangleq \phi_r(\mathsf{Y}_1,\dots,\mathsf{Y}_r)$ is a non-negative $\mathfrak{Y}_r$-measurable random value, and $\Phi_r \triangleq \frac{\phi_r}{\mathbf{1}\Xi_{1,r}^\top\pi}$, then*

$$\mathbf{E}\left\{\mathbf{I}_{A_r^s}(\omega)\Phi_r\right\} = \int_{\mathbb{R}^M}\cdots\int_{\mathbb{R}^M}\phi_r(y_1,\dots,y_r)dy_r\dots dy_1. \tag{A18}$$

**Proof of Lemma A1.** We consider a non-negative integrable function $\phi_1 = \phi_1(y) : \mathbb{R}^M \to \mathbb{R}_+$ and a $\mathfrak{Y}_1$-measurable random value

$$\Phi_1 \triangleq \frac{\phi_1(\mathsf{Y}_1)}{\mathbf{1}\xi_1^\top(\mathsf{Y}_1)\pi} = \frac{\phi_1(\mathsf{Y}_1)}{\sum_{i,j=1}^N\sum_{m=0}^s\int_{\mathcal{D}}\mathcal{N}(\mathsf{Y}_1,fu,\sum_{p=1}^N u^pG_p)\rho^{i,j,m}(du)\pi_i}. \tag{A19}$$

We find $\mathbf{E}\left\{\mathbf{I}_{a_1^s}(\omega)\Phi_1\right\}$:

$$\mathbf{E}\left\{\mathbf{I}_{a_1^s}(\omega)\Phi_1\right\} = \int_{\mathbb{R}^M}\int_{\mathcal{D}}\frac{\phi_1(y)\sum_{k,\ell=1}^N\sum_{n=0}^s\mathcal{N}(y,fv,\sum_{q=1}^N v^qG_q)\rho^{k,\ell,n}(dv)\pi_k}{\sum_{i,j=1}^N\sum_{m=0}^s\int_{\mathcal{D}}\mathcal{N}(y,fu,\sum_{p=1}^N u^pG_p)\rho^{i,j,m}(du)\pi_i}dy =$$
$$= \int_{\mathbb{R}^M}\phi_1(y)\frac{\sum_{k,\ell=1}^N\sum_{n=0}^s\int_{\mathcal{D}}\mathcal{N}(y,fv,\sum_{q=1}^N v^qG_q)\rho^{k,\ell,n}(dv)\pi_k}{\sum_{i,j=1}^N\sum_{m=0}^s\int_{\mathcal{D}}\mathcal{N}(y,fu,\sum_{p=1}^N u^pG_p)\rho^{i,j,m}(du)\pi_i}dy = \int_{\mathbb{R}^M}\phi_1(y)dy. \tag{A20}$$

Let us consider a non-negative integrable function $\phi_2 = \phi_1(y_1,y_2) : \mathbb{R}^{2M} \to \mathbb{R}_+$ and a $\mathfrak{Y}_2$-measurable random value

$$\Phi_2 \triangleq \frac{\phi_1(\mathsf{Y}_1,\mathsf{Y}_2)}{\mathbf{1}\Xi_{1,2}^\top(\mathsf{Y}_1,\mathsf{Y}_2)\pi} =$$
$$= \frac{\phi_2(\mathsf{Y}_1,\mathsf{Y}_2)}{\sum\limits_{i,i_2,j=1}^N\sum\limits_{m_1,m_2=0}^s\int_{\mathcal{D}}\int_{\mathcal{D}}\mathcal{N}(\mathsf{Y}_1,fu_1,\sum\limits_{p_1=1}^N u^{p_1}G_{p_1})\mathcal{N}(\mathsf{Y}_2,fu_2,\sum\limits_{p_2=1}^N u^{p_2}G_{p_2})\rho^{i,i_2,m_1}(du_1)\rho^{i_2,j,m_2}(du_2)\pi_i}.$$

We find $\mathbf{E}\left\{\mathbf{I}_{A_2^s}(\omega)\Phi_2\right\}$:

$$\mathbf{E}\left\{\mathbf{I}_{A_2^s}(\omega)\Phi_2\right\} = \int_{\mathbb{R}^M}\int_{\mathbb{R}^M}\phi_2(y_1,y_2)\times$$

$$\times\frac{\sum\limits_{k,k_2,\ell=1}^{N}\sum\limits_{n_1,n_2=0}^{s}\int_{\mathcal{D}}\int_{\mathcal{D}}\mathcal{N}(y_1,fv_1,\sum\limits_{q_1=1}^{N}v^{q_1}G_{q_1})\mathcal{N}(y_2,fv_2,\sum\limits_{q_2=1}^{N}v^{q_2}G_{q_2})\rho^{k,k_2,n_1}(dv_1)\rho^{k_2,\ell,n_2}(dv_2)\pi_k}{\sum\limits_{i,i_2,j=1}^{N}\sum\limits_{m_1,m_2=0}^{s}\int_{\mathcal{D}}\int_{\mathcal{D}}\mathcal{N}(y_1,fu_1,\sum\limits_{p_1=1}^{N}u^{p_1}G_{p_1})\mathcal{N}(y_2,fu_2,\sum\limits_{p_2=1}^{N}u^{p_2}G_{p_2})\rho^{i,i_2,m_1}(du_1)\rho^{i_2,j,m_2}(du_2)\pi_i}dy_2dy_1 =$$

$$= \int_{\mathbb{R}^M}\int_{\mathbb{R}^M}\phi_2(y_1,y_2)dy_2dy_1.$$

The correctness of the Lemma assertion in the general case of $\mathbf{E}\left\{\mathbf{I}_{A_r^s}(\omega)\Phi_r\right\}$ can be verified similarly. Lemma A1 is proved. □

Let us define an upper estimate for the norm of $\Delta_r = \widetilde{X}_r - \overline{X}_r$. From the definitions of $\Xi$, $\Psi$ and $\Gamma$ it follows that

$$\Gamma_{1,r} \triangleq \Psi_{1,r} - \Xi_{1,r} = \sum_{t=1}^{r}\Psi_{1,t-1}\gamma_t\Psi_{t+1,r}. \tag{A21}$$

Making the same inferences as for $\Delta_1$, we can deduce that

$$\|\Delta_r\|_1 \leqslant \frac{1}{\mathbf{1}\Xi_{1,r}^\top\pi}\|[\Gamma_{1,r}^\top\pi\mathbf{1} - \mathbf{1}\Gamma_{1,r}^\top\pi I]\|_1 \leqslant 2\sum_{t=1}^{r}\frac{1}{\mathbf{1}\Xi_{1,r}^\top\pi}\mathbf{1}\Psi_{t+1,r}^\top\overline{\gamma}_t^\top\Psi_{1,t-1}^\top\pi. \tag{A22}$$

To estimate the contribution of each summand in (A22) we use (A18). To simplify derivation we consider the case $r = 3$, function $\phi(y_1,y_2,y_3) : \mathbb{R}^{3M} \to \mathbb{R}_+$

$$\phi(y_1,y_2,y_3) = \mathbf{1}\psi^\top(y_3)\overline{\gamma}^\top(y_2)\psi^\top(y_1)\pi$$

and the $\mathfrak{Y}_3$-measurable random value $\Phi \triangleq \frac{\phi(Y_1,Y_2,Y_3)}{\mathbf{1}\Xi_{1,3}^\top(Y_1,Y_2,Y_3)\pi}$. Let us estimate from above the mathematical expectation

$$\mathbf{E}\left\{\mathbf{I}_{A_3^s}(\omega)\Phi\right\} = \int_{\mathbb{R}^M}\int_{\mathbb{R}^M}\int_{\mathbb{R}^M}\sum_{i,j,k,m=1}^{N}\pi_i\psi^{ij}(y_1)\overline{\gamma}^{jk}(y_2)\psi^{km}(y_3)dy_3dy_2dy_1 =$$

$$= \sum_{i,j,k=1}^{N}\pi_i\sum_{\ell=1}^{L}\varrho_\ell^{ij}\int_{\mathbb{R}^M}\overline{\gamma}^{jk}(y_2)dy_2\sum_{m=1}^{N}\sum_{n=1}^{L}\varrho_n^{km} = Q\sum_{i,j=1}^{N}\pi_i\sum_{\ell=1}^{L}\varrho_\ell^{ij}\sum_{k=1}^{N}\int_{\mathbb{R}^M}\overline{\gamma}^{jk}(y_2)dy_2 \leqslant$$

$$\leqslant Q\delta\sum_{i=1}^{N}\pi_i\sum_{j=1}^{N}\sum_{\ell=1}^{L}\varrho_\ell^{ij} \leqslant Q^2\delta.$$

Acting in the same way, we can prove that for arbitrary $r \geqslant 2$ the inequality

$$\mathbf{E}\left\{\mathbf{I}_{A_r^s}(\omega)\frac{\mathbf{1}\Psi_{t+1,r}^\top\overline{\gamma}_t^\top\Psi_{1,t-1}^\top\pi}{\mathbf{1}\Xi_{1,r}^\top\pi}\right\} \leqslant Q^{r-1}\delta$$

is valid for all $r$ summands in the RHS of (A22). Finally $\mathbf{E}\left\{\mathbf{I}_{A_r^s}(\omega)\|\Delta_r\|_1\right\} \leqslant 2rQ^{r-1}\delta$, and the correctness of (42) follows from the fact that the latter inequality is valid for arbitrary $\pi \in \Pi$. Theorem 2 is proved. □

## References


1. Wonham, W.M. Some Applications of Stochastic Differential Equations to Optimal Nonlinear Filtering. *J. Soc. Ind. Appl. Math. Series A Control* **1964**, *2*, 347–369, doi:10.1137/0302028. [CrossRef]
2. Kalman, R.E.; Bucy, R.S. New results in linear filtering and prediction theory. *Trans. ASME Ser. D J. Basic Eng.* **1961**, 95–108. [CrossRef]
3. Rabiner, L.R. A tutorial on hidden Markov models and selected applications in speech recognition. *Proc. IEEE* **1989**, *77*, 257–286, doi:10.1109/5.18626. [CrossRef]
4. Ephraim, Y.; Merhav, N. Hidden Markov processes. *IEEE Trans. Inf. Theory* **2002**, *48*, 1518–1569, doi:10.1109/TIT.2002.1003838. [CrossRef]
5. Cappé, O.; Moulines, E.; Ryden, T. *Inference in Hidden Markov Models*; Springer: Berlin/Heidelberg, Germany 2005. [CrossRef]
6. Elliott, R.J.; Moore, J.B.; Aggoun, L. *Hidden Markov Models: Estimation and Control*; Springer: New York, NY, USA, 1995. [CrossRef]
7. McLane, P.J. Optimal linear filtering for linear systems with state-dependent noise. *Int. J. Control* **1969**, *10*, 41–51, doi:10.1080/00207176908905798.
8. Dragan, V.; Aberkane, S. $\mathcal{H}_2$-optimal filtering for continuous-time periodic linear stochastic systems with state-dependent noise. *Syst. Control Lett.* **2014**, *66*, 35–42, doi:10.1016/j.sysconle.2013.12.020.
9. Dragan, V.; Morozan, T.; Stoica, A. *Mathematical Methods in Robust Control of Discrete-Time Linear Stochastic Systems*; Springer: New York, NY, USA, 2010. [CrossRef]
10. Liptser, R.; Shiryaev, A. *Statistics of Random Processes II: Applications*; Springer: Berlin/Heidelberg, Germany, 2001. [CrossRef]
11. Takeuchi, Y.; Akashi, H. Least-squares state estimation of systems with state-dependent observation noise. *Automatica* **1985**, *21*, 303 – 313, doi:10.1016/0005-1098(85)90063-9. [CrossRef]
12. Joannides, M.; LeGland, F. Nonlinear filtering with continuous time perfect observations and noninformative quadratic variation. In Proceedings of the 36th IEEE Conference on Decision and Control, San Diego, CA, USA, 10–12 December 1997; Volume 2, pp. 1645–1650, doi:10.1109/CDC.1997.657750.
13. Borisov, A. Optimal filtering in systems with degenerate noise in the observations. *Autom. Remote Control* **1998**, *59*, 1526–1537.
14. Crisan, D.; Kouritzin, M.; Xiong, J. Nonlinear filtering with signal dependent observation noise. *Electron. J. Probab.* **2009**, *14*, 1863–1883, doi:10.1214/EJP.v14-687. [CrossRef]
15. Kushner, H. *Probability Methods for Approximations in Stochastic Control and for Elliptic Equations*; Academic Press: New York, NY, USA, 1977. [CrossRef]
16. Kushner, H.J.; Dupuis, P.G. *Numerical Methods for Stochastic Control Problems in Continuous Time*; Springer: Berlin/Heidelberg, Germany, 1992.
17. Ito, K.; Rozovskii, B. Approximation of the Kushner Equation for Nonlinear Filtering. *SIAM J. Control Optim.* **2000**, *38*, 893–915, doi:10.1137/S0363012998344270. [CrossRef]
18. Clark, J. The design of robust approximations to the stochastic differential equations of nonlinear filtering. *Commun. Syst. Random Proc. Theory* **1978**, *25*, 721–734.
19. Malcolm, W.P.; Elliott, R.J.; van der Hoek, J. On the numerical stability of time-discretised state estimation via Clark transformations. In *42nd IEEE International Conference on Decision and Control*; IEEE: Piscataway, NJ, USA, 2003; Volume 2, pp. 1406–1412, doi:10.1109/CDC.2003.1272807.
20. Yin, G.; Zhang, Q.; Liu, Y. Discrete-time approximation of Wonham filters. *J. Control Theory Appl.* **2004**, *2*, 1–10. doi:10.1007/s11768-004-0017-7. [CrossRef]
21. Borisov, A.V. Wonham Filtering by Observations with Multiplicative Noises. *Autom. Remote Control* **2018**, *79*, 39–50, doi:10.1134/S0005117918010046.
22. Borisov, A.V.; Semenikhin, K.V. State Estimation by Continuous-Time Observations in Multiplicative Noise. *IFAC Pap. OnLine* **2017**, *50*, 1601–1606, doi:10.1016/j.ifacol.2017.08.316. [CrossRef]
23. Liptser, R.; Shiryaev, A. *Theory of Martingales*; Mathematics and its Applications; Springer: Dortrecht, The Netherlands, 1989. [CrossRef]
24. Stoyanov, J. *Counterexamples in Probability*; Wiley: Hoboken, NJ, USA, 1997. [CrossRef]
25. Kolmogorov, A.; Fomin, S. *Elements of the Theory of Functions and Functional Analysis*; Dover: Mineola, NY, USA, 1999. [CrossRef]

26. Platen, E.; Bruti-Liberati, N. *Numerical Solution of Stochastic Differential Equations with Jumps in Finance*; Springer: Berlin/Heidelberg, Germany, 2010. doi:10.1007/978-3-642-13694-8. [CrossRef]
27. Bertsekas, D.P.; Shreve, S.E. *Stochastic Optimal Control: The Discrete-Time Case*; Academic Press: New York, NY, USA, 1978.
28. Zolotarev, V. Metric Distances in Spaces of Random Variables and Their Distributions. *Math. USSR-Sbornik* **1976**, *30*, 373–401, doi:10.1070/sm1976v030n03abeh002280.
29. Zolotarev, V. Limit Theorems as Stability Theorems. *Theory Prob. Appl.* **1989**, *34*, 153–163, doi:10.1137/1134006.
30. Borovkov, A. *Asymptotic Methods in Queuing Theory*; John Wiley & Sons: Hoboken, NJ, USA, 1984. [CrossRef]
31. Liptser, R.; Shiryaev, A. *Statistics of Random Processes: I. General Theory*; Springer: Berlin/Heidelberg, Germany 2001. [CrossRef]

Σ mathematics

MDPI

*Article*

# On the Fractional Wave Equation

**Francesco Iafrate and Enzo Orsingher ***

Dipartimento di Scienze Statistiche, Sapienza, University of Rome, 00185 Rome, Italy; francesco.iafrate@uniroma1.it

* Correspondence: enzo.orsingher@uniroma1.it

Received: 13 May 2020; Accepted: 27 May 2020; Published: 31 May 2020

**Abstract:** In this paper we study the time-fractional wave equation of order $1 < \nu < 2$ and give a probabilistic interpretation of its solution. In the case $0 < \nu < 1$, $d = 1$, the solution can be interpreted as a time-changed Brownian motion, while for $1 < \nu < 2$ it coincides with the density of a symmetric stable process of order $2/\nu$. We give here an interpretation of the fractional wave equation for $d > 1$ in terms of laws of stable $d-$dimensional processes. We give a hint at the case of a fractional wave equation for $\nu > 2$ and also at space-time fractional wave equations.

**Keywords:** Hankel contours; multivariate stable processes; contour integrals; fractional laplacian

---

## 1. Introduction

In this paper we study in detail the solution of the time-fractional equation

$$\frac{\partial^\nu u}{\partial t^\nu} = c^2 \sum_{j=1}^{d} \frac{\partial^2 u}{\partial x_j^2} \tag{1}$$

for $1 < \nu < 2$ under the initial conditions

$$\begin{cases} u(x,0) = \delta(x) \\ u_t(x,0) = 0. \end{cases} \tag{2}$$

The time-fractional derivative is hereafter understood in the Caputo sense:

$$\frac{\partial^\nu u}{\partial t^\nu} = \frac{1}{\Gamma(m-\nu)} \int_0^t \frac{\partial^m}{\partial t^m} u(x,s)\,(t-s)^{m-\nu-1}\,\mathrm{d}s \qquad m-1 < \nu < m. \tag{3}$$

We first prove that the Fourier transform of the solution of the Cauchy problem (1) and (2) is

$$\mathcal{U}(\gamma_1, \dots, \gamma_d, t) = \mathcal{U}(\gamma, t) = E_{\nu,1}(-c^2\|\gamma\|^2 t^\nu) \tag{4}$$

where

$$E_{\nu,1}(x) = \sum_{k=0}^{\infty} \frac{x^k}{\Gamma(\nu k + 1)} \qquad x \in \mathbb{R} \tag{5}$$

is the one-parameter Mittag-Leffler function, first introduced in [1]. The representation of (4) as a contour integral on the Hankel path $H_a$

$$E_{\nu,1}(-c^2\|\gamma\|^2 t^\nu) = \frac{1}{2\pi i} \int_{H_a} \frac{e^w w^{\nu-1}}{w^\nu + t^\nu c^2 \|\gamma\|^2}\,\mathrm{d}w \tag{6}$$

permits us to obtain a representation of (6) as

$$E_{\nu,1}(-c^2\|\gamma\|^2t^\nu) = \frac{\sin \pi\nu}{\pi}\int_0^\infty \frac{z^{\nu-1}e^{-tzc^{2/\nu}\|\gamma\|^{2/\nu}}}{z^{2\nu}+2z^\nu\cos\pi\nu+1}\mathrm{d}z + \frac{1}{\nu}\left[e^{c^{2/\nu}\|\gamma\|^{2/\nu}te^{i\pi/\nu}} + e^{c^{2/\nu}\|\gamma\|^{2/\nu}te^{-i\pi/\nu}}\right]. \tag{7}$$

Some details about the representation (6) and the Hankel path can be found in [2]. For $d = 1$, $1 < \nu < 2$ the inversion of (7) is presented in [3] with the conclusion that the solution of (1) is the distribution of a stable symmetric process of order $2/\nu$.

We here show that for $d > 1, 1 < \nu < 2$ the solution can be expressed in terms of the law of a $d-$dimensional stable process $S_\alpha(t)$ with a suitable choice of the measure $\Gamma$ appearing in

$$\mathbb{E}e^{i\gamma\cdot S_\alpha(t)} = e^{-t\int_{\mathbb{S}^{d-1}}\|\gamma\cdot s\|^\alpha\left(1-i\,\mathrm{sign}(\gamma\cdot s)\tan\frac{\pi\alpha}{2}\right)\Gamma(\mathrm{d}s)}. \tag{8}$$

In particular, for $\Gamma$ uniform on the upper and lower hemispheres of $\mathbb{S}^{d-1} = \{s \in \mathbb{R}^d : \|s\| = 1\}$, we prove that (8) yields the characteristic functions in square brackets of Formula (7). We give also the explicit forms of $u(x,t)$ of the solution of (1) in terms of Bessel functions $J_{\frac{d}{2}-1}(\rho\|x\|)$, which for $d = 1$ can be reduced to Fujita's result. Some results concerning wave equations of fractional type can be found, e.g., in [4].

## 2. The Fractional Wave Equation

In this note we present some relationship between stable processes (and their inverses) with fractional equations. Stable processes are studied in depth in the monograph [5]. Some simple and well known results state that a symmetric stable process $S_\alpha(t), 0 < \alpha \leq 2$ with characteristic function

$$\mathbb{E}e^{i\gamma S_\alpha(t)} = e^{-|\gamma|^\alpha t} \qquad \gamma \in \mathbb{R}, t > 0 \tag{9}$$

has distribution $p_\alpha(x,t), x \in \mathbb{R}, t > 0$, satisfying the fractional equation

$$\frac{\partial p}{\partial t} = \frac{\partial^\alpha}{\partial|x|^\alpha}p \tag{10}$$

where $\frac{\partial^\alpha}{\partial|x|^\alpha}$ is the Riesz fractional derivative usually defined as

$$\frac{\partial^\alpha}{\partial|x|^\alpha}f(x) = \frac{1}{2\Gamma(m-\alpha)\cos(\pi\alpha/2)}\frac{d^m}{\mathrm{d}x^m}\int_{-\infty}^{+\infty}\frac{f(y)}{|x-y|^{\alpha+1-m}}\mathrm{d}y \qquad m-1<\alpha<m \tag{11}$$

with Fourier transform

$$\int_{-\infty}^{+\infty} e^{i\gamma x}\frac{\partial^\alpha}{\partial|x|^\alpha}f(x)\mathrm{d}x = -|\gamma|^\alpha\int_{-\infty}^{+\infty}e^{i\gamma x}f(x)\mathrm{d}x. \tag{12}$$

For the $d$-dimensional isotropic stable process $S_\alpha(t) = (S_\alpha^1(t),\dots,S_\alpha^d(t))$ with characteristic function,

$$\mathbb{E}e^{i\gamma S_\alpha^d(t)} = e^{-\|\gamma\|^\alpha t} \qquad \gamma \in \mathbb{R}^d, t > 0 \tag{13}$$

The corresponding probability law $p_\alpha(x_1,\dots,x_d,t) = p_\alpha(x,t)$ satisfies the equation

$$\frac{\partial p}{\partial t} = -(-\Delta)^\alpha p \tag{14}$$

where $-(-\Delta)^\alpha$ is the fractional Laplacian defined as the operator such that

$$-(-\Delta)^{\alpha} f(x) = \frac{1}{(2\pi)^d} \int_{\mathbb{R}^d} e^{-i(\gamma,x)} \|\gamma\|^{\alpha} \hat{f}(\gamma) \mathrm{d}\gamma\,, \qquad f \in Dom((-\Delta)^{\alpha}) \tag{15}$$

where $\hat{f}(\gamma)$ is the Fourier transform of a function $f(x), x \in \mathbb{R}^d$ and the domain of the operator $Dom((-\Delta)^{\alpha})$ is

$$Dom((-\Delta)^{\alpha}) = \left\{ f \in L^1_{loc}(R^n) : \int_{\mathbb{R}^n} |\hat{f}(\gamma)|^2 (1 + \|\gamma\|^{2\alpha})\, \mathrm{d}\gamma < \infty \right\}$$

(on this point see for example [6]). The connection between fractional operators and stochastic processes is explored, e.g., in [7]. A detailed comparison of the several possible definitions of the fractional Laplacian can be found in [8]. For the time-fractional equation (see [9]),

$$\begin{cases} \frac{\partial^{\nu} p}{\partial t^{\nu}} = c^2 \frac{\partial^2 p}{\partial x^2} & 0 < \nu \leq 2,\, x \in \mathbb{R},\, t > 0 \\ u(x,0) = \delta(x) \\ u_t(x,0) = 0, \end{cases} \tag{16}$$

we have that the solution of the Cauchy problem is explicitly given by

$$u(x,t) = \frac{1}{2ct^{\nu/2}} W_{-\nu/2, 1-\nu/2}\left(-\frac{|x|}{ct^{\nu}}\right) \tag{17}$$

where

$$W_{\alpha,\beta}(x) = \sum_{k=0}^{\infty} \frac{x^k}{\Gamma(\alpha k + \beta)} \qquad \alpha > -1, b > 0, x \in \mathbb{R}$$

is the Wright function. The $d-$dimensional counterpart of (16) is

$$\begin{cases} \frac{\partial^{\nu} p}{\partial t^{\nu}} = c^2 \Delta u & 0 < \nu \leq 2,\, x \in \mathbb{R}^d,\, t > 0 \\ u(x,0) = \delta(x) \\ u_t(x,0) = 0. \end{cases} \tag{18}$$

Some details about time-fractional derivatives can be found in [10]. For $0 < \nu < 1$ the solution of (18) corresponds to the distribution of the vector process

$$B\left(L_{\nu}(t)\right), \quad t \geq 0 \tag{19}$$

where $B(t) = (B_1(t), \ldots, B_d(t))$ is the $d-$dimensional Brownian motion and $L_{\nu}(t)$ is the inverse of the stable subordinator $H_{\nu}(t)$ (see [11]).

In the more general case

$$\begin{cases} \frac{\partial^{\nu} p}{\partial t^{\nu}} = -c^2 (-\Delta)^{\alpha} u & 0 < \nu \leq 2,\, 0 < \alpha \leq 1,\, x \in \mathbb{R}^d,\, t > 0 \\ u(x,0) = \delta(x) \\ u_t(x,0) = 0. \end{cases} \tag{20}$$

The solution of the Cauchy problem (20) is the probability density of the process

$$S_{\alpha}(L_{\nu}(t)), \; t \geq 0 \tag{21}$$

where $S^{\alpha}(t) = (S^1_{\alpha}(t), \ldots, S^d_{\alpha}(t))$ is an isotropic stable process (see [11]). We here consider the case where in (18) and (20) the order of the fractional derivative is $1 < \nu \leq 2$.We start first with (18) and observe that the Laplace–Fourier transform of the solution $u_{\alpha}(x,t)$ is

$$\int_0^\infty e^{-\lambda t} \int_{\mathbb{R}^d} e^{i\gamma\cdot x} u(x,t) \mathrm{d}x = \frac{\lambda^{\nu-1}}{\lambda^\nu + c^2 \|\gamma\|^2} \tag{22}$$

and the Fourier transform reads

$$\int_{\mathbb{R}^d} e^{i\gamma\cdot x} u(x,t)\, \mathrm{d}x = E_{\nu,1}(-c^2\|\gamma\|^2 t^\nu). \tag{23}$$

The Mittag-Leffler function $E_{\nu,1}(-c^2\|\gamma\|^2 t^\nu)$ can be represented as a contour integral on the Hankel path as

$$E_{\nu,1}(-c^2\|\gamma\|^2 t^\nu) = \frac{1}{2\pi i} \int_{H_a} \frac{e^w w^{\nu-1}}{w^\nu + t^\nu c^2 \|\gamma\|^2} \mathrm{d}w \tag{24}$$

where $Ha$ is the contour in the complex plane represented in Figure 1.

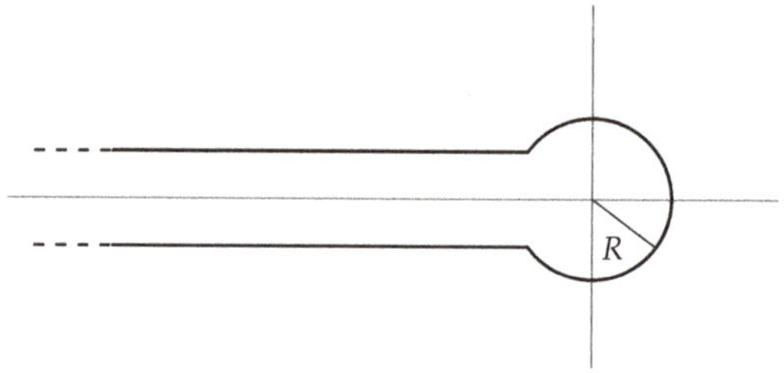


**Figure 1.** Hankel path in the complex plane.

The representation (24) is a consequence of the integral representation of the inverse of the Gamma function

$$\frac{1}{\Gamma(\nu)} = \frac{1}{2\pi i} \int_{Ha} e^w w^{-\nu} \mathrm{d}w.$$

The integral in (24) can be developed by inserting a ring of radius $\epsilon < R$.

The contour $C$ is composed by the circumferences $C_R$ and $C_\epsilon$ with two segments joining $Re^{-i\pi}$ with $\epsilon e^{-i\pi}$ and $Re^{i\pi}$ with $\epsilon e^{i\pi}$, and is run counterclockwise. See Figure 2.

In order to evaluate

$$\frac{1}{2\pi i} \int_C \frac{e^w w^{\nu-1}}{w^\nu + t^\nu c^2 \|\gamma\|^2} \mathrm{d}w \tag{25}$$

we perform the transformation $w^\nu = z^{2m}$ for $2m - 1 < \nu < 2m$.

The contour of Figure 2 after the transformation $w^\nu = z^{2m}$ takes the form shown in Figure 3.

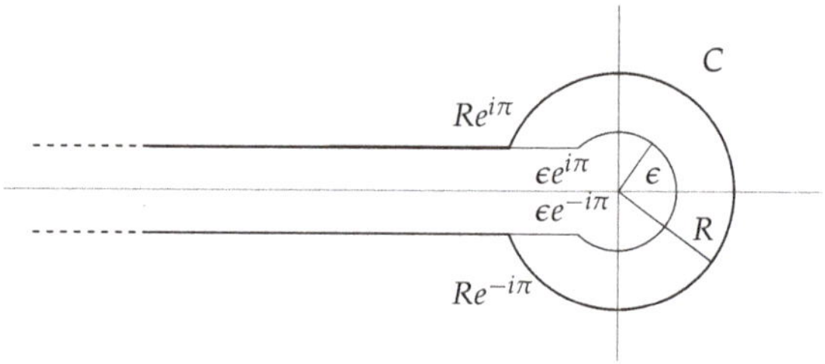


**Figure 2.** Representation of the contour $C$.

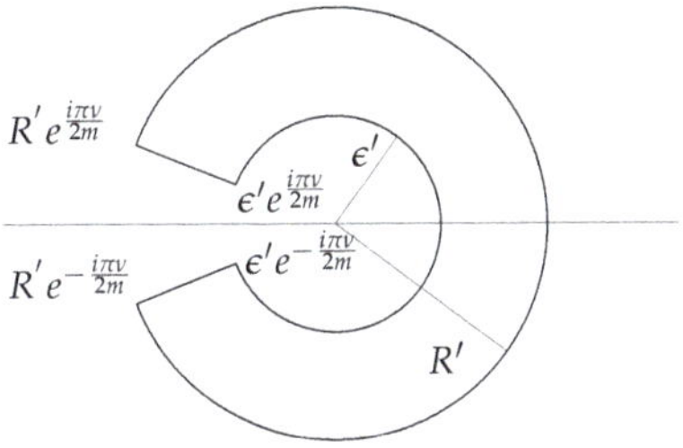


**Figure 3.** Representation of the contour $C'$, with $R' = R^{\frac{\nu}{2m}}$ and $\epsilon' = \epsilon^{\frac{\nu}{2m}}$.

Therefore, the horizontal segments of Figure 2 are rotated by an angle of amplitude $\pm\pi\nu/2m$ and the radii are subject to contraction or dilation according to the value of $\nu$. The integral on $C'$ thus obtained from (25) is

$$\frac{1}{2\pi i}\int_C \frac{e^w w^{\nu-1}}{w^\nu + t^\nu c^2\|\gamma\|^2}\,\mathrm{d}w = \frac{2m}{2\pi\nu i}\int_{C'} \frac{e^{z^{2m/\nu}} z^{2m-1}}{z^{2m} + t^\nu c^2\|\gamma\|^2}\,\mathrm{d}z. \tag{26}$$

The integral on the right side of (26) can be evaluated by means of the Cauchy residue theorem. The function

$$f(z) = \frac{e^{z^{2m/\nu}} z^{2m-1}}{z^{2m} + t^\nu c^2\|\gamma\|^2} \qquad z \in \mathbb{C} \tag{27}$$

has $2m$ poles at points $z_k = e^{i\pi\frac{(2k+1)}{2m}}(c^2\|\gamma\|^2 t^\nu)^{\frac{1}{2m}}$ for $0 \le k \le 2m-1$. It is easy to show that the residues of (27) at the poles $z_k$ are given by

$$\lim_{z\to z_k}(z - z_k)\frac{e^{z^{2m/\nu}} z^{2m-1}}{z^{2m} + t^\nu c^2\|\gamma\|^2} = \frac{e^{z_k^{2m/\nu}}}{2m}. \tag{28}$$

Thus the integral (26) can be written as

$$\frac{1}{2\pi i}\int_{C'} f(z)\,\mathrm{d}z = \frac{1}{2m}\sum_{k=0}^{2m-1} e^{z_k^{2m/\nu}} \qquad \text{where } z_k = e^{i\pi\frac{2k+1}{2m}}(c^2\|\gamma\|^2 t^\nu)^{\frac{1}{2m}} \tag{29}$$

By adding the contribution of the segments $(Re^{-i\pi}, \epsilon e^{-i\pi})$ and $(Re^{i\pi}, \epsilon e^{i\pi})$ for $R \to \infty$ and $\epsilon \to 0$ we obtain

$$E_{\nu,1}(-c^2\|\gamma\|^2 t^\nu) = \frac{\sin\pi\nu}{\pi}\int_0^\infty \frac{z^{\nu-1}e^{-tzc^{2/\nu}\|\gamma\|^{2/\nu}}}{z^{2\nu} + 2z^\nu\cos\pi\nu + 1}\mathrm{d}z + \frac{2m}{\nu}\left[\frac{1}{2m}\sum_{k=0}^{2m-1} e^{z_k^{2m/\nu}}\right]. \tag{30}$$

For $m = 1$ we must distinguish the cases $0 < \nu < 1$ where

$$E_{\nu,1}(-c^2\|\gamma\|^2 t^\nu) = \frac{\sin\pi\nu}{\pi}\int_0^\infty \frac{z^{\nu-1}e^{-tzc^{2/\nu}\|\gamma\|^{2/\nu}}}{z^{2\nu} + 2z^\nu\cos\pi\nu + 1}\mathrm{d}z \tag{31}$$

and $1 < \nu < 2$, where

$$\begin{aligned} E_{\nu,1}(-c^2\|\gamma\|^2 t^\nu) = {} & \frac{\sin\pi\nu}{\pi}\int_0^\infty \frac{z^{\nu-1}e^{-tzc^{2/\nu}\|\gamma\|^{2/\nu}}}{z^{2\nu} + 2z^\nu\cos\pi\nu + 1}\mathrm{d}z \\ & + \frac{1}{\nu}\left[e^{c^{2/\nu}\|\gamma\|^{2/\nu}te^{i\pi/\nu}} + e^{c^{2/\nu}\|\gamma\|^{2/\nu}te^{-i\pi/\nu}}\right]. \end{aligned} \tag{32}$$

In order to simplify the formulas involved in the analysis we take $c = 1$.

For $m = 2$ we have the subcases $2 < \nu < 3$ and $3 < \nu < 4$. In the first case the contour integral of Figure 3 involves two poles and thus yields two additional terms in the representation of the Mittag-Leffler function (32). In the second case we have the contribution of four poles in the contour integral of Figure 3, so that for $3 < \nu < 4$

$$E_{\nu,1}(-c^2\|\gamma\|^2 t^\nu) = \frac{\sin \pi\nu}{\pi} \int_0^\infty \frac{z^{\nu-1} e^{-tzc^{2/\nu}\|\gamma\|^{2/\nu}}}{z^{2\nu} + 2z^\nu \cos \pi\nu + 1} \mathrm{d}z + \frac{1}{\nu}\left[e^{\|\gamma\|^{2/\nu} t e^{i\pi/4\nu}} + e^{\|\gamma\|^{2/\nu} t e^{i3\pi/4\nu}} + e^{\|\gamma\|^{2/\nu} t e^{i5\pi/4\nu}} + e^{\|\gamma\|^{2/\nu} t e^{i7\pi/4\nu}}\right]. \tag{33}$$

The contours for $2 < \nu < 3$ and $3 < \nu < 4$ are depicted below (Figure 4).

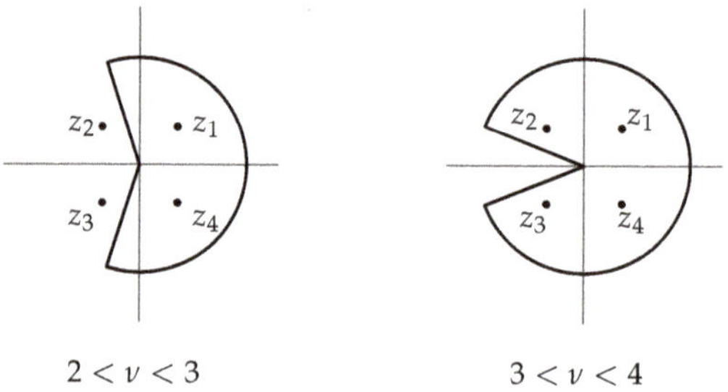


**Figure 4.** Representation of the contour $C'$ for $m = 2$. The dots indicate the poles of $f(z)$.

The substantial difference between the cases $1 < \nu < 2$ and $\nu > 2$ is that in the first case we have that $\Re(e^{\pm i\pi/\nu})$ is negative and the contribution of the poles correspond to the characteristic function of stable processes, whereas for $2 < \nu < 3$, $\Re(e^{i\pi/\nu})$ and $\Re(e^{7i\pi/\nu})$ are positive and thus are not characteristic functions of random variables. Let us now concentrate our attention on the integrals in Equations (31) and (32) (which is also true in the general case for $\nu > 2$). If we write

$$\begin{aligned} &\frac{\sin \pi\nu}{\pi} \int_0^\infty \frac{z^{\nu-1} e^{-tz\|\gamma\|^{2/\nu}}}{z^{2\nu} + 2z^\nu \cos \pi\nu + 1} \mathrm{d}z \\ &= \frac{\sin \pi\nu}{\pi\nu} \int_0^\infty \frac{e^{-tw^{1/\nu}\|\gamma\|^{2/\nu}}}{w^2 + 2w\cos \pi\nu + 1} \mathrm{d}w \\ &= \frac{\sin \pi\nu}{\pi\nu} \int_0^\infty \frac{e^{-tw^{1/\nu}\|\gamma\|^{2/\nu}}}{(w + \cos \pi\nu)^2 + \sin^2 \pi\nu} \mathrm{d}w \\ &= \mathbb{E} e^{-t\|\gamma\|^{2/\nu} W^{1/\nu}} \end{aligned} \tag{34}$$

where $W$ is a non-negative r.v. with density

$$f(w) = \frac{\sin \pi\nu}{\pi\nu} \frac{\mathrm{d}w}{(w + \cos \pi\nu)^2 + \sin^2 \pi\nu} \qquad w > 0,\ 0 < \nu < 1$$

Note that for $1 < \nu < 2$ the function (2) is negative on $(0, \infty)$. We note also that the r.v. $W_\nu, 0 < \nu < 1$ with density

$$P(W_\nu \in \mathrm{d}w)/\mathrm{d}w = \frac{\sin \pi\nu}{\pi} \frac{w^{\nu-1}}{1 + w^{2\nu} + 2w^\nu \cos \pi\nu} \qquad w > 0 \tag{35}$$

appearing in (35) has the same distribution as the ratio of two independent stable subordinators of degree $0 < \nu < 1$.

We now give the inverse Fourier transform of (34) for $0 < \nu < 1$, $d = 1$.

$$
\begin{aligned}
p_\nu(x,t) &= \frac{1}{2\pi}\int_{-\infty}^{\infty} e^{-i\gamma x} E_{\nu,1}(-\gamma^2 t^\nu)\mathrm{d}\gamma \\
&= \frac{1}{2\pi}\int_{-\infty}^{\infty} e^{-i\gamma x}\frac{\sin\pi\nu}{\pi}\int_0^\infty \frac{z^{\nu-1}e^{-tz\gamma^{2/\nu}}}{z^{2\nu}+2z^\nu\cos\pi\nu+1}\mathrm{d}z\,\mathrm{d}\gamma = (tz\gamma^{\frac{2}{\nu}} = w) \\
&= \frac{1}{2\pi}\int_{-\infty}^{\infty} e^{-i\gamma x}\frac{\sin\pi\nu}{\pi}\int_0^\infty \frac{e^{-w}w^{\nu-1}t^\nu\gamma^2}{w^{2\nu}+t^{2\nu}\gamma^4+2t^\nu\gamma^2 w^\nu\cos\pi\nu}\mathrm{d}w\,\mathrm{d}\gamma.
\end{aligned} \tag{36}
$$

We start by evaluating the following integral

$$
\begin{aligned}
&\frac{1}{2\pi}\int_{-\infty}^{\infty} e^{-i\gamma x}\frac{t^\nu\gamma^2}{w^{2\nu}+t^{2\nu}\gamma^4+2t^\nu\gamma^2w^\nu\cos\pi\nu}\mathrm{d}\gamma \\
&= \frac{1}{2\pi t^{\nu/2}}\int_{-\infty}^{\infty}\frac{e^{-ix\gamma' t^{-\nu/2}}\gamma'^2}{\gamma'^4+2\gamma'^2w^\nu\cos\pi\nu+w^{2\nu}}\mathrm{d}\gamma' \\
&= \frac{1}{2\pi(wt)^{\nu/2}}\int_{-\infty}^{\infty}\frac{e^{-ix\gamma(w/t)^{\nu/2}}\gamma^2}{1+\gamma^4+2\gamma^2\cos\pi\nu}\mathrm{d}\gamma \\
&= \frac{1}{2\pi(wt)^{\nu/2}}\int_{-\infty}^{\infty}\frac{e^{-i\gamma A}\gamma^2}{1+\gamma^4+2\gamma^2\cos\pi\nu}\mathrm{d}\gamma
\end{aligned}
$$

where $A = (w/t)^{\nu/2}x$.

We must now evaluate the integral of

$$
f(z) = \frac{z^2e^{-izA}}{z^4+2z^2\cos\pi\nu+1} \qquad z\in\mathbb{C}
$$

on a suitable contour $C_R$. The four roots of $z^4+2z^2\cos\pi\nu+1=0$ are

$$
\begin{cases}
z_1 = e^{\frac{i\pi\nu}{2}-i\frac{\pi}{2}} = \sin\frac{\pi\nu}{2} - i\cos\frac{\pi\nu}{2} \\
z_2 = e^{i\frac{\pi\nu}{2}+i\frac{\pi}{2}} = -\sin\frac{\pi\nu}{2} + i\cos\frac{\pi\nu}{2} \\
z_3 = e^{-i\frac{\pi\nu}{2}-i\frac{\pi}{2}} = -\sin\frac{\pi\nu}{2} - i\cos\frac{\pi\nu}{2} \\
z_4 = e^{-i\frac{\pi\nu}{2}+i\frac{\pi}{2}} = \sin\frac{\pi\nu}{2} + i\cos\frac{\pi\nu}{2}
\end{cases}
$$

and are located in C as in Figure 5, because $1<\nu<2$.

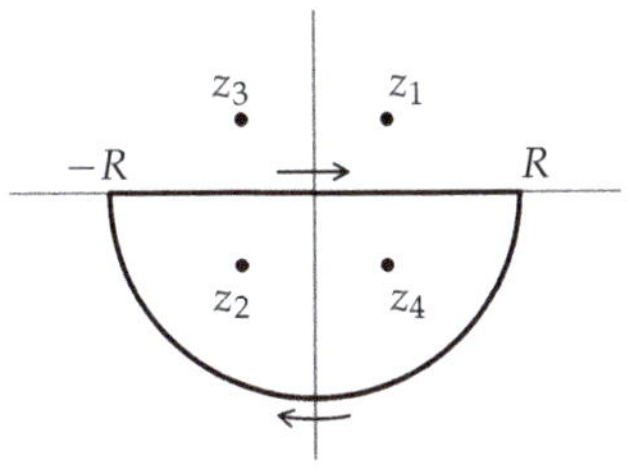


**Figure 5.** Integration contour for $x>0$.

We observe that $e^{-izA} = e^{-i(u+iv)x(w/t)^{\nu/2}} = e^{vA}e^{-iuA}$ for $x>0$ and $v<0$, the curvilinear integral on the half-circle $Re^{i\theta}$, $0\le\theta\le\pi$ tends to zero as $R\to\infty$. By the residue theorem we thus have

$$
\int_{C_R} f(z)\,\mathrm{d}z = -2\pi i(R_{z_2}+R_{z_4}) \tag{37}
$$

The minus sign is due to the fact that the contour in Figure 5 is run clockwise.

The residues $R_{z_2}$ and $R_{z_4}$ have the following values

$$R_{z_2} = -\frac{e^{i\frac{\pi\nu}{2}} e^{Ae^{i\pi\nu/2}}}{4\sin\pi\nu} \qquad R_{z_4} = \frac{e^{-i\frac{\pi\nu}{2}} e^{Ae^{-i\pi\nu/2}}}{4\sin\pi\nu} \tag{38}$$

and thus

$$\begin{aligned} \int_{C_R} f(z)\,\mathrm{d}z &= -\frac{2\pi i}{2^2\sin\pi\nu}\left(e^{-i\frac{\pi\nu}{2}} e^{Ae^{-i\pi\nu/2}} - e^{i\frac{\pi\nu}{2}} e^{Ae^{i\pi\nu/2}}\right) \\ &= -\frac{\pi}{\sin\pi\nu} e^{A\cos\frac{\pi\nu}{2}} \sin\left(\frac{\pi\nu}{2} + A\sin\frac{\pi\nu}{2}\right) \end{aligned} \tag{39}$$

For $x < 0$, the integration of $f(z)$ must be performed on the contour of Figure 6 and

$$\int_{C_R} f(z)\,\mathrm{d}z = -2\pi i(R_{z_1} + R_{z_3}) \tag{40}$$

the sign being in this case positive because the path is run counterclockwise.

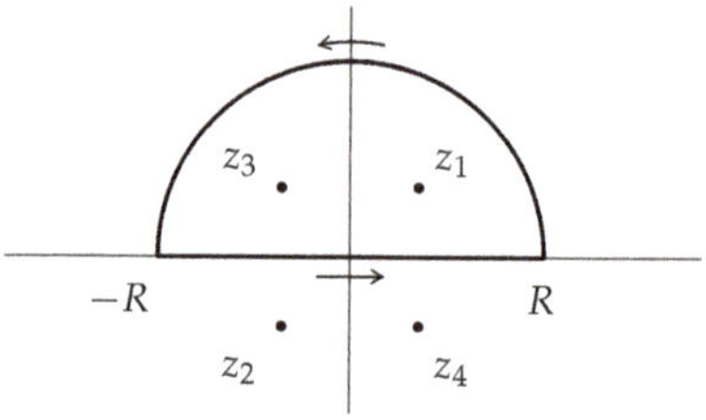


**Figure 6.** Integration contour for $x < 0$.

The residues in this case are

$$R_{z_1} = \frac{e^{i\pi\nu/2} e^{-Ae^{i\pi\nu/2}}}{4\sin\pi\nu} \qquad R_{z_3} = -\frac{e^{-i\pi\nu/2} e^{-Ae^{-i\pi\nu/2}}}{4\sin\pi\nu}$$

The integral (40), therefore, takes the form

$$\begin{aligned} \int_{C_R} f(z)\,\mathrm{d}z &= \frac{2\pi i}{4\sin\pi\nu}\left(e^{i\pi\nu/2} e^{-Ae^{i\pi\nu/2}} - e^{-i\pi\nu/2} e^{-Ae^{-i\pi\nu/2}}\right) \\ &= -\frac{\pi}{\sin\pi\nu} e^{-A\cos\frac{\pi\nu}{2}} \sin\left(\frac{\pi\nu}{2} - A\sin\frac{\pi\nu}{2}\right) \end{aligned} \tag{41}$$

In conclusion we have that

$$\begin{aligned} &\int_{-\infty}^{\infty} \frac{e^{-ix\gamma(w/t)^{\nu/2}}\gamma^2}{1+\gamma^4+2\gamma^2\cos\pi\nu}\mathrm{d}\gamma \\ &= -\frac{\pi}{\sin\pi\nu} e^{|x|\cos\frac{\pi\nu}{2}(w/t)^{\nu/2}} \sin\left(\frac{\pi\nu}{2} + |x|\left(\frac{w}{t}\right)^{\frac{\nu}{2}} \sin\frac{\pi\nu}{2}\right). \end{aligned} \tag{42}$$

We now consider the integration with respect to $w$ in (36). This leads to the evaluation of the following integral

$$\frac{1}{2\pi}\int_0^\infty e^{-w} w^{\nu-1} \frac{\sin\pi\nu}{\pi(wt)^{\nu/2}} \left[-\frac{\pi}{\sin\pi\nu} e^{|x|\cos\frac{\pi\nu}{2}(w/t)^{\nu/2}} \sin\left(\frac{\pi\nu}{2}+|x|\left(\frac{w}{t}\right)^{\frac{\nu}{2}}\sin\frac{\pi\nu}{2}\right)\right] dw \tag{43}$$
$$= -\frac{1}{2\pi t^{\frac{\nu}{2}}}\int_0^\infty e^{-w} w^{\frac{\nu}{2}-1} e^{|x|\cos\frac{\pi\nu}{2}(\frac{w}{t})^{\nu/2}} \sin\left(\frac{\pi\nu}{2}+|x|\left(\frac{w}{t}\right)^{\frac{\nu}{2}}\sin\frac{\pi\nu}{2}\right) dw$$
$$= -\frac{1}{\pi\nu t^{\frac{\nu}{2}}}\int_0^\infty e^{-tw^{\frac{2}{\nu}}} (w^{\frac{2}{\nu}} t)^{\frac{\nu}{2}-1} \frac{2}{\nu} t w^{\frac{2}{\nu}-1} e^{w|x|\cos\frac{\pi\nu}{2}} \sin\left(\frac{\pi\nu}{2}+w|x|\sin\frac{\pi\nu}{2}\right) dw$$
$$= -\frac{1}{\pi\nu}\int_0^\infty e^{-tw^{\frac{2}{\nu}}} e^{w|x|\cos\frac{\pi\nu}{2}} \sin\left(\frac{\pi\nu}{2}+w|x|\sin\frac{\pi\nu}{2}\right) dw$$

The last step of (43) can be developed as follows

$$-\frac{1}{\pi\nu}\int_0^\infty e^{-tw^{2/\nu}} e^{w|x|\cos\frac{\pi\nu}{2}} \left[e^{i\left(\frac{\pi\nu}{2}+w|x|\sin\frac{\pi\nu}{2}\right)} - e^{-i\left(\frac{\pi\nu}{2}+w|x|\sin\frac{\pi\nu}{2}\right)}\right] \frac{dw}{2i} \tag{44}$$
$$= -\frac{e^{i\pi\nu/2}}{2i\pi\nu}\int_0^\infty e^{-tw^{2/\nu}} e^{w|x|e^{i\pi\nu/2}} dw + \frac{e^{-i\pi\nu/2}}{2i\pi\nu}\int_0^\infty e^{-tw^{2/\nu}} e^{w|x|e^{-i\pi\nu/2}} dw$$

We evaluate the first integral in (44) by taking the contour integral of the function

$$f(z) = e^{-tz^{2/\nu}+|x|ze^{i\pi\nu 2}} \qquad z \in \mathbb{C}$$

along the path $C_R$ depicted in Figure 7.

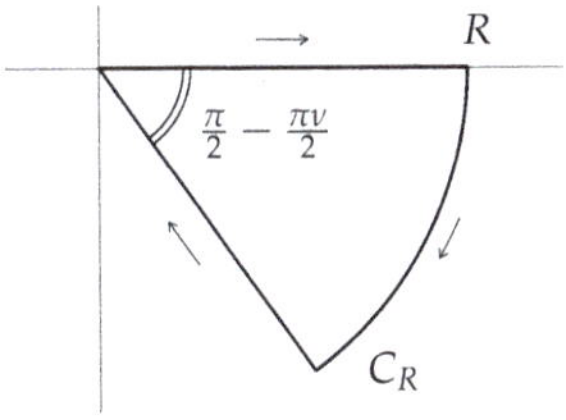


**Figure 7.** Path $C_R$ corresponding to the change of variables $z' = e^{i\frac{\pi}{2}(1-\nu)}z$ in the first integral of (44).

By the Cauchy theorem we have that

$$\int_{C_R} f(z)\,\mathrm{d}z = \int_0^R f(w)\,\mathrm{d}w + \int_C f(Re^{i\theta})\,\mathrm{d}\theta + \int_R^0 f(ze^{i\frac{\pi}{2}-i\frac{\pi\nu}{2}})e^{i\frac{\pi}{2}-i\frac{\pi\nu}{2}}\,\mathrm{d}z = 0 \tag{45}$$

The integral on the arc $C$ tends to zero because

$$\left|\int_C f(Re^{i\theta})\,\mathrm{d}\theta\right| \leq \int_C |f(Re^{i\theta})|\,\mathrm{d}\theta = \int_{\frac{(1-\nu)\pi}{2}}^0 e^{-tR^{\frac{2}{\nu}}\cos\frac{2\theta}{\nu}} e^{|x|R\cos\left(\theta+\frac{\pi\nu}{2}\right)}\,\mathrm{d}\theta \tag{46}$$

Since $\pi > \theta + \frac{\pi\nu}{2} > \frac{\pi}{2}$ as $\theta \in (\frac{(1-\nu)\pi}{2}, 0)$, the exponent of the second factor of (46) is negative as well as the first one because $\cos(2\theta/\nu)$ for $\theta$ ranging in the same interval.

We thus conclude that (46) converges to zero as $R \to \infty$ and thus (45) yields

$$\int_0^\infty e^{-tw^{2/\nu}} e^{w|x|e^{i\frac{\pi\nu}{2}}}\,\mathrm{d}w = \int_0^\infty e^{i\frac{\pi}{2}-i\frac{\pi\nu}{2}} e^{-t\left(ze^{i\frac{\pi}{2}-i\frac{\pi\nu}{2}}\right)^{2/\nu}+i|x|z}\,\mathrm{d}z \tag{47}$$
$$= ie^{-i\frac{\pi\nu}{2}}\int_0^\infty e^{i|x|z}e^{tz^{2/\nu}e^{i\pi/\nu}}\,\mathrm{d}z$$

In order to evaluate the second integral of (44) we integrate

$$f(z) = e^{-tz^{2/\nu}+|x|ze^{-i\pi\nu 2}} \qquad z \in \mathbb{C}$$

along the contour of Figure 8.

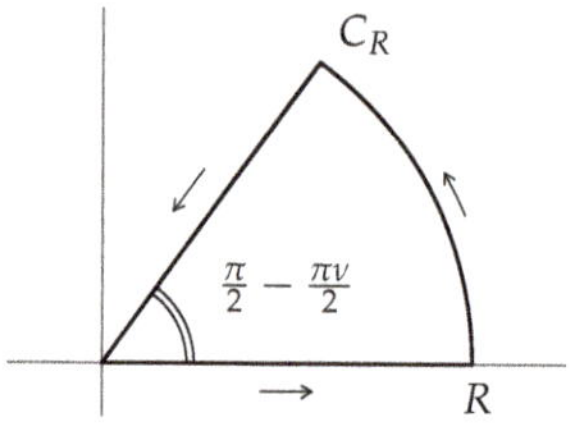


**Figure 8.** Path $C_R$ corresponding to the change of variables $z' = e^{i\frac{\pi}{2}(\nu-1)}z$ in the second integral of (44).

By performing the same steps as above we obtain

$$\begin{aligned}\int_0^\infty e^{-tw^{2/\nu}} e^{w|x|e^{-i\frac{\pi\nu}{2}}}\,\mathrm{d}w &= \int_0^\infty e^{i\frac{\pi\nu}{2}-i\frac{\pi}{2}} e^{-t\left(ze^{i\frac{\pi\nu}{2}-i\frac{\pi}{2}}\right)^{2/\nu}+|x|ze^{i\frac{\pi\nu}{2}-i\frac{\pi}{2}}e^{-i\frac{\pi\nu}{2}}}\,\mathrm{d}z \\ &= -ie^{i\frac{\pi\nu}{2}}\int_0^\infty e^{-i|x|z} e^{tz^{2/\nu}e^{-i\pi/\nu}}\,\mathrm{d}z\end{aligned} \tag{48}$$

in view of (47) and (48) the integral (44) becomes

$$\begin{aligned}&-\frac{1}{2\pi\nu}\int_0^\infty e^{i|x|z}e^{tz^{2/\nu}e^{i\pi/\nu}}\,\mathrm{d}z - \frac{1}{2\pi\nu}\int_0^\infty e^{-i|x|z}e^{tz^{2/\nu}e^{-i\pi/\nu}}\,\mathrm{d}z \\ &= -\frac{1}{2\pi\nu}\int_0^\infty e^{i|x|z}e^{tz^{2/\nu}e^{i\pi/\nu}}\,\mathrm{d}z - \frac{1}{2\pi\nu}\int_{-\infty}^0 e^{i|x|z}e^{t(-z)^{2/\nu}e^{-i\pi/\nu}}\,\mathrm{d}z \\ &= -\frac{1}{2\pi\nu}\int_{-\infty}^{+\infty} e^{i|x|z}e^{t|z|^{2/\nu}e^{\frac{i\pi}{\nu}\mathrm{sign}z}}\,\mathrm{d}z \\ &= -\frac{1}{2\pi\nu}\int_{-\infty}^{+\infty} e^{-i|x|z}e^{t|z|^{\frac{2}{\nu}}e^{-\frac{i\pi}{\nu}\mathrm{sign}z}}\,\mathrm{d}z\end{aligned}$$

From (32), $d = 1$, we conclude that

$$\begin{aligned}p_\nu(x,t) &= \frac{1}{2\pi}\int_{-\infty}^{+\infty} e^{-i\gamma x}E_{\nu,1}(-\gamma^2 t^\nu)\,\mathrm{d}\gamma \\ &= \frac{1}{2\pi\nu}\left[-\int_{-\infty}^{+\infty} e^{-i|x|\gamma}e^{t|\gamma|^{2/\nu}e^{-\frac{i\pi}{\nu}\mathrm{sign}\gamma}}\,\mathrm{d}\gamma + \int_{-\infty}^{+\infty} e^{-ix\gamma}e^{t|\gamma|^{2/\nu}e^{\frac{i\pi}{\nu}\mathrm{sign}\gamma}}\,\mathrm{d}\gamma + \int_{-\infty}^{+\infty} e^{-ix\gamma}e^{t|\gamma|^{2/\nu}e^{-\frac{i\pi}{\nu}\mathrm{sign}\gamma}}\,\mathrm{d}\gamma\right] \\ &= \frac{1}{2\pi\nu}\int_{-\infty}^{+\infty} e^{i|x|\gamma}e^{t|\gamma|^{2/\nu}e^{-\frac{i\pi}{\nu}\mathrm{sign}\gamma}}\,\mathrm{d}\gamma\end{aligned} \tag{49}$$

**Remark 1.** *The function $h(\gamma, \frac{2}{\nu}) = e^{t|\gamma|^{2/\nu}e^{-\frac{i\pi}{\nu}\mathrm{sign}\gamma}} = e^{t|\gamma|^{\frac{2}{\nu}}\cos\frac{\pi}{\nu}(1-i\mathrm{sign}\gamma\tan\frac{\pi}{\nu})}$ is the characteristic function of a stable random variable of order $1 < 2/\nu < 2$ with symmetry parameter $\beta = 1$. Many details about the properties of such densities can be found in [12]. The function $p_{\frac{2}{\nu}}(x,t) = \frac{1}{2\pi}\int_{-\infty}^{+\infty} e^{-i\gamma x}h(\gamma, 2/\nu)\mathrm{d}\gamma$ is unimodal with a positive maximal point and is such that $\int_0^\infty p_{\frac{2}{\nu}}(x,t) = \mathrm{d}x = \nu/2$. Analogously, the function $h(\gamma, -2/\nu)$ is the characteristic function of a negatively skewed random variable. This implies that the function $p_\nu$ can be seen as the superposition of the densities of stable random variables with index $\beta = \pm 1$, conditional to be respectively positive or negative.*

## 3. The Multidimensional Case for $1 < \nu < 2$

The Cauchy problem

$$\begin{cases}\frac{\partial^\nu u}{\partial t^\nu} = \sum_{j=1}^d \frac{\partial^2 u}{\partial x_j^2} & x \in \mathbb{R}^d,\ t > 0 \\ u(x,0) = \delta(x) \\ u_t(x,0) = 0.\end{cases} \tag{50}$$

has solution with Fourier transform

$$\int_{\mathbb{R}^d} e^{i\gamma\cdot x}u(x,t)\,\mathrm{d}x = E_{\nu,1}(-\|\gamma\|^2 t^\nu) \tag{51}$$
$$= \frac{\sin \pi\nu}{\pi}\int_0^\infty \frac{z^{\nu-1}e^{-tz\|\gamma\|^{2/\nu}}}{z^{2\nu}+2z^\nu \cos\pi\nu+1}\mathrm{d}z + \frac{1}{\nu}\left[e^{\|\gamma\|^{2/\nu}te^{i\pi/\nu}} + e^{\|\gamma\|^{2/\nu}te^{-i\pi/\nu}}\right].$$

as shown in the analysis presented above. Thus, the solution to the Cauchy problem (50) reads

$$u(x,t) = \frac{1}{(2\pi)^d}\int_{\mathbb{R}^d} e^{-i\gamma\cdot x} \tag{52}$$
$$\times\left\{\frac{\sin\pi\nu}{\pi}\int_0^\infty \frac{z^{\nu-1}e^{-tz\|\gamma\|^{2/\nu}}}{z^{2\nu}+2z^\nu\cos\pi\nu+1}\mathrm{d}z + \frac{1}{\nu}\left[e^{\|\gamma\|^{2/\nu}te^{i\pi/\nu}} + e^{\|\gamma\|^{2/\nu}te^{-i\pi/\nu}}\right]\right\}\mathrm{d}\gamma$$

We must therefore evaluate the following three $d-$dimensional integrals, the first one being a function of $z$.

$$\int_{\mathbb{R}^d} e^{-i\gamma\cdot x}e^{-t\|\gamma\|^{2/\nu}z}\mathrm{d}\gamma,\qquad \int_{\mathbb{R}^d} e^{-i\gamma\cdot x}e^{\|\gamma\|^{2/\nu}te^{i\pi/\nu}}\mathrm{d}\gamma,\qquad \int_{\mathbb{R}^d} e^{-i\gamma\cdot x}e^{\|\gamma\|^{2/\nu}te^{-i\pi/\nu}}\mathrm{d}\gamma \tag{53}$$

Since the three integrals (53) are substantially similar, we restrict ourselves to the evaluation of the first one. In spherical coordinates we have that

$$\int_{\mathbb{R}^d} e^{-i\gamma\cdot x}e^{-t\|\gamma\|^{\frac{2}{\nu}}z}\mathrm{d}\gamma \tag{54}$$
$$= \int_0^\infty \rho^{d-1}\mathrm{d}\rho\int_0^\pi \mathrm{d}\theta_1\cdots\int_0^\pi \mathrm{d}\theta_{d-2}\int_0^{2\pi}\mathrm{d}\phi\,\sin\theta_1^{d-2}\cdots\sin\theta_{d-2}$$
$$\times\, e^{-i\rho(x_d\sin\theta_1\cdots\sin\theta_{d-2}\sin\phi+x_{d-1}\sin\theta_1\cdots\sin\theta_{d-2}\cos\phi+\cdots+x_2\sin\theta_1\cos\theta_2+x_1\cos\theta_1)}e^{-t\rho^{2/\nu}z}$$
$$= \int_0^\infty \rho^{d-1}e^{-t\rho^{2/\nu}z}\frac{(2\pi)^{\frac{d}{2}}J_{\frac{d}{2}-1}(\rho\|x\|)}{(\rho\|x\|)^{\frac{d}{2}-1}}\mathrm{d}\rho$$

The last step is the hyperspherical integral

$$\int_{\{\gamma_1,\ldots\gamma_d:\sum_{j=1}^d\gamma_j^2=\rho^2\}} e^{-i\sum_{j=1}^d x_j\gamma_j}\mathrm{d}\gamma_1\ldots\mathrm{d}\gamma_d \tag{55}$$
$$= \int_0^\pi \mathrm{d}\theta_1\cdots\int_0^\pi\mathrm{d}\theta_{d-2}\int_0^{2\pi}\mathrm{d}\phi\,\sin\theta_1^{d-2}\cdots\sin\theta_{d-2}$$
$$\times\, e^{-i\rho(x_d\sin\theta_1\cdots\sin\theta_{d-2}\sin\phi+x_{d-1}\sin\theta_1\cdots\sin\theta_{d-2}\cos\phi+\cdots+x_2\sin\theta_1\cos\theta_2+x_1\cos\theta_1)}$$
$$= \frac{(2\pi)^{\frac{d}{2}}J_{\frac{d}{2}-1}(\rho\|x\|)}{(\rho\|x\|)^{\frac{d}{2}-1}}$$

which is proven in detail in [13], Formula (2.151).

By inserting (54) into (52) we have that

$$u(x,t) = \frac{1}{(2\pi)^d}\int_0^\infty \rho^{d-1}\frac{(2\pi)^{\frac{d}{2}} J_{\frac{d}{2}-1}(\rho\|x\|)}{(\rho\|x\|)^{\frac{d}{2}-1}}\mathrm{d}\rho \tag{56}$$
$$\times\left\{\frac{\sin\pi\nu}{\pi}\int_0^\infty \frac{z^{\nu-1}e^{-t\rho^{2/\nu}z}}{z^{2\nu}+2z^{\nu}\cos\pi\nu+1}\mathrm{d}z + \frac{1}{\nu}\left[e^{\rho^{2/\nu}te^{i\pi/\nu}} + e^{\rho^{2/\nu}te^{-i\pi/\nu}}\right]\right\}$$
$$= \frac{1}{(2\pi)^{\frac{d}{2}}\|x\|^{\frac{d}{2}-1}}\int_0^\infty \rho^{\frac{d}{2}} J_{\frac{d}{2}-1}(\rho\|x\|)\mathrm{d}\rho$$
$$\times\left\{\frac{\sin\pi\nu}{\pi}\int_0^\infty \frac{z^{\nu-1}e^{-t\rho^{2/\nu}z}}{z^{2\nu}+2z^{\nu}\cos\pi\nu+1}\mathrm{d}z + \frac{1}{\nu}\left[e^{\rho^{2/\nu}te^{i\pi/\nu}} + e^{\rho^{2/\nu}te^{-i\pi/\nu}}\right]\right\}$$

Note that the integral in $z$ after the change of variable $z^{\nu} = z'$ becomes

$$\frac{\sin\pi\nu}{\pi\nu}\int_0^\infty \frac{e^{-t\rho^{2/\nu}z^{1/\nu}}}{z^2+2z\cos\pi\nu+1}\mathrm{d}z = \frac{1}{\nu}\int_0^\infty e^{-t\rho^{2/\nu}z^{1/\nu}}\left[\frac{1}{\pi}\frac{\sin\pi\nu}{(z-\cos\pi\nu)^2+\sin^2\pi\nu}\right]\mathrm{d}z \tag{57}$$

since $1 < \nu < 2$ the function

$$f(z) = \frac{1}{\pi}\frac{\sin\pi\nu}{(z-\cos\pi\nu)^2+\sin^2\pi\nu} \tag{58}$$

has the form shown in Figure 9.

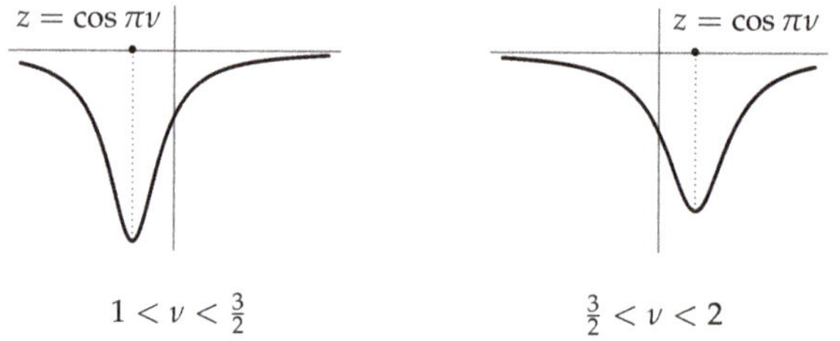


**Figure 9.** Plot of the function $f$ given in (58).

In conclusion

$$\frac{1}{\nu}\int_0^\infty f(z)\,\mathrm{d}z = \begin{cases} -1 & \frac{3}{2} < \nu < 2 \\ -1+\frac{1}{\nu} & 1 < \nu < \frac{3}{2}. \end{cases} \tag{59}$$

We recall now the definition of an $\alpha-$stable $d-$dimensional process $S^{\alpha}(t) = (S_1^{\alpha}(t),\ldots,S_d^{\alpha}(t))$, $0 < \alpha < 2$.

Its characteristic function has the following form [14]

$$\mathbb{E}e^{i\gamma\cdot S^{\alpha}(t)} = \begin{cases} e^{-t\int_{\mathbb{S}^{d-1}}|\gamma\cdot s|^{\alpha}(1-i\,\mathrm{sign}(\gamma\cdot s)\tan\frac{\pi\alpha}{2})\Gamma(\mathrm{d}s)+i\gamma\cdot\mu} & \alpha\neq 1 \\ e^{-t\int_{\mathbb{S}^{d-1}}|\gamma\cdot s|(1-i\frac{2}{\pi}\mathrm{sign}(\gamma\cdot s)\log(\gamma\cdot s))\Gamma(\mathrm{d}s)+i\gamma\cdot\mu} & \alpha = 1 \end{cases} \tag{60}$$

where $\mu\in\mathbb{R}^d$, $\Gamma$ is a finite measure on the sphere $\mathbb{S}^{d-1} = \{s\in\mathbb{R}^d : \|s\| = 1\}$.

Since $\|\gamma\cdot s\| = \|\gamma\|\|s\|\cos\theta = \|\gamma\|\cos\theta$, where $s\in\mathbb{S}^{d-1}$ so that $\|s\| = 1$. Furthermore sign $(\gamma\cdot s) = \mathrm{sign}\cos\theta$. We can assume $\gamma$ oriented through the north pole of $\mathbb{S}^{d-1}$ and thus $\theta$ can be viewed as the latitude of vector $s$, as shown in Figure 10.

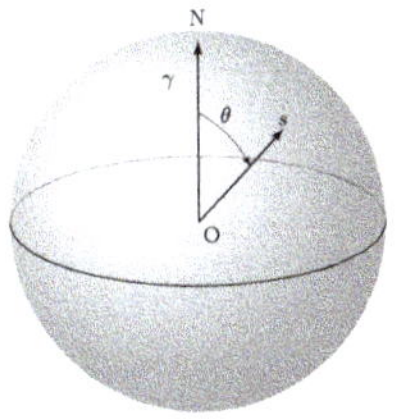


**Figure 10.** Domain of integration in (60), upon suitable rotation of the axis of the sphere.

We take the case $1 < \alpha < 2$ and rewrite the characteristic function as

$$\begin{aligned} \mathbb{E}e^{i\gamma\cdot S^{\alpha}(t)} &= e^{-t\|\gamma\|^{\alpha}\int_{\mathbb{S}^{d-1}}\cos^{\alpha}\theta(\cos\frac{\pi\alpha}{2}-i\,\mathrm{sign}(\cos\theta)\sin\frac{\pi\alpha}{2})\frac{\Gamma(\mathrm{d}s)}{\cos\frac{\pi\alpha}{2}}+i\gamma\cdot\mu} \\ &= e^{-t\|\gamma\|^{\alpha}\sigma^{\alpha}\int_{\mathbb{S}^{d-1}}\cos^{\alpha}\theta e^{-i\frac{\pi\alpha}{2}\mathrm{sign}(\cos\theta)}\Gamma(\mathrm{d}s)+i\gamma\cdot\mu} \end{aligned} \tag{61}$$

where $\sigma^{\alpha} = \frac{1}{\cos\frac{\pi\alpha}{2}}$.

For simplicity we assume $d = 3$ and suppose that $\Gamma$ is a uniformly distributed measure on the upper hemisphere of the unit sphere. Thus

$$\Gamma(\mathrm{d}s) = \frac{\sin\theta}{2\pi}\,\mathrm{d}\theta\mathrm{d}\phi \qquad 0 < \theta < \frac{\pi}{2},\, 0 < \phi < 2\pi. \tag{62}$$

with $\mathrm{sign}(\cos\theta) = 1$. For $\mu = 0$ the integral in (61) becomes

$$\frac{e^{-i\frac{\pi\alpha}{2}}}{2\pi}\int_{\mathbb{S}^2}\cos^{\alpha}\sin\theta\,\mathrm{d}\phi\mathrm{d}\theta = \frac{e^{-i\frac{\pi\alpha}{2}}}{2\pi}\int_0^{\frac{\pi}{2}}\sin\theta\cos^{\alpha}\theta\int_0^{2\pi}\mathrm{d}\phi = \frac{e^{-i\frac{\pi\alpha}{2}}}{\alpha+1} \tag{63}$$

The characteristic function (61) turns out to be

$$\mathbb{E}e^{i\gamma\cdot S^{\alpha}(t)} = e^{-t\|\gamma\|^{\alpha}\frac{1}{\cos\frac{\pi\alpha}{2}}\frac{e^{-i\frac{\pi\alpha}{2}}}{\alpha+1}} = e^{t\|\gamma\|^{\frac{2}{\nu}}\frac{1}{|\cos\frac{\pi}{\nu}|}\frac{e^{-i\frac{\pi}{\nu}}}{\alpha+1}} = e^{t\|\gamma\|^{\frac{2}{\nu}}\frac{1}{|\cos\frac{\pi}{\nu}|}e^{-i\frac{\pi}{\nu}}\sigma^{\nu}}$$

Since $\alpha = 2/\nu$ and $1 < \nu < 2$, we have $\pi > \pi/\nu > \pi/2$ so $\cos\frac{\alpha\pi}{2} = \cos\frac{\pi}{\nu}$ is negative.

If $\Gamma$ is distributed on the lower hemisphere $\mathrm{sign}(\cos\theta) = -1$, and in the same way we have that

$$\mathbb{E}e^{i\gamma\cdot S^{\alpha}(t)} = e^{-t\|\gamma\|^{\alpha}\frac{1}{\cos\frac{\pi\alpha}{2}}\frac{e^{i\frac{\pi\alpha}{2}}}{\alpha+1}} = e^{t\|\gamma\|^{\frac{2}{\nu}}\frac{1}{|\cos\frac{\pi}{\nu}|}\frac{e^{i\frac{\pi}{\nu}}}{\alpha+1}} = e^{t\|\gamma\|^{\frac{2}{\nu}}\frac{1}{|\cos\frac{\pi}{\nu}|}e^{i\frac{\pi}{\nu}}\sigma^{\nu}}$$

The situation in the space $\mathbb{S}^{d-1}$, $d > 3$ is quite similar with the integral in (61) evaluated in hyperspherical coordinates.

**Author Contributions:** Conceptualization, F.I. and E.O.; Writing—original draft, F.I. and E.O.; Writing—review & editing, F.I. and E.O. All authors have read and agreed to the published version of the manuscript.

**Funding:** This research received no external funding.

**Conflicts of Interest:** The authors declare no conflict of interest.

## References

1. Mittag-Leffler, G.M. Sur la nouvelle fonction $E_{\alpha}(x)$. *CR Acad. Sci. Paris* **1903**, *137*, 554–558.
2. Erdélyi, A.; Magnus, W.; Oberhettinger, F.; Tricomi, F. *Higher Transcendental Functions*; McGraw-Hill Book Company: New York, NY, USA, 1955; Volume III.
3. Fujita, Y. Integrodifferential equation which interpolates the heat equation and the wave equation. *Osaka J. Math.* **1990**, *27*, 309–321.
4. Mainardi, F. The time fractional diffusion-wave equation. *Radiophys. Quantum Electron.* **1995**, *38*, 13–24. [CrossRef]

5. Zolotarev, V.M. *One-Dimensional Stable Distributions*; American Mathematical Soc.: Providence, RI, USA, 1986; Volume 65.
6. D'Ovidio, M.; Orsingher, E.; Toaldo, B. Time-changed processes governed by space-time fractional telegraph equations. *Stoch. Anal. Appl.* **2014**, *32*, 1009–1045. [CrossRef]
7. Kolokol'cov, V.N. *Markov Processes, Semigroups, and Generators*; Walter de Gruyter: Berlin, Germany, 2011; Volume 38.
8. Kwaśnicki, M. Ten equivalent definitions of the fractional laplace operator. *Fract. Calc. Appl. Anal.* **2017**, *20*, 7–51. [CrossRef]
9. Wyss, W. The fractional diffusion equation. *J. Math. Phys.* **1986**, *27*, 2782–2785. [CrossRef]
10. Podlubny, I. *Fractional Differential Equations: An Introduction to Fractional Derivatives, Fractional Differential Equations, to Methods of Their Solution and Some of Their Applications*; Mathematics in Science and Engineering, Elsevier Science: Amsterdam, The Netherlands, 1998.
11. Orsingher, E.; Toaldo, B. Space–Time Fractional Equations and the Related Stable Processes at Random Time. *J. Theor. Probab.* **2017**, *30*, 1–26. [CrossRef]
12. Lukacs, E. *Characteristic Functions*; Charles Griffin & Co., Ltd.: Glasgow, Scotland, 1970.
13. Orsingher, E.; De Gregorio, A. Random flights in higher spaces. *J. Theor. Probab.* **2007**, *20*, 769–806. [CrossRef]
14. Samorodnitsky, G.; Taqqu, M. *Stable Non-Gaussian Random Processes: Stochastic Models with Infinite Variance*; Stochastic Modeling Series, Taylor & Francis: Abingdon, UK, 1994.

mathematics
MDPI

Article

# Probability Models and Statistical Tests for Extreme Precipitation Based on Generalized Negative Binomial Distributions

Victor Korolev [1,2,3,4] and Andrey Gorshenin [1,2,3,*]

1 Moscow Center for Fundamental and Applied Mathematics, Lomonosov Moscow State University, 119991 Moscow, Russia; vkorolev@cs.msu.ru
2 Faculty of Computational Mathematics and Cybernetics, Lomonosov Moscow State University, 119991 Moscow, Russia
3 Federal Research Center "Computer Science and Control" of the Russian Academy of Sciences, 119333 Moscow, Russia
4 Department of Mathematics, School of Science, Hangzhou Dianzi University, Hangzhou 310018, China
* Correspondence: agorshenin@frccsc.ru

Received: 4 April 2020; Accepted: 14 April 2020; Published: 16 April 2020

**Abstract:** Mathematical models are proposed for statistical regularities of maximum daily precipitation within a wet period and total precipitation volume per wet period. The proposed models are based on the generalized negative binomial (GNB) distribution of the duration of a wet period. The GNB distribution is a mixed Poisson distribution, the mixing distribution being generalized gamma (GG). The GNB distribution demonstrates excellent fit with real data of durations of wet periods measured in days. By means of limit theorems for statistics constructed from samples with random sizes having the GNB distribution, asymptotic approximations are proposed for the distributions of maximum daily precipitation volume within a wet period and total precipitation volume for a wet period. It is shown that the exponent power parameter in the mixing GG distribution matches slow global climate trends. The bounds for the accuracy of the proposed approximations are presented. Several tests for daily precipitation, total precipitation volume and precipitation intensities to be abnormally extremal are proposed and compared to the traditional PoT-method. The results of the application of this test to real data are presented.

**Keywords:** precipitation; limit theorems; statistical test; generalized negative binomial distribution; generalized gamma distribution; asymptotic approximations; extreme order statistics; random sample size

**MSC:** 60F05; 62G30; 62E20; 62P12; 65C20

---

## 1. Introduction

In this paper, we continue the research we started in [1,2]. We develop the mathematical models for statistical regularities in precipitation proposed in the papers mentioned above. We consider the models for the statistical regularities in the duration of a wet period, maximum daily precipitation within a wet period and total precipitation volume per wet period. The base for the models is the generalized negative binomial (GNB) introduced in the recent paper [3]. The GNB distribution is a mixed Poisson distribution, the mixing distribution being generalized gamma (GG). The results of fitting the GNB distribution to real data are presented and demonstrate excellent concordance of the GNB model with the empirical distribution of the duration of wet periods measured in days. Based on this GNB model, asymptotic approximations are proposed for the distributions of the maximum daily precipitation volume within a wet period and of the total precipitation volume for a wet period.

The asymptotic distribution of the maximum daily precipitation volume within a wet period turns out to be a tempered scale mixture of the gamma distribution in which the scale factor has the Weibull distribution, whereas the asymptotic approximation for the total precipitation volume for a wet period turns out to be the GG distribution. These asymptotic approximations are deduced using limit theorems for statistics constructed from samples with random sizes having the GNB distribution. The bounds for the accuracy of the proposed approximations are discussed theoretically and illustrated statistically. The proposed approximations appear to be very accurate. Based on these models, two approaches are proposed to the definition of abnormally extremal precipitation. These approaches can be regarded as a further development of those proposed in [2].

The importance of the problem of modeling statistical regularities in extreme precipitation is indisputable. Understanding climate variability and trends at relatively large time horizons is of crucial importance for long-range business, say, agricultural projects and forecasting of risks of water floods, dry spells and other natural disasters. Modeling regularities and trends in heavy and extreme daily precipitation is important for understanding climate variability and change at relatively small or medium time horizons. However, these models are much more uncertain as compared to those derived for mean precipitation or total precipitation during a wet period. In [4], a detailed review of this phenomenon is presented and it is noted that, at least for the European continent, most results hint at a growing intensity of heavy precipitation over the last decades.

In [2], we proposed a rather reasonable approach to the unambiguous (algorithmic) determination of extreme or abnormally heavy total precipitation for a wet period. This approach was based on the NB model for the duration of wet periods measured in days, and, as a consequence, on the distribution of the total precipitation volume during a wet period. This approach has some advantages. First, estimates of the parameters of the total precipitation are weakly affected by the accuracy of the daily records and are less sensitive to missing values. Second, the corresponding mathematical models are theoretically based on limit theorems of probability theorems that yield unambiguous asymptotic approximations, which are used as adequate mathematical models. Third, this approach gives an unambiguous algorithm for the determination of extreme or abnormally heavy total precipitation that does not involve statistical significance problems owing to the low occurrence of such (relatively rare) events.

The problem of the construction of a statistical test for the precipitation volume to be abnormally large can be mathematically formalized as follows. Let $m \geqslant 2$ be a natural number and consider a sample of $m$ positive observations $X_1, X_2, \ldots, X_m$. With finite $m$, among $X_i$'s there is always an extreme observation, say, $X_1$, such that $X_1 \geqslant X_i$, $i = 1, 2, \ldots, m$. Two cases are possible: (i) $X_1$ is a 'typical' observation and its extreme character is conditioned by purely stochastic circumstances (there must be an extreme observation within a finite homogeneous sample) and (ii) $X_1$ is abnormally large so that it is an 'outlier' and its extreme character is due to some exogenous factors.

To construct a test for distinguishing between these two cases for abnormally extreme daily precipitation, we use the fact that the distribution of the maximum daily precipitation per wet period is a tempered scale mixture of the gamma distribution in which the scale factor has the Weibull distribution. According to this model, a daily precipitation volume is considered to be abnormally extremal if it exceeds a certain (pre-defined) quantile of this distribution.

As regards testing for anomalous extremeness of total precipitation volume during a wet period, we use the GG distribution as the model of statistical regularities of its behavior. The theoretical grounds for this model are provided by the law of large numbers for random sums in which the number of summands has the GNB distribution. It turns out that, as compared to the ordinary negative binomial (NB) model (see [2]), the additional exponent power parameter in the corresponding GG distribution matches slow global climate trends. Hence, the hypothesis that the total precipitation volume during a certain wet period is abnormally large can be re-formulated as the homogeneity hypothesis of a sample from the GG distribution. Two equivalent tests are proposed for testing this hypothesis. One of them is based on the beta distribution whereas the second is based on the

Snedecor–Fisher distribution. Both of these tests deal with the relative contribution of the total precipitation volume for a wet period to the considered set (sample) of successive wet periods. Within the second approach, it is possible to introduce the notions of relatively abnormal and absolutely abnormal precipitation volumes. These tests are scale-free and depend only on the easily estimated shape parameter of the GNB distribution and the time-scale parameter determining the denominator in the fractional contribution of a wet period under consideration. The tests appeared to be applicable not only to total precipitation volumes over wet periods but also to the precipitation intensities (the ratios of total precipitation volumes per wet periods to the durations of the corresponding wet periods measured in days).

## 2. Generalized Negative Binomial Model for the Duration of Wet Periods

The main results of this paper strongly rely on the GNB model, a wide and flexible family of discrete distributions that are mixed Poisson laws with the mixing GG distribution. Namely, we say that a random variable $N_{r,\gamma,\mu}$ ($r > 0$, $\gamma \in \mathbb{R}$ and $\mu > 0$) has the generalized negative binomial distribution, if

$$\mathbb{P}(N_{r,\gamma,\mu} = k) = \frac{1}{k!}\int_0^\infty e^{-z} z^k g^*(z; r, \gamma, \mu)dz, \quad k = 0, 1, 2..., \tag{1}$$

where $g^*(z; r, \gamma, \mu)$ is the density of GG distribution:

$$g^*(x; r, \gamma, \mu) = \frac{|\gamma|\mu^r}{\Gamma(r)} x^{\gamma r - 1} e^{-\mu x^\gamma}, \quad x \geqslant 0, \tag{2}$$

with $\gamma \in \mathbb{R}$, $\mu > 0$, $r > 0$. The GNB distributions seem to be very promising in the statistical description of many real phenomena, being very convenient and almost universal models.

It is necessary to explain why this combination of the mixed and mixing distributions is considered. First of all, the Poisson kernel is used as mixed for the following reasons. Pure Poisson processes can be regarded as the best models of stationary (time-homogeneous) chaotic flows of events [5]. Recall that the attractiveness of a Poisson process as a model of homogeneous discrete stochastic chaos is due to at least two circumstances. First, Poisson processes are point processes characterized by the time intervals between successive points that are independent random variables (r.v.'s) with one and the same exponential distribution, and, as is well known, the exponential distribution possesses the maximum differential entropy among all absolutely continuous distributions concentrated on the nonnegative half-line with finite expectations, whereas the entropy is a natural and convenient measure of uncertainty. Second, the points forming the Poisson process are uniformly distributed along the time axis in the sense that for any finite time interval $[t_1, t_2]$, $t_1 < t_2$, the conditional joint distribution of the points of the Poisson process that fall into the interval $[t_1, t_2]$ under the condition that the number of such points is fixed and equals, say, $n$, coincides with the joint distribution of the order statistics constructed from an independent sample of size $n$ from the uniform distribution on $[t_1, t_2]$, whereas the uniform distribution possesses the maximum differential entropy among all absolutely continuous distributions concentrated on finite intervals and very well corresponds to the conventional impression of an absolutely unpredictable random variable (see, e g., [5,6]). However, in actual practice, as a rule, the parameters of the chaotic stochastic processes are influenced by poorly predictable «extrinsic» factors, which can be regarded as stochastic so that most reasonable probabilistic models of non-stationary (time-non-homogeneous) chaotic point processes are doubly stochastic Poisson processes, also called Cox processes (see, e.g., [5,7,8]). These processes are defined as Poisson processes with stochastic intensities. Such processes proved to be adequate models in insurance [5,7,8], financial mathematics [9], physics [10] and many other fields. Their one-dimensional distributions are mixed Poisson.

In order to have a flexible model of a mixing distribution that is "responsible" for the description of statistical regularities of the manifestation of external stochastic factors, we suggest to use the GG distributions defined by the density (2). The class of GG distributions was first described as a unitary family in 1962 by E. Stacy [11] as the class of probability distributions simultaneously containing both Weibull and gamma distributions. The family of GG distributions contains practically all the most popular absolutely continuous distributions concentrated on the non-negative half-line. In particular, the family of GG distributions contains:

- The gamma distribution ($\gamma = 1$) and its special cases
  - The exponential distribution ($\gamma = 1, r = 1$),
  - The Erlang distribution ($\gamma = 1, r \in \mathbb{N}$),
  - The chi-square distribution ($\gamma = 1, \mu = \frac{1}{2}$);
- The Nakagami distribution ($\gamma = 2$);
- The half-normal (folded normal) distribution (the distribution of the maximum of a standard Wiener process on the interval $[0,1]$) ($\gamma = 2, r = \frac{1}{2}$);
- The Rayleigh distribution ($\gamma = 2, r = 1$);
- The chi-distribution ($\gamma = 2, \mu = 1/\sqrt{2}$);
- The Maxwell distribution (the distribution of the absolute values of the velocities of molecules in a dilute gas) ($\gamma = 2, r = \frac{3}{2}$);
- The Weibull–Gnedenko distribution (the extreme value distribution of type III) ($r = 1, \gamma > 0$);
- The (folded) exponential power distribution ($\gamma > 0, r = \frac{1}{\gamma}$);
- The inverse gamma distribution ($\gamma = -1$) and its special case
  - The Lévy distribution (the one-sided stable distribution with the characteristic exponent $\frac{1}{2}$ – the distribution of the first hit time of the unit level by the Brownian motion) ($\gamma = -1$, $r = \frac{1}{2}$);
- The Fréchet distribution (the extreme value distribution of type II) ($r = 1, \gamma < 0$)

and other laws. The limit point of the class of GG-distributions is

- The log-normal distribution ($r \to \infty$).

GG distributions are widely applied in many practical problems. There are dozens of papers dealing with the application of GG-distributions as models of regularities observed in practice. Apparently, the popularity of GG-distributions is due to the fact that most of them can serve as adequate asymptotic approximations, since all the representatives of the class of GG-distributions listed above appear as limit laws in various limit theorems of probability theory in rather simple limit schemes. Below we will formulate a general limit theorem (an analog of the law of large numbers) for random sums of independent r.v.'s in which the GG-distributions are limit laws. It is worth noting that the GG distribution and its limit cases give a general form of the exponential distribution of rank 1 for the scale parameter.

In [1], the data registered in so climatically different points as Potsdam (Brandenburg, Germany) and Elista (Kalmykia, Russia) was analyzed, and it was demonstrated that the fluctuations of the numbers of successive wet days with very high confidence fit the NB distribution with shape parameters $r = 0.847$ and $r = 0.876$, respectively. In the same paper, a schematic attempt was undertaken to explain this phenomenon by the fact that NB distributions can be represented as mixed Poisson laws with mixing gamma distributions whereas, as it already has been mentioned, the Poisson distribution is the best model for the discrete stochastic chaos and the mixing distribution accumulates the stochastic influence of factors that can be assumed exogenous with respect to the local system under consideration.

The NB distributions are special cases of the GNB distributions. This family of discrete distributions is very wide and embraces Poisson distributions (as limit points corresponding to a degenerate mixing distribution), NB (Polya) distributions including geometric distributions (corresponding to the gamma mixing distribution, see [12]), Sichel distributions (corresponding to the inverse gamma mixing distributions, see [13,14]), Weibull–Poisson distributions (corresponding to the Weibull mixing distributions, see [15]) and many other types supplying descriptive statistics with many flexible models. More examples of mixed Poisson laws can be found in [8,16].

It is quite natural to expect that, having introduced one more free parameter into the pure negative binomial model, namely, the power parameter in the exponent of the original gamma mixing distribution, instead of the negative binomial model one might obtain a more flexible GNB model that provides an even better fit with the statistical data of the durations of wet days. The analysis of the real data shows that this is indeed so.

In Figures 1 and 2 there are the histograms constructed from real data of 3323 wet periods in Potsdam and 2937 wet periods in Elista. On the same pictures, there are the graphs of the fitted NB distribution (that is, the GNB distribution with $\gamma = 1$) and the fitted GNB distribution with additionally adjusted scale and power parameters. For vividness, in the GNB model, the value of the shape parameter $r$ was taken the same as that obtained for the NB model and equal to 0.876 for Elista and 0.847 for Potsdam. For the "fine tuning" of the GNB models with these fixed values of $r$, the minimization of the $\ell_1$-norm of the difference between the histogram and the fitted GNB model was used. In Appendix A, the Algorithm A1 of for the computation of GNB probabilities by the minimization of the $\ell_1$, $\ell_2$ and $\ell_\infty$-norms of the difference between the histogram and the fitted GNB model is presented.

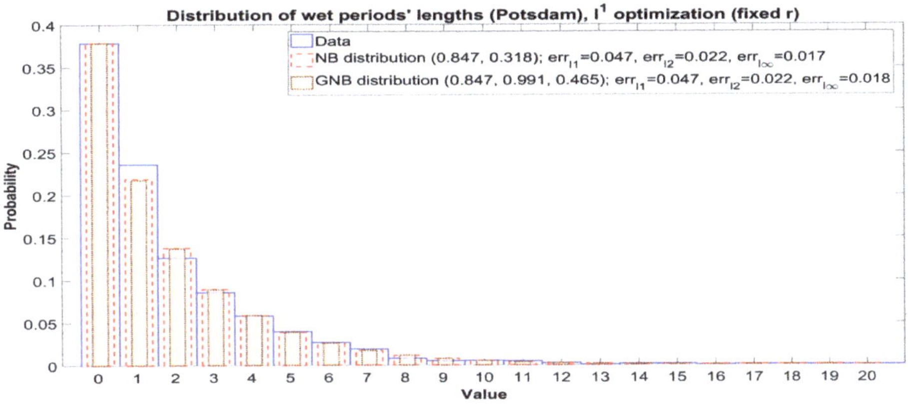


**Figure 1.** The histograms constructed from real data of 3320 wet periods in Potsdam and the fitted negative binomial (NB) and generalized negative binomial (GNB) models, $\ell_1$-distance minimization.

The analytic and asymptotic properties of the GNB distributions were studied in [3]. In particular, it was shown in that paper that the GNB distribution with shape parameter and exponent power parameter less than one is actually mixed geometric. The mixed geometric distributions were introduced and studied in [17] (also see [15,18]). A mixed geometric distribution can be interpreted in terms of the Bernoulli trials as follows. First, as a result of some "preliminary" experiment the value of some r.v. taking values in $[0, 1]$ is determined, which is then used as the probability of success in the sequence of Bernoulli trials in which the original "unconditional" mixed Poisson r.v. is nothing else than the "conditionally" geometrically distributed r.v. having the sense of the number of trials up to the first failure. This makes it possible to assume that the sequence of wet/dry days is not independent but is conditionally independent and the random probability of success is determined by some outer stochastic factors. As such, we can consider the seasonality or the type of the cause of a wet period. So, since the GG-distribution is a more general and, hence, a more flexible model than the "pure" gamma distribution, there arises a hope that the GNB distribution could provide an even better goodness of fit to the statistical regularities in the duration of wet periods than the "pure" NB binomial distribution.

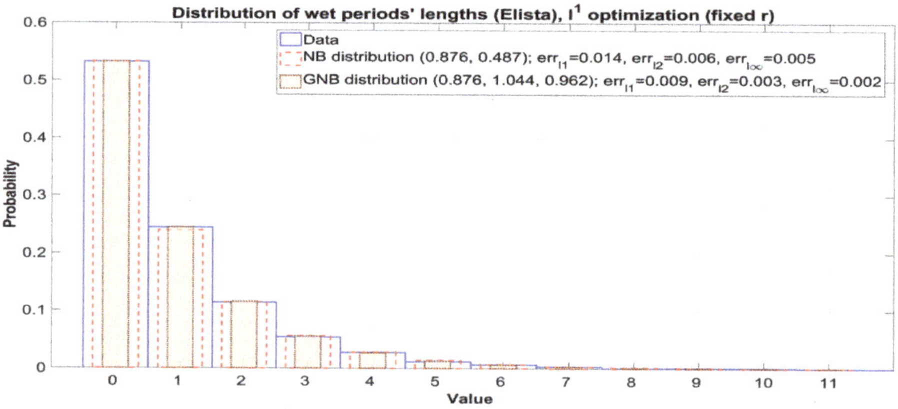


**Figure 2.** The histograms constructed from real data of 2937 wet periods in Elista and the fitted NB and GNB models, $\ell_1$-distance minimization.

## 3. Notation, Definitions and Mathematical Preliminaries

In the paper, conventional notation is used. The symbols $\stackrel{d}{=}$ and $\Longrightarrow$ denote the coincidence of distributions and convergence in distribution, respectively.

In what follows, for brevity and convenience, the results will be presented in terms of r.v.'s with the corresponding distributions. It will be assumed that all the r.v.'s are defined on the same probability space $(\Omega, \mathfrak{F}, \mathsf{P})$.

An r.v. having the gamma distribution with shape parameter $r > 0$ and scale parameter $\mu > 0$ will be denoted $G_{r,\mu}$,

$$\mathsf{P}(G_{r,\mu} < x) = \int_0^x g(z; r, \mu)dz, \text{ with } g(x; r, \mu) = \frac{\mu^r}{\Gamma(r)} x^{r-1} e^{-\mu x}, \ x \geqslant 0,$$

where $\Gamma(r)$ is Euler's gamma-function, $\Gamma(r) = \int_0^\infty x^{r-1} e^{-x} dx$, $r > 0$.

In this notation, obviously, $G_{1,1}$ is an r.v. with the standard exponential distribution: $\mathsf{P}(G_{1,1} < x) = [1 - e^{-x}]\mathbf{1}(x \geqslant 0)$ (here and in what follows $\mathbf{1}(A)$ is the indicator function of a set $A$).

A GG-distribution is the absolutely continuous distribution defined by the density (Equation (2)). The distribution function (d.f.) corresponding to the density $g^*(x; r, \gamma, \mu)$ will be denoted $F^*(x; r, \gamma, \mu)$.

The properties of GG-distributions are described in [11,19]. An r.v. with the density $g^*(x; r, \gamma, \mu)$ will be denoted $\overline{G}_{r,\gamma,\mu}$. It can be easily made sure that

$$\overline{G}_{r,\gamma,\mu} \stackrel{d}{=} G_{r,\mu}^{1/\gamma}, \tag{3}$$

and hence,

$$(\overline{G}_{r,\gamma,\mu})^\gamma \stackrel{d}{=} G_{r,\mu} \tag{4}$$

For convenience, for an r.v. with the Weibull distribution, a particular case of GG-distributions corresponding to the density $g^*(x; 1, \gamma, 1)$ and the d.f. $[1 - e^{-x^\gamma}]\mathbf{1}(x \geqslant 0)$ with $\gamma > 0$, we will use a special notation $W_\alpha$, that is, $W_\gamma \stackrel{d}{=} \overline{G}_{1,\gamma,1}$. Thus, $G_{1,1} \stackrel{d}{=} W_1$. The density $g^*(x; 1, \alpha, 1)$ with $\alpha < 0$ defines the Fréchet or inverse Weibull distribution. It is easy to see that

$$W_1^{1/\gamma} \stackrel{d}{=} W_\gamma. \tag{5}$$

An r.v. $N_{r,p}$ is said to have the negative binomial (NB) distribution with parameters $r > 0$ (shape) and $p \in (0,1)$ (success probability), if

$$\mathsf{P}(N_{r,p} = k) = \frac{\Gamma(r+k)}{k!\Gamma(r)} \cdot p^r(1-p)^k, \quad k = 0,1,2,...$$

A particular case of the NB distribution corresponding to the value $r = 1$ is the geometric distribution. Let $p \in (0,1)$ and let $N_{1,p}$ be the r.v. having the geometric distribution with parameter $p$:

$$\mathsf{P}(N_{1,p} = k) = p(1-p)^k, \quad k = 0,1,2,...$$

This means that for any $m \in \mathbb{N}$

$$\mathsf{P}(N_{1,p} \geqslant m) = \textstyle\sum_{k=m}^{\infty} p(1-p)^k = (1-p)^m.$$

Let $Y$ be an r.v. taking values in the interval $(0,1)$. Moreover, let for all $p \in (0,1)$ the r.v. $Y$ and the geometrically distributed r.v. $N_{1,p}$ be independent. Let $M = N_{1,Y}$, that is, $M(\omega) = N_{1,Y(\omega)}(\omega)$ for any $\omega \in \Omega$. The distribution

$$\mathsf{P}(M \geqslant m) = \int_0^1 (1-y)^m d\mathsf{P}(Y < y), \quad m \in \mathbb{N},$$

of the r.v. $M$ will be called $Y$-mixed geometric [17].

It is well known that the negative binomial distribution is a mixed Poisson distribution with the gamma mixing distribution [12] (also see [20]): for any $r > 0$, $p \in (0,1)$ and $k \in \{0\} \bigcup \mathbb{N}$ we have

$$\frac{\Gamma(r+k)}{k!\Gamma(r)} \cdot p^r(1-p)^k = \frac{1}{k!}\int_0^{\infty} e^{-z} z^k g(z;r,\mu)dz, \tag{6}$$

where $\mu = p/(1-p)$.

The d.f. and the density of a strictly stable distribution with the characteristic exponent $\alpha$ and shape parameter $\theta$ defined by the characteristic function (ch.f.)

$$\mathfrak{f}(t;\alpha,\theta) = \exp\big\{-|t|^{\alpha}\exp\{-\tfrac{1}{2}i\pi\theta\alpha \mathrm{sign} t\}\big\}, \quad t \in \mathbb{R},$$

where $0 < \alpha \leqslant 2$, $|\theta| \leqslant \min\{1, \frac{2}{\alpha} - 1\}$, will be respectively denoted $F(x;\alpha,\theta)$ and $f(x;\alpha,\theta)$ (see, e.g., [21]). An r.v. with the d.f. $F(x;\alpha,\theta)$ will be denoted $S_{\alpha,\theta}$.

To symmetric, strictly stable distributions, there corresponds the value $\theta = 0$. To one-sided strictly stable distributions concentrated on the nonnegative halfline, there correspond the values $\theta = 1$ and $0 < \alpha \leqslant 1$. The pairs $\alpha = 1$, $\theta = \pm 1$ correspond to the distributions degenerate in $\pm 1$, respectively. All the other strictly stable distributions are absolutely continuous. Stable densities cannot explicitly be represented via elementary functions with four exceptions: the normal distribution ($\alpha = 2$, $\theta = 0$), the Cauchy distribution ($\alpha = 1$, $\theta = 0$), the Lévy distribution ($\alpha = \frac{1}{2}$, $\theta = 1$) and the distribution symmetric to the Lévy law ($\alpha = \frac{1}{2}$, $\theta = -1$). Expressions of stable densities in terms of the Fox functions (generalized Meijer G-functions) can be found in [22,23].

The standard normal d.f. will be denoted $\Phi(x)$,

$$\Phi(x) = \frac{1}{\sqrt{2\pi}}\int_{-\infty}^{x} e^{-y^2/2}dy, \quad y \in \mathbb{R}.$$

An r.v. with the d.f. $\Phi(x)$ will be denoted $X$. The folded or half- normal distribution is the distribution of the r.v. $|X|$. It can be easily verified that $S_{2,0} \stackrel{d}{=} \sqrt{2}X$.

In [24,25], it was demonstrated that if $\gamma \in (0,1]$, then

$$W_\gamma \stackrel{d}{=} W_1 \cdot S_{\gamma,1}^{-1} \quad (7)$$

with the r.v.'s on the right-hand side being independent.

For $r \in (0,1)$ let $G_{r,1}$ and $G_{1-r,1}$ be independent gamma-distributed r.v.'s. Let $\mu > 0$. Introduce the r.v.

$$Z_{r,\mu} = \frac{\mu(G_{r,1} + G_{1-r,1})}{G_{r,1}} \stackrel{d}{=} \mu Z_{r,1} \stackrel{d}{=} \mu\left(1 + \tfrac{1-r}{r} Q_{1-r,r}\right), \quad (8)$$

where $Q_{1-r,r}$ is the r.v. with the Snedecor–Fisher distribution defined by the probability density

$$q(x;1-r,r) = \frac{(1-r)^{1-r} r^r}{\Gamma(1-r)\Gamma(r)} \cdot \frac{1}{x^r[r + (1-r)x]}, \quad x \geqslant 0. \quad (9)$$

In the paper [26], it was shown that any gamma distribution with shape parameter no greater than one is mixed exponential. For convenience, we formulate this result as the following lemma.

**Lemma 1** ([26]). *The density of a gamma distribution $g(x;r,\mu)$ with $0 < r < 1$ can be represented as*

$$g(x;r,\mu) = \int_0^\infty z e^{-zx} p(z;r,\mu) dz,$$

*where*

$$p(z;r,\mu) = \frac{\mu^r}{\Gamma(1-r)\Gamma(r)} \cdot \frac{\mathbf{1}(z \geqslant \mu)}{(z-\mu)^r z}$$

*is the density of the r.v. $Z_{r,\mu}$ introduced above. In other words, if $0 < r < 1$, then*

$$G_{r,\mu} \stackrel{d}{=} \frac{W_1}{Z_{r,\mu}}, \quad (10)$$

*where the random variables $W_1$ and $Z_{r,\mu}$ are independent. Moreover, a gamma distribution with shape parameter $r > 1$ cannot be represented as a mixed exponential distribution.*

Let $r > 0$, $\gamma \in \mathbb{R}$ and $\mu > 0$. Let the r.v. $N_{r,\gamma,\mu}$ have the GNB distribution. Its d.f. will be denoted $F_{GNB}(x;r,\gamma,\mu)$.

Along with the arguments given above in favor of the adequacy of the GNB models for the duration of wet periods based on their definition as mixed Poisson distributions, this effect can also be explained (at least in part) by their one more important property of being mixed geometric formulated as the following theorem.

**Theorem 1** ([3]). *If $r \in (0,1]$, $\gamma \in (0,1]$ and $\mu > 0$, then a GNB distribution is a $Y_{r,\gamma,\mu}$-mixed geometric distribution:*

$$\mathsf{P}(N_{r,\gamma,\mu} = k) = \int_0^1 y(1-y)^k d\mathsf{P}(Y_{r,\gamma,\mu} < y), \quad k = 0,1,2..., \quad (11)$$

*where*

$$Y_{r,\gamma,\mu} \stackrel{d}{=} \frac{S_{\gamma,1} Z_{r,\mu}^{1/\gamma}}{1 + S_{\gamma,1} Z_{r,\mu}^{1/\gamma}} \stackrel{d}{=} \frac{\mu^{1/\gamma} S_{\gamma,1} (G_{r,1} + G_{1-r,1})^{1/\gamma}}{G_{r,1}^{1/\gamma} + \mu^{1/\gamma} S_{\gamma,1} (G_{r,1} + G_{1-r,1})^{1/\gamma}}, \quad (12)$$

*where the r.v.'s $S_{\gamma,1}$ and $Z_{\mu,r}$ or $S_{\gamma,1}$, $G_{r,1}$ and $G_{1-r,1}$ are independent.*

## 4. The Asymptotic Approximation to the Probability Distribution of Extremal Daily Precipitation within a Wet Period

In this section, the probability distribution of extremal daily precipitation within a wet period will be deduced as an asymptotic approximation. We will require some auxiliary statements formulated as lemmas.

The following asymptotic property of the GNB distribution will play the fundamental role in the construction of asymptotic approximations to the distributions of extreme daily precipitation within a wet period and the total precipitation volume per wet period and the corresponding statistical tests for precipitation to be abnormally heavy.

**Lemma 2** ([3]). *For $r > 0$, $\gamma \in \mathbb{R}$, $\mu > 0$ let $N_{r,\gamma,\mu}$ be an r.v. with the GNB distribution. We have*

$$\mu^{1/\gamma} N_{r,\gamma,\mu} \Longrightarrow \overline{G}_{r,\gamma,1} \stackrel{d}{=} G_{r,1}^{1/\gamma} \tag{13}$$

*as $\mu \to 0$. If, moreover, $r \in (0,1]$ and $\gamma \in (0,1]$, then the limit law can be represented as*

$$\overline{G}_{r,\gamma,1} \stackrel{d}{=} \frac{W_1}{S_{\gamma,1} Z_{r,1}^{1/\gamma}} \stackrel{d}{=} \frac{W_1^{1/\gamma}}{Z_{r,1}^{1/\gamma}} \stackrel{d}{=} \left( \frac{W_1 G_{r,1}}{G_{r,1} + G_{1-r,1}} \right)^{1/\gamma} \stackrel{d}{=} W_1^{1/\gamma} \cdot \left(1 + \tfrac{1-r}{r} Q_{1-r,r}\right)^{-1/\gamma}, \tag{14}$$

*where the r.v.'s $W_1$, $S_{\gamma,1}$ and $Z_{r,1}$ are independent as well as the r.v.'s $W_1$ and $Z_{r,1}$, or the r.v.'s $W_1$, $G_{r,1}$ and $G_{1-r,1}$, and the r.v. $Q_{1-r,r}$ has the Snedecor–Fisher distribution with parameters $1-r$ and $r$, see Equation (9).*

Let $\mu > 0$, $\gamma > 0$. Instead of an infinitesimal parameter $\mu$, in order to construct asymptotic approximations with "large" sample size, introduce an auxiliary "infinitely large" parameter $n \in \mathbb{N}$ and assume that $\mu = \mu_n = \mu n^{-\gamma}$. It can be easily made sure that in this case

$$\overline{G}_{r,\gamma,\mu/n^\gamma} \stackrel{d}{=} n\overline{G}_{r,\gamma,\mu}. \tag{15}$$

Then for $r > 0$, $\mu > 0$ for any $n \in \mathbb{N}$, we have

$$n^{-1}\overline{G}_{r,\gamma,\lambda/n^\gamma} \stackrel{d}{=} \overline{G}_{r,\gamma,\lambda} \stackrel{d}{=} \lambda^{-1/\gamma}\overline{G}_{r,\gamma,1} \stackrel{d}{=} \lambda^{-1/\gamma} G_{r,1}^{1/\gamma}. \tag{16}$$

The standard Poisson process (the Poisson process with unit intensity) will be denoted $P(t)$, $t \geqslant 0$.

**Lemma 3** ([27]). *Let $\Lambda_1, \Lambda_2, \ldots$ be a sequence of positive r.v.'s such that for any $n \in \mathbb{N}$ the r.v. $\Lambda_n$ is independent of the standard Poisson process $P(t)$, $t \geqslant 0$. The convergence*

$$n^{-1}P(\Lambda_n) \Longrightarrow \Lambda$$

*as $n \to \infty$ to some nonnegative r.v. $\Lambda$ takes place if and only if*

$$n^{-1}\Lambda_n \Longrightarrow \Lambda, \quad n \to \infty. \tag{17}$$

Lemma 3 can be regarded as a special case of the following result. Consider a sequence of r.v.'s $W_1, W_2, \ldots$ Let $N_1, N_2, \ldots$ be natural-valued r.v.'s such that for every $n \in \mathbb{N}$ the r.v. $N_n$ is independent of the sequence $W_1, W_2, \ldots$ In the following statement, the convergence is meant as $n \to \infty$.

**Lemma 4** ([28,29]). *Assume that there exists an infinitely increasing (convergent to zero) sequence of positive numbers $\{b_n\}_{n \geqslant 1}$ and an r.v. $W$ such that*

$$b_n^{-1} W_n \Longrightarrow W.$$

*If there exist an infinitely increasing (convergent to zero) sequence of positive numbers* $\{d_n\}_{n\geqslant 1}$ *and an r.v. N such that*

$$d_n^{-1} b_{N_n} \Longrightarrow N, \tag{18}$$

*then*

$$d_n^{-1} W_{N_n} \Longrightarrow W \cdot N, \tag{19}$$

*where the r.v.'s on the right-hand side of Equation (19) are independent. If, in addition,* $N_n \longrightarrow \infty$ *in probability and the family of scale mixtures of the d.f. of the r.v. W is identifiable, then Condition (18) is not only sufficient for Equation (19), but is necessary as well.*

Consider a sequence of independent identically distributed (i.i.d.) r.v.'s $X_1, X_2, \ldots$. Let $N_1, N_2, \ldots$ be a sequence of natural-valued r.v.'s such that for each $n \in \mathbb{N}$ the r.v. $N_n$ is independent of the sequence $X_1, X_2, \ldots$. Denote $M_n = \max\{X_1, \ldots, X_{N_n}\}$.

**Lemma 5** ([30]). *Let* $\Lambda_1, \Lambda_2, \ldots$ *be a sequence of positive r.v.'s such that for each* $n \in \mathbb{N}$ *the r.v.* $\Lambda_n$ *is independent of the Poisson process* $P(t)$, $t \geqslant 0$. *Let* $N_n = P(\Lambda_n)$. *Assume that there exists a nonnegative r.v.* $\Lambda$ *such that Convergence (17) takes place. Let* $X_1, X_2, \ldots$ *be i.i.d. r.v.'s a common d.f.* $F(x)$. *Assume also that* $\sup\{x : F(x) < 1\} = \infty$ *and there exists a number* $\alpha > 0$ *such that for each* $x > 0$

$$\lim_{y\to\infty} \frac{1 - F(xy)}{1 - F(y)} = x^{-\alpha}. \tag{20}$$

*Then*

$$\lim_{n\to\infty} \sup_{x\geqslant 0} \left| \mathsf{P}\left( \frac{M_n}{F^{-1}(1 - \frac{1}{n})} < x \right) - \int_0^\infty e^{-zx^{-\alpha}} d\mathsf{P}(\Lambda < z) \right| = 0.$$

Now we turn to the main results of this section. The principal role in our reasoning will be played by Lemma 5. In order to justify its applicability, we need to make sure that the daily precipitation volumes satisfy Condition (20). A thorough statistical analysis shows that, although being rather adequate and, in general, acceptable model, the traditional gamma distribution (used, e.g., in [4]) is not the best model for statistical regularities in daily precipitation. The analysis of meteorological data (daily precipitation volumes) registered over 60 years at two geographic points with a very different climate: Potsdam (Brandenburg, Germany) with a mild climate influenced by the closeness to the ocean with warm Gulfstream flow and Elista (Kalmykia, Russia) with a radically continental climate convincingly suggests the Pareto-type model for the distribution of daily precipitation volumes, see Figures 3 and 4. For comparison, on these figures, the graphs of the best gamma-densities there are also presented. It can be seen that the gamma model fits the histograms in a noticeably worse way than the Pareto distribution.

**Theorem 2.** *Let* $n \in \mathbb{N}$, $\gamma > 0$, $\mu > 0$ *and let* $N_{r,\gamma,\mu_n}$ *be an r.v. with the GNB distribution with parameters* $r > 0$, $\gamma > 0$ *and* $\mu_n = \mu/n^\gamma$. *Let* $X_1, X_2, \ldots$ *be i.i.d. r.v.'s with a common d.f.* $F(x)$. *Assume that* $\mathrm{rext}(F) = \infty$ *and there exists a number* $\alpha > 0$ *such that Relation (20) holds for any* $x > 0$. *Then*

$$\lim_{n\to\infty} \sup_{x\geqslant 0} \left| \mathsf{P}\left( \frac{\max\{X_1, \ldots, X_{N_{r,\gamma,\mu_n}}\}}{F^{-1}(1 - \frac{1}{n})} < x \right) - F(x; r, \alpha, \gamma, \mu) \right| = 0,$$

*where*

$$F(x; r, \alpha, \gamma, \mu) = \int_0^\infty e^{-\lambda x^{-\alpha}} g^*(\lambda; r, \gamma, \mu) d\lambda \equiv \mathsf{P}(M_{r,\alpha,\gamma,\mu} < x), \quad x \in \mathbb{R}.$$

*The limit r.v.* $M_{r,\alpha,\gamma,\mu}$ *admits the following product representations:*

$$M_{r,\alpha,\gamma,\mu} \stackrel{d}{=} \frac{\overline{G}_{r,\alpha\gamma,\mu}}{W_\alpha} \stackrel{d}{=} \left( \frac{\overline{G}_{r,\gamma,\mu}}{W_1} \right)^{1/\alpha} \stackrel{d}{=} \mu^{-1/\alpha\gamma} \left( \frac{G_{r,1}}{W_\gamma} \right)^{1/\alpha\gamma} \tag{21}$$

*and in each term, the involved random variables are independent.*

**Proof.** By definition, the GNB distribution is a mixed Poisson distribution with the GG mixing distribution. So, $N_{r,\gamma,\mu_n} \stackrel{d}{=} P(\overline{G}_{r,\gamma,\mu_n})$. Therefore, from Equation (16), Lemma 3 with $\Lambda_n = \overline{G}_{r,\gamma,\mu_n}$ and Lemma 5 with the account of the absolute continuity of the limit distribution it immediately follows that

$$\lim_{n\to\infty}\sup_{x\geqslant 0}\left|\mathsf{P}\left(\frac{\max\{X_1,\ldots,X_{N_{r,\gamma,\mu_n}}\}}{F^{-1}(1-\frac{1}{n})}<x\right)-\int_0^\infty e^{-zx^{-\alpha}}g^*(z;r,\gamma,\mu)dz\right|=0.$$

Since the Fréchet (inverse Weibull) d.f. $e^{-x^{-\alpha}}$ with $\alpha > 0$ corresponds to the r.v. $W_\alpha^{-1}$, it is easy to make sure

$$F(x;r,\alpha,\gamma,\mu)\equiv\int_0^\infty e^{-zx^{-\alpha}}g^*(z;r,\gamma,\mu)dz=\mathsf{P}\left(\frac{\overline{G}^{1/\alpha}_{r,\gamma,\mu}}{W_\alpha}<x\right).$$

Moreover, using relation $\overline{G}_{r,\gamma,\mu} \stackrel{d}{=} G^{1/\gamma}_{r,\mu}$, it is easy to see that

$$\frac{\overline{G}^{1/\alpha}_{r,\gamma,\mu}}{W_\alpha}\stackrel{d}{=}\frac{\overline{G}_{r,\alpha\gamma,\mu}}{W_\alpha}\stackrel{d}{=}\left(\frac{\overline{G}_{r,\gamma,\mu}}{W_1}\right)^{1/\alpha}\stackrel{d}{=}\mu^{-1/\alpha\gamma}\left(\frac{G_{r,1}}{W_\gamma}\right)^{1/\alpha\gamma},$$

where in each term the involved random variables are independent. The theorem is proved. □

If $\gamma = 1$, then the limit distribution $F(x;r,\alpha,1,\mu)$ corresponds to the results of [31,32].

**Theorem 3.** *The distribution of the r.v. $M_{r,\alpha,\gamma,\mu}$ admits the following representations.*

*(i)* *If $r \in (0,1]$, it is the scale mixture of the distribution of the ratio of two independent Weibull-distributed r.v.'s:*

$$M_{r,\alpha,\gamma,\mu}\stackrel{d}{=}(\mu Z_{r,1})^{-1/\alpha\gamma}\cdot\frac{W_{\alpha\gamma}}{W_\gamma},$$

*where all the involved random variables are independent and the r.v. $Z_{r,1}$ is defined in Equation (8).*

*(ii)* *If $\gamma \in (0,1]$, it is the scale mixture of the tempered Snedecor–Fisher distribution with parameters $r$ and 1:*

$$M_{r,\alpha,\gamma,\mu}\stackrel{d}{=}\left(\frac{S_{\gamma,1}}{\mu r}\cdot Q_{r,1}\right)^{1/\alpha\gamma},$$

*where $S_{\gamma,1}$ is a positive strictly stable r.v. with characteristic exponent $\gamma$ independent of the r.v. $Q_{r,1}$ with the Snedecor–Fisher distribution in Equation (9) with parameters $r$ and 1.*

*(iii)* *If $\gamma \in (0,1]$ and $r \in (0,1]$, it is the scale mixture of the Pareto laws:*

$$M_{r,\alpha,\gamma,\mu}\stackrel{d}{=}\Pi_\alpha\left(S_{\gamma,1}Z^{1/\gamma}_{r,1}\right)^{-1/\alpha},$$

*where $\mathbb{P}(\Pi_\alpha > x) = (x^\alpha + 1)^{-1}$, $x \geqslant 0$.*

*(iv)* *If $r \in (0,1]$ and $\alpha\gamma \in (0,1]$, it is the scale mixture of the folded normal laws:*

$$M_{r,\alpha,\gamma,\mu}\stackrel{d}{=}|X|\cdot\frac{\sqrt{2W_1}}{\mu^{1/\alpha\gamma}W_\alpha S_{\alpha\gamma,1}Z^{1/\alpha\gamma}_{r,1}},$$

*where all the involved r.v.'s are independent.*

**Proof.** To prove (i) it suffices to consider the rightmost term in Equation (21), apply relations $W_1^{1/\gamma} \stackrel{d}{=} W_\gamma$ and $G_{r,\mu} \stackrel{d}{=} \frac{W_1}{Z_{r,\mu}}$ (here $0 < r < 1$ and the r.v.'s $W_1$ and $Z_{r,\mu}$ are independent (for details, see Lemma 1).

To prove (ii) it suffices to transform the rightmost term in (21) with the account of representation in Equation (7) and use the definition of the Snedecor–Fisher distribution as the distribution of the ratio of two independent gamma-distributed r.v.'s (see, e.g., Section 27 in [33]).

To prove (iii) it suffices to transform the second term in Equation (21) with the account of Equation (14) and notice that the distribution of the ratio of two independent exponentially distributed r.v.'s coincides with that of the random variable $\Pi_1$.

To prove (iv) it suffices to transform the second term in (21) with the account of (14) and notice that $W_1 \stackrel{d}{=} |X|\sqrt{2W_1}$ with the r.v.'s on the right-hand side being independent (see, e.g., [25]). The theorem is proved. □

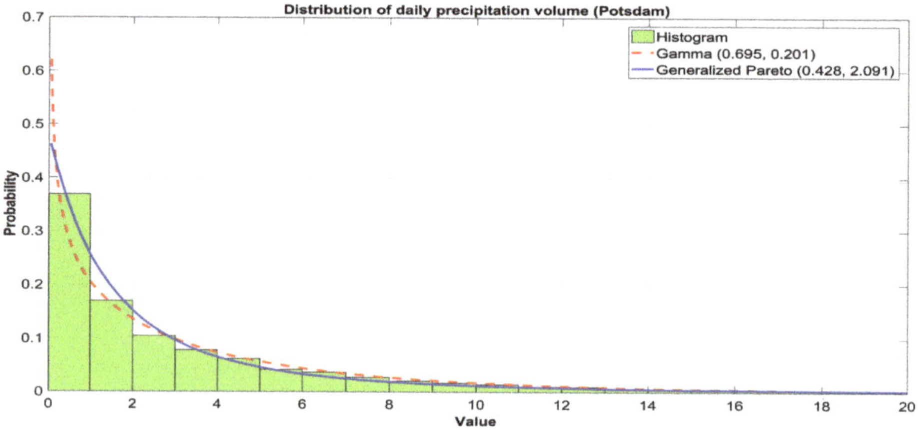


**Figure 3.** The histogram of daily precipitation volumes in Potsdam and the fitted Pareto and gamma distributions.

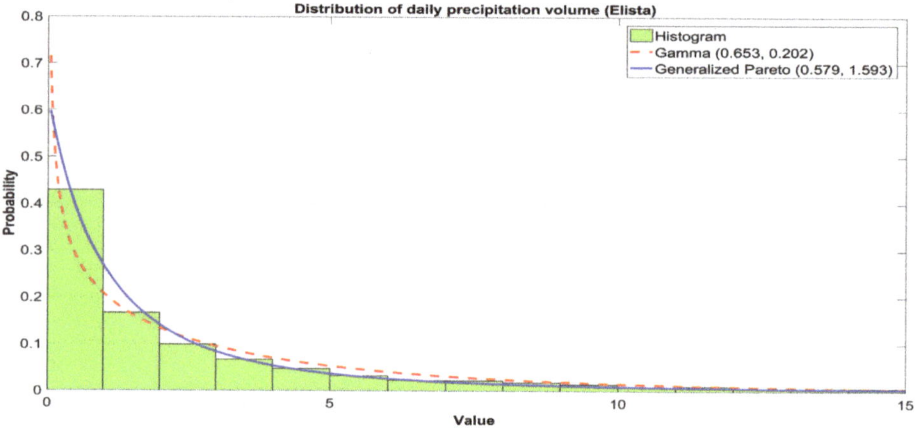


**Figure 4.** The histogram of daily precipitation volumes in Elista and the fitted Pareto and gamma distributions.

The product representations for the random value $M_{r,\alpha,\gamma,\mu}$ established in Theorem 3 can be used for computer simulation.

**Theorem 4.** *If $r \in (0,1]$, $\mu > 0$ and $\alpha\gamma \in (0,1]$, then the d.f. $F(x;r,\alpha,\gamma,\mu)$ is mixed exponential:*

$$1 - F(x;r,\alpha,\gamma,\mu) = \int_0^\infty e^{-ux} dA(u), \quad x \geqslant 0,$$

*where $A(u) = \mathsf{P}(\mu^{1/\alpha\gamma} W_\alpha S_{\alpha\gamma,1} Z_{r,1}^{1/\alpha\gamma} < u)$, $u \geqslant 0$, and all the involved r.v.'s are independent.*

**Proof.** To prove this statement, it suffices to transform the second term in Equation (21) with the account of Equation (14) and obtain

$$M_{r,\alpha,\gamma,\mu} \stackrel{d}{=} \frac{W_1}{\mu^{1/\alpha\gamma} W_\alpha S_{\alpha\gamma,1} Z_{r,1}^{1/\alpha\gamma}}.$$

□

**Corollary 1.** *Let $r \in (0,1]$, $\alpha\gamma \in (0,1]$, $\mu > 0$. Then the distribution function $F(x;r,\alpha,\gamma,\mu)$ is infinitely divisible.*

**Proof.** This statement immediately follows from Theorem 3 and the result of [34] stating that the product of two independent non-negative r.v.'s is infinitely divisible, if one of the two is exponentially distributed. □

It is possible to deduce explicit expressions for the moments of the r.v. $M_{r,\alpha,\gamma,\mu}$.

**Theorem 5.** *Let $0 < \delta < \alpha$. Then*

$$\mathbb{E} M_{r,\alpha,\gamma,\mu}^{\delta} = \frac{\Gamma(r + \frac{\delta}{\alpha\gamma})\Gamma(1 - \frac{\delta}{\alpha})}{\mu^{\delta/\alpha\gamma}\Gamma(r)}.$$

**Proof.** From Equation (14) it follows that $\mathbb{E} M_{r,\alpha,\gamma,\mu}^{\delta} = \mu^{-\delta/\alpha\gamma} \mathbb{E} G_{r,1}^{\delta/\alpha\gamma} \cdot \mathbb{E} W_1^{-\delta/\alpha}$. It is easy to verify that $\mathbb{E} G_{r,1}^{\delta/\alpha\gamma} = \Gamma(r + \frac{\delta}{\alpha\gamma})/\Gamma(r)$, $\mathbb{E} W_1^{-\delta/\alpha} = \Gamma(1 - \frac{\delta}{\alpha})$. Hence follows the desired result. □

Consider the bounds for the rate of convergence in Theorem 2. For this purpose, we will use one more auxiliary statement.

**Lemma 6.** *Let $\lambda > 0$, $X_1, X_2, \ldots$ be i.i.d. r.v.'s with a common d.f. $F(x)$, $P(t)$ be the standard Poisson process independent of $X_1, X_2, \ldots$. Assume that there exists a d.f. $H(x)$ such that for any $x \in \mathbb{R}$*

$$\lim_{n\to\infty} \mathsf{P}\left( \frac{1}{F^{-1}(1 - \frac{1}{n})} \max_{1 \leqslant k \leqslant n} X_k < x \right) = H(x). \tag{22}$$

*Then for any $n \in \mathbb{N}$*

$$\left| \mathsf{P}\left( \frac{1}{F^{-1}(1 - \frac{1}{n})} \max_{1 \leqslant k \leqslant P(n\lambda)} X_k < x \right) - H^\lambda(x) \right| \leqslant \left| n\left[1 - F\left(xF^{-1}(1 - \tfrac{1}{n})\right)\right] - \log H(x) \right| \lambda H^\lambda(x).$$

**Proof.** This statement is a special case of Corollary 2 in [35]. □

**Theorem 6.** *Let $n \in \mathbb{N}$, $\gamma > 0$, $\mu > 0$ and let $N_{r,\gamma,\mu_n}$ be an r.v. with the GNB distribution with parameters $r > 0$, $\gamma > 0$ and $\mu_n = \mu/n^\gamma$. Let $X_1, X_2, \ldots$ be i.i.d. r.v.'s with the common Pareto d.f.*

$$F(x) = 1 - \frac{c}{ax^\alpha + c}, \quad x \geqslant 0,$$

*with* $a, c, \alpha > 0$,

$$F(x; r, \alpha, \gamma, \mu) = \int_0^\infty e^{-\lambda x^{-\alpha}} g^*(\lambda; r, \gamma, \mu) d\lambda, \quad x \in \mathbb{R}.$$

*Then for any* $x \in \mathbb{R}$

$$\left| \mathsf{P}\left( \left[ \frac{a}{c(n-1)} \right]^{1/\gamma} \max_{1 \leqslant k \leqslant N_{r,\gamma,\mu_n}} X_k < x \right) - F(x; r, \alpha, \gamma, \mu) \right| \leqslant$$

$$\leqslant \left| \frac{x^\alpha - 1}{x^\alpha(n-1)+1} \right| \cdot \int_0^\infty \lambda e^{-\lambda x^{-\alpha}} g^*(\lambda; r, \gamma, \mu) d\lambda \leqslant \left| \frac{x^\alpha - 1}{x^\alpha(n-1)+1} \right| \cdot \frac{\Gamma(r + \frac{1}{\gamma})}{\mu^{1/\gamma}\Gamma(r)}.$$

**Proof.** First of all, check Condition (20). We have

$$\frac{1 - F(xy)}{1 - F(y)} = \frac{ay^\alpha + c}{ax^\alpha y^\alpha + c} \longrightarrow x^{-\alpha}$$

as $y \to \infty$, that is, Condition (20) holds implying Equation (22) with $H(x) = e^{-x^{-\alpha}}$ in accordance with the classical theory of extremes (see, e.g., [36]). Second, note that in the case under consideration $F^{-1}(1 - \frac{1}{n}) = [\frac{c(n-1)}{a}]^{1/\alpha}$ so that $F(xF^{-1}(1 - \frac{1}{n})) = 1 - [x^\alpha(n-1)+1]^{-1}$ and

$$n\left[1 - F\left(xF^{-1}\left(1 - \frac{1}{n}\right)\right)\right] - \log H(x) = \frac{n}{x^\alpha(n-1)+1} - \frac{1}{x^\alpha} = \frac{x^\alpha - 1}{x^\alpha(n-1)+1}.$$

Third, from Equation (15) it follows that $N_{r,\gamma,\mu/n^\gamma} \stackrel{d}{=} P(n\overline{G}_{r,\gamma,\mu})$ with independent $P(t)$ and $\overline{G}_{r,\gamma,\mu}$. Therefore, by Lemma 6 we have

$$\left| \mathsf{P}\left( \left[ \frac{a}{c(n-1)} \right]^{1/\gamma} \max_{1 \leqslant k \leqslant N_{r,\gamma,\mu_n}} X_k < x \right) - F(x; r, \alpha, \gamma, \mu) \right| \leqslant$$

$$\leqslant \int_0^\infty \left| \mathsf{P}\left( \left[ \frac{a}{c(n-1)} \right]^{1/\gamma} \max_{1 \leqslant k \leqslant P(n\lambda)} X_k < x \right) - e^{-\lambda x^{-\alpha}} \right| g^*(\lambda; r, \gamma, \mu) d\lambda \leqslant$$

$$\leqslant \left| \frac{x^\alpha - 1}{x^\alpha(n-1)+1} \right| \cdot \int_0^\infty \lambda e^{-\lambda x^{-\alpha}} g^*(\lambda; r, \gamma, \mu) d\lambda \leqslant \left| \frac{x^\alpha - 1}{x^\alpha(n-1)+1} \right| \cdot \mathsf{E}\overline{G}_{r,\gamma,\mu} =$$

$$= \left| \frac{x^\alpha - 1}{x^\alpha(n-1)+1} \right| \cdot \frac{\Gamma(r + \frac{1}{\gamma})}{\mu^{1/\gamma}\Gamma(r)}.$$

The theorem is proved. □

Actually, Theorem 6 states that the rate of convergence in Theorem 2 is $O(\mu_n^{1/\gamma})$ as $\mu_n \to 0$.

The results of this section serve as a theoretical base for the construction of a test for abnormally extreme daily precipitation. The distribution of the maximum daily precipitation per wet period can be assumed to be a tempered scale mixture of the gamma distribution in which the scale factor has the Weibull distribution. According to the typical construction of a test, a daily precipitation volume is considered to be abnormally extremal, if it exceeds a certain (pre-defined) quantile of this distribution. A detailed description of this test and algorithm of estimation of the parameters of the distribution mentioned above deserve a separate study as well as its application to real data.

## 5. The Asymptotic Approximation to the Probability Distribution of Total Precipitation over a Wet Period. Generalized R ényi Theorem for Gnb Random Sums

As far ago as in the 1950s, being interested in modeling rare events, A. Rényi studied rarefaction of renewal point processes and proved his famous theorem on convergence of rarefied renewal processes to the Poisson process [37,38]. The Rényi theorem states that the distribution of a geometric sum (i.e., a sum of a random number of i.i.d. r.v.'s in which the number of summands is a r.v. with the geometric distribution independent of the summands) normalized by its expectation converges to the exponential law as the expectation of the sum infinitely increases. The normalization of a sum by its expectation is typical for laws of large numbers. Therefore, the Rényi theorem can be regarded as the law of large numbers for geometric sums. A general law of large numbers for random sums of independent identically distributed (i.i.d.) random variables (r.v.'s) was proved in [28]. It was demonstrated there that the distribution of a random sum normalized by its expectation converges to some distribution, if and only if the distribution of the random index (the number of summands) converges to the same distribution (up to a scale parameter) under the same normalization. In [3] the law of large numbers for GNB random sums was proved. However, a direct application of this result to modeling the probability distribution of total precipitation over a wet period is hampered by the following very interesting practical observation.

One might have expected that successive daily precipitation volumes $X_1, X_2, \ldots$ satisfy the classical law of large numbers, that is, the arithmetic mean $\frac{1}{n}(X_1 + \ldots + X_n)$ converges to some number $a$ almost surely as $n$ infinitely grows, as it was done in [2]. However, a thorough analysis of real data shows that this not quite so. In Figure 5, there are the graphs of the averaged daily precipitation volumes in Potsdam and Elista demonstrating the slowly decreasing trend for Potsdam and slowly increasing trend for Elista.

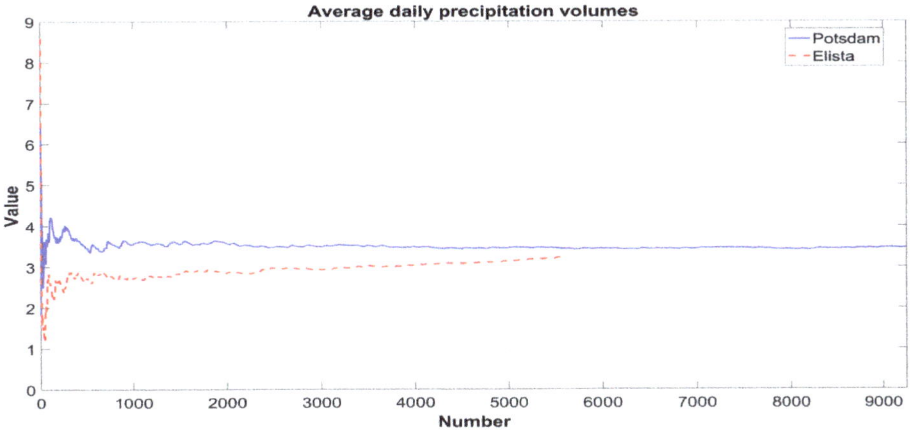


**Figure 5.** Stabilization of the cumulative averages of daily precipitation volumes as $n$ grows in Potsdam (continuous line) and Elista (dash line).

This means that, in order to match the stabilization of the averages at some level $a$, it is required to normalize the sum $X_1 + \ldots + X_n$ not by $n$, but by a somewhat more complicated function of $n$ that can match the influence of slow global trends. As such, a function of $n$, consider a power function $n^\beta$ with $\beta > 0$ and assume that not necessarily i.i.d. r.v.'s $X_1, X_2, \ldots$ satisfy the condition

$$\frac{1}{n^\beta} \sum_{j=1}^{n} X_j \Longrightarrow a \in (0, \infty) \tag{23}$$

as $n \to \infty$. The parameters $a$ and $\beta$ can be rather reliably estimated by the least squares technique.

Let $X_1, X_2, \ldots, X_n$ be the observed values of successive nonzero daily precipitation volumes, $n \in \mathbb{N}$ be the total number of available observations. For a natural $k = 1, \ldots, n$ denote $s_k = X_1 + \ldots + X_k$.

If Condition (23) holds, then for $k$ large enough ($1 \leqslant m \leqslant k \leqslant n$), the following estimates of the parameters $a$ and $\beta$ in Relation (23) can be used:

$$\widetilde{a} = \exp\left\{ \frac{\sum_{k=m}^{n} \log s_k \cdot \sum_{k=m}^{n} (\log k)^2 - \sum_{k=m}^{n} \log k \cdot \sum_{k=m}^{n} \left( \log k \cdot \log s_k \right)}{(n-m+1) \sum_{k=m}^{n} (\log k)^2 - \left( \sum_{k=m}^{n} \log k \right)^2} \right\}, \tag{24}$$

$$\widetilde{\beta} = \frac{\sum_{k=m}^{n} \log s_k - (n-m+1) \log \widetilde{a}}{\sum_{k=m}^{n} \log k}. \tag{25}$$

Indeed, if Condition (23) holds, the following approximate equality can be written:

$$\frac{T_k}{k^{\beta}} \approx a \Leftrightarrow -\beta \log k + \log T_k \approx \log a.$$

Therefore, the estimates of the parameters $a$ and $\beta$ can be found as the solution of the least squares problem

$$\sum_{k=m}^{n} (\log T_k - \beta \log k - \log a)^2 \longrightarrow \min_{\beta, \log a}.$$

This solution can be found explicitly and has the form

$$\widetilde{\log a} = \frac{\sum_{k=m}^{n} \log T_k \cdot \sum_{k=m}^{n} (\log k)^2 - \sum_{k=m}^{n} \log k \cdot \sum_{k=m}^{n} \left( \log k \cdot \log T_k \right)}{(n-m+1) \sum_{k=m}^{n} (\log k)^2 - \left( \sum_{k=m}^{n} \log k \right)^2},$$

$$\widetilde{\beta} = \frac{\sum_{k=m}^{n} \log T_k - (n-m+1) \widetilde{\log a}}{\sum_{k=m}^{n} \log k},$$

that leads to Formulas (24) and (25). This least squares method for estimation of $a$ and $\beta$ is realized by Algorithm A2 (see Appendix A).

The application of Equation (23) to real data from Potsdam and Elista with $a$ and $\beta$ estimated by Equations (24) and (25) is illustrated in Figure 6. It can be seen that the cumulative averages stabilize at the level $a = 4.087$ with $\beta = 0.981$ for Potsdam and at the level $a = 0.96$ with $\beta = 1.146$ for Elista.

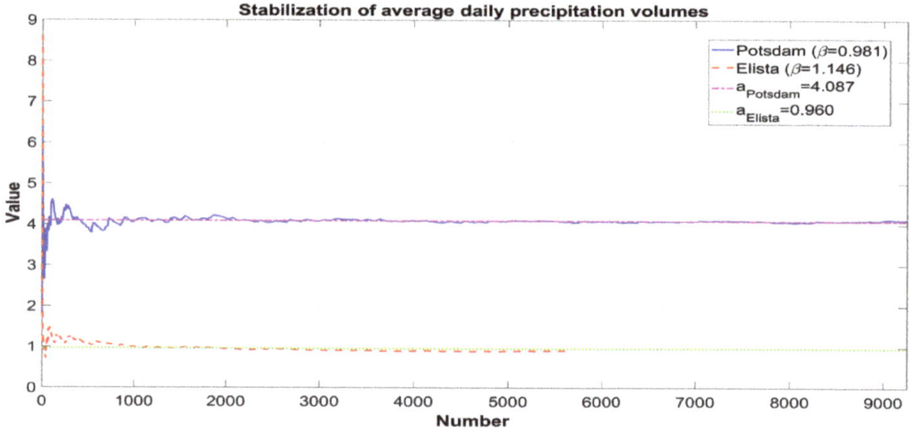


**Figure 6.** Stabilization of the cumulative averages of daily precipitation volumes as $n$ grows with $\beta = 1.139$ for Potsdam (solid line) and with $\beta = 0.981$ for Elista (dashed line).

So, to construct the asymptotic approximation to the probability distribution of total precipitation over a wet period, we should prove a generalized Rényi theorem for GNB random sums improving an analogous statement proved in [3]. It must be especially noted that in the following theorem, the r.v.'s $X_1, X_2, \ldots$ are not assumed to be i.i.d.

**Theorem 7.** *Assume that the nonzero daily precipitation volumes $X_1, X_2, \ldots$ satisfy Condition (23) with some $\beta > 0$ and $a > 0$. Let the numbers $r > 0$, $\gamma$ and $\mu > 0$ be arbitrary. For each $n \in \mathbb{N}$, let the r.v. $N_{r,\gamma,\mu_n}$ have the GNB distribution with parameters $r$, $\gamma$ and $\mu_n = \mu/n^\gamma$. Assume that for each $n \in \mathbb{N}$ the r.v. $N_{r,\gamma,\mu_n}$ is independent of the sequence $X_1, X_2, \ldots$ Then*

$$\frac{a\mu^{\beta/\gamma}}{n^\beta}\sum_{j=1}^{N_{r,\gamma,\mu/n^\gamma}} X_j \Longrightarrow \overline{G}_{r,\gamma/\beta,1} \stackrel{d}{=} G_{r,1}^{\beta/\gamma}$$

*as $n \to \infty$.*

**Proof.** The proof is based on Lemma 4 and Equation (16). From Equation (16) it follows that

$$\frac{\mu^{1/\gamma}}{n} \cdot N_{r,\gamma,\mu/n^\gamma} \Longrightarrow \overline{G}_{r,\gamma,1} \tag{26}$$

as $n \to \infty$. By virtue of Condition (23), in Lemma 4 let $b_n = n^\beta/a$. As $N_n$ in Lemma 4 take $N_{r,\gamma,\mu/n^\gamma}$. Then $b_{N_n} = \frac{1}{a}N^\beta_{r,\gamma,\mu/n^\gamma}$. From Equation (26) it follows that, as $n \to \infty$,

$$\tfrac{1}{a}N^\beta_{r,\gamma,\mu/n^\gamma} \cdot \frac{\mu^{\beta/\gamma}}{n^\beta} \Longrightarrow \tfrac{1}{a}\overline{G}^\beta_{r,\gamma,1} \stackrel{d}{=} \tfrac{1}{a}\overline{G}_{r,\gamma/\beta,1} \stackrel{d}{=} \tfrac{1}{a}G_{r,1}^{\beta/\gamma}. \tag{27}$$

Therefore, as $d_n$ we can take $d_n = n^\beta/\mu^{\beta/\gamma}$. So, using Equation (27) in the role of Equation (18) in Lemma 4, we obtain Equation (19) in the form

$$\frac{\mu^{\beta/\gamma}}{n^\beta}\sum_{j=1}^{N_{r,\gamma,\mu/n^\gamma}} X_j \Longrightarrow \tfrac{1}{a}\overline{G}_{r,\gamma/\beta,1} \stackrel{d}{=} \tfrac{1}{a}G_{r,1}^{\beta/\gamma}, \tag{28}$$

whence follows the desired result. The theorem is proved. □

Theorem 7 presents a good tool for the account of the parameters $\beta$ and $\gamma$ characterizing the deviation from traditional NB and arithmetic mean models due to the influence of possible (slow) global trends. If in Theorem 7 $r = \gamma = \beta = 1$, then we obtain a version of the Rényi theorem [39] generalized to non-identically distributed and not necessarily independent summands. If in Theorem 7 $\beta = 1$, then we obtain the law of large numbers for GNB random sums (see [3]). If in Theorem 7 $\gamma = 1$, then we obtain the law of large numbers for NB random sums modified for the case $\beta \neq 1$.

Therefore, if daily precipitation volumes $X_1, X_2, \ldots$ (of course, being non-identically distributed and not independent), with the account of the excellent fit of the GNB model for the duration of a wet period (see Figure 1), with rather small $\mu$, the GG distribution can be regarded as an adequate and theoretically well-based model for the total precipitation volume over a (long enough) wet period.

As regards the bounds for the rate of convergence in Theorem 7, consider a special case of $\beta = 1$ and i.i.d. $X_1, X_2, \ldots$

As a measure of the distance between probability distributions, consider the $\zeta$-metric proposed by V. M. Zolotarev in [40,41] (also see [42], p. 44). Let $s > 0$. There exists a unique representation of the number $s$ as $s = m + \alpha$ where $m$ is an integer and $0 < \alpha \leqslant 1$. By $\mathcal{F}_s$ we denote the set of all real-valued bounded functions $f$ on $\mathbb{R}$ that are $m$ times differentiable and $|f^{(m)}(x) - f^{(m)}(y)| \leqslant |x-y|^\alpha$. Let $X$ and $Y$ be two r.v.'s in which the distribution functions will be denoted $F_X(x)$ and $F_Y(x)$, respectively.

The $\zeta$-metric $\zeta_s(X, Y) \equiv \zeta_s(F_X, F_Y)$ in the space of probability distributions is defined by the equality $\zeta_s(X, Y) = \sup\{|\mathsf{E}(f(X) - f(Y))| : f \in \mathcal{F}_s\}$. In particular,

$$\zeta_1(X, Y) = \int_{\mathbb{R}} |F_X(x) - F_Y(x)| dx.$$

In [43], it was shown that in the case $\beta = 1$ and i.i.d. $X_1, X_2, \ldots$ for $1 \leqslant s \leqslant 2$ we have

$$\zeta_s\Big(\frac{\mu^{1/\gamma}}{na}\sum\nolimits_{j=1}^{N_{r,\gamma,\mu/n^\gamma}} X_j,\, \overline{G}_{r,\gamma,1}\Big) \leqslant \frac{(\mathsf{E}X_1^2)^{s/2}}{n^{s/2}|\mathsf{E}X_1|^s} \cdot \frac{\Gamma(1+\gamma)\Gamma(r+\frac{s}{2\gamma})}{\Gamma(1+s)\Gamma(r)}.$$

In particular,

$$\zeta_2\Big(\frac{\mu^{1/\gamma}}{na}\sum\nolimits_{j=1}^{N_{r,\gamma,\mu/n^\gamma}} X_j,\, \overline{G}_{r,\gamma,\mu}\Big) \leqslant \frac{\mathsf{E}X_1^2}{2n(\mathsf{E}X_1)^2} \cdot \frac{\Gamma(r+\frac{1}{\gamma})}{\Gamma(r)}.$$

The results presented above justify the GG models for the probability distribution of total precipitation volume over a wet period improving the models considered in [2]. Statistical tests for the detection of anomalously extreme total volumes will be considered below.

## 6. Statistical Tests for Anomalously Extreme Total Precipitation Volumes

Now we turn to the construction of the tests for the total precipitation volume during a wet period to be abnormally large.

In what follows, based on the results of the preceding section, we will assume that the total precipitation volume during a wet period has the GG distribution with some parameters $r > 0$, $\gamma > 0$ and $\mu > 0$.

Let $m \in \mathbb{N}$ and $\overline{G}_{r,\gamma,\mu}^{(1)}, \overline{G}_{r,\gamma,\mu}^{(2)}, \ldots, \overline{G}_{r,\gamma,\mu}^{(m)}$ be independent r.v.'s having the same GG distribution with parameters $r > 0$, $\gamma$ and $\mu > 0$. Also, let $G_{r,\mu}^{(1)}, G_{r,\mu}^{(2)}, \ldots, G_{r,\mu}^{(m)}$ be i.i.d. r.v.'s having the same gamma distribution with parameters $r > 0$ and $\mu > 0$.

The base for the first step in the construction of the desired test is the following obvious conclusion: if the r.v.'s $\overline{G}_{r,\gamma,\mu}^{(1)}, \overline{G}_{r,\gamma,\mu}^{(2)}, \ldots, \overline{G}_{r,\gamma,\mu}^{(m)}$ are identically distributed (that is, the sample $\overline{G}_{r,\gamma,\mu}^{(1)}, \overline{G}_{r,\gamma,\mu}^{(2)}, \ldots, \overline{G}_{r,\gamma,\mu}^{(m)}$ is homogeneous), then the r.v.'s $(\overline{G}_{r,\gamma,\mu}^{(1)})^\gamma, (\overline{G}_{r,\gamma,\mu}^{(2)})^\gamma, \ldots, (\overline{G}_{r,\gamma,\mu}^{(m)})^\gamma$ are also identically distributed (that is, the sample $(\overline{G}_{r,\gamma,\mu}^{(1)})^\gamma, (\overline{G}_{r,\gamma,\mu}^{(2)})^\gamma, \ldots, (\overline{G}_{r,\gamma,\mu}^{(m)})^\gamma$ is homogeneous.

Consider the relative contribution of the r.v. $(\overline{G}_{r,\gamma,\mu}^{(1)})^\gamma$ to the sum $(\overline{G}_{r,\gamma,\mu}^{(1)})^\gamma + (\overline{G}_{r,\gamma,\mu}^{(2)})^\gamma + \ldots + (\overline{G}_{r,\gamma,\mu}^{(m)})^\gamma$:

$$R = \frac{(\overline{G}_{r,\gamma,\mu}^{(1)})^\gamma}{(\overline{G}_{r,\gamma,\mu}^{(1)})^\gamma + (\overline{G}_{r,\gamma,\mu}^{(2)})^\gamma + \ldots + (\overline{G}_{r,\gamma,\mu}^{(m)})^\gamma}. \tag{29}$$

From Equation (4), it obviously follows that

$$R \stackrel{d}{=} \frac{G_{r,\mu}^{(1)}}{G_{r,\mu}^{(1)} + G_{r,\mu}^{(2)} + \ldots + G_{r,\mu}^{(m)}} \stackrel{d}{=} \frac{G_{r,1}^{(1)}}{G_{r,1}^{(1)} + G_{r,1}^{(2)} + \ldots + G_{r,1}^{(m)}} \stackrel{d}{=} R^*$$

(see Equation(29)).

So, the r.v. $R$ characterizes the relative precipitation volume for one (long enough) wet period with respect to the total precipitation volume registered for $m$ wet periods.

Note that

$$R = \Big(1 + \frac{1}{G_{r,\mu}^{(1)}}(G_{r,\mu}^{(2)} + \ldots + G_{r,\mu}^{(m)})\Big)^{-1} \stackrel{d}{=} \Big(1 + \frac{G_{(m-1)r,\mu}}{G_{r,\mu}}\Big)^{-1},$$

where the gamma-distributed r.v.'s on the right hand side are independent. The distribution of the r.v. $R$ was described in [2] where it was demonstrated that $R \stackrel{d}{=} \left(1 + \frac{k}{r}Q_{k,r}\right)^{-1}$ where $Q_{k,r}$ is the r.v. having the Snedecor–Fisher distribution determined for $k > 0$, $r > 0$ by the Lebesgue density

$$f_{k,r}(x) = \frac{\Gamma(k+r)}{\Gamma(k)\Gamma(r)}\left(\frac{k}{r}\right)^k \frac{x^{k-1}}{(1+\frac{k}{r}x)^{k+r}}, \quad x \geqslant 0. \tag{30}$$

It should be noted that the particular value of the scale parameter is insignificant. For convenience, it is assumed equal to one. It can be easily made sure by standard calculation using Equation (30), the distribution of the r.v. $R$ is determined by the density

$$p(x;k,r) = \frac{\Gamma(k+r)}{\Gamma(r)\Gamma(k)}(1-x)^{k-1}x^{r-1}, \quad 0 \leqslant x \leqslant 1,$$

that is, it is the beta distribution with parameters $k = (m-1)r$ and $r$.

Then the test for the homogeneity of an independent sample of size $m$ consisting of the GG distributed observations of total precipitation volumes during $m$ wet periods with known $\gamma$ based on the r.v. $R$ looks as follows. Let $V_1, \ldots, V_m$ be the total precipitation volumes during $m$ wet periods and, moreover, $V_1 \geqslant V_j$ for all $j \geqslant 2$. Calculate the quantity

$$SR = \frac{V_1^\gamma}{V_1^\gamma + \ldots + V_m^\gamma}$$

($SR$ means «Sample $R$»). From what was said above, it follows that under the hypothesis $H_0$: «the precipitation volume $V_1$ under consideration is not abnormally large» the r.v. $SR$ has the beta distribution with parameters $k = (m-1)r$ and $r$. Let $\varepsilon \in (0,1)$ be a small number, $\beta_{k,r}(1-\varepsilon)$ be the $(1-\varepsilon)$-quantile of the beta distribution with parameters $k = (m-1)r$ and $r$. If $SR > \beta_{k,r}(1-\varepsilon)$, then the hypothesis $H_0$ must be rejected, that is, the volume $V_1$ of precipitation during one wet period must be regarded as abnormally large. Moreover, the probability of erroneous rejection of $H_0$ is equal to $\varepsilon$.

Instead of $R$, the quantity

$$R_0 = \frac{(m-1)(\overline{G}^{(1)}_{r,\gamma,\mu})^\gamma}{(\overline{G}^{(2)}_{r,\gamma,\mu})^\gamma + \ldots + (\overline{G}^{(m)}_{r,\gamma,\mu})^\gamma} \stackrel{d}{=} \frac{(m-1)G^{(1)}_{r,\mu}}{G^{(2)}_{r,\mu} + \ldots + G^{(m)}_{r,\mu}} \stackrel{d}{=} \frac{k}{r}\frac{G_{r,\mu}}{G_{k,\mu}} \stackrel{d}{=} \frac{k}{r}\frac{G_{r,1}}{G_{k,1}} \stackrel{d}{=} Q_{r,k}$$

can be considered. Then, as is easily seen, the r.v.'s $R$ and $R_0$ are related by the one-to-one correspondence

$$R = \frac{R_0}{m-1+R_0} \quad \text{or} \quad R_0 = \frac{(m-1)R}{1-R},$$

so that the homogeneity test for a sample from the GG distribution equivalent to the one described above and, correspondingly, the test for a precipitation volume during a wet period to be abnormally large, can be based on the r.v. $R_0$, which has the Snedecor–Fisher distribution with parameters $r$ and $k = (m-1)r$.

Namely, again let $V_1, \ldots, V_m$ be the total precipitation volumes during $m$ wet periods and, moreover, $V_1 \geqslant V_j$ for all $j \geqslant 2$. Calculate the quantity

$$SR_{GG} = \frac{(m-1)V_1^\gamma}{V_2^\gamma + \ldots + V_m^\gamma}. \tag{31}$$

($SR_0$ means «Sample $R_0$»). From what was said above, it follows that under the hypothesis $H_0$: «the precipitation volume $V_1$ under consideration is not abnormally large» the r.v. $SR$ has the Snedecor–Fisher distribution with parameters $r$ and $k = (m-1)r$. Let $\varepsilon \in (0,1)$ be a small number, $q_{r,k}(1-\varepsilon)$ be the $(1-\varepsilon)$-quantile of the Snedecor–Fisher distribution with parameters $r$

and $k = (m-1)r$. If $SR_0 > q_{r,k}(1-\varepsilon)$, then the hypothesis $H_0$ must be rejected, that is, the volume $V_1$ of precipitation during one wet period must be regarded as abnormally large. Moreover, the probability of erroneous rejection of $H_0$ is equal to $\varepsilon$.

Let $l$ be a natural number, $1 \leqslant l < m$. It is worth noting that, unlike the test based on the statistic $R$, the test based on $R_0$ can be modified for testing the hypothesis $H_0'$: «the precipitation volumes $V_{i_1}, V_{i_2}, \ldots, V_{i_l}$ do not make an abnormally large cumulative contribution to the total precipitation volume $V_1 + \ldots + V_m$». For this purpose denote

$$T_l^\gamma = V_{i_1}^\gamma + V_{i_2}^\gamma + \ldots + V_{i_l}^\gamma, \quad T^\gamma = V_1^\gamma + V_2^\gamma + \ldots + V_m^\gamma$$

and consider the quantity

$$SR_0' = \frac{(m-l)T_l^\gamma}{l(T^\gamma - T_l^\gamma)}.$$

In the same way as it was done above, it is easy to make sure that

$$SR_0' \stackrel{d}{=} \frac{(m-l)G_{lr,l}}{lG_{(m-l)r,1}} \stackrel{d}{=} Q_{lr,(m-l)r}.$$

Let $\varepsilon \in (0,1)$ be a small number, $q_{lr,(m-1)r}(1-\varepsilon)$ be the $(1-\varepsilon)$-quantile of the Snedecor–Fisher distribution with parameters $lr$ and $k = (m-l)r$. If $SR_0' > q_{lr,(m-l)r}(1-\varepsilon)$, then the hypothesis $H_0'$ must be rejected, that is, the cumulative contribution of the precipitation volumes $V_{i_1}, V_{i_2}, \ldots, V_{i_l}$ into the total precipitation volume $V_1 + \ldots + V_m$ must be regarded as abnormally large. Moreover, the probability of erroneous rejection of $H_0'$ is equal to $\varepsilon$.

## 7. Comparison of Tests for Anomalously Extreme Precipitation Volumes Based on Gamma and Gg Distributions

In this section, the results of the application of the test based on the statistic $R$ in Equation (29) to the analysis of the time series of daily precipitation observed in Potsdam and Elista from 1950 to 2007 are considered and compared with similar results for the case of gamma distributed total precipitation volumes during wet periods [2].

The results of the application of the tests for a total precipitation volume during one wet period to be abnormally large based on GG and gamma models in the moving mode are shown in Figures 7 and 8 (Potsdam) and Figures 9 and 10 (Elista).

If $m$ is the window width (the number of observations in a moving window). A fixed sample point falls in exactly $m$ windows. One of the following cases can occur for a fixed observation:

- Absolute (abs) extreme, if at all $m$ windows it is recognized as abnormally extreme;
- Intermediate (int) extreme, if it is recognized as abnormally extreme for at least half of windows containing it;
- Relative (rel) extreme, if it is recognized as abnormally extreme for at least one window;
- Not extremal, if it is not recognized as abnormally extreme for all windows.

Algorithm A3 (see Appendix A) realizes the method based on the statistic $R$ described above.

For the sake of vividness on these figures the time horizon equals 90 and 360 days and the significance level $\alpha$ of the tests is 0.01. The absolutely, intermediate and relatively abnormal precipitation volumes are marked by downward-pointing triangles, circles and squares, respectively, for the test based on the gamma model, whereas the corresponding test based on the statistic $R$ based on the GG distribution are marked by upward-pointing triangles, diamonds and right-pointing triangles, respectively. It is worth noting that MATLAB's notations are used here for these markers.

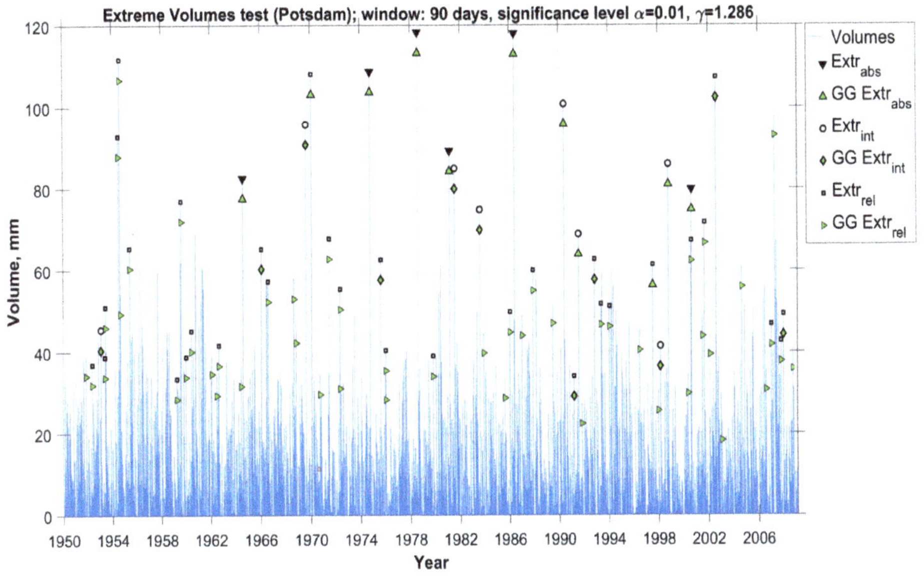


**Figure 7.** Abnormal precipitation volumes (Potsdam, 90 days).

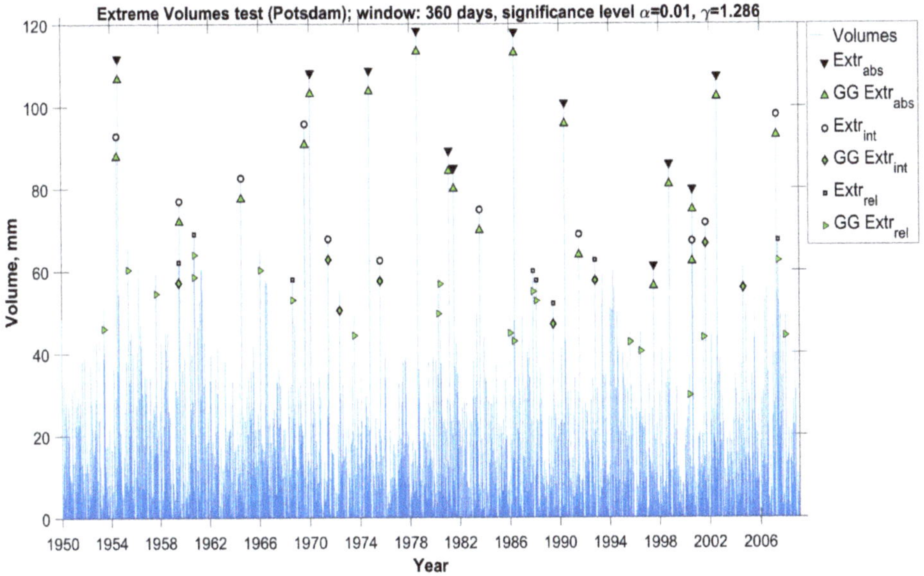


**Figure 8.** Abnormal precipitation volumes (Potsdam, 360 days).

Figures 7–10 demonstrate non-trivial values of the parameter $\gamma$, that is, $\gamma \neq 1$. For Potsdam $\gamma = 1.286$, whereas for Elista $\gamma$ equals 1.279. At the same time, the results of the two methods are quite close, although the approach based on the GG distribution demonstrates a higher quality of determining potentially extreme observations. The same conclusions are valid for smaller window sizes.

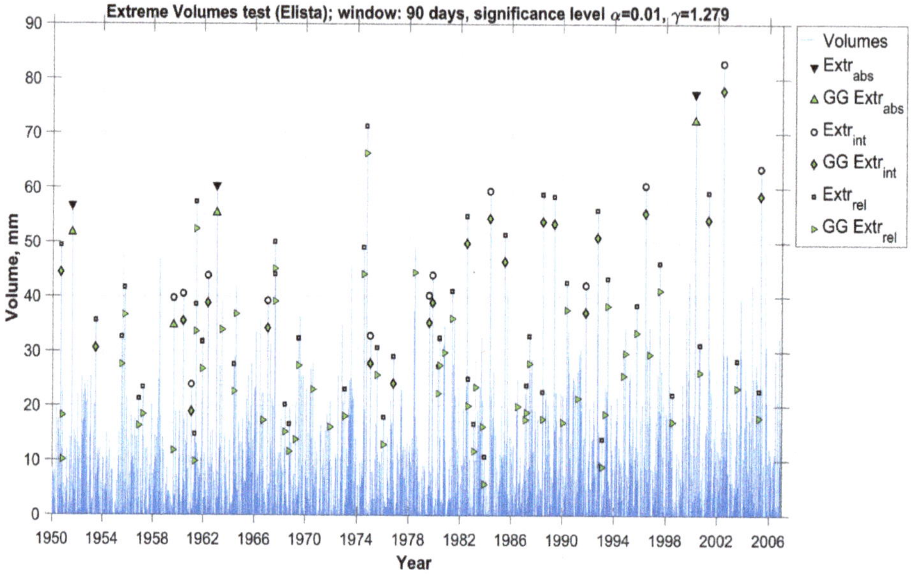


**Figure 9.** Abnormal precipitation volumes (Elista, 90 days).

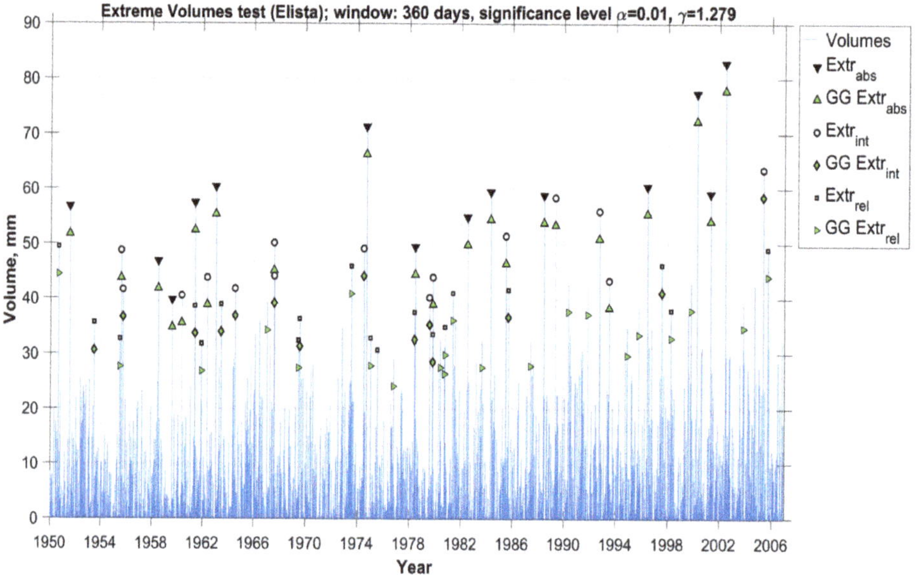


**Figure 10.** Abnormal precipitation volumes (Elista, 360 days).

## 8. Comparison of GG-Based Statistical Test and Peaks over Threshold Methodology for Extreme Precipitation Intensities

One important precipitation indicator is the precipitation intensity that is defined as the ratio of the total precipitation volume over a wet period to the duration of this wet period measured in days. The extreme precipitation volumes and intensities are relevant to various problems of climatology and hydrometeorology (see, for example, [44–47]). Traditionally, these phenomena are investigated for different geographical regions or countries [48,49]. In particular, the issue of determining threshold values, the excess of which leads to the extreme events, for example, in daily rainfalls or their intensities, is the key point of the study. Precipitation intensities are important not only for forecasting floods but also for solving problems such as runoff and soil erosion [50,51]. It can be explained by the contemporary climate change scenarios that predict a significant increase in the frequency of high intensity rainfall events, primarily in the dry areas. Moreover, precipitation can induce shallow landslides [52,53] and debris flows [54].

Statistical analysis of real data shows that the probability distribution of the precipitation intensity can be approximated by the gamma distribution with very high accuracy. In [55], some theoretical arguments were presented to justify the gamma model for the distribution of precipitation intensities. So, the statistical approach described in Section 6 and in [2] can be also used for identification of abnormally large intensities. For the analysis, the precipitation intensities in Potsdam and Elista (verified samples without missing values) are used as the initial data. This section presents a comparison of a non-parametric approach based on the extreme value theory as well as modified Peaks over Threshold (PoT) methodology [56] with the parametric approach that significantly involves testing parametric statistical hypotheses to determine extreme intensities of wet periods (see Section 6). The classical version of PoT [57] is quite popular for solving a wide range of climatic problems. In particular, the following results can be mentioned: the inverse Weibull distribution as an extreme wind speed model [58], a time-dependent versions of the PoT model for severe storm waves [59] and daily temperatures [60], probability model for rainfalls of high magnitude [61], analysis of precipitation extremes in a changing climate [62]. Most applications of the extreme value theory assume stationarity, but it is well-known that real events are not stationary. So, the generalized results analogous to Theorem 7 are required. All the numerical methods are implemented as a MATLAB program. Algorithm A4 demonstrates the method based on the PoT and GG-test (see Appendix A).

Figures 11 and 12 present the results obtained by the modified PoT algorithm in which the Weibull distribution is considered as the distribution of time between extreme events. Starting from the maximum threshold value that coincides with the maximum of the analyzed data, the hypothesis that the time intervals between the moments of excess of a certain threshold have the Weibull distribution is tested. The corresponding $P$-value is saved, and the threshold is shifted down by a certain (small) step (in this case, 0.01). It is worth noting that a similar procedure was suggested in [56] for precipitation volumes under the assumption that the time intervals between excesses have the exponential distribution. For a given significance level (in both cases, $\alpha$ is chosen as 0.01), the corresponding hypothesis is not rejected for all thresholds for which the $P$-values are located to the right of the red vertical line in the upper graphs in Figures 11 and 12. The lower graphs show the parameters of the fitted Weibull distribution.

On Figures 13 and 14 the results of the test (see Section 6) for both the GG and usual gamma distribution are compared with those of the PoT method based on the exponential and Weibull distributions for the intensities in Potsdam and Elista. The following notation is used:

- The thresholds with the indices low correspond to the minimum levels at which the hypothesis of exponential or Weibull distribution is not rejected (the lowest point to the right of the red line on the upper graphs in Figures 11 and 12);
- The thresholds with the indices maxval correspond to the maximum $P$-value (the rightmost point in the upper graphs);
- The thresholds with the indices high correspond to the upper level, when the corresponding hypothesis is not rejected (the highest point to the right of the red line on the graphs).

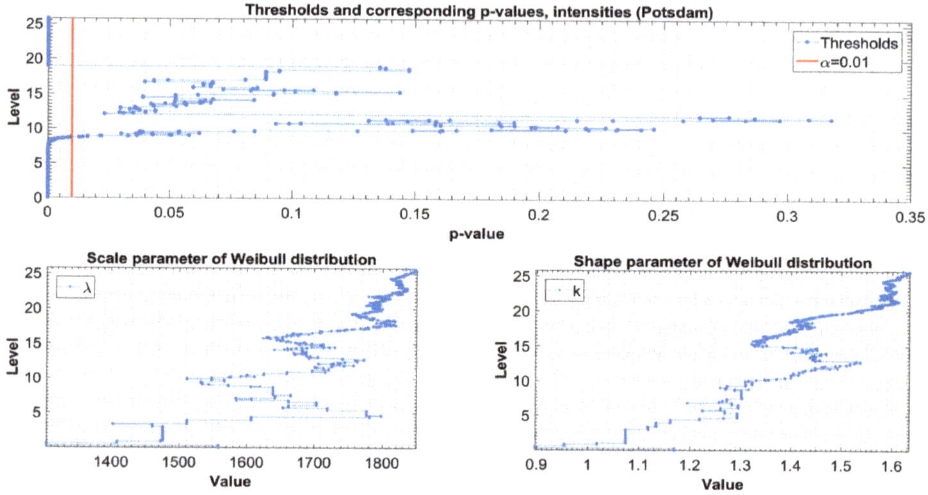


**Figure 11.** P-values that correspond to the various thresholds (Potsdam).

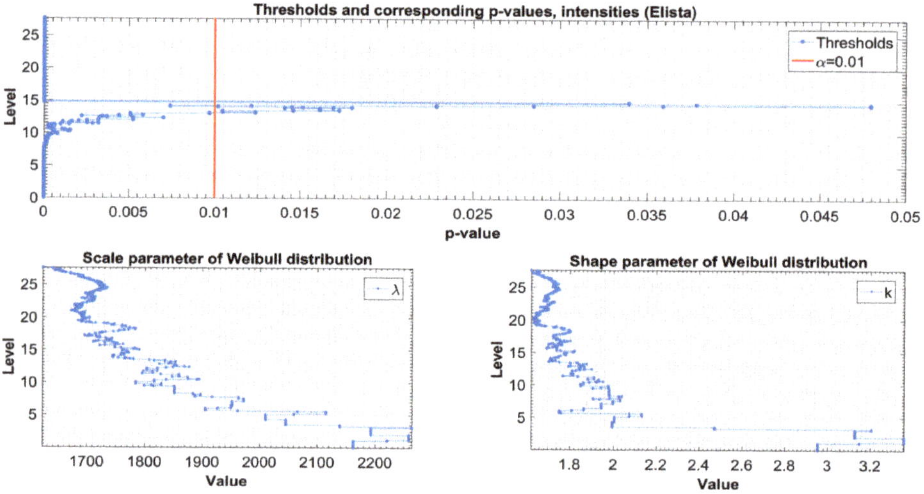


**Figure 12.** P-values that correspond to the various thresholds (Elista).

The green-filled downward-pointing triangles mark the intensities, which are classified as absolutely abnormal based on the GG test (see Section 6). The black upward-pointing triangles correspond to the decision based on the classical gamma distribution test (that is, $\gamma = 1$, see (31)). The circles denote intermediate extreme observations, and the squares mark relatively extreme ones. This classification is described in Section 7. It is worth noting that for the GG test in Potsdam the value of $\gamma$ is 1.0775 and for Elista $\gamma = 1.1257$.

For Potsdam, the results of gamma and GG tests are good and close. In addition, the PoT method is also effective in the case where the threshold is chosen with the maximum $P$-value. However, for Elista, with less rainfalls with lower intensities, the results are quite different. Indeed, the decisions of the PoT method are close for the exponential and Weibull cases (the thresholds differ by only 0.29).

However, a statistical test based on the gamma distribution identifies only four intensities as absolutely extreme, while the GG test identifies more absolute extremes, including those below the thresholds mentioned above.

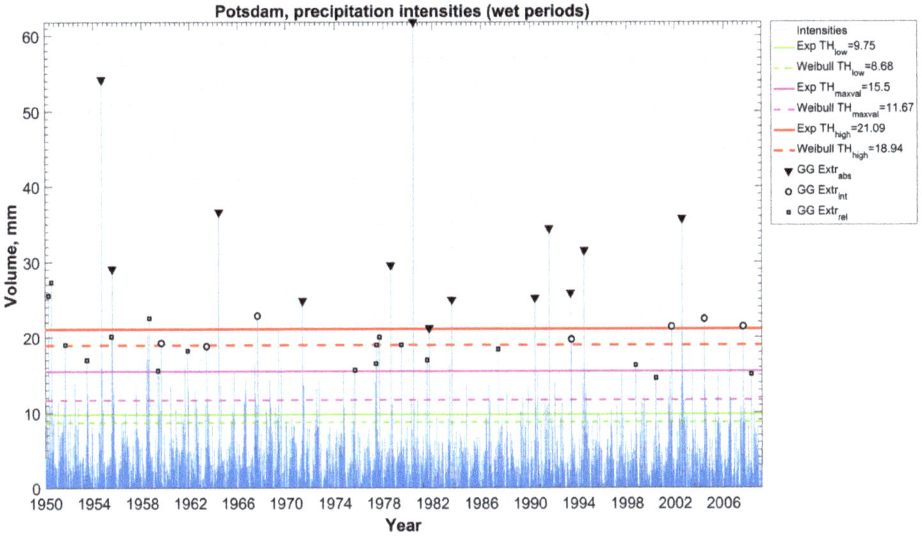


**Figure 13.** Comparison of statistical tests andpPeaks over threshold methodology for extreme precipitation intensities (Potsdam).

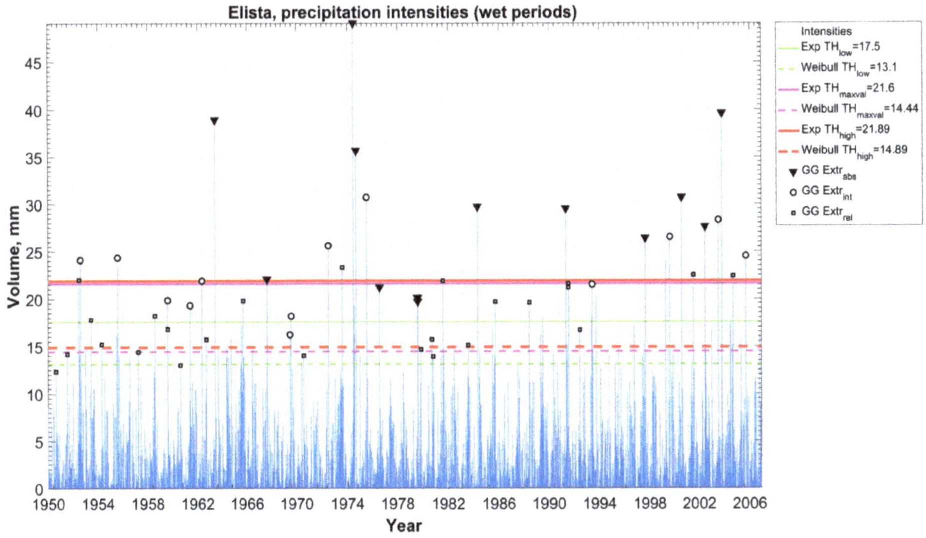


**Figure 14.** Comparison of statistical tests and peaks over threshold methodology for extreme precipitation intensities (Elista).

## 9. Conclusions and Discussion

In the paper, asymptotic models for some precipitation characteristics based on GNB distributions were considered. Also, a statistical test based on the GG distribution was proposed for the determination of the type of precipitation extremes. The GG and GNB distributions are not quite widespread, so the methods for the estimation of their parameters are, as a rule, not implemented in standard statistical packages. Therefore, the implementation of appropriate procedures requires the creation of specialized software solutions, for example, based on the functional approach, as it was done in the study described in this paper using the MATLAB programming language. However, as was demonstrated in the paper, the results of fitting such distributions to real data turned out to be better as compared to conventional models. Therefore, for processing spatial meteorological data from a large number of stations, the proposed methods and models can be effectively implemented as high-performance computing services.

**Author Contributions:** Conceptualization, V.K., A.G.; formal analysis, V.K., A.G.; funding acquisition, A.G.; investigation, A.G., V.K.; methodology, V.K., A.G.; project administration, V.K., A.G.; resources, A.G.; software, A.G.; supervision, V.K.; validation, A.G.; visualization, A.G.; writing—original draft, V.K., A.G.; writing—review and editing, A.G., V.K. All authors have read and agreed to the published version of the manuscript.

**Funding:** The results presented in Section 8 were obtained by Andrey Gorshenin to meet the research goals of grant No 18-71-00156 of the Russian Science Foundation.

**Acknowledgments:** Authors thank the reviewers for their valuable comments that helped to improve the presentation of the material.

**Conflicts of Interest:** The authors declare no conflict of interest.

## Appendix A. Algorithms

This section presents the above-described algorithms in the form pseudo-code. All of these algorithms have been implemented using Matlab programming language without any tools specific to this development environment. They can be successfully implemented, for example, with the Python programming language taking into account some minor changes.

**Algorithm A1.** GNB approximations

1: LOAD(Data);   // Loading initial data (Potsdam, Elista)
2: $\alpha$=0.05;   // Significance level
3: WP=WETPERIODS(Data);   // Finding wet periods in Data
4: r=NBFIT(WP-1, $\alpha$);   // Finding parameter r
5: [$\gamma_{l1}, \mu_{l1}, err_{l1}$]=GNBAPPROX($l_1$, r);   // GNB approximation based on $\ell_1$-distance minimization
6: [$\gamma_{l2}, \mu_{l2}, err_{l2}$]=GNBAPPROX($l_2$, r);   // GNB approximation based on $\ell_2$-distance minimization
7: [$\gamma_{l\infty}, \mu_{l\infty}, err_{l\infty}$]=GNBAPPROX($l_\infty$, r);   // GNB approximation based on $\ell_\infty$-distance minimization
  // Plotting initial histograms and GNB approximations for all cases
8: HISTOGRAMS($\gamma_{l1}, \mu_{l1}, err_{l1}, \gamma_{l2}, \mu_{l2}, err_{l2}, \gamma_{l\infty}, \mu_{l\infty}, err_{l\infty}$);

**Algorithm A2.** Stabilization of averages

1: LOAD(Potsdam, Elista);   // Loading initial data
2: m=3000;   // The value can be chosen empirically
3: DATAPREPROCESSING(Potsdam, Elista);
4: [$\beta_P, a_P$]=StabParams(Potsdam);   // Search for stabilization parameters using Formulas (24) and (25)
5: [$\beta_E, a_E$]=StabParams(Elista);
6: PLOTAVERAGES(Potsdam, Elista, $\beta_P, a_P, \beta_E, a_E$);   // Drawing results

| Algorithm A3. Statistical test for extreme volumes |
|---|

1: α=0.01;                                                                      // Significance level
2: LOAD(Data);                                                // Loading initial data (Potsdam, Elista)
                                  // Correspondence between astronomical time and sample elements
3: window=DAYSTOOBSERVATIONS(Days);
4: Vols=VOLUMES(Data);                                          // Volumes obtained from raw data
5: γ=GAMMAFIT(Vols, α);                                                   // Finding parameter γ
6: [SR, $SR_{GG}$]=GGSTATISTICS(Vols,γ);         // Finding values of statistics based on Formula (31)
7: [$Extr_{abs}$, GG $Extr_{abs}$, $Extr_{int}$, GG $Extr_{int}$, $Extr_{rel}$, GG $Extr_{rel}$]=GGTEST(SR, $SR_{GG}$);
                                 // Plotting initial data and decisions based on statistics SR and $SR_{GG}$
8: PLOTEXTREMES($Extr_{abs}$, GG $Extr_{abs}$, $Extr_{int}$, GGExt$r_{int}$, $Extr_{rel}$, GG $Extr_{rel}$);

| Algorithm A4. PoT and GG-based test for intensities |
|---|

1: α=0.01;                                                                      // Significance level
2: LOAD(Data);                                                // Loading initial data (Potsdam, Elista)
3: Ints=INTENSITIES(Data);                                    // Intensities obtained from raw data
4: [GG $Extr_{abs}$, GG $Extr_{int}$, GG $Extr_{rel}$]=GGTEST(Ints,$SR_{GG}$);
                                                                                     // Modified PoT
5: level=MAX(Ints);                             // Initial PoT level (threshold) equal to maximum data value
6: L=LENGTH(Ints);
7: k=0;
                                          // Determining the dependence of the level on the p-value
8: **while** level⩾0 **do**
9:     I=FIND(Ints>level);
                            // Minimum sufficient number of elements in the sample exceeding the threshold
10:    **if** LENGTH(I)<MinNum) **then**
11:        level=level-step;
12:        CONTINUE;
13:    **end if**
14:    [ExpParam(k),ExpPval(k)] = FITEXP(I,α);
15:    [WeibullParams(k),WeibullPval(k)] = FITWEIBULL(I,α);
16:    k++;
17:    level=level-step;
18: **end while**
19: PLOTEXTREMES(GG $Extr_{abs}$, GGExt$r_{int}$, GG $Extr_{rel}$);                  // Plotting intensity extremes
20: PLOTTHRESHOLDS( );                                                         // Plotting thresholds

## References

1. Korolev, V.; Gorshenin, A.; Gulev, S.; Belyaev, K.; Grusho, A. Statistical Analysis of Precipitation Events. *AIP Conf. Proc.* **2017**, *1863*, 090011. [CrossRef]
2. Korolev, V.; Gorshenin, A.; Belyaev, K. Statistical tests for extreme precipitation volumes. *Mathematics* **2019**, *7*, 648. [CrossRef]
3. Korolev, V.; Zeifman, A. Generalized negative binomial distributions as mixed geometric laws and related limit theorems. *Lith. Math. J.* **2019**, *59*, 1461–1466. [CrossRef]
4. Zolina, O.; Simmer, C.; Belyaev, K.; Kapala, A.; Gulev, S.; Koltermann, P. Changes in the duration of European wet and dry spells during the last 60 years. *J. Clim.* **2013**, *26*, 2022–2047. [CrossRef]
5. Bening, V.; Korolev, V. *Generalized Poisson Models and Their Applications in Insurance and Finance*; VSP: Utrecht, The Netherlands, 2002.

6. Gnedenko, B.; Korolev, V. *Random Summation: Limit Theorems and Applications*; CRC Press: Boca Raton, FL, USA, 1996.
7. Grandell, J. *Doubly Stochastic Poisson Processes*; Lecture Notes Mathematics; Springer: Berlin/Heidelberg, Germany; New York, NY, USA, 1976; Volume 529.
8. Grandell, J. *Mixed Poisson Processes*; Chapman and Hall: London, UK, 1997.
9. Korolev, V.; Chertok, A.; Korchagin, A.; Zeifman, A. Modeling high-frequency order flow imbalance by functional limit theorems for two-sided risk processes. *Appl. Math. Comput.* **2015**, *253*, 224–241. [CrossRef]
10. Korolev, V.; Skvortsova, N. *Stochastic Models of Structural Plasma Turbulence*; VSP: Utrecht, The Netherlands, 2006.
11. Stacy, E. A generalization of the gamma distribution. *Ann. Math. Stat.* **1962**, *38*, 1187–1192. [CrossRef]
12. Greenwood, M.; Yule, G. An inquiry into the nature of frequency-distributions of multiple happenings, etc. *J. R. Stat. Soc.* **1920**, *83*, 255–279. [CrossRef]
13. Holla, M. On a Poisson-inverse Gaussian distribution. *Metrika* **1967**, *11*, 115–121. [CrossRef]
14. Sichel, H. On a family of discrete distributions particular suited to represent long tailed frequency data. In Proceedings of the 3rd Symposium on Mathematical Statistics, Pretoria, South Africa, 19–22 July 1971; pp. 51–97.
15. Korolev, V.; Korchagin, A.; Zeifman, A. Poisson theorem for the scheme of Bernoulli trials with random probability of success and a discrete analog of the Weibull distribution. *Informat. Appl.* **2016**, *10*, 11–20. [CrossRef]
16. Steutel, F.; van Harn, K. *Infinite Divisibility of Probability Distributions on the Real Line*; Marcel Dekker: New York, NY, USA, 2004.
17. Korolev, V. Limit distributions for doubly stochastically rarefied renewal processes and their properties. *Theory Probab. Appl.* **2016**, *61*, 753–773. [CrossRef]
18. Korolev, V.; Korchagin, A.; Zeifman, A. On doubly stochastic rarefaction of renewal processes. *AIP Conf. Proc.* **2017**, *1863*, 090010. [CrossRef]
19. Zaks, L.; Korolev, V. Generalized variance gamma distributions as limit laws for random sums. *Informat. Appl.* **2013**, *7*, 105–115. [CrossRef]
20. Korolev, V.; Bening, V.; Shorgin, S. *Mathematical Foundations of Risk Theory*, 2nd ed.; FIZMATLIT: Moscow, Russia, 2011.
21. Zolotarev, V. *One-Dimensional Stable Distributions*; American Mathematical Society: Providence, RI, USA, 1986.
22. Schneider, W. *Stable Distributions: Fox Function Representationand Generalization*; Springer: Berlin, Germany, 1986; pp. 497–511.
23. Uchaikin, V.; Zolotarev, V. *Infinite Divisibility of Probability Distributions on the Real Line*; VSP: Utrecht, The Netherlands, 1999.
24. Shanbhag, D.; Sreehari, M. On certain self-decomposable distributions. *Z. Für Wahrscheinlichkeitstheorie Und Verwandte Geb.* **1977**, *38*, 217–222. [CrossRef]
25. Korolev, V. Product representations for random variables with the Weibull distributions and their applications. *J. Math. Sci.* **2016**, *218*, 298–313. [CrossRef]
26. Gleser, L. The gamma distribution as a mixture of exponential distributions. *Am. Stat.* **1989**, *43*, 115–117. [CrossRef]
27. Korolev, V. On convergence of distributions of compound Cox processes to stable laws. *Theory Probab. Appl.* **1999**, *43*, 644–650. [CrossRef]
28. Korolev, V. Convergence of random sequences with independent random indexes. I. *Theory Probab. Appl.* **1994**, *39*, 313–333. [CrossRef]
29. Korolev, V. Convergence of random sequences with independent random indexes. II. *Theory Probab. Appl.* **1995**, *40*, 770–772. [CrossRef]
30. Korolev, V.; Gorshenin, A.; Sokolov, I. Max-compound Cox processes. I. *J. Math. Sci.* **2019**, *237*, 789–803. [CrossRef]
31. Korolev, V.; Gorshenin, A. The probability distribution of extreme precipitation. *Dokl. Earth Sci.* **2017**, *477*, 1461–1466. [CrossRef]
32. Gorshenin, A.; Korolev, V. Scale mixtures of Frechet distributions as asymptotic approximations of extreme precipitation. *J. Math. Sci.* **2018**, *234*, 886–903. [CrossRef]

33. Johnson, N.; Kot, S.; Balakrishnan, N. *Continuous Univariate Distributions*, 2nd ed.; Wiley: New York, NY, USA, 1995; Volume 2.
34. Goldie, C. A class of infinitely divisible distributions. *Math. Proc. Camb. Philos. Soc.* **1967**, *63*, 1141–1143. [CrossRef]
35. Korolev, V.; Gorshenin, A.; Sokolov, I. Max-compound Cox processes. III. *arXiv* **2020**, arXiv:1912.02237v2.
36. Galambos, J. *The Asymptotic Theory of Extreme Order Statistics*; Wiley: New York, NY, USA, 1978.
37. Rényi, A. A Poisson-folyamat egy jellemzese. *Magy. Tud. Acad. Mat. Kut. Int. Közl.* **1956**, *1*, 519–527. [CrossRef]
38. Rényi, A. On an extremal property of the Poisson process. *Ann. Inst. Stat. Math.* **1964**, *16*, 129–133. [CrossRef]
39. Kalashnikov, V. *Geometric Sums: Bounds for Rare Events with Applications*; Kluwer Academic Publishers: Dordrecht, The Netherlands, 1997.
40. Zolotarev, V. Approximation of Distributions of Sums of Independent Random Variables with Values in Infinite-Dimensional Spaces. *Theory Probab. Appl.* **1976**, *21*, 721–737. [CrossRef]
41. Zolotarev, V. Ideal Metrics in the Problem of Approximating Distributions of Sums of Independent Random Variables. *Theory Probab. Appl.* **1977**, *22*, 433–449. [CrossRef]
42. Zolotarev, V. *Modern Theory of Summation of Random Variables*; VSP: Utrecht, The Netherlands, 1997.
43. Korolev, V.; Zeifman, A. Bounds for convergence rate in laws of large numbers for mixed Poisson random sums. *arXiv* **2020**, arXiv:2003.12495.
44. Cheng, L.; AghaKouchak, A. Nonstationary Precipitation Intensity-Duration-Frequency Curves for Infrastructure Design in a Changing Climate. *Sci. Rep.* **2014**, *4*, 7093. [CrossRef]
45. Wasko, C.; Sharma, A. Steeper temporal distribution of rain intensity at higher temperatures within Australian storms. *Nat. Geosci.* **2015**, *8*, 527–U166. [CrossRef]
46. Mo, C.; Ruan, Y.; He, J.; Jin, J.; Liu, P.; Sun, G. Frequency analysis of precipitation extremes under climate change. *Int. J. Climatol.* **2019**, *39*, 1373–1387. [CrossRef]
47. Donat, M.; Angelil, O.; Ukkola, A. Intensification of precipitation extremes in the world's humid and water-limited regions. *Environ. Res. Lett.* **2019**, *14*. [CrossRef]
48. Groisman, P.; Knight, R.; Karl, T. Changes in Intense Precipitation over the Central United States. *J. Hydrometeorol.* **2012**, *13*, 47–66. [CrossRef]
49. Xu, C.; Qiao, Y.; Jian, M. Interdecadal Change in the Intensity of Interannual Variation of Spring Precipitation over Southern China and Possible Reasons. *J. Clim.* **2013**, *32*, 5865–5881. [CrossRef]
50. Ziadat, F.; Taimeh, A. Effect of Rainfall Intensity, Slope, Land Use and Antecedent Soil Moisture on Soil Erosion in an Arid Environment. *Land Degrad. Dev.* **2013**, *24*, 582–590. [CrossRef]
51. Jomaa, S.; Barry, D.; Rode, M.; Sander, G.; Parlange, J. Linear scaling of precipitation-driven soil erosion in laboratory flumes. *Catena* **2017**, *152*, 285–291. [CrossRef]
52. Bezak, N.; Auflic, M.J.; Mikos, M. Application of hydrological modelling for temporal prediction of rainfall-induced shallow landslides. *Landslides* **2019**, *16*, 1273–1283. [CrossRef]
53. Bliznak, V.; Kaspar, M.; Muller, M.; Zacharov, P. Sub-daily temporal reconstruction of extreme precipitation events using NWP model simulations. *Atmos. Res.* **2019**, *224*, 65–80. [CrossRef]
54. Huang, W.; Nychka, D.; Zhang, H. Estimating precipitation extremes using the log-histospline. *Environmetrics* **2019**, *30*. [CrossRef]
55. Martinez-Villalobos, C.; Neelin, J. Why Do Precipitation Intensities Tend to Follow Gamma Distributions? *J. Atmos. Sci.* **2019**, *76*, 3611–3631. [CrossRef]
56. Gorshenin, A.; Korolev, V. Determining the extremes of precipitation volumes based on a modified "Peaks over Threshold". *Inform. I Ee Primen.* **2018**, *12*, 16–24. [CrossRef]
57. Leadbetter, M. On a basis for "Peaks over Threshold" modeling. *Stat. Probab. Lett.* **1991**, *12*, 357–362. [CrossRef]
58. Simiu, E.; Heckert, N. Extreme wind distribution tails: A "peaks over threshold" approach. *J. Struct. Eng.-ASCE* **1996**, *122*, 539–547.:5(539). [CrossRef]
59. Mendez, F.; Menendez, M.; Luceno, A.; Losada, I. Estimation of the long-term variability of extreme significant wave height using a time-dependent Peak Over Threshold (POT) model. *J. Geophys. Res.-Ocean.* **2006**, *111*, C07024. [CrossRef]

60. Kyselý, J.; Picek, J.; Beranova, R. Estimating extremes in climate change simulations using the peaks-over-threshold method with a non-stationary threshold. *Glob. Planet. Chang.* **2010**, *72*, 55–68. [CrossRef]
61. Begueria, S.; Angulo-Martinez, M.; Vicente-Serrano, S.; Lopez-Moreno, I.; El-Kenawy, A. Assessing trends in extreme precipitation events intensity and magnitude using non-stationary peaks-over-threshold analysis: A case study in northeast Spain from 1930 to 2006. *Int. J. Climatol.* **2011**, *31*, 2102–2114. [CrossRef]
62. Roth, M.; Buishand, T.; Jongbloed, G.; Tank, A.; van Zanten, J. A regional peaks-over-threshold model in a nonstationary climate. *Water Resour. Res.* **2012**, *48*, W11533. [CrossRef]

mathematics

MDPI

*Article*

# Rates of Convergence in Laplace's Integrals and Sums and Conditional Central Limit Theorems

**Vassili N. Kolokoltsov**

Department of Statistics, University of Warwick, Coventry CV4 7AL, UK; v.kolokoltsov@warwick.ac.uk

Received: 28 February 2020; Accepted: 30 March 2020; Published: 1 April 2020

**Abstract:** We obtained the exact estimates for the error terms in Laplace's integrals and sums implying the corresponding estimates for the related laws of large number and central limit theorems including the large deviations approximation.

**Keywords:** integrals and sums; rates of convergence; conditional law of large numbers; conditional central limit theorem

## 1. Introduction

The Laplace integrals find applications in numerous problems of mathematics and applied science, and the literature on these integrals is abundant. For example, let us mention the applications in statistical physics, see e.g., [1] or Lecture 5 in [2], in the pattern analysis [3], in the large deviation theory [4–6], where it is sometimes referred to as the Laplace–Varadhan method, in the analysis of Weibullian chaos [7], in the asymptotic methods for large excursion probabilities [8], in the asymptotic analysis of stochastic processes [9], and in the calculation of the tunneling effects in quantum mechanics and quantum fields, see [10,11]. It can be used to essentially simplify Maslov's type derivation of the Gibbs, Bose–Einstein and Pareto distribution [12]. An infinite-dimensional version and a non-commutative versions of the Laplace approximations were developed recently in [13,14], respectively.

The majority of research on this topic is devoted to the asymptotic expansions, or even, following the general approach to large deviation of Varadhan, just to the logarithmic asymptotics, see also [15]. In the present paper, following the recent trend for the searching of the best constants for the error term in the central-limit-type results, see [16] and references therein, we are interested in exact estimates for the main error term of the Laplace approximation. This approach to Laplace integrals was initiated by the author in book [9] (Appendix B), where the stress was on the integrals with complex phase. Here we aimed at making these asymptotic more precise for real phase including the most general case of both exponent and the pre-exponential term in the integral depending on the parameter (which is crucial for the applications to the classical conditional large numbers (LLN) that we have in mind here), and stressing two new applications, to the sums instead of integrals (Laplace–Varadhan asymptotics) and to the conditional law of large numbers (LLN) and central limit theorems (CLT) of large deviations.

The content of the paper is as follows. In Section 2 we obtained the estimates for the error term in Laplace approximation with minimum of the phase in the interior of the domain of integration improving slightly on estimates from [9], and in Section 3 we derived the resulting LLN and CLT results. In Sections 4 and 5 the same program was carried out for the case of phase minima occurring in the border of the domain. In Section 6 we derived the analogous results for the case of sums, rather than integrals. In Section 7 we show how our results can be applied to the conditional LLN and CLT of large deviations.

## 2. Phase Minimum Inside the Domain of Integration

Here we present the estimates of the remainder in the asymptotic formula for the Laplace integrals with the critical point of the phase lying in the interior of the domain of integration, adapting and streamlining the arguments of [9].

Consider the integral

$$I(N) = \int_{\Omega} f(x,N) \exp\{-NS(x,N)\}\, dx, \quad N \geq N_0 > 0, \tag{1}$$

where $\Omega$ is an open bounded subset of the Euclidean space $\mathbf{R}^d$, equipped with the Euclidean norm $|.|$, with Euclidean volume $|\Omega|$, the amplitude $f$ and the phase $S$ are continuous real functions of $x \in \Omega, N \geq N_0$.

**Remark 1.** *The assumption that $\Omega$ is bounded is not essential, but simplifies explicit estimates for the error terms. One should think of $\Omega$ as a bounded subset of the full domain of integration containing all minimum points of $S(.,N)$. If $f$ is integrable outside $\Omega$, the integral of $f(x,N)\exp\{-NS(x,N)\}$ over $\mathbf{R}^d \setminus \Omega$ will be exponentially small as compared with Equation (1).*

Recall that the $k$th order derivative

$$\phi^{(k)}(x) = \frac{\partial^k \phi}{\partial x^k}$$

of a real function $\phi$ on $\mathbf{R}^d$ can be viewed as the multi-linear map

$$\phi^{(k)}(x)[v] = \frac{\partial^k \phi}{\partial x^k}(x)[v] = \sum_{i_1,\cdots,i_k=1}^{d} \frac{\partial^k \phi(x)}{\partial x_{i_1}\cdots \partial x_{i_k}} v_{i_1}\cdots v_{i_k}, \quad v \in \mathbf{R}^d.$$

The second derivative will be written as usual in the matrix form

$$\phi''(x)[v] = \left(\frac{\partial^2 \phi}{\partial x^2}(x)v, v\right).$$

We shall denote by $\|\phi^{(k)}(x)\|$ the corresponding norm defined as the lowest constant for which the estimate

$$|\phi^{(k)}(x)[v]| \leq \|\phi^{(k)}(x)\| |v|^k$$

holds for all $v$.

**Remark 2.** *It is a standard way to define norms of multi-linear mappings, see e.g., [17]. However, as all norms on finite-dimensional spaces are equivalent, the choice of a norm is not very essential here.*

Let us make now the following assumptions on the functions $f$ and $S$:
(C1) $f(x,N))$ is a Lipshitz continuous function of $x$ with

$$f_0 = \sup_{x\in\Omega,\, N>N_0} |f(x,N)| < \infty, \quad f_1 = \sup_{x\neq y,\, N>N_0} \frac{|f(x,N)-f(y,N)|}{|x-y|} < \infty;$$

(C2) $S(x,N)$ is a thrice continuously differentiable function in $x$ such that

$$S_3 = \sup_{x\in\Omega, N\geq N_0} \|\frac{\partial^3 S(x,N)}{\partial x^3}\| < \infty;$$

and

$$\Lambda_m|\xi|^2 \leq \left(\frac{\partial^2 S}{\partial x^2}(x,N)\xi,\xi\right) \leq \Lambda_M|\xi|^2$$

for all $x \in \Omega$, $N \geq N_0$, $\xi \in \mathbf{R}^d$, with positive constants $\Lambda_m, \Lambda_M$; the latter condition can be concisely written as

$$\Lambda_m \leq \frac{\partial^2 S}{\partial x^2}(x,N) \leq \Lambda_M,$$

where the usual ordering on symmetric matrices is used;

(C3) For any $N \geq N_0$ there exists a unique point $x(N)$ of global minimum of $S(.,N)$ in $\Omega$, and the ball

$$U(N) = \{x : |x - x(N)| < N^{-1/3}\} \tag{2}$$

is contained in $\Omega$. Let us denote by $D_N$ the matrix of the second derivatives of $S$ at $x(N)$, that is

$$D_N = \frac{\partial^2 S}{\partial x^2}(x(N),N). \tag{3}$$

Notice that from convexity of $S$ in $\Omega$ and Assumption (C3) it follows that

$$S_{min}(N) = \inf\{S(x,N) : x \in \Omega \setminus U(N)\} = \min\{S(x) : x \in \partial U(N)\}. \tag{4}$$

Our approach to the study of the Laplace integral $I(N)$ is based on its decomposition

$$I(N) = I'(N) + I''(N),$$

with

$$I'(N) = \int_{U(N)} f(x,N)\exp\{-NS(x,N)\}\,dx, \quad I''(N) = \int_{\Omega\setminus U(N)} f(x,N)\exp\{-NS(x,N)\}\,dx. \tag{5}$$

**Remark 3.** *In the proof below one can use $U(N) = \{x : |x - x(N)| < N^{-\varkappa}\}$ instead of Equations (2) with $1/3 \leq \varkappa < 1/2$, the lower bounds coming from the estimate of $I_1$ below, and the upper bound from the estimate of $I_3$ below.*

**Proposition 1.** *Under Assumptions (C1)–(C3),*

$$I(N) = \exp\{-NS(x(N),N)\}\left(\frac{2\pi}{N}\right)^{d/2}\left[\frac{f(x(N),N)}{\sqrt{\det D_N}} + \frac{\omega(N)}{\sqrt{N}}\right] + \omega^{exp}(N), \tag{6}$$

*where $\omega(N)$ is a bounded function depending on $\Lambda_m, f_0, f_1, S_3, d$, and $\omega^{exp}(N)$ is exponentially small, compared to the main term. Explicitly*

$$|\omega(N)| \leq d\Lambda_m^{-(1+d)/2}\left[f_1 + \frac{d+1}{6\Lambda_m} f_0 S_3 e^{S_3/6}\right] \tag{7}$$

$$|\omega^{exp}(N)| \leq f_0 \exp\{-NS(x(N),N)\}\exp\{-\Lambda_m N^{1/3}/2\}$$
$$\times\left[|\Omega| + \frac{(2\pi)^{d/2}N^{-d/3}}{\Lambda_m N^{1/3}}\left(\frac{1}{\Gamma(d/2)} + \frac{2^{d/2}}{2\Lambda_m N^{1/3}}\right)\right]. \tag{8}$$

**Proof.** From the Taylor formula for functions on $\mathbf{R}$

$$g(t) = g(0) + g'(0)t + \int_0^t (t-s)g''(s)ds$$

it follows that

$$S(x,N) - S(x(N),N) = S(x(N) + t(x - x(N)),N)|_{t=0}^{t=1}$$

$$= \int_0^1 (1-\tau)\left(\frac{\partial^2 S}{\partial x^2}(x(N) + \tau(x - x(N)),N)(x - x(N)), x - x(N)\right) d\tau. \quad (9)$$

Consequently, for $x \in \partial U(N)$ we have by Assumption (C2) that

$$S(x,N) - S(x(N),N) \geq \frac{1}{2}\Lambda_m |x - x(N)|^2 = \frac{1}{2}\Lambda_m N^{-2/3} \quad (10)$$

It follows then from Equation (4) that

$$S_{min}(N) = \inf\{S(x,N) : x \in \Omega \setminus U(N)\} \geq S(x(N),N) + \frac{1}{2}\Lambda_m N^{-2/3}, \quad (11)$$

so that

$$I''(N) \leq \exp\{-NS_{min}(N)\} \int_\Omega f(x,N)\,dx \leq f_0|\Omega| \exp\{-\Lambda_m N^{1/3}/2\} \exp\{-NS(x(N),N)\}. \quad (12)$$

To go further we shall need the Taylor expansion of $S$ up to the third order. Namely, from Equation (9) we deduce the expansion

$$S(x,N) - S(x(N),N) = \frac{1}{2}(D_N(x - x(N)), x - x(N)) + \sigma(x,N), \quad (13)$$

where, due to the equation $\int_0^1 (1-\tau)\tau\,d\tau = 1/6$,

$$|\sigma(x,N)| \leq \frac{1}{6} S_3 |x - x(N)|^3. \quad (14)$$

Turning to $I'(N)$ we further decompose it into the four integrals

$$I'(N) = \exp\{-NS(x(N),N)\}(I_1(N) + I_2(N) + I_3(N) + I_4(N)) \quad (15)$$

with

$$I_1(N) = \int_{U(N)} f(x,N) \exp\{-\frac{N}{2}(D_N(x - x(N)), x - x(N))\}(e^{-N\sigma(x,N)} - 1)dx,$$

$$I_2(N) = \int_{U(N)} (f(x,N) - f(x(N),N) \exp\{-\frac{N}{2}(D_N(x - x(N)), x - x(N))\}\,dx,$$

$$I_3(N) = f(x(N),N) \int_{\mathbf{R}^d \setminus U(N)} \exp\{-\frac{N}{2}(D_N(x - x(N)), x - x(N))\}\,dx,$$

$$I_4(N) = f(x(N),N) \int_{\mathbf{R}^d} \exp\{-\frac{N}{2}(D_N(x - x(N)), x - x(N))\}\,dx.$$

It follows from Equation (14) that, for $x \in U(N)$, $N|\sigma(x,N)| \leq S_3/6$. Using Equation (14) again and the trivial estimate $|e^t - 1| \leq |t|e^{|t|}$, we conclude that, for $x \in U(N)$,

$$|e^{-N\sigma(x,N)} - 1| \leq \frac{1}{6} e^{S_3/6} N S_3 |x - x(N)|^3. \quad (16)$$

Consequently,

$$|I_1(N)| \leq \frac{1}{6} e^{S_3/6} S_3 f_0 N \int_{\mathbf{R}^d} |y|^3 \exp\{-N\Lambda_m |y|^2/2\}.$$

From the standard integral

$$\int_{\mathbf{R}^d} |y|^k \exp\{-\alpha|y|^2\}\, dy = \pi^{d/2}\alpha^{-(k+d)/2}\frac{\Gamma((k+d)/2)}{\Gamma(d/2)}, \tag{17}$$

we deduce that

$$|I_1(N)| \le \frac{1}{6}\pi^{d/2}\frac{\Gamma((3+d)/2)}{\Gamma(d/2)}2^{(3+d)/2}\Lambda_m^{-(3+d)/2} f_0 S_3 e^{S_3/6} N^{-(d+1)/2}. \tag{18}$$

Next,

$$|I_2(N)| \le f_1 \int_{\mathbf{R}^d} |y| \exp\{-N\Lambda_m|y|^2/2\},$$

or, using Equation (17) with $k = 1$,

$$|I_2(N)| \le \pi^{d/2}\frac{\Gamma((1+d)/2)}{\Gamma(d/2)}2^{(1+d)/2}\Lambda_m^{-(1+d)/2} f_1 N^{-(d+1)/2}. \tag{19}$$

Next,

$$|I_3(N)| \le f_0 \int_{\{y:|y|\ge N^{-1/3}\}} \exp\{-N\Lambda_m|y|^2/2\}dy$$

$$= f_0 \exp\{-\Lambda_m N^{1/3}/2\} \int_{N^{-1/3}}^{\infty} \exp\{-N\Lambda_m(r^2 - N^{-2/3})/2\}|S^{d-1}|r^{d-1}\, dr,$$

where

$$|S^{d-1}| = 2\frac{\pi^{d/2}}{\Gamma(d/2)}$$

is the area of the unit sphere in $\mathbf{R}^d$. Changing $r$ to $z$ so that

$$z = N\Lambda_m(r^2 - N^{-2/3})/2 \Longleftrightarrow r^2 = N^{-2/3}\left(1 + \frac{2z}{\Lambda_m N^{1/3}}\right),$$

and thus $dz = N\Lambda_m r\, dr$, the last integral rewrites as

$$\frac{f_0}{\Lambda_m} N^{-(d+1)/3} \exp\{-\Lambda_m N^{1/3}/2\} \int_0^{\infty} e^{-z}\left(1 + \frac{2z}{N^{1/3}\Lambda_m}\right)^{(d-2)/2} |S^{d-1}|\, dz,$$

so that, using the inequality $(1+\omega) \le 2^n(1+\omega^n)$,

$$|I_3(N)| \le \frac{f_0}{\Lambda_m} N^{-(d+1)/3} \exp\{-\Lambda_m N^{1/3}/2\}\frac{\pi^{d/2}}{\Gamma(d/2)}2^{d/2}\left[1 + \left(\frac{2}{\Lambda_m N^{1/3}}\right)^{(d-2)/2}\Gamma\left(\frac{d}{2}\right)\right]. \tag{20}$$

**Remark 4.** *For $d = 1$ we get simply*

$$|I_3(N)| \le \frac{2f_0}{\Lambda_m} N^{-(d+1)/3} \exp\{-\Lambda_m N^{1/3}/2\},$$

*and for $d = 2$ the same with $2\pi$ instead of 2.*

Finally $I_4$ is calculated explicitly giving the main term of asymptotics:

$$I_4(N) = f(x(N), N)\left(\frac{2\pi}{N}\right)^{d/2} (\det D_N)^{-1/2}.$$

Summarizing the estimates for all integrals involved and performing elementary simplifications, in particular using $\Gamma((d+1)/2) < d\Gamma(d/2)/\sqrt{2}$ and $\Gamma(1+\alpha) = \alpha\Gamma(\alpha)$, yields estimate Equation (7). □

**Proposition 2.** *Under (C1)–(C3) assume additionally that S is four times differentiable and f has a Lipschitz continuous first derivatives with respect to x with*

$$S_4 = \sup_{x\in\Omega, N\geq N_0} \|\frac{\partial^4 S(x,N)}{\partial x^4}\| < \infty, \quad f_2 = \sup_{x\neq y, N\geq N_0} \frac{|\frac{\partial f}{\partial x}(x,N) - \frac{\partial f}{\partial x}(y,N)|}{|x-y|} < \infty.$$

*Then*

$$I(N) = \exp\{-NS(x(N),N)\}\left(\frac{2\pi}{N}\right)^{d/2}\left[\frac{f(x(N),N)}{\sqrt{\det D_N}} + \frac{\omega(N)}{N}\right] + \omega^{exp}(N), \tag{21}$$

*where the exponentially small term $\omega^{exp}$ has exactly the same estimate as in the previous Proposition and $\omega(N)$ is a bounded function depending on $\Lambda_m, f_0, f_1, f_2, S_3, S_4, d$. Explicitly,*

$$|\omega(N)| \leq \max(1, \Lambda_m^{-3-d/2})[f_0 S_3^2 d^3 e^{S_3/6} + f_0 S_4 d^2 + f_2 d + f_2 S_3 d^3 + f_1 S_3 d^2]. \tag{22}$$

**Remark 5.** *The key difference in the error term here is the denominator N instead of $\sqrt{N}$ in Equation (6).*

**Proof.** We again decompose $I(N)$ in the sum $I(N) = I'(N) + I''(N)$ with $I'(N), I''(N)$ given by Equation (5) and estimate $I''(N)$ by Equation (12). Estimation of $I'(N)$ needs more careful analysis using further terms of the Taylor expansion of $S$ and $f$. Namely we decompose it first as

$$I'(N) = \exp\{-NS(x(N),N)\}(I_1(N) + I_2(N)) \tag{23}$$

with

$$I_1(N) = \int_{U(N)} f(x,N)\exp\{-\frac{N}{2}(D_N(x-x(N)), x-x(N))\}[e^{-N\sigma(x,N)} - 1 + N\sigma(x,N)]\, dx,$$

$$I_2(N) = \int_{U(N)} f(x,N)\exp\{-\frac{N}{2}(D_N(x-x(N)), x-x(N))\}[1 - N\sigma(x,N)]\, dx.$$

From Equation (14) we get

$$|e^{-N\sigma(x,N)} - 1 + N\sigma(x,N)| \leq \frac{1}{2}(N|\sigma(x,N)|)^2 e^{N|\sigma(x,N)|} \leq \frac{1}{2}N^2(S_3/6)^2|x-x(N)|^6 e^{S_3/6}.$$

Consequently,

$$|I_1(N)| \leq \frac{f_0 S_3^2}{72} e^{S_3/6} N^2 \int_{\mathbf{R}^d} |y|^6 \exp\{-N\Lambda_m |y|^2/2\}.$$

From Equation (17) with $k = 6$ we deduce that

$$|I_1(N)| \leq \frac{f_0 S_3^2}{72} e^{S_3/6} \pi^{d/2} \frac{\Gamma((6+d)/2)}{\Gamma(d/2)} \left(\frac{2}{\Lambda_m}\right)^{(6+d)/2} N^{-(d+2)/2}$$

$$= \frac{f_0 S_3^2}{72} e^{S_3/6} (2\pi)^{d/2} \frac{d(d+2)(d+4)}{\Lambda_m^{3+d/2}} N^{-(d+2)/2}. \tag{24}$$

To evaluate $I_2(N)$ we use the Taylor expansion of $S$ to the fourth order yielding

$$\sigma(x,N) = \frac{1}{6}\frac{\partial^3 S}{\partial x^3}(x(N),N)[x-x(N)] + \tilde{\sigma}(x,N)$$

with

$$|\tilde{\sigma}(x,N)| \leq \frac{1}{24} S_4 |x-x(N)|^4.$$

Consequently, $I_2(N)$ can be represented as $I_2(N) = J_1(N) + J_2(N)$ with

$$J_1(N) = -N \int_{U(N)} f(x,N) \exp\{-\frac{N}{2}(D_N(x-x(N)), x-x(N))\}\tilde{\sigma}(x,N)\,dx,$$

$$J_2(N) = \int_{U(N)} f(x,N) \exp\{-\frac{N}{2}(D_N(x-x(N)), x-x(N))\} \left[1 - \frac{N}{6}\frac{\partial^3 S}{\partial x^3}(x(N),N)[x-x(N)]\right] dx.$$

Using the estimate for $\tilde{\sigma}$ we obtain

$$|J_1(N)| \le \frac{1}{24} N f_0 S_4 \int_{\mathbf{R}^d} \exp\{-N\Lambda_m |y|^2/2\}|y|^4\,dy$$

$$= \frac{1}{24} f_0 S_4 \pi^{d/2} \frac{\Gamma((4+d)/2)}{\Gamma(d/2)} \left(\frac{2}{\Lambda_m}\right)^{(4+d)/2} N^{-(d+2)/2}$$

$$= \frac{1}{24} f_0 S_4 (2\pi)^{d/2} \frac{d}{2}\left(\frac{d}{2}+1\right) \frac{4}{\Lambda_m^{(4+d)/2}} N^{-(d+2)/2}.$$

To evaluate $J_2$ we expand $f$ in Taylor series writing

$$f(x,N) = f(x(N),N) + \left(\frac{\partial f}{\partial x}(x(N),N), x-x(N)\right)$$

$$+[f(x,N) - f(x(N),N) - \left(\frac{\partial f}{\partial x}(x(N),N), x-x(N)\right)].$$

Substituting this in $J_2$ and using the fact that the integral of an odd function over a ball centered at the origin vanishes, we get

$$J_2(N) = J_{21}(N) + J_{22}(N) + J_{23}(N)$$

with

$$J_{21}(N) = \int_{U(N)} [f(x,N) - f(x(N),N) - \left(\frac{\partial f}{\partial x}(x(N),N), x-x(N)\right)]$$

$$\times \exp\{-\frac{N}{2}(D_N(x-x(N)), x-x(N))\} \left[1 - \frac{N}{6}\frac{\partial^3 S}{\partial x^3}(x(N),N)[x-x(N)]\right] dx,$$

$$J_{22}(N) = -\int_{U(N)} \frac{N}{6} \left(\frac{\partial f}{\partial x}(x(N),N), x-x(N)\right) \frac{\partial^3 S}{\partial x^3}(x(N),N)[x-x(N)]$$

$$\times \exp\{-\frac{N}{2}(D_N(x-x(N)), x-x(N))\}\,dx,$$

$$J_{23}(N) = \int_{U(N)} f(x(N),N) \exp\{-\frac{N}{2}(D_N(x-x(N)), x-x(N))\}\,dx.$$

The first two integrals are estimated as above, that is

$$|J_{21}(N)| \le \int_{\mathbf{R}^d} \frac{1}{2} f_2 |y|^2 \left(1 + \frac{NS_3}{6}|y|^3\right) \exp\{-N\Lambda_m|y|^2/2\}\,dy$$

$$= \frac{1}{2} f_2 \pi^{d/2} \left[\frac{d}{2}\left(\frac{2}{\Lambda_m}\right)^{(2+d)/2} N^{-(d+2)/2} + \frac{S_3}{6}\left(\frac{2}{\Lambda_m}\right)^{(5+d)/2} \frac{\Gamma((5+d)/2)}{\Gamma(d/2)} N^{-(d+3)/2}\right],$$

and

$$|J_{22}(N)| \le \int_{\mathbf{R}^d} \frac{1}{6} N f_1 S_3 |y|^4 \exp\{-N\Lambda_m|y|^2/2\}\,dy$$

$$= \frac{1}{6} f_1 S_3 \pi^{d/2} \left(\frac{2}{\Lambda_m}\right)^{(4+d)/2} N^{-(d+2)/2} \frac{\Gamma((4+d)/2)}{\Gamma(d/2)}.$$

Finally, $J_{23}(N)$ was estimated in Proposition 1 by representing it as the difference between the integral over the whole space $\mathbf{R}^d$ and the integral over $\mathbf{R}^d \setminus U(N)$, the first term yielding the main term of the asymptotics and the second one being exponentially small. Exponentially small terms are exactly the same as in the previous Proposition. Summarizing the estimates obtained and slightly simplifying, yields Equation (22). □

## 3. LLN and CLT for Internal Minima of the Phase

**Theorem 1.** *Let $\Omega$ be a bounded open subset of $\mathbf{R}^d$ and $f(x,N)$, $S(x,N)$ be continuous functions on $\Omega \times [N_0, \infty)$ satisfying conditions of Proposition 1. Assume that $f(x,N)$ is strictly positive and the sequence of global minima $x(N)$ converges, as $N \to \infty$, to a point $x_0$ belonging to the interior of $\Omega$.*

*Let $\xi_N$ denote a $\Omega$-valued random variable having density $\phi_N(x)$ that is proportional to $f(x,N)\exp\{-NS(x,N)\}$, that is*

$$\phi_N(x) = f(x,N)\exp\{-NS(x,N)\} \left(\int_\Omega f(x,N)\exp\{-NS(x,N)\}\,dx\right)^{-1}.$$

*(i) Then $\xi_N$ weakly converge to $x_0$. More explicitly, for a smooth $g$, one has*

$$|\mathbf{E}g(\xi_N) - g(x_0)| \le \left(\frac{c_1}{\sqrt{N}} + |x(N) - x_0|\right) \|g\|_{C^1(\Omega)} \tag{25}$$

*with a constant $c_1$ depending on $f_0, \Lambda_m, S_3, d$ and $f_m = \min_{x\in\Omega} f(x)$, which can be explicitly derived from Equations (7) and (8).*

*(ii) If additionally $S$ satisfies the conditions of Proposition 2, then*

$$|\mathbf{E}g(\xi_N) - g(x_0)| \le \frac{c_2}{N}\|g\|_{C^2(\Omega)} + |x(N) - x_0| \|g\|_{C^1(\Omega)}, \tag{26}$$

*with a constant $c_2$ depending on $f_0, f_1, \Lambda_m, S_3, S_4, d$ and $f_m$.*

**Proof.** From Propositions 1 and 2 we conclude that

$$|\mathbf{E}g(\xi_N) - g(x(N))| \le \frac{c_1}{\sqrt{N}}\|g\|_{C^1(\Omega)} \tag{27}$$

and

$$|\mathbf{E}g(\xi_N) - g(x(N))| \le \frac{c_2}{N}\|g\|_{C^2(\Omega)} \tag{28}$$

in cases (i) and (ii) respectively. The estimates of Equations (25) and (26) are then obtained from the triangle inequality. □

Next we were interested in the convergence of the normalized fluctuations of $\xi_N$ around $x_0$, namely, of the random variables

$$\eta_N = \sqrt{N}(\xi_N - x_0). \tag{29}$$

To simplify the formulas below we shall assume that $f(x,N) = 1$, but everything remains valid under general $f$ satisfying the assumptions above,

To analyze the fluctuations, we use their moment generating functions

$$M_N(p) = \mathbf{E}\exp\{(p, \eta_N)\} = \frac{\int_\Omega \exp\{-NS(x,N) + \sqrt{N}(p, x - x_0)\}\,dx}{\int_\Omega \exp\{-NS(x,N)\}\,dx} \tag{30}$$

for $p \in \mathbf{R}^d$.

The numerator in Equation (30) can be written in the form of Equation (1) as

$$I(p) = \int_\Omega \exp\{-N\left(S(x,N) - \frac{1}{\sqrt{N}}(p, x - x_0)\right)\}\,dx = \int_\Omega \exp\{-NS^*(x,N)\}\,dx$$

where the new phase is

$$S^*(x,N) = S(x,N) - \frac{1}{\sqrt{N}}(p, x - x_0).$$

To shorten the notations, we shall denote by primes the derivatives of $S$ or $S^*$ with respect to the variable $x$. $S^*$ is also convex, as $S$ is, and has the same derivatives of order 2 and higher as $S$. To apply the Laplace method we need to find its point of global minimum, which coincides with its (unique) critical point, that we denote by $x^* = x^*(p,N)$ and that solves the equation

$$(S^*)'(x^*, N) = 0 \Longleftrightarrow S'(x^*, N) = p/\sqrt{N}. \tag{31}$$

As a preliminary step to proving our CLT let us perform some elementary analysis of this equation proving its well posedness and finding its dependence on $N$ in the first approximation. We shall need the following elementary result.

**Lemma 1.** *Let $S(x)$ be a smooth convex function in $\mathbf{R}^d$ s.t. $S''(x) \geq \Lambda_m$ everywhere and $S'(x_0) = 0$. Then for any $K$ the mapping $z \mapsto S'(x_0 + z)$ is a diffeomorphism of the ball $B_K = \{z : |z| \leq K\}$ on its image and this image contains the ball $B_{K\Lambda_m}$.*

**Proof.** Injectivity is straightforward from convexity. Let us prove the last statement, that is, that for any $y \in B_{K\Lambda_m}$ there exists $z \in B_K$ such that $S'(x_0 + z) = y$. For any $\alpha > 0$, this claim is equivalent to the existence of a fixed point for a mapping

$$\Phi(z) = z - \alpha(S'(x_0 + z) - y) = z - \alpha \int_0^1 S''(x_0 + sz)z\,ds + \alpha y$$

in $B_K$. By the famous fixed point principle, to show the existence of a fixed point, it is sufficient to show that $\Phi$ maps $B_K$ to itself, that is, $\|\Phi(z)\| \leq K$ whenever $\|z\| \leq K$. Let

$$\Lambda_M = \sup_{z \in B_k} \|S''(x_0 + z)\|$$

and take $\alpha = 1/\Lambda_M$. Then the symmetric matrix $B = 1 - \alpha \int_0^1 S''(x_0 + sz)\,ds$ is such that $0 \leq B \leq 1 - \Lambda_m/\Lambda_M$ for all $z \in B_K$. Hence, if $z \in B_K$ we have

$$\|\Phi(z)\| \leq (1 - \frac{\Lambda_m}{\Lambda_M})K + \frac{\|y\|}{\Lambda_M}.$$

Hence, the inequality $\|\Phi(z)\| \leq K$ is fulfilled whenever $\|y\| \leq \Lambda_m K$, as was claimed. □

Thus the image of the set $U(N)$ contains the ball of radius $\Lambda_m N^{-1/3}$, so that for every $y : |y| \leq \Lambda_m N^{-1/3}$ there exists a unique $x \in U(N)$ such that $S'(x) = y$.

On the other hand, for any $K$ we can take $N_1 = \max(N_0, (K/\Lambda_m)^6)$, which is such that

$$\frac{p}{\sqrt{N}} < \Lambda_m N^{-1/3}$$

for all $N > N_1$ and $|p| \leq K$. Consequently, by Lemma 1, for such $p$ and $N$, there exists a unique solution $x^* = x^*(p, N)$ of Equation (31) in $\Omega$, and $x^* \in U(N)$, i.e.,

$$|x^* - x(N)| \leq N^{-1/3}. \tag{32}$$

Next, expanding $S'(x, N)$ in the Taylor series around $x(N)$ (where $S'(x(N), N) = 0$), we find from Equation (31) that

$$|S''(x(N), N)(x^* - x(N)) - \frac{p}{\sqrt{N}}| \leq S_3|x^* - x(N)|^2, \tag{33}$$

and thus

$$|D_N(x^* - x(N)) - \frac{p}{\sqrt{N}}| \leq S_3 N^{-2/3} \tag{34}$$

(recall that we denote $D_N = S''(x(N), N)$).

This allows us to improve the preliminary estimate of Equation (32) and to obtain

$$|x^* - x(N)| \leq D_N^{-1}\left(\frac{p}{\sqrt{N}} + \frac{S_3}{N^{2/3}}\right) \leq \frac{|p| + S_3}{\Lambda_m \sqrt{N}}. \tag{35}$$

Hence from Equation (33) we get

$$|D_N(x^* - x(N)) - \frac{p}{\sqrt{N}}| \leq \frac{S_3(|p| + S_3)^2}{\Lambda_m^2 N}. \tag{36}$$

Finally we conclude that

$$x^*(p, N) = x(N) + \frac{1}{\sqrt{N}} D_N^{-1} p + \frac{\epsilon}{N} \tag{37}$$

with

$$|\epsilon| \leq \frac{S_3(|p| + S_3)^2}{\Lambda_m^3}. \tag{38}$$

We can now prove a convergence result that can be called the *CLT for Laplace integrals.*

**Theorem 2.** *Under the assumption of Theorem 1 (i), assume additionally that $x(N)$ converges to $x_0$ quickly enough, that is*

$$|x(N) - x_0| \leq cN^{-\delta - 1/2} \tag{39}$$

*with positive constants $c, \delta$. Then the fluctuations $\eta_N = \sqrt{N}(\xi_N - x_0)$ converge weakly to a centered Gaussian random variable with the moment generating function*

$$M(p) = \exp\{\frac{1}{2}(p, D_N^{-1} p)\}. \tag{40}$$

**Proof.** We show that the moment generating functions of the fluctuations $\eta_N$ given by Equation (30) converge, as $N \to \infty$, to the function $M(p)$, the convergence being uniform on bounded subsets of $p$. By the well known characterization of weak convergence this will apply the weak convergence of the random fluctuations $\eta_N$.

Applying Proposition 1 to the numerator and denominator of the r.h.s. of Equation (30) we get, for $N > N_0$,

$$M_N(p) = \frac{\exp\{-NS^*(x^*(p, N), N)\}}{\exp\{-NS(x(N), N)\}} \frac{\sqrt{\det D_N}}{\sqrt{\det S''(x^*(p, N), N)}} \left(1 + \frac{\omega(x, N, p)}{\sqrt{N}}\right), \tag{41}$$

where $\omega$ is a bounded function, with a bound, depending on $S_3, \Lambda_m, p, d$, that can be found explicitly from Equation (7).

We have

$$S(x^*(p,N),N) = S\left(x(N) + \frac{1}{\sqrt{N}} D_N^{-1} p + \frac{\epsilon}{N}, N\right)$$

$$= S(X(N),N) + \frac{1}{2}\left(D_N(\frac{1}{\sqrt{N}} D_N^{-1} p + \frac{\epsilon}{N}), \frac{1}{\sqrt{N}} D_N^{-1} p + \frac{\epsilon}{N}\right) + \phi N^{-3/2},$$

with

$$|\phi| \le S_3 \left| D_N^{-1} p + \frac{\epsilon}{\sqrt{N}} \right|^3,$$

and consequently

$$S(x^*(p,N),N) = S(X(N),N) + \frac{1}{2N}(p, D_N^{-1} p) + \frac{1}{N^{3/2}}(p,\epsilon) + \frac{1}{2N^2}(D_N\epsilon,\epsilon) + \phi N^{-3/2}.$$

Therefore,

$$S^*(x^*(p,N),N) = S(x^*(p,N),N) - \frac{1}{\sqrt{N}}\left(p, x(N) + \frac{1}{\sqrt{N}} D_N^{-1} p + \frac{\epsilon}{N} - x_0\right)$$

$$= S(X(N),N) - \frac{1}{2N}(p, D_N^{-1} p) + \frac{1}{2N^2}(D_N\epsilon,\epsilon) + \frac{\phi}{N^{3/2}} - \frac{1}{\sqrt{N}}(p, x(N) - x_0).$$

Using Equation (63) we conclude that

$$\left| N[S(x(N),N) - S^*(x^*(p,N),N)] - \frac{1}{2}(p, D_N^{-1} p) \right| \le c\left(N^{-1/2} + N^{-\delta}\right),$$

where the constant $c$ depends on $p, S_3, \Lambda_m, \Lambda_M, d$.

Next, from Equation (35) we get

$$\|S''(x(N),N) - S''(x^*(p,N),N)\| \le S_3 \frac{|p| + S_3}{\Lambda_m \sqrt{N}},$$

so that

$$\left| \frac{\sqrt{\det D_N}}{\sqrt{\det S''(x^*(p,N),N)}} - 1 \right| \le \frac{c}{\sqrt{N}} \tag{42}$$

with another constant $c$ depending on $p, S_3, \Lambda_m, \Lambda_M, d$. Consequently, we deduce from Equation (41) that

$$M_N(p) = \exp\{\frac{1}{2}(p, D_N^{-1} p) + \frac{c(N,p)}{\sqrt{N}}\}\left(1 + \frac{\omega(N,p)}{\sqrt{N}}\right) \tag{43}$$

with some functions $c, \omega$, which are bounded on bounded subsets of $p$, implying the required convergence of the functions $M_N(p)$. □

## 4. Phase Minimum on the Border of the Domain of Integration

Here we present the estimates of the remainder in the asymptotic formula for the Laplace integrals with the critical point of the phase lying on the boundary of the domain of integration.

Let us start with a simple one-dimensional result, which is version of the well known Watson lemma. The proof can be performed as above by decomposing the domain of integration $[0, a]$ into the two intervals: $[0, N^{-1/2}]$ and $[N^{-1/2}, a]$. We omit the detail of the proof.

**Lemma 2.** *Let $S(x,N)$ and $f(x,N)$ be two continuous functions on the domain $\{x \in [0,a], N \ge 1\}$ with $a > 0$. Let $f$ be continuously differentiable and $S$ be twice continuously differentiable with respect to $x$, with the uniform bounds*

$$|S''(x,N)| \le s_2, \quad |f(x,N)| \le f_0, \quad |f'(x,N)| \le f_1,$$

*and the lower bound*

$$S'(x,N) \geq s_1,$$

*with some strictly positive constants* $s_1, s_2, f_0, f_1$, *where by primes we denote derivatives with respect to* $x$. *Then, for the Laplace integral*

$$I(N) = \int_0^a \exp\{-NS(x,N)\} f(x,N)\, dx,$$

*we have the asymptotic expression*

$$I(N) = \frac{\exp\{-NS(0,N)\}}{NS'(0,N)} \left( f(0,N) + \frac{\omega(N)}{N} \right) + \omega^{exp}(N), \tag{44}$$

*where*

$$|\omega(N)| \leq \frac{f_1}{S'(0,N)} + \frac{f_0}{(S'(0,N))^2} s_2 e^{s_2/2},$$

$$|\omega^{exp}(N)| \leq 2af_0 \exp\{-NS(0,N)\} \exp\{-s_1\sqrt{N}\}.$$

**Remark 6.** *One can obtain similar result by decomposing* $[0,a] = [0, N^{-\gamma}] \cup [N^{-\gamma}, a]$ *for any* $\gamma \in [1/2, 1)$, *in which case the exponentially small term will get the estimate*

$$|\omega^{exp}(x,N)| \leq 2af_0 \exp\{-NS(0,N)\} \exp\{-s_1 N^{1-\gamma}\}.$$

*This also shows that Lemma 2 remains essentially valid for small* $a$ *of order* $a = N^{-\gamma}$, $\gamma < 1$, *which is used in the proof of the next result.*

Let us turn to the general case. Namely, assume $\Omega$ is a bounded open set in $\mathbf{R}^{d+1}$. The coordinates in $\mathbf{R}^{d+1}$ will be denoted $(x,y)$ with $x \in \mathbf{R}, y \in \mathbf{R}^d$. Let

$$\Omega_+ = \{(x,y) \in \Omega : x \geq \psi(y)\}, \tag{45}$$

with some smooth function $\psi$. It will be convenient to introduce the sections of $\Omega$ as the sets

$$\Omega(x) = \{y : (x,y) \in \Omega\}.$$

We are interested in the asymptotics of the Laplace integral

$$I(N) = \int_{\Omega_+} f(x,y,N) \exp\{-NS(x,y,N)\}\, dxdy, \quad N > N_0, \tag{46}$$

with continuous functions $f$ and $S$ referred to as the amplitude and phase respectively.

Let us first discuss the case of $\Omega_+$ with a plane boundary, that is with $\psi(Y) = 0$, or equivalently with

$$\Omega_+ = \{(x,y) \in \Omega : x \geq 0\}. \tag{47}$$

We shall assume the following:

(C1′) $f(x,y,N)$ is a continuously differentiable function on $\Omega_+$ (up to the border) with

$$f_0 = \sup_{(x,y)\in\Omega_+, N\geq N_0} |f(x,y,N)| < \infty, \quad f_1 = \sup_{(x,y)\in\Omega_+, N\geq N_0} \left( |\frac{\partial f}{\partial x}| + |\frac{\partial f}{\partial y}| \right) < \infty;$$

(C2′) $S(x,y,N)$ is thrice continuously differentiable function of $x$ and $y$ such that

$$\frac{\partial^2 S}{\partial y^2}(x,y,N) \geq \Lambda_m$$

(where $\geq$ is the usual order on symmetric matrices) and

$$\frac{\partial S}{\partial x}(x,y,N) \geq g_m$$

with positive constants $\Lambda_m, g_m$, and

$$S_2 = \sup_{(x,y)\in\Omega_+, N\geq N_0} \max\left( \|\frac{\partial^2 S}{\partial x^2}\|, \|\frac{\partial^2 S}{\partial x \partial y}\|, \|\frac{\partial^2 S}{\partial y^2}\| \right) < \infty,$$

$$S_3 = \sup_{(x,y)\in\Omega_+, N\geq N_0} \max\left( \|\frac{\partial^3 S}{\partial x^3}\|, \|\frac{\partial^3 S}{\partial x^2 \partial y}\|, \|\frac{\partial^3 S}{\partial x \partial y^2}\|, \|\frac{\partial^3 S}{\partial y^3}\| \right) < \infty.$$

**Remark 7.** *As was noted above, the norms of higher derivatives in the estimates that we are using are their norms as multi-linear operators. For instance,* $\|\frac{\partial^2 S(x,y,N)}{\partial x \partial y}\|$ *is the minimum of constants* $\alpha$ *such that*

$$\left| \sum_{j=1}^{d} \frac{\partial^2 S(x,y,N)}{\partial x \partial y_j} x y_j \right| \leq \alpha |x||y|.$$

(C3') For any $N > N_0$, there exists a unique point of global minimum of $S$ in $\Omega_+$, this point lies on the boundary $\{x = 0\}$, i.e., it has coordinates $(0, y(N))$ with some $y(N) \in R^d$, and the box

$$U(N) = \{(x,y) : x \in [0, N^{-2/3}], |y - y(N)| \leq N^{-1/3}\} \tag{48}$$

is contained in $\Omega_+$. We shall also use the sections

$$U(x,N) = \{y : (x,y) \in U(N)\}.$$

Let us denote by $D_N$ the matrix of the second derivatives of $S$ as a function of $y$ at $(0, y(N), N)$, and by $g_N$ the gradient of $S$ as a function of $x$ at $(0, y(N), N)$, that is

$$D_N = \frac{\partial^2 S}{\partial y^2}(0, y(N), N), \quad g_N = \frac{\partial S}{\partial x}(0, y(N), N). \tag{49}$$

The approach of our analysis is to decompose the integral $I(N)$ into the sum of two integrals

$$I(N) = I'(N) + I''(N),$$

over the sets $\{x \leq N^{-2/3}\}$ and $\{x > N^{-2/3}\}$, to represent the first integral as the double integral, so that

$$I''(N) = \int_{\Omega \cap \{(x,y): x > N^{-2/3}\}} f(x,y,N) \exp\{-NS(x,y,N)\}\, dxdy, \tag{50}$$

$$I'(N) = \int_0^{N^{-2/3}} I(x,N) dx, \quad I(x,N) = \int_{\Omega(x)} f(x,y,N) \exp\{-NS(x,y,N)\}\, dy, \tag{51}$$

and to use Proposition 1 for the estimation of $I(x,N)$, $x \in [0, N^{-2/3}]$, and finally Lemma 2 to estimate $I'(N)$.

**Theorem 3.** *Under the assumptions (C1')–(C3'), the formula*

$$I(N) = \exp\{-NS(0, y(N), N)\} \left(\frac{2\pi}{N}\right)^{d/2} \frac{1}{N} \left[ \frac{f(0, y(N))}{g_N \sqrt{\det D_N}} + \frac{\omega(N)}{\sqrt{N}} \right] + \omega^{exp}(N) \tag{52}$$

*holds for $\Omega_+$ from Equation (47) and $N > N_1 = \max(N_0, (2S_2/\Lambda_m)^3)$, where $\omega^{exp}(N)$ is an exponentially small term and*

$$|\omega(N)| \leq \frac{1}{g_m \Lambda_m^{d/2}} \left[ f_1 \left( 1 + \frac{d}{\Lambda_m} \right) + f_0 d \max(S_3, S_2) e^{\max(S_3, S_2)} \left( 1 + \frac{1}{\Lambda_m^2} + \frac{1}{g_m^2} \right) \right]. \tag{53}$$

**Proof.** Integral $I''(N)$ from Equation (50) yields clearly an exponentially small contribution, similar to the integral $I''(N)$ in Proposition 1, so we omit the details here.

To calculate $I(x, N)$ we have to know critical points of the phase $S(x, y, N)$ as a function of $y$, that is the solutions $y^*(x, N)$ of the equation

$$\frac{\partial S}{\partial y}(x, y^*(x, N), N) = 0. \tag{54}$$

As $S$ is convex in $y$, the solution is unique, if it exists. Proceeding as in Lemma 1, that is, searching for a fixed point of the mapping

$$z \mapsto z - \frac{\partial S}{\partial y}(x, y(N) + z, N),$$

we find that there exists a unique solution $y^*(x, N)$ of Equation (54) whenever

$$S_2 < \Lambda_m N^{1/3} \Longleftrightarrow N > N_1 \tag{55}$$

such that

$$|y^*(x, N) - y(N)| \leq N^{-1/3}. \tag{56}$$

Next, using the Taylor expansion of $\partial S / \partial y$ around the point $(0, y(N), N)$ we get that

$$0 = \frac{\partial S}{\partial y}(x, y^*(x, N), N)$$

$$= \frac{\partial^2 S}{\partial y \partial x}(0, y(N), N) x + \frac{\partial^2 S}{\partial y^2}(0, y(N), N)(y^*(x, N) - y(N)) + \phi(x, y, N) \tag{57}$$

with

$$\phi(x, y, N) \leq 2S_3(|x|^2 + |y^*(x, N) - y(N)|^2) \leq 4S_3 N^{-2/3}.$$

This implies

$$y^*(x, N) - y(N) = -D_N^{-1} \left( \frac{\partial^2 S}{\partial y \partial x}(0, y(N), N) x + \phi(x, y, N) \right),$$

so that

$$|y^*(x, N) - y(N)| \leq \frac{S_2 + 4S_3}{\Lambda_m} N^{-2/3}, \tag{58}$$

which is an essential improvement as compared with the initial estimate of Equation (56). It ensures that the distance from $y^*(x, N)$ to the border of $U(x, N)$ is of order $N^{-1/3}$, so that Proposition 1 can in fact be applied to the integral $I(x, N)$ leading to

$$I(x, N) = \omega^{exp}(x, y, N)$$

$$+ \exp\{-NS(x, y^*(x, N), N)\} \left( \frac{2\pi}{N} \right)^{d/2} \left[ \frac{f(x, y^*(x, N))}{\left( \det \frac{\partial S^2}{\partial y^2}(x, y^*(x, N), N) \right)^{1/2}} + \frac{\omega(x, y, N)}{\sqrt{N}} \right], \tag{59}$$

where $\omega^{exp}$ is exponentially small compared to the main term and

$$|\omega(x,y,N)| \le d\Lambda_m^{-(1+d)/2}\left[f_1 + \frac{d+1}{6\Lambda_m} f_0 S_3 e^{S_3/6}\right].$$

In order to apply Lemma 2 we need to get lower and upper bounds to the quantities

$$\frac{\partial}{\partial x} S(x, y^*(x,N), N) \text{ and } \left|\frac{\partial}{\partial x}\left(\det \frac{\partial S^2}{\partial y^2}(x, y^*(x,N), N)\right)^{-1/2}\right|,$$

respectively.

We have

$$\frac{\partial}{\partial x} S(x, y^*(x,N), N) = \frac{\partial S}{\partial x}(x, y^*(x,N), N) + \frac{\partial S}{\partial y}(x, y^*(x,N), N)\frac{\partial y^*}{\partial x}(x,N).$$

But the second term vanishes. Hence

$$\frac{\partial}{\partial x} S(x, y^*(x,N), N) = \frac{\partial S}{\partial x}(x, y^*(x,N), N) \ge g_m.$$

Next, differentiating Equation (54) with respect to $y$ we obtain

$$\frac{\partial y^*}{\partial x}(x,N) = -\left[\frac{\partial^2 S}{\partial y^2}(x, y^*(x,N), N)\right]^{-1}\frac{\partial^2 S}{\partial x \partial y}(x, y^*(x,N), N),$$

implying the estimate

$$\|\frac{\partial y^*}{\partial x}(x,N)\| \le \frac{S_2}{\Lambda_m}. \tag{60}$$

Consequently, using the formula for the differentiation of the determinant of invertible symmetric matrices,

$$(\det A)' = \det A \operatorname{tr}(A' A^{-1}),$$

we can estimate

$$\left|\frac{\partial}{\partial x}\left(\det \frac{\partial S^2}{\partial y^2}(x, y^*(x,N), N)\right)^{-1/2}\right| \le \frac{dS_3}{2\Lambda_m^2}\left(\det \frac{\partial S^2}{\partial y^2}(x, y^*(x,N), N)\right)^{-1/2}.$$

Hence Lemma 2 can be applied to the calculation of $I'(N)$ given by Equations (51) and (59) yielding Equation (52). □

**Remark 8.** *Arguing as in Proposition 2, one can improve the estimate of the remainder term in Equation (52) to be of order $N^{-1}$, by assuming more regularity on $S$ and $f$.*

The general case of Equation (45) can be directly reduced to the case of $\Omega_+$ from Equation (47). In fact, changing coordinates $(x,y)$ to $(z,y)$ with $z = x - \psi(y)$ we get that $\Omega_+$ turns to $\tilde{\Omega}_+ = \{(z,y) : z \ge 0\}$. Making this change of the variable of integration in $I(N)$ yields

$$I(N) = \int_{\tilde{\Omega}_+} \tilde{f}(z,y,N) \exp\{-N\tilde{S}(z,y,N)\}\, dxdy, \quad N > N_0,$$

with $\tilde{S}(z,y,N) = S(z+\psi(y),y,N)$, $\tilde{f}(z,y,N) = f(z+\psi(y),y,N)$. Assuming that these functions satisfy the conditions of Theorem 3 we obtain

$$I(N) = \exp\{-NS(\psi(y(N)),y(N),N)\}\left(\frac{2\pi}{N}\right)^{d/2}\frac{1}{N}\left[\frac{f(\psi(y(N)),y(N))}{g_N\sqrt{\det\tilde{D}_N}}+\frac{\tilde{\omega}(N)}{\sqrt{N}}\right]+\tilde{\omega}^{exp}(N), \quad (61)$$

where

$$\tilde{D}_N = \frac{\partial^2\tilde{S}}{\partial y^2}(0,y(N),N) = \left(\frac{\partial^2 S}{\partial y^2}+\frac{\partial S}{\partial x}\frac{\partial^2\psi}{\partial y^2}+\frac{\partial^2 S}{\partial x^2}\frac{\partial\psi}{\partial y}\right)(\psi(y(N)),y(N),N)$$

and with similar change in the constants appearing in $\tilde{\omega}(N)$ and $\tilde{\omega}^{exp}(N)$.

## 5. LLN and CLT for Minima on the Boundary

The results on weak convergence of random variables with exponential densities given above for the case of the phase having minimum in the interior of the domain can be now recast for the case of the phase having minimum on the boundary of the domain of integration. The following statements are proved by literally the same argument as Theorems 1 and 2. We omit details.

**Theorem 4.** *Let $\Omega$ be a bounded open set in $\mathbf{R}_+^{d+1}$ with coordinates $(x,y)$, $x \in \mathbf{R}, y \in \mathbf{R}^d$, and let*

$$\Omega_+ = \{(x,y) \in \Omega : x \geq 0\}.$$

*Let the functions $f(x,y,N)$, $S(x,y,N)$ be a continuous functions on $\Omega_+ \times [1,\infty)$ satisfying condition (C1')- (C3') from Theorem 3. Assume moreover that $f$ is bounded below by a positive constants and that the sequence of global minima $(0,y(N))$ converges, as $N \to \infty$, to a point $(0,y_0)$ belonging to the interior of $\Omega$.*

*Let $(\xi_N^x, \xi_N^y)$ denote a $\Omega_+$-valued random variable having density $\phi_N(x,y)$ that is proportional to $f(x,y,N)\exp\{-NS(x,y,N)\}$, that is*

$$\phi_N(x,y) = f(x,y,N)\exp\{-NS(x,y,N)\}\left(\int_{\Omega_+} f(x,y,N)\exp\{-NS(x,y,N)\}\,dxdy\right)^{-1}.$$

*Then $(\xi_N^x, \xi_N^y)$ weakly converge to a constant $(0,y_0)$. More explicitly, for a smooth $g$, one has*

$$|\mathbf{E}g(\xi_N^x,\xi_N^y) - g(0,y_0)| \leq \left(\frac{c}{\sqrt{N}} + |y(N)-y_0|\right)\|g\|_{C^1(\Omega)} \quad (62)$$

*with a constant $c$ depending only on $S$ (actually on the bounds for the derivatives of $S$ up to the third order).*

**Theorem 5.** *Under the assumptions of Theorem 4 assume additionally that*

$$|y(N) - y_0| \leq cN^{-\delta-1/2}. \quad (63)$$

*Then the fluctuations $(\eta_N^x, \eta_N^y) = (N\xi_N^x, \sqrt{N}(\xi_n^y - y_0))$ converge weakly to a $(d+1)$-dimensional random vector such that its last coordinates form a centered Gaussian random vector with the moment generating function*

$$M(p) = \exp\{\frac{1}{2}(p, D_N^{-1}p)\}, \quad (64)$$

*and the first coordinate is independent and represents a $g_0$- exponential random variable. The rates of convergence with all explicit constants are obtained directly from Theorem 3.*

## 6. Laplace Sums with Error Estimates

It is more or less straightforward to modify the above results to the of sums rather than integrals. Namely, instead of the integral $I(N)$ from Equation (1) let us consider the sum

$$\Sigma(N) = \frac{1}{N^d} \sum_{k=(k_1,\cdots,k_d):x_k=k/N\in\Omega} f(x_k)\exp\{-NS(x_k,N), \quad N>1, \tag{65}$$

where $\Omega$ is an open polyhedron of the Euclidean space $\mathbf{R}^d$, with Euclidean volume $|\Omega|$, the amplitude $f$ and the phase $S$ are continuous real functions.

**Theorem 6.** *Under the assumptions of Proposition 1,*

$$\Sigma(N) = \exp\{-NS(x(N),N)\}\left(\frac{2\pi}{N}\right)^{d/2}\left[\frac{f(x(N),N)}{\sqrt{\det D_N}} + \frac{\tilde{\omega}(N)}{\sqrt{N}}\right] + \omega^{exp}(N), \tag{66}$$

*where*

$$|\tilde{\omega}(N)| \le |\omega(N)| + (f_0+f_1)C(\Lambda_m,\Lambda_M,S_3),$$

*and where $\omega(N)$ and $\omega^{exp}(N)$ are the same as in Proposition 1 and $C(\Lambda_M,S_3)$ is yet another constant depending on $\Lambda_m,\Lambda_M,S_3$.*

**Proof.** We use the well known (and easy to prove) fact (a simplified version of the Euler–Maclorin formula) that

$$|\int_\Omega g(x) - \frac{1}{N^d}\sum_{k=(k_1,\cdots,k_d):x_k=k/N\in\Omega} g(x_k)| \le \frac{1}{N}\int_\Omega |g'(x)|\,dx. \tag{67}$$

Consequently,

$$|\Sigma(N)-I(N)| \le \frac{1}{N}\int |f'(x)|\exp\{-NS(x,N)\}\,dx + \int f(x)|S'(x,N)|\exp\{-NS(x,N)\}\,dx, \tag{68}$$

where $I(N)$ is from Equation (1). The first integral on the r.h.s. of Equation (68) is clearly of order $1/N$, as compared with the main term of $I(N)$ given in Proposition 1. The pre-exponential term in the second integral vanishes at the critical point $(x(N),N)$ of $S(x,N)$. Hence the required estimate for the second integral is obtained directly from Proposition 1. □

Now all LLN and CLT results obtained above for continuous distributions can be reformulated and proved straightforwardly for the case of discrete random variables taking values in the lattice $\{x_k = k/N \in \Omega\}$ with probabilities proportional to $f(x_k)\exp\{-NS(x_k,N)\}$.

## 7. Application to LLN and CLT of Large Deviations

Conditional LLN (conditioned on the sums of the corresponding random variables to stay in a certain prescribed domain, usually some linear subspace or a convex set) are well developed in probability (see e.g., [2,18] for two different contexts). The results above can be used to supply exact estimates for the error terms in these approximations. To illustrate this statement in the most transparent way let us start with the classical multidimensional local theorem of large deviations as given in [4] (that extends earlier results of [6]). Namely, let $\xi,\xi_1,\xi_2,\cdots$ be a sequence of independent identically distributed $\mathbf{R}^k$-valued random vectors. Assume that the set $O$ of vectors $\lambda\in\mathbf{R}^k$ such that the moment generating function $v(\lambda) = \mathbf{E}e^{(\lambda,\xi)}$ is well defined has a nonempty interior $O^0$. It is well known (and easy to see) that the functions $v$ and $\ln v$ are convex and the sets $O^0$ and its closure $\bar{O}^0 = \bar{O}$

are convex. The function $\psi(\alpha) = \inf[\ln v(\lambda) - (\alpha, \lambda)]$ is called the entropy and it is concave. Moreover, the infimum in its definition is attained, so that there exists $\lambda(\alpha) \in O$ such that

$$\psi(\alpha) = \inf_{\lambda}[\ln v(\lambda) - (\alpha, \lambda)] = \ln v(\lambda(\alpha)) - (\alpha, \lambda(\alpha)),$$

and the function $\lambda(\alpha)$ is a diffeomorphism of $O^0$ onto some open domain $\Omega$ in $\mathbf{R}^k$. Assume that the random variable $\xi$ has a bounded probability density $p(x)$, and define the family of distributions $P_\alpha$ with the densities

$$\pi_\alpha(x) = \exp\{(\lambda(\alpha), x) - \psi(\alpha)\}p(x).$$

Let $p_n(x)$ be the density of the averaged sum $S_n/n = (\xi_1 + \cdots + \xi_n)/n$.

Theorem 1 of [4] states (though we formulate it equivalently in terms of the density of $S_n/n$, rather than $S_n$ as is done in [4]) that if $\Phi$ is any compact set in $\Omega$, then

$$p_n(\alpha) = \frac{n^{k/2} e^{n\psi(\alpha)}}{(2\pi)^{k/2} \det(M(\alpha))^{1/2}} \left(1 + \sum_{j=1}^{s} c_j(\alpha) n^{-j} + O(n^{-s})\right), \quad (69)$$

where $s$ is arbitrary, the estimate is uniform for $\alpha \in \Phi$, $M(\alpha)$ is the matrix of the second moments of the distributions $P_\alpha$, the coefficients $c_j(\alpha)$ depend only on $2j+2$ moments of $P_\alpha$ and are uniformly bounded in $\Phi$.

The densities of Equation (69) are exactly of the type dealt with in our Theorems 1, 2, and 4, and Equation (5). Thus, these theorems are applied directly for finding the rates of convergence for LLN and CLT for the sums of independent variables when $S_n/n$ is reduced to some convex bounded set with smooth boundary or a linear subspace. These conditional versions of LLN may be applied even if $\mathbf{E}\xi$ is not defined, so that the usual LLN does not hold.

When the random variable $\xi$ has values in a lattice, a version with sums, that is Theorem 6, should be applied to get the rates of convergence in the corresponding laws of large numbers.

**Funding:** FRC CSC RAS, Supported by the RFFI grant 18-07-01405.

**Conflicts of Interest:** The author declares no conflicts of interest.

## References

1. Albeverio, S.; Kondratiev, Y.; Kozicki, Y.; Röckner, M. *The Statistical Mechanics of Quantum Lattice Systems. A Path Integral Approach*; European Mathematical Society: Zürich, Switzerland, 2009.
2. Minlos, R.A. Lectures on stastistical physics. *Uspehi Mat. Nauk* **1968**, *23*, 133–190.
3. Grenander, U. *Lectures in Pattern Theory*; Springer: New York, NY, USA, 1981; Volume III.
4. Borovkov, A.A.; Rogozin, B.A. On the Multidimensional Central Limit Theorem. *Teor. Veroyatn. Primen.* **1965**, *10*, 61–69.
5. Del Moral, P.; Zajic, T. A note on the Laplace-Varadhan integral lemma. *Bernoulli* **2003**, *9*, 49–65. [CrossRef]
6. Richter, W. Multidimensional Local Limit Theorems for Large Deviations. *Teor. Veroyatn. Primen.* **1958**, *3*, 107–114.
7. Korshunov, D.A.; Piterbarg, V.I.; Hashorva, E. On the Asymptotic Laplace Method and Its Application to Random Chaos. *Mat. Zametki* **2015**, *97*, 868–883. [CrossRef]
8. Piterbarg, V.; Simonova, I. Asymptotic expansions for probabilities of large excursions of nonstationary Gaussian processes. *Mat. Zametki* **1984**, *35*, 909–920.
9. Kolokoltsov, V.N. *Semiclassical Analysis for Diffusions and Stochastic Processes*; Springer: Berlin/Heidelberg, Germany, 2000; Volume 1724 .
10. Dobrokhotov, S.Y.; Kolokoltsov, V.N. Double-well Splitting of the Low Energy Levels of the Schrödinger Operator for Multidimensional $\phi^4$-models on Tori. *J. Math. Phys.* **1995**, *36*, 1038–1053. [CrossRef]

11. Kolokoltsov, V.N. On the asymptotics of the low lying eigenvalues and eigenfunctions of the Schrödinger operator. In *Doklady Akademii Nauk*; Mezhdunarodnaya Kniga: Moscow, Russia, 1993; Volume 328, pp. 649–653.
12. Maslov, V.P. On a general theorem of set theory leading to the Gibbs, Bose-Einstein, and Pareto distributions as well as to the Zipf-Mandelbrot law for the stock market. *Math. Notes* **2005**, *78*, 807–813. [CrossRef]
13. Albeverio, S.; Steblovskaya, V. Asymptotics of Gaussian Integrals in infinite dimensions. *Infnite Dimens. Anal. Quantum Probab. Relat. Top.* **2019**, *22*, 1950004. [CrossRef]
14. De Roeck, W.; Maes, C.; Netocný, K.; Rey-Bellet, L. A note on the non-commutative Laplace-Varadhan integral lemma. *Rev. Math. Phys.* **2010**, *22*, 839–858. [CrossRef]
15. Maslov, V.P.; Chebotarev, A.M. On the second term of the logarithmic asymptotics of functional integrals. *Itogi Nauki i Tehkniki Teor. Veroyatnosti* **1982**, *19*, 127–154.
16. Korolev, V.Y.; Shevtsova, I.G. An upper bound for the absolute constant in the Berry-Esseen inequality. *Teor. Veroyatn. Primen.* **2009**, *54*, 671–695. [CrossRef]
17. Kolokoltsov, V.N. *Differential Equations on Measures and Functional Spaces*; Springer International Publishing: Basel, Switzerland, 2019.
18. Vasicek, O.A. A conditional law of large numbers. *Ann. Probab.* **1980**, *8*, 142–147. [CrossRef]

mathematics

MDPI

*Article*

# Sensitivity Analysis and Simulation of a Multiserver Queueing System with Mixed Service Time Distribution

**Evsey Morozov [1,2,3,†], Michele Pagano [4,†] and Irina Peshkova [1,*,†,‡] and Alexander Rumyantsev [1,2,†]**

1 Department of Applied Mathematics and Cybernetics, Petrozavodsk State University, 185035 Petrozavodsk, Russia; emorozov@karelia.ru (E.M.); ar0@krc.karelia.ru (A.R.)

2 Institute of Applied Mathematical Research, Karelian Research Centre, Russian Academy of Sciences, 185910 Petrozavodsk, Russia

3 Moscow Center for Fundamental and Applied Mathematics, Moscow State University, 119991 Moscow, Russia

4 Department of Information Engineering, University of Pisa, 56126 Pisa, Italy; michele.pagano@iet.unipi.it

* Correspondence: iaminova@petrsu.ru; Tel.: +7-8142-71-3261

† These authors contributed equally to this work.

‡ Current address: Lenina Str., 33, 185910 Petrozavodsk, Russia.

Received: 9 June 2020; Accepted: 24 July 2020; Published: 3 August 2020

**Abstract:** The motivation of mixing distributions in communication/queueing systems modeling is that some input data (e.g., service time in queueing models) may follow several distinct distributions in a single input flow. In this paper, we study the sensitivity of performance measures on proximity of the service time distributions of a multiserver system model with two-component Pareto mixture distribution of service times. The theoretical results are illustrated by numerical simulation of the $M/G/c$ systems while using the perfect sampling approach.

**Keywords:** pareto mixture distribution; multiserver system; uniform distance; perfect simulation

## 1. Introduction

Mixtures of distributions arise in complex stochastic systems and they are extensively used for statistical analysis in many real fields, such as lifetime modeling, ageing or failure processes, engineering reliability [1], and survival theory [2], where data are assumed to be heterogeneous. The application of the mixture of distributions in the modeling of queueing systems is often induced by diverse structure of the customers in the system, e.g., by various service time requirements of multiple classes of customers that arrive into the system (for instance, the transmission time of IP datagrams with different lengths), or by the noisy/biased measurements that induce the so-called contaminated distributions. Ignoring such a diversity at the modeling phase may lead to significant deviation of system performance at practical implementation phase as compared to the modelled values. This motivates various types of analysis, including the analysis of continuity, robustness, monotonicity, stability, and sensitivity. In this regard, we mention the fundamental result obtained for telecommunication system models by B. A. Sevast'yanov [3], and the basic monographs [4,5].

The authors would like to use this opportunity to pay tribute to Professor Vladimir Zolotarev and to note his outstanding role as the founder of the International Seminar on Stability Problems for Stochastic Models. One of the authors had a great pleasure to communicate with Professor Zolotarev over many years, and all of the authors actively participated in the seminar he has founded. In the context of this paper, it is especially appropriate to emphasize an important role of Professor Zolotarev in the study of the stability and monotonicty of queuieng processes, see [6–8].

Information flows in modern telecommunications and computing systems have the form of a superposition of some sequential-parallel structures [9]. Ranging from small personal devices up to large scale high-performance computing systems, all of these may be modeled as multiserver queueing systems. Thus, it is highly important to study the performance of such systems and, in particular, the sensitivity of stationary performance indexes with respect to the variability of input parameters. However, direct output analysis of queueing systems is often tricky (see e.g., [10,11]), and explicit expressions for the distributions of steady-state performance indexes of a multiserver system are, in general, hardly available and, beyond classic models, known in some special cases only. In some cases, the analysis may be performed by obtaining asymptotic upper bounds, as in the paper [12], or studying the continuity of the process, as in [8], or stochastic stability of the queueing process, like in [4,13], or by means of simulation. In the present paper, we utilize the latter approach.

This paper is dedicated to sensitivity analysis of a steady-state performance index of a multiserver system with respect to service time distribution having the form of the so-called finite mixture [14]. However, instead of studying the direct parametric sensitivity, we focus on a more delicate analysis of the (combined) effect of the service time distribution on the steady-state performance estimate. That is, we compare the basic system to a disturbed one, using a sensitivity measure (Kolmogorov–Smirnov distance) both for the service time distributions, and for the steady-state performance estimate (queue size). The service time distribution perturbation is performed by changing the mixing coefficient and parameters defining the mixture components. We formalize this at the end of Section 3.

In general, the output distributions are hardly analytically available and, in this case, we must be able to obtain the steady-state performance indexes by simulation. As a basic model, we consider the classical $M/G/c$ model, where the steady-state distribution of the vector workload process is unknown as well; however, it can be estimated by means of the recently developed method of regenerative perfect simulation [15]. In more detail, as the target (perturbed) service time distribution we take the two-component mixture of Type-II Pareto distributions with support on the positive axis, which is known as Lomax distributions, as well as two-component exponential (hyperexponential) distribution. Such a choice also allows for obtaining some analytical expressions. Our interest to Pareto distribution is caused by the heavy tailed property of this distribution that is frequently observed in models of file size and flow duration [16].

This paper continues the study performed in [17] in the context of monotonicity. The key idea of the present paper is to study qualitatively the sensitivity of the steady-state distribution of the system performance index (steady-state queue size) to the variability of service time distribution by means of simulation. We also apply the auxiliary results on the failure rates comparison, which allows us to characterize the monotonicity of some stationary performance measures.

The structure of the paper is as follows. In Section 2, we introduce the two-component mixture of distributions and discuss some properties that are used in the subsequent analysis. Subsequently, we define the uniform distance between the mixture and the corresponding parent distribution. In Section 3, some known stochastic monotonicity properties of the multiserver system are collected, which further are specified for the considered mixture distributions. In Section 4, we describe the perfect sampling algorithm that is then used to sample from exact (but unknown) steady-state distribution of a multiserver queue $M/G/c$. The results of simulation are presented in Section 5. We study the sensitivity of the steady-state queue size distribution with respect to (w.r.t.) the shape parameters of mixture and the mixing coefficient and illustrate stochastic monotonocity of the system performance. The discussion of the simulation results finalizes the paper in the concluding Section 6.

## 2. Two-Component Mixture Distributions

The goal of this Section is to derive the uniform distance between the two-component mixture distribution and its parent distribution. First, we introduce the two-component mixture, and then give a few properties, including the stochastic monotonicity. This property is further used to obtain the monotonicity of the corresponding output queueing process.

Let $X_i$ be independent random variables having mean $\mathrm{E}X_i$, density $f_i$, tail distribution function (d.f.) $\overline{F}_i(x) = 1 - F_i(x)$, and failure rate

$$r_i(x) = \frac{f_i(x)}{\overline{F}_i(x)}, \ i = 1,2,$$

defined for such $x$ that $\overline{F}_i(x) > 0$. We assume that $F_1 \neq F_2$ to avoid trivial case. Let $I$ be a Bernoulli random variable independent of $X_i$, with success probability $P(I = 1) = p$. Subsequently, it is called the random variable

$$X_M = IX_1 + (1 - I)X_2,$$

has the two-component mixture distribution [18] (we use the index $M$ to denote the mixture). The mean $\mathrm{E}X_M$ and density, $f_M$ of $X_M$ equal, respectively,

$$\mathrm{E}X_M = p\mathrm{E}X_1 + (1 - p)\mathrm{E}X_2, \tag{1}$$

$$f_M(x) = pf_1(x) + (1 - p)f_2(x), \tag{2}$$

and it is easy to see that the tail distribution is

$$\overline{F}_M(x) = p\overline{F}_1(x) + (1 - p)\overline{F}_2(x). \tag{3}$$

Note that the d.f. $F_i$ may belong to the same family of distributions but have other parameters. In reliability analysis, such a mixture may be interpreted as a contaminated distribution [19], where $1 - p$ is, as a rule, small enough. $F_1$ is called the parent distribution and $F_2$ is the contaminating distribution. In this Section, we focus on the distance between the mixture and its parent distribution.

A straightforward analysis shows that the failure rate of the mixture has the following form [20]:

$$r_M(x) = \frac{pf_1(x) + (1 - p)f_2(x)}{p\overline{F}_1(x) + (1 - p)\overline{F}_2(x)} = a(x)r_1(x) + (1 - a(x))r_2(x), \tag{4}$$

where

$$a(x) = \frac{p\overline{F}_1(x)}{p\overline{F}_1(x) + (1 - p)\overline{F}_2(x)}, \quad x \geq 0.$$

In particular, it follows from Equation (4) that

$$r_M(x) \geq \min(r_1(x), r_2(x)), \quad x \geq 0. \tag{5}$$

It is worth mentioning that the mixture preserves the monotonicity of failure rate in the following way: if both rates $r_i(x)$ are non-increasing, that is d.f.'s $F_i(x)$ are decreasing failure rate distributions (DFR), then the mixture $F_M(x)$ is DFR distribution as well [21]. Indeed, one can check that

$$r'_M(x) = a(x)r'_1(x) + (1 - a(x))r'_2(x) - a(x)(1 - a(x))(r_1(x) - r_2(x))^2, \tag{6}$$

also see [1]. Subsequently, if $r'_i(x) < 0$, $i = 1,2$, it follows from Equation (6) that $r'_M(x) < 0$, since $a(x) \in [0,1]$ for any $x \in (0,\infty)$. In particular, it follows from Equation (6) that the mixture of two exponential distributions is DFR (note that the exponential distribution has constant failure rate).

Another example of a DFR distribution is the Type-II Pareto distribution, denoted by $Pareto(\alpha_i, x_0)$, having d.f. (see e.g., [22])

$$F_i(x) = 1 - \left(\frac{x_0}{x_0 + x}\right)^{\alpha_i}, \ x \geq 0, \ x_0 > 0, \ \alpha_i > 0, \ i = 1,2.$$

The failure rate of $Pareto(\alpha_i, x_0)$ equals

$$r_i(x) = \frac{\alpha_i}{x_0 + x}, \; x \geq 0, \;\; i = 1, 2, \tag{7}$$

and it is monotonically decreases to 0 as $x \to \infty$. As has been noted above, the two-component mixture of Pareto distributions is DFR distribution. However, the failure rate of finite mixtures, in general, is a complicated function [20].

The uniform distance between distributions $F$ and $G$, defined as [12]

$$\Delta(F, G) = \sup_x |F(x) - G(x)|, \tag{8}$$

is a recognised measure, which is actively used in the sensitivity analysis [12]. It is easy to see that the uniform distance $\Delta(F_M, F_1)$ between the mixture distribution Equation (3) and its parent distribution is

$$\Delta(F_M, F_1) = \sup_{x \geq 0} |pF_1(x) + (1-p)F_2(x) - F_1(x)| = (1-p) \sup_{x \geq 0} |F_1(x) - F_2(x)|. \tag{9}$$

Note that, if the densities $f_i$ exist, and there exists $x^*$ that delivers the supremum in Equation (9),

$$\Delta(F_M, F_1) = |F_1(x^*) - F_2(x^*)|,$$

then $x^*$ satisfies the equality

$$f_1(x^*) = f_2(x^*). \tag{10}$$

By definition of the failure rates, $r_i$, it then follows that

$$r_1(x^*)\overline{F}_1(x^*) = r_2(x^*)\overline{F}_2(x^*).$$

Thus, expression Equation (9) can be written in the following convenient form

$$\Delta(F_M, F_1) = (1-p)\frac{|r_2(x^*) - r_1(x^*)|}{r_2(x^*)}\overline{F}_1(x^*) = (1-p)\frac{|r_1(x^*) - r_2(x^*)|}{r_1(x^*)}\overline{F}_2(x^*). \tag{11}$$

Note that Equation (11) allows obtaining the following upper bound for the distance $\Delta(F_M, F_1)$:

$$\Delta(F_M, F_1) \leq (1-p)\frac{|r_2(x^*) - r_1(x^*)|}{r_1(x^*)} =: \delta(x^*). \tag{12}$$

In particular, for the hyperexponential distribution, that is for two-component mixture of exponential distributions with densities $f_i(x) = \lambda_i e^{-\lambda_i x}, i = 1, 2$, it follows from Equation (10), that

$$x^* = \frac{\log \lambda_1 - \log \lambda_2}{\lambda_1 - \lambda_2},$$

and in this case expression Equation (11) becomes

$$\Delta(F_M, F_1) = (1-p)\frac{|\lambda_2 - \lambda_1|}{\lambda_2}\left(\frac{\lambda_1}{\lambda_2}\right)^{-\frac{\lambda_1}{\lambda_1 - \lambda_2}} \leq (1-p)\frac{|\lambda_2 - \lambda_1|}{\lambda_2}. \tag{13}$$

Note that the last inequality in Equation (13) is a particular case of Equation (12). Expression Equation (13) is consistent with a more general result for the so-called univariate scale mixture $X_M$ having form [2]

$$X_M \stackrel{d}{=} \frac{X_1}{Y}, \tag{14}$$

with d.f.

$$\hat{F}_M(x) = \int_0^\infty F_1(\theta x) dG(\theta),$$

where $F_1$ is the parent distribution of the random variable $X_1$ and $G$ is the distribution of a mixing random variable $Y \geq 0$. It is clear that the transformation Equation (14) is a scale change, and if $Y \in \{y_1, \ldots, y_m\}$ is a discrete random variable, then Equation (14) becomes

$$X_M = \sum_{i=1}^{m} \frac{1}{y_i} I(Y = y_i) X_1, \tag{15}$$

which is a finite mixture, where $I$ is an indicator function. The aforementioned general result for the univariate scale mixture states that if $F_1$ is an exponential d.f., then an upper bound for the uniform distance may be obtained as follows [2,23]:

$$\Delta(\hat{F}_M, F_1) \leq \mathrm{E}|Y - 1|. \tag{16}$$

To show that Equation (16) indeed coincides with Equation (13) for the two-component scale mixture case, let $Y$ have point masses at 1 and $\lambda_2/\lambda_1$ with probabilities $p$ and $1 - p$, respectively. It immediately follows from Equations (15) and (16) that

$$\mathrm{E}(Y - 1) = (1 - p)\frac{|\lambda_1 - \lambda_2|}{\lambda_2}.$$

Now, we return to the two-component $Pareto(\alpha_i, x_0)$ mixture $F_M$. It follows from Equation (11) that in this case

$$\Delta(F_M, F_1) = (1 - p)\frac{|\alpha_1 - \alpha_2|}{\alpha_2} \left(\frac{\alpha_2}{\alpha_1}\right)^{\frac{\alpha_1}{\alpha_1 - \alpha_2}}, \tag{17}$$

where the value $x^*$ satisfying equality Equation (10) equals

$$x^* = x_0 \left( \left(\frac{\alpha_1}{\alpha_2}\right)^{\frac{1}{\alpha_1 - \alpha_2}} - 1 \right).$$

Note that the r.h.s. of Equation (17) is similar to the r.h.s. of Equation (13). This similarity is caused by the specific shape of the failure rate of the distribution $Pareto(\alpha_i, x_0)$. Moreover, in such a case, the quantity $\delta(x^*)$ defined in Equation (12) does not depend on $x^*$ and, thus, for Pareto mixture, it readily follows from Equation (12) that

$$\Delta(F_M, F_1) \leq (1 - p)\frac{|\alpha_1 - \alpha_2|}{\alpha_2} =: \delta(\alpha_1, \alpha_2). \tag{18}$$

In Figure 1, to illustrate the dependence of the uniform distance on the parameter $\alpha_2$ of the contaminating distribution, we depict $\Delta(F_M, F_1)$ jointly with $\delta(\alpha_1, \alpha_2)$ for fixed $\alpha_1 = 2$ and $p = 0.9$ by varying $\alpha_2$ in the interval (1, 5).

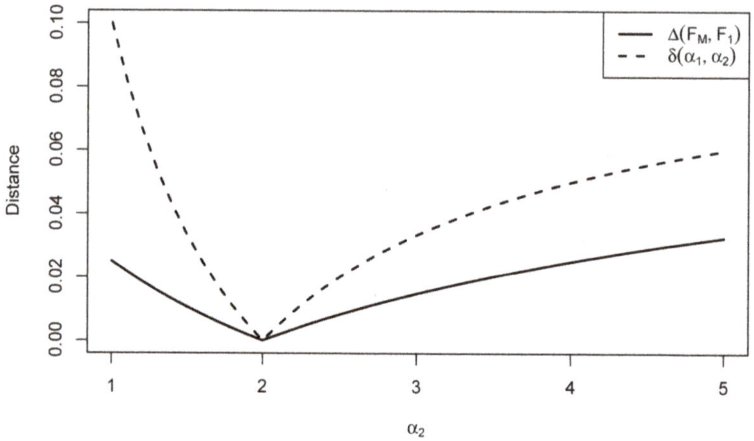


**Figure 1.** The distance $\Delta(F_M, F_1)$ with mixing parameter $p = 0.9$ and an upper bound $\delta(\alpha_1, \alpha_2)$ vs. parameter $\alpha_2$.

## 3. Multiserver System Sensitivity

In this Section, we formalize our main goal for the numerical experiments conducted and discussed in Section 5. We then demonstrate how stochastic and failure rate ordering can be applied to multiserver systems with mixed service time distribution. The numerical experiments equipped with the stochastic comparison technique not only allow for obtaining the absolute value, but also characterizing the monotonicity of performance indexes.

Consider a classical First-Come-First-Served (FCFS) $c$-server $M/G/c$ queueing system that is fed by a Poisson input with rate $\lambda$, arrival instants $\{t_i,\, i \geq 1\}$ with $t_1 = 0$, independent and identically distributed (iid) interarrival times $T_i = t_{i+1} - t_i$ and iid service times $\{S_i,\, i \geq 1\}$. Note that $\lambda = 1/\mathrm{E}T$, where $T$ is generic interarrival time. Now, we consider the $c$-dimensional vector of the remaining workload process in such a system,

$$W_i = (W_{i,1}, \ldots, W_{i,c}),$$

where $W_{i,k}$ is the $k$th smallest component of the vector which is observed by the $i$th arrival [24]. Thus, the vector components are kept in ascending order,

$$W_{i,1} \leq \cdots \leq W_{i,c},$$

and the quantity $W_{i,j}$, "observed" by the arriving customer $i$, equals the unfinished work which must be done by server $j$ provided no new work arrives after arrival instant $t_i$ of customer $i$; $j = 1, \ldots, c$. If there are no idle servers upon arrival of customer $i$, then s/he waits in a common infinite capacity queue until the server with minimal work, $W_{i,1}$, becomes free. It is easy to see that $W_{i,1}$ is the waiting time of customer $i$ which starts being served at time $t_i + W_{i,1}$. It is well-known that the workload vector sequence follows the celebrated stochastic Kiefer–Wolfowitz recursion [25]:

$$W_{i+1} = R(W_i + e_1 S_i - \mathbf{1} T_i)^+, \tag{19}$$

where $e_1 = (1, 0, \ldots, 0)$ and $\mathbf{1}$ is the vector of ones, operator $R$ puts the components in an ascending order, and operation $(\cdot)^+ = \max(0, \cdot)$ is applied componentwise (we omit the sub-index for a generic element of a sequence). In what follows, we assume that the stability condition holds [25],

$$\rho := \lambda \mathrm{E}S < c. \tag{20}$$

Define the departure instant of customer $i$ by $d_i = t_i + W_{i,1} + S_i$. Now define the process

$$Q_n = \sum_{j\geq 1, j\neq n} I(t_j \leq t_n < d_j), \tag{21}$$

counting the queue size (number of customers in the system) at the arrival instant $t_n$. Under condition Equation (20), $Q_n$ converges in distribution, as $n \to \infty$, to the steady-state queue size $Q$, with stationary distribution

$$\pi_n = \mathrm{P}(Q = n),\ n \geq 0.$$

Note that when service times $S_i$ are exponential, the steady-state queue size distribution, $\pi_n, n \geq 0$, is well known [26]:

$$\pi_n = \begin{cases} \left(\sum_{k=0}^{c-1} \frac{\rho^k}{k!} + \frac{\rho^c}{(c-1)!(c-\rho)}\right)^{-1}, & n = 0, \\ \pi_0 \frac{\rho^n}{n!}, & 1 \leq n \leq c, \\ \pi_0 \frac{\rho^n}{c! c^{n-c}}, & n > c. \end{cases} \tag{22}$$

The operators $R(\cdot)$ and $(\cdot)^+$ in Equation (19) preserve ordering, and it allows for us to establish the monotonicity of the workload sequence in the multiserver system in the case when the driving sequences $\{T_n^{(i)}, S_n^{(i)}, n \geq 1\}$, $i = 1,2$ satisfy stochastic order. We recall that the stochastic order $X_2 \leq_{st} X_1$ between two random variables $X_1, X_2$ means that the tail d.f.'s satisfy inequality

$$\mathrm{P}(X_2 > x) \leq \mathrm{P}(X_1 > x),\ \ x \geq 0. \tag{23}$$

In is known [27] that, in two $c$-server systems with stochastically ordered input sequences, $T^{(2)} \geq_{st} T^{(1)}$ and $S^{(2)} \leq_{st} S^{(1)}$, the workload sequences $\{W_n^{(i)}\}, i = 1,2$, are (componentwise) ordered in the following way

$$W_n^{(2)} \leq_{st} W_n^{(1)}\ \ n \geq 1. \tag{24}$$

It also holds for the steady-state workloads:

$$W^{(2)} \leq_{st} W^{(1)}.$$

If the input in both systems is the same, which is $T^{(2)} =_{st} T^{(1)}$, then the the queue length process at the arrival instants satisfy similar ordering both in path-wise sense and in steady-state [27]

$$Q^{(2)} \leq_{st} Q^{(1)}. \tag{25}$$

The stochastic ordering $\leq_{st}$ can be transformed into the ordering with probability 1 by the coupling technique [28]. In the context of this work, it is worth mentioning that the sufficient condition for the stochastic ordering $S^{(2)} \leq_{st} S^{(1)}$ is the failure rate ordering [29]:

$$r_2(x) \geq r_1(x),\ x \geq 0, \tag{26}$$

where $r_i$ is the failure rate of r.v. $S^{(i)}$, $i = 1,2$. We summarize the discussion in the following lemma which is a straightforward result of [27].

**Lemma 1.** *Consider two $c$-server systems with stochastically equivalent input, $T^{(2)} =_{st} T^{(1)}$, and failure rate ordered service time distributions, $r_2(x) \geq r_1(x)$, $x \geq 0$. Subsequently, Equation (25) holds.*

Now, we consider two $M/G/c$ queueing systems, denoted by $\Sigma^{(1)}$ and $\Sigma^{(M)}$, fed by (stochastically) identical Poisson process with rate $\lambda$. Let the first system $\Sigma^{(1)}$ have the service time distribution $F_1$. We refer below to the first system as being basic. In the second (contaminated) system $\Sigma^{(M)}$, we use service time distribution $F_M$ defined by Equation (3), with the same $F_1$ and some $F_2$ and $p \in (0,1)$. Let now $Q^{(1)}$ (with d.f. $F_{Q^{(1)}}$) be the steady-state queue size in the first system. Define similarly $Q^{(M)}$ and $F_{Q^{(M)}}$ for the system $\Sigma^{(M)}$. We are interested in studying the sensitivity of the uniform distance

$$\Delta(F_{Q^{(M)}}, F_{Q^{(1)}}) = \sup_{x \geq 0} |F_{Q^{(1)}}(x) - F_{Q^{(M)}}(x)|. \tag{27}$$

More formally, we study the effect of $\Delta(F_M, F_1)$ given by Equation (9), on the steady-state performance $\Delta(F_{Q^{(M)}}, F_{Q^{(1)}})$ defined in Equation (27), by varying the mixing coefficient $p$ and parameters defining the mixture components $F_1$ and $F_2$. However, since the distributions $F_{Q^{(M)}}$ and $F_{Q^{(1)}}$ are not available explicitly in general, we use simulation to obtain the corresponding estimates. As such, we study a combined effect of the service time distribution on the steady-state performance estimate.

The generic service time $S^{(M)}$ in the contaminated system $\Sigma^{(M)}$ has a two-component mixture d.f. $F_M$ and, thus, it follows from Equation (5) that the conditions of Lemma 1 are satisfied, since M, where $r_M$ is the failure rate of $S^{(M)}$. In particular, this means that the basic system $\Sigma^{(1)}$ is heavier loaded than the contaminated system $\Sigma^{(M)}$. It then follows from Lemma 1 and Equation (23) that the difference $F_{Q^{(1)}}(x) - F_{Q^{(M)}}(x)$ (see Equation (27)) is negative for all $x \geq 0$. In Section 5, we study the distance Equation (27) numerically.

## 4. Exact Steady-State Simulation by Regenerative Approach

In general, there are no closed form expressions for the steady-state distribution of the queue length and vector workload process in an $M/G/c$ system. Although a number of approximations exist [30–33], in general the accuracy of such methods is a point of discussion [34], especially when the service times distribution is heavy-tailed. Thus, to study the sensitivity we need to rely on simulation. A contribution of this work is that unlike classical discrete-event simulation (crude Monte-Carlo), which always has the so-called transient (warm-up) period during which an influence of initial conditions exists, we use the perfect simulation technique that allows exact sampling from the (unknown) steady-state distribution. In what follows, we rely on the regenerative approach designed for the $M/G/c$ system in the work [15] (although there are recently developed more sophisticated techniques based on backward coupling, for instant [35], which are valid for a more general $G/G/c$ system). Below, we outline the approach from [15].

This approach uses the so-called a Random Assignment (RA) system $M/G/c$ as a majorant for the original $M/G/c$ system. In the RA system, each new customer is assigned to arbitrary server randomly (that is with probability $1/c$). As a result, the remaining workload in server $j$ that customer $n$ meets, denoted by $V_{n,j}$, satisfies recursion

$$V_{n+1,j} = [V_{n,j} + I(U_n = j)S_n - T_n]^+, \quad j = 1, \ldots, c, \tag{28}$$

where iid random variables $\{U_n\}$ are uniformly distributed over $\{1, \ldots, c\}$, and $I(U_n = j) = 1$ means that customer $n$ is routed to server $j$. The RA system is indeed is a collection of $M/G/1$ systems, each with Poisson input with rate $\lambda/c$. As a result, in each, such a system the stationary workload, $D$, is distributed in accordance with the following version of the Pollaczek–Khintchine formula [15]

$$D = \sum_{i=1}^{L} S_i^{(e)}, \tag{29}$$

where $L$ has geometric distribution

$$P(L = k) = \left(\frac{\rho}{c}\right)^k \left(1 - \frac{\rho}{c}\right), \tag{30}$$

and $S^{(e)}$ has the so-called equilibrium (integrated tail) distribution,

$$\mathrm{P}\left(S^{(e)} > x\right) = \frac{1}{\mathrm{E}S}\int_x^\infty \overline{F}_S(t)dt, \quad (31)$$

where $F_S$ is the d.f. of original service time $S$. It is well-known that both workload process and queue size process in the RA system dominate the corresponding process in the original $M/G/c$ system [5,24,27]. Applying coupling, this dominance holds with probability (w.p.) 1. In particular, the regenerations of RA system (the instants when customers meet totally idle system) are also regeneration instants of the original system $M/G/c$. These results then are used to sample from the steady-state distribution of the RA system as follows:

1. sample the values $L_i, i = 1, \dots, c$ according to geometric distribution Equation (30);
2. sample $S_1^{(e)}, \dots, S_{L_i}^{(e)}, i = 1, \dots, c$ according to integrated tail distribution Equation (31); and,
3. construct the (stationary) components $D_i$ for $i = 1, \dots, c$, by formula Equation (29).

Subsequently, starting from the steady-state vector $V_1 = (D_1, \dots, D_c)$ containing iid components, the recursion Equation (28) is applied to each separate queue in the RA system until the event

$$V_{\tau_e} = (V_{\tau_e,1}, \dots, V_{\tau_e,c}) = \mathbf{0}$$

happens at the (arrival) instant of some customer $\tau_e$. Thus, $\tau_e$ is the length of equilibrium (steady-state) remaining regeneration period. Note that by construction, at each step of this recursion, the workload vector has steady-state distribution in the RA system. Omitting unnecessary details, the remaining steps of algorithm are as follows [15]:

1. sample stochastic copies $V^{(k)} = (V_1^{(k)}, V_2^{(k)}, \dots)$, $k = 1, 2, \dots$ of the sequence of workload vectors using recursion Equation (28); each sequence starts with $V_1^{(k)} = 0$ and lasts until the event $V_{\tau(k)}^{(k)} = 0$ happens at some instant $\tau(k)$; note that $\{\tau(k)\}$ are iid random variables distributed as a generic regeneration period $\tau$ of RA system;
2. repeat previous step until the event $\tau(j) > \tau_e$ happens in some sample $V^{(j)} = (V_1^{(j)}, V_2^{(j)}, \dots)$; and,
3. the value $V_{\tau_e}^{(j)}$ of the workload vector $V^{(j)}$ at instant $\tau_e$, has the target steady-state distribution of the workload in the original $M/G/c$ system.

We note that, although this approach allows to sample exactly from the steady-state distribution, the regeneration period in the dominated RA system can be very large in practice, and, thus, can lead to unacceptable long simulation. For further details on perfect sampling, see [15,35–37].

Now we explain how to sample from the equilibrium distribution of a two-component mixture. Let $\overline{F}_M$ be the tail of a two-component mixture Equation (3). Subsequently, it follows from Equation (3) that

$$\overline{F}_M^{(e)}(x) = \frac{1}{\mathrm{E}X_M}\int_x^\infty \overline{F}_M(u)du = \frac{p\mathrm{E}X_1}{\mathrm{E}X_M}\overline{F}_1^{(e)}(x) + \frac{(1-p)\mathrm{E}X_2}{\mathrm{E}X_M}\overline{F}_2^{(e)}(x). \quad (32)$$

It is clearly seen from Equation (32) that the equilibrium distribution of a mixture is itself a two-component mixture of equilibrium distributions of the components. Thus, to sample from the equilibrium distribution (32), we sample from $F_1^{(e)}$ w.p. $q = p\mathrm{E}X_1/\mathrm{E}X_M$, and sample from $F_2^{(e)}$ w.p. $1-q$. Finally, note that, as easy to see, if the original distributions are $Pareto(\alpha_i, x_0)$, then $\overline{F}_i^{(e)}$ are also $Pareto(\alpha_i - 1, x_0), i = 1, 2$ (also see [38]).

## 5. Simulation Results

As a sanity check of the perfect sampling $M/G/c$ model, we validate the algorithm via the $M/M/c$ system having input rate $\lambda = 7.5$, service rate $\mu = 1.5$, $c = 10$ servers, and $\rho = \lambda/\mu = 5$.

We run $N = 5000$ samples from steady-state process using perfect simulation and build the empirical queue size distribution vs. theoretical values that were obtained from Equation (22). We depict the results of validation on Figure 2. Note that the uniform distance between the empirical and theoretical distributions is 0.0091.

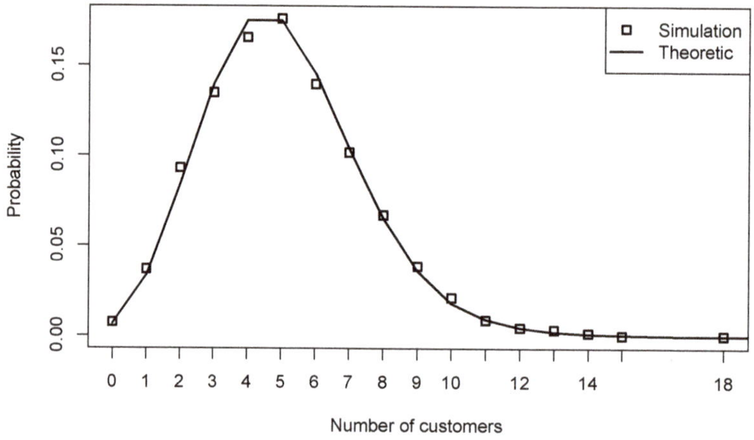


**Figure 2.** Theoretical distribution of the steady-state queue size in an $M/M/10$ system vs.empirical distribution ($N = 5000$ samples), with input rate $\lambda = 7.5$, service rate $\mu = 1.5$. The uniform distance between the theoretical and estimated queue size distributions equals $\Delta = 0.0091$.

### 5.1. Experiment 1: Hyperexponential Case

Now, we step away from the basic Markovian case, $M/M/c$, having service time distribution $F_1(x) = 1 - e^{-\mu_1 x}$, by introducing a contaminated system $M/G/c$ with generic service time, $S^{(M)}$, having two-state hyperexponential distribution, $H_2$, with $F_i(x) = 1 - e^{-\mu_i x}$, and mixing coefficient $p \in (0, 1)$. Note that such a case has computationally tractable solution, see [39]. However, we use the perfect sampling algorithm to check the accuracy of the sensitivity analysis. We fix

$$\mu_1 = 2, \; c = \lambda = 5, \; p = 0.7,$$

and vary $\mu_2$ over range $(2, 8]$ with step 0.4. We obtain the empirical queue size d.f., $\hat{F}_{Q^{(1)}}$, in the basic, and $\hat{F}_{Q^{(M)}}$ in the contaminated system, and construct Equation (27) for each combination of the parameters while using $N = 10{,}000$ samples from the steady-state distribution. The linear dependence of $\Delta(\hat{F}_{Q^{(M)}}, \hat{F}_{Q^{(1)}})$ on $\Delta(F_M, F_1)$ is clearly seen in Figure 3.

### 5.2. Experiment 2a: Pareto Case, Sensitivity to Mixing Parameter

In the following experiments, we use an $M/G/c$ system with $c = 4$, load $\rho = 0.5$, and $Pareto(\alpha_1, 1)$ service time d.f., with $\alpha_1 = 2.1$ as the basic system for comparison. The input rate of the basic system is taken as $\lambda = \rho c(\alpha_1 - 1)$ so as to guarantee the desired load $\rho = 0.5$. Note that, to the best of our knowledge, there is no explicit expression for the steady-state queue size in such a system, and thus simulation is used to obtain the corresponding estimates of the steady-state queue size d.f. To obtain such an estimate, $N = 10{,}000$ samples from the corresponding steady-state distribution are obtained by the perfect sampling technique described in Section 4.

In the first experiment, we study the steady-state queue size distribution sensitivity to the mixing parameter, $p$. The mixing coefficient is iterated over the discrete values $p = 0.95, 0.9, \ldots, 0.25$, and the empirical steady-state queue size d.f., $F_{Q_p^{(M)}}$, is constructed for the disturbed system with mixture service time d.f., $F_M$ given in Equation (3), consisting of $Pareto(\alpha_1, 1)$ and $Pareto(\alpha_2, 1)$ with mixing parameter $p$, where $\alpha_2 = 4.9$. The input rate $\lambda$ is fixed at the level $\lambda = 2.2$, so as to guarantee the load $\rho = 0.5$ in the basic system. Note that the parameter $p$ is varied in such a way that the mixing proportion of $Pareto(\alpha_2, 1)$ distribution becomes larger with smaller $p$, and dominates the $Pareto(\alpha_1, 1)$, for $p < 0.5$. Finally, we plot the values $\Delta(F_M, F_1)$ vs. $\Delta(\hat{F}_{Q_p^{(M)}}, \hat{F}_{Q^{(1)}})$ for the values $p$ given. The results are depicted in Figure 4. Note that the dependence of the distance is approximately linear in mixing probability, $p$.

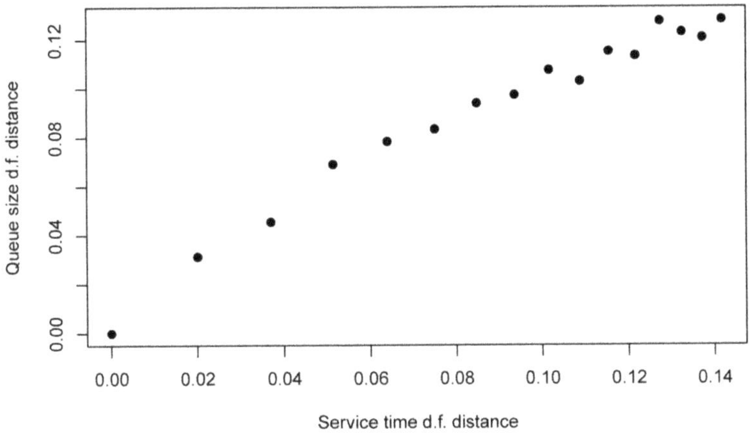


**Figure 3.** Distance, $\Delta(\hat{F}_{Q^{(M)}}, \hat{F}_{Q^{(1)}})$, between the empirical queue size d.f. in a basic $M/M/5$ system with input rate $\lambda = 5$, service rate $\mu_1 = 2$, compared to a contaminated $M/H_2/5$ system with input rate $\lambda = 5$ and hyperexponential service times being a mixture with $\mu_1 = 2$ and $\mu_2 = 2, 2.4, \ldots, 8$, $p = 0.7$, obtained from $N = 10{,}000$ samples, vs. service time d.f. distance, $\Delta(F_M, F_1)$.

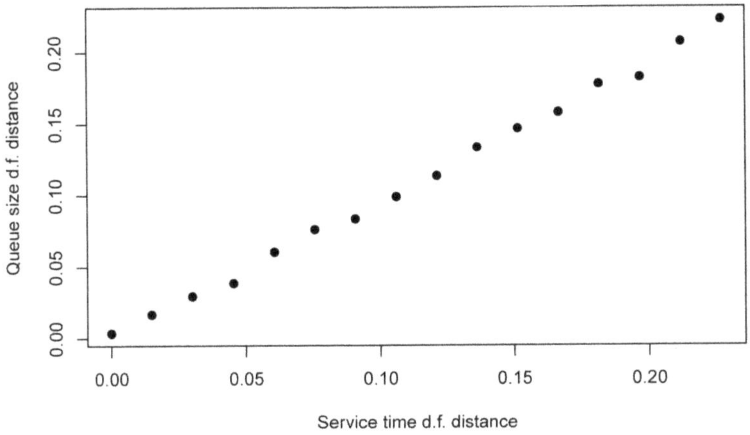


**Figure 4.** Distance between the empirical queue size d.f. in a basic $M/G/c$ system with $c = 4$, $\rho = 0.5$, $F_1$ being $Pareto(2.1, 1)$ service time d.f. and $\lambda = 2.2$, and system with a mixture, $F_M$ of $Pareto(2.1, 1)$ and $Pareto(4.9, 1)$ service time d.f. vs. the distance between $F_1$ and $F_M$, for varying $p = 1, 0.95, \ldots, 0.25$.

### *5.3. Experiment 2b: Pareto Case, Sensitivity to Contaminating Distribution*

In the following experiment, we study the sensitivity of the steady-state queue size distribution on the parameter $\alpha_2$ of the mixture. Now $p = 0.7$ is fixed, and $\alpha_2$ is iterated over the discrete set $\alpha_2 \in \{2.1, 2.3, \ldots, 4.9\}$, ceteris paribus. As in the previous experiment, we build the empirical steady-state queue size distribution of the basic system, $\hat{F}_{Q^{(1)}}$, by exact sampling from steady state using the method described in Section 4. We plot the values $\Delta(F_M, F_1)$ vs. $\Delta(\hat{F}_{Q^{(M)}_{\alpha_2}}, \hat{F}_{Q^{(1)}})$ for the given values of $\alpha_2$ as a parametric functions of $\alpha_2$. The results are depicted in Figure 5, where, unlike the previous scenarios, the nonlinear dependence on $\alpha_2$ is clear.

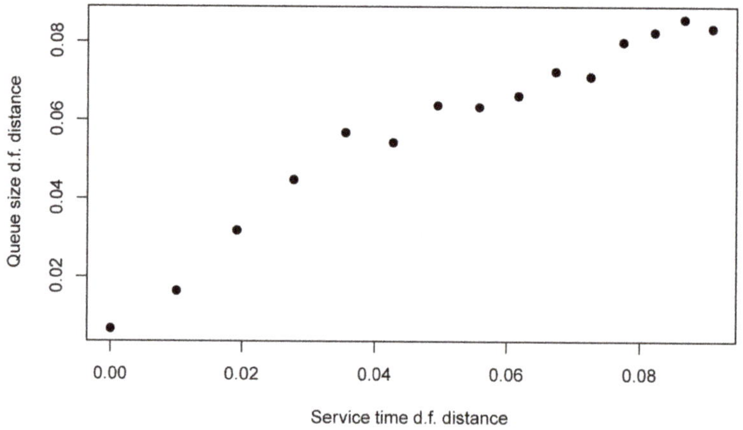


**Figure 5.** Distance between the empirical queue size d.f. in a basic $M/G/c$ system with $c = 4$, $\rho = 0.5$, $F_1$ being $Pareto(2.1, 1)$ service time d.f. and $\lambda = 2.2$, and system with a mixture, $F_M$ of $Pareto(2.1, 1)$ and $Pareto(\alpha_2, 1)$ service time d.f. vs. the distance between $F_1$ and $F_M$, for fixed $p = 0.7$ and varying $\alpha_2 = 2.1, 2.3, \ldots, 4.9$.

Note that the non-linear dependence of $\Delta(F_{Q^{(M)}}, F_{Q^{(1)}})$ on $\alpha_2$ may be caused by the non-linear dependence of the distance of service time distributions, $\Delta(F_M, F_1)$, on $\alpha_2$, see Figure 1. Moreover, the mean service time, $S^{(M)}$, also differs from mean service time of the basic system, which causes appropriate changes in the load, $\rho$, in the disturbed system, see Figure 6.

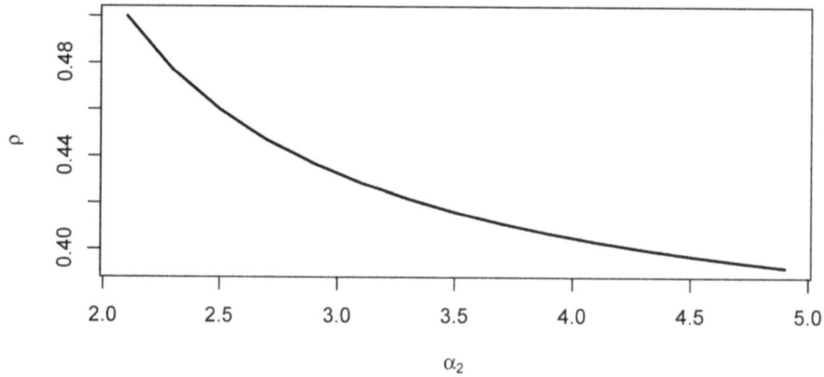


**Figure 6.** Dependence of the system load, $\rho$, on the parameter $\alpha_2 = 2.1, 2.3, \ldots, 4.9$ of the mixture distribution in an $M/G/c$ system with $c = 4$, $\lambda = 2.2$, mixture, $F_m$ of $Pareto(2.1, 1)$ and $Pareto(\alpha_2, 1)$ service time d.f. with mixing coefficient $p = 0.7$.

Using the results of Experiment 2b, we illustrate the stochastic monotonicity property Equation (25) for selected values of parameter $\alpha_2$. Figure 7 depicts the results.

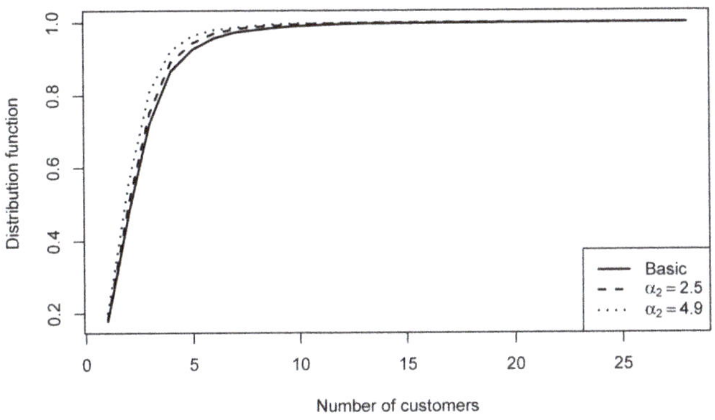


**Figure 7.** Stochastic monotonicity of the system output, in terms of steady-state queue size d.f., on the parameter $\alpha_2 = 2.1, 2.5, 4.9$ of the mixture distribution in an $M/G/c$ system with $c = 4$, $\lambda = 2.2$, mixture, $F_M$ of $Pareto(2.1, 1)$ and $Pareto(\alpha_2, 1)$ service time d.f. with mixing coefficient $p = 0.7$.

## 5.4. Experiment 2c: Pareto Case, Constant Load

In the final experiment, we study the joint effect of both the parent and the contaminating (Pareto) distributions. To do so, we change $\alpha_1 = 2.1, 2.5, \ldots, 4.9$, vary $\alpha_2 = \alpha_1, \alpha_1 + 0.4, \ldots, 4.9$. To mitigate the effect of changing load illustrated by Figure 6, we simultaneously change the parameter $\lambda$, so as to guarantee constant load $\rho = 0.5$ for all systems, keeping $p = 0.7, c = 4$ constant. The comparison is done to the system with the parent distribution of $\alpha_1 = 2.1$ of the service times. Each point is obtained then by $N = 10{,}000$ samples by the perfect sampling technique. Figure 8 depicts the results, where the color reflects the parent distribution parameter, $\alpha_1$, and size of a dot is proportional to $\alpha_2$. With increasing distance of the parent distribution from the contaminating distribution, the distance changes in a linear manner. Moreover, increased $\alpha_1$ changes the starting point (which in all lines corresponds to the parent distribution with parameter $\alpha_1$), and increasing $\alpha_2$ for fixed $\alpha_1$ increases the distance both in the input (for the mixture) and performance index (queue size distribution distance). Interestingly, for the lower line that corresponds to the fixed $\alpha_1 = 2.1$ and varying $\alpha_2$, there seems to be a slightly negative slope, which is likely to be the result of an increasing variance and, hence, decreasing accuracy. However, this effect might be interesting to study separately in the future.

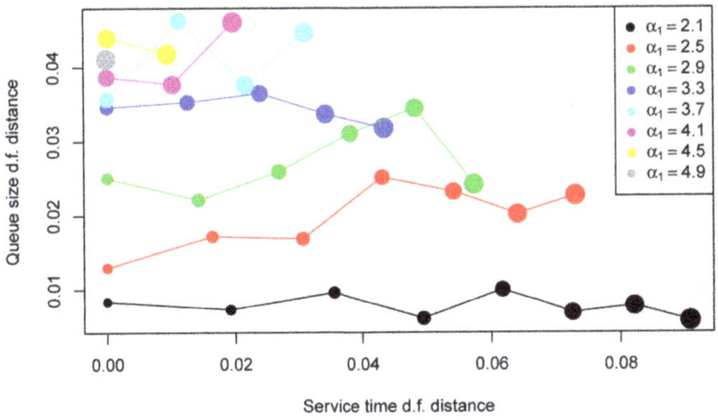


**Figure 8.** Distance between the empirical queue size d.f. in a basic $M/G/c$ system with $c = 4$, $F_1$ being $Pareto(\alpha_1, 1)$ service time d.f., and system with a mixture, $F_M$ of $Pareto(\alpha_1, 1)$ and $Pareto(\alpha_2, 1)$ service time d.f. vs. the distance between $F_1$ and $F_M$, for fixed $p = 0.7$, fixed $\rho = 0.5$, varying $\alpha_1 = 2.1, 2.4, \ldots, 4.9$ (color), varying $\alpha_2 = \alpha_1, \alpha_1 + 0.4, \ldots, 4.9$ (dot size), and varying $\lambda$, so as to fix the load, $\rho$.

Finally, we note that, to speedup the computation, we used parallel computation of the uniform distance for various system configurations using the resources of the High-Performance Datacenter of Karelian Research Centre of Russian Academy of Sciences.

## 6. Conclusions and Discussion

In this paper, the effect of the service time distribution perturbation on the steady-state performance measures of a multiserver queueing system is studied. The explicit form for the sensitivity measure (Kolmogorov-Smirnov distance) between the service time distribution functions was obtained, and the performance estimates were obtained by the regenerative perfect simulation technique. The simulation results outline the qualitative nature of the sensitivity, which is, in most cases, linear (possibly after appropriate scaling of the input rate to guarantee the constant load).

The approach to sensitivity analysis that is presented in this paper can be applied to more sophisticated, and more practically oriented systems, such as the simultaneous service multiserver system [40], which would result, though, in an increased dimension of the system state. However, we note that the steady-state exact sampling by regenerative simulation has several serious drawbacks. First, the average working time of the algorithm may be infinite [36], e.g., in a system with large number of servers (which indeed depends on the regenerative cycle length). This problem can be solved either by the coupling-from-the-past technique [35] (which, although, is rather technically tricky), or by non traditional regenerative techniques, such as the artificial regeneration [41] or regenerative envelopes [40]. Finally, the study may be extended to larger classes of service time distributions. At that, we leave these as opportunities for future research.

**Author Contributions:** Conceptualization, writing—original draft preparation, I.P.; simulation and visualization, A.R.; writing—review and editing, M.P.; supervision, project administration—E.M. All authors have read and agreed to the published version of the manuscript.

**Funding:** The research is supported by Russian Foundation for Basic Research, projects No. 19-57-45022, 19-07-00303, 18-07-00156, 18-07-00147.

**Conflicts of Interest:** The authors declare no conflict of interest.

## Abbreviations

The following abbreviations are used in this manuscript:

| | |
|---|---|
| d.f. | distribution function |
| w.p. | with probability |

## References

1. Al-Hussaini, E.K.; Sultan, K.S. Reliability and hazard based on finite mixture models. In *Handbook of Statistics*; Elsevier: Amsterdam, The Netherlands, 2001; Volume 20, pp. 139–183. [CrossRef]
2. Shaked, M.; Spizzichino, F. Mixtures and monotonicity of failure rate functions. In *Handbook of Statistics*; Elsevier: Amsterdam, The Netherlands, 2001; Volume 20, pp. 185–198. [CrossRef]
3. Sevast'yanov, B.A. An Ergodic Theorem for Markov Processes and Its Application to Telephone Systems with Refusals. *Theory Probab. Its Appl.* **1957**, *2*, 104–112. [CrossRef]
4. Kalashnikov, V.V. Stability Analysis of in Queueing Problems by a Method of Trial Functions. *Theory Probab. Its Appl.* **1977**, *22*, 86–103. [CrossRef]
5. Müller, A.; Stoyan, D. *Comparison Methods for Stochastic Models and Risks*; Wiley Series in Probability and Statistics; Wiley: Hoboken, NJ, USA, 2002.
6. Zolotarev, V.M. On the stochastic continuity of the queuing systems of type G | G | 1. *Theory Probab. Its Appl.* **1977**, *21*, 250–269. [CrossRef]
7. Zolotarev, V.M. Quantitative estimates for the continuity property of queueing systems of type G | G | ∞. *Theory Probab. Its Appl.* **1978**, *22*, 679–691. [CrossRef]
8. Zolotarev, V.M. Qualitative Estimates in Problems of Continuity of Queuing Systems. *Theory Probab. Its Appl.* **1975**, *20*, 211–213. [CrossRef]

9. Batrakova, D.; Korolev, V.; Shorgin, S. A new method for the probabilistic and statistical analysis of information flows in telecommunication networks. *Inform. Appl.* **2007**, *1*, 40–53.
10. Daley, D.J. Queueing Output Processes. *Adv. Appl. Probab.* **1976**, *8*, 395. [CrossRef]
11. Daley, D.J. Revisiting queueing output processes: A point process viewpoint. *Queueing Syst.* **2011**, *68*, 395–405. [CrossRef]
12. Korolev, V.Y.E.; Krylov, V.A.; Kuz'min, V.Y.E. Stability of finite mixtures of generalized Gamma-distributions with respect to disturbance of parameters. *Inform. Appl.* **2011**, *5*, 31–38.
13. Kalashnikov, V.V.; Tsitsiashvili, G.S. Stability analysis of queueing systems. *J. Sov. Math.* **1981**, *17*, 2238–2255. [CrossRef]
14. McLachlan, G.J.; Lee, S.X.; Rathnayake, S.I. Finite Mixture Models. *Annu. Rev. Stat. Its Appl.* **2019**, *6*, 355–378. [CrossRef]
15. Sigman, K. Exact simulation of the stationary distribution of the FIFO M/G/c queue: the general case for $\rho < c$. *Queueing Syst.* **2012**, *70*, 37–43. [CrossRef]
16. Feitelson, D.G. *Workload Modeling for Computer Dystems Performance Evaluation*; Cambridge University Press: Cambridge, UK, 2014.
17. Morozov, E.; Peshkova, I.; Rumyantsev, A. On Failure Rate Comparison of Finite Multiserver Systems. In *Distributed Computer and Communication Networks*; Vishnevskiy, V.M., Samouylov, K.E., Kozyrev, D.V., Eds.; Springer International Publishing: Cham, Germany, 2019; Volume 11965, pp. 419–431._32. [CrossRef]
18. Marshall, A.W.; Olkin, I. *Life Distributions: Structure of Nonparametric, Semiparametric, and Parametric Families*; Springer Series in Statistics; Springer: New York, NY, USA; London, UK, 2007.
19. Goldstein, M. Contamination Distributions. In *The Annals of Statistics*; Institute of Mathematical Statistics: Beachwood, OH, USA, 1982; Volume 10, pp. 174–183.
20. Block, H.W. The Failure Rates of Mixtures. In *Advances in Distribution Theory, Order Statistics, and Inference*; Balakrishnan, N., Sarabia, J.M., Castillo, E., Eds.; Birkhäuser Boston: Boston, MA, USA, 2006; pp. 267–277._17. [CrossRef]
21. Barlow, R.E.; Proschan, F. *Mathematical Theory of Reliability*; Classics in Applied Mathematics; Society for Industrial and Applied Mathematics: Philadelphia, PA, USA, 1996. [CrossRef]
22. Goldie, C.M.; Klüppelberg, C. Subexponential Distributions. In *A Practical Guide to Heavy Tails: Statistical Techniques and Applications*; Birkhauser Boston Inc.: Cambridge, MA, USA, 1998; pp. 435–459.
23. Shaked, M. Bounds on the Distance of a Mixture from Its Parent Distribution. *J. Appl. Probab.* **1981**, *18*, 853–863. [CrossRef]
24. Asmussen, S. *Applied Probability and Queues*; Springer: New York, NY, USA, 2003.
25. Kiefer, J.D.; Wolfowitz, J. On the theory of queues with many servers. *Trans. Am. Math. Soc.* **1955**, *78*, 1–18. [CrossRef]
26. Kleinrock, L. *Theory, Volume 1, Queueing Systems*; Wiley-Interscience: Hoboken, NJ, USA, 1975.
27. Whitt, W. Comparing counting processes and queues. *Adv. Appl. Probab.* **1981**, *13*, 207–220. [CrossRef]
28. Thorrison, H. *Coupling, Stationarity, and Regeneration*; Springer: New York, NY, USA, 2000.
29. Shaked, M.; Shanthikumar, J.G. *Stochastic Orders*; Springer Series in Statistics; Springer: New York, NY, USA, 2007.
30. Whitt, W. Approximations for the GI/G/M Queue. *Prod. Oper. Manag.* **1993**, *2*, 114–161. [CrossRef]
31. Van Hoorn, M.; Tijms, H. Approximations for the waiting time distribution of the M/G/c queue. *Perform. Eval.* **1982**, *2*, 22–28. [CrossRef]
32. Ma, B.N.W.; Mark, J.W. Approximation of the Mean Queue Length of an M/G/c Queueing System. *Oper. Res.* **1995**, *43*, 158–165. [CrossRef]
33. Kimura, T. Approximations for multi-server queues: system interpolations. *Queueing Syst.* **1994**, *17*, 347–382. [CrossRef]
34. Gupta, V.; Harchol-Balter, M.; Dai, J.G.; Zwart, B. On the inapproximability of M/G/K: Why two moments of job size distribution are not enough. *Queueing Syst.* **2010**, *64*, 5–48. [CrossRef]
35. Blanchet, J.; Pei, Y.; Sigman, K. Exact sampling for some multi-dimensional queueing models with renewal input. *Adv. Appl. Probab.* **2019**, *51*, 1179–1208. [CrossRef]
36. Xiong, Y. Perfect and Nearly Perfect Sampling of Work-Conserving Queues. Ph.D. Thesis, The School of Graduate and Postdoctoral Studies, The University of Western Ontario, London, ON, Canada, 2015.
37. Blanchet, J.; Dong, J.; Pei, Y. Perfect Sampling of GI/GI/c Queues. *arXiv* **2015**, arXiv: 1508.02262.

38. Nair, N.U.; Preeth, M. On some properties of equilibrium distributions of order n. *Stat. Methods Appl.* **2009**, *18*, 453–464. [CrossRef]
39. de Smit, J.H. A numerical solution for the multi-server queue with hyper-exponential service times. *Oper. Res. Lett.* **1983**, *2*, 217–224. [CrossRef]
40. Morozov, E.; Peshkova, I.; Rumyantsev, A. On Regenerative Envelopes for Cluster Model Simulation. In Proceedings of the Distributed Computer and Communication Networks: 19th International Conference, DCCN 2016, Moscow, Russia, 21–25 November 2016; Vishnevskiy, V.M., Samouylov, K.E., Kozyrev, D.V., Eds.; Springer International Publishing: Cham, Germany, 2016; pp. 222–230._20. [CrossRef]
41. Rumyantsev, A.; Peshkova, I. Artificial Regeneration Based Regenerative Estimation of Multiserver System with Multiple Vacations Policy. In *Information Technologies and Mathematical Modelling. Queueing Theory and Applications*; Dudin, A., Nazarov, A., Moiseev, A., Eds; Springer International Publishing: Berlin/Heidelberg, Germany, 2019; Volume 1109, pp. 38–50._4. [CrossRef]

mathematics

MDPI

Article

# Statistical Indicators of the Scientific Publications Importance: A Stochastic Model and Critical Look †

**Lev B. Klebanov [1,*], Yulia V. Kuvaeva [2] and Zeev E. Volkovich [3]**

1 Department of Probability and Mathematical Statistics, Charles University, 116 36 Prague, Czech Republic
2 Department of Finance, Money Circulation and Credit, Ural State University of Economics, 620144 Yekaterinburg, Russia; ykuvaeva1974@mail.ru
3 Software Engineering Department, ORT Braude College, Karmiel 21982, Israel; vlvolkov@braude.ac.il
* Correspondence: lev.klebanov@mff.cuni.cz

† We dedicate this work to the blessed memory of Vladimir Mikhailovich Zolotarev, to whom we owe our interest in heavy-tailed distributions.

Received: 31 March 2020; Accepted: 25 April 2020; Published: 3 May 2020

**Abstract:** A model of scientific citation distribution is given. We apply it to understand the role of the Hirsch index as an indicator of scientific publication importance in Mathematics and some related fields. The proposed model is based on a generalization of such well-known distributions as geometric and Sibuya laws. Real data analysis of the Hirsch index and corresponding citation numbers is given.

**Keywords:** citation distribution; Hirsch index; geometric distribution; Sibuya distribution

## 1. Introduction

In theory, a rather large number of indexes are proposed, which supposedly measure the significance of the scientific publications of an author. Among the most popular of them should be noted:

(i1) the total number of citations of a particular author [1–3];
(i2) Hirsch index of the author [4] (see also [5]).

It is these two indexes that we consider in the proposed work.

The definition of the numerical value of the index (i1) is clear from its name.

Recall the definition of the Hirsch index (see [4]). The Hirsch Index $h$ is the number of articles that have been cited at least $h$ times each. This index was introduced in [4], where its properties were explained. In our opinion, these do not correspond to the index purpose. However, we dwell on the description of both the positive and negative sides of the Hirsch index after constructing citation models for scientific articles. One of them has already been stated by us in preprint [6].

## 2. Citation Model Construction

We now turn to the construction of the author's citation model. It will be considered as a composite of two models. The first of it describes the process of publishing an article by one author which will be cited, and the second describes the process of citing such an article.

Let us make some assumptions, which we discuss later.

**Assumption 1.** *Let the probability of rejection or non citing of the manuscript be $q$ and the decisions on publication of different manuscripts are taken independently.*

Then it is clear that the probability that the scientist will have exactly $k$ cited papers equals $q(1-q)^k$, $k = 0, 1, \ldots$. In other words, the number of publications of a scientist has a geometric distribution with parameter $q$. This distribution supposes that the number of an author publications may be arbitrarily large. However, $(1-q)^k$ tends to zero rather fast as $k \to \infty$ and, therefore, the mean value of the number of publications is not too large. The generating function of this distribution has the form

$$Q(z) = \frac{q}{1-(1-q)z}. \tag{1}$$

Of course, here we assume that all the journals to which the author sends manuscripts have the same review system, i.e., all of them accept the manuscripts of this author with the same probability $1-q$. More realistic is the situation with a random parameter $q$:

$$Q(z) = \int_0^1 \frac{q}{1-(1-q)z} d\,\Xi(q),$$

where $\Xi$ is a probability distribution on $[0,1]$ interval and then $\mathbb{P}\{X = k\} = \mathbb{E}(q(1-q)^n)$.

Let us go back to (1). How large may be the time spent by a scientist to publish a corresponding number of papers? Of course, this time is a random variable $T$ and we are interested in its distribution. The usual assumption on the working time is its exponential distribution with parameter $\lambda = \mathbb{E}T$ and the Laplace transform $\varphi(t) = 1/(1+\lambda t)$. Suppose that times needed for the publication of $j$-th paper is $T_j$, and $T_1, T_2, \ldots$ are independent and identically distributed as $T$ random variables. Then the time needed for all publications has the Laplace transform

$$\sum_{k=1}^{\infty} \varphi^k(t) q(1-q)^{k-1} = \frac{1}{1+\lambda t/q},$$

i.e., it has exponential distribution with the parameter $\lambda/q$.

It is natural to assume that each cited publication will produce some number of citations. Of course, the likelihood that the article will be quoted again depends on the number of previous citations.

**Assumption 2.** *Assume the probability that an article having $k-1$ ($k \geq 1$) citations will not have new quotes equalling $p/k^\gamma$ where $p$ is the probability that the article will not be quoted for the first time. The parameter $\gamma$ is responsible for the speed of convergence of the rejection probability to zero.*

Consequently, the likelihood that the article will be quoted exactly k times equals $p/k^\gamma \prod_{j=1}^{k-1}(1 - p/j^\gamma)$. For the case of $\gamma = 1$, the generating probability function for the number of citations of this article is $1-(1-z)^p$. The corresponding distribution function is named after Sibuya [7]. Below we consider the case of arbitrary positive $\gamma$. The corresponding study has general mathematical interest. Therefore, we provide it in a number of sections below.

## 3. Distribution of Citation Number of a Paper

Let us consider an ordered sequence of experiments $\{\mathcal{E}_n;\ n = 1, 2, \ldots\}$, where an event $A$ may appear in each of the experiments with the probability $p_n$. Define a random variable $X$ as the number of the first experiment in which $A$ appears. We suppose that $X$ is an improper random variable in the sense that it may take infinite value (that is, the event $A$ will never appear). For the case $\mathbb{P}\{X = \infty\} = 0$ we say that $X$ is a proper random variable. It is clear that, since we define any product from 1 to 0 to be 1,

$$\mathbb{P}\{X = n\} = p_n \cdot \prod_{k=1}^{n-1} (1 - p_k) \tag{2}$$

and

$$\mathbb{P}\{X=\infty\} = \lim_{n\to\infty} \prod_{k=1}^{n-1} (1-p_k).$$

Particular cases are:

1. The probabilities $p_n = p$ are constant. So (2) is

$$\mathbb{P}\{X=n\} = p \cdot (1-p)^{n-1}, \quad \mathbb{P}\{X=\infty\} = 0 \tag{3}$$

corresponding to the classical geometric distribution. Its tail is

$$\mathbb{P}\{X \geq n\} = (1-p)^{n-1}, \quad m = 1, 2, \ldots$$

Clearly, the tail and probabilities (3) decrease exponentially fast as $n$ tends to infinity.

2. The probabilities are given by $p_n = p/n$, where p is a number from the interval $(0,1)$. Equation (3) is transformed to

$$\mathbb{P}\{X=n\} = \frac{p}{n} \cdot \prod_{k=1}^{n-1} (1-\frac{p}{k}). \tag{4}$$

According to (4) X is a proper random variable and has, in this case, the Sibuya distribution with parameter $p \in (0,1)$ with the following tail

$$\mathbb{P}\{X \geq n\} = \frac{\Gamma(n-p)}{\Gamma(n) \cdot \Gamma(1-p)} \sim \frac{1}{\Gamma(1-p) \cdot n^p}$$

having heavy power asymptotic for $n \to \infty$. Such the distribution does not have a finite mean value. It is not difficult to see that

$$\mathbb{P}\{X=n\} \sim p/(n^{p+1} \cdot \Gamma(1-p)), \quad n \to \infty.$$

The presented distributions can be respected as a kind of "extreme points" from the perspective of the tail behavior for proper random variable X. Hence, it is natural to study roughly speaking the cases "happening between them"; namely to consider, for example, the situations when $p_n = p/n^\gamma$, with $p \in (0,1)$ and $\gamma > 0$. As it was mentioned above, the parameter $\gamma$ is responsible for the speed of convergence of the rejection probability to zero.

## 4. Main Result on Citation Number Distribution

The research subject is in the asymptotic behavior of the probabilities (2) for $p_n = p/n^\gamma$ with $\gamma \geq 0$. Additionally, to the discussed earlier values of $\gamma = 0$ or $\gamma = 1$, we distinguish the following two cases:

(A) $0 < \gamma < 1$;
(B) $\gamma > 1$.

Let us consider the case (A). We have

$$\mathbb{P}\{X=n\} = \frac{p}{n^\gamma} \cdot \prod_{k=1}^{n-1} (1-\frac{p}{k^\gamma}). \tag{5}$$

Consider the product from right-hand-side of (5) in more details.

$$\prod_{k=1}^{n-1} (1-\frac{p}{k^\gamma}) = \exp\Big\{\sum_{k=1}^{n-1} \log(1-p/k^\gamma)\Big\} = \exp\Big\{-\sum_{k=1}^{n-1}\sum_{j=1}^{\infty} \frac{p^j}{jk^{\gamma j}}\Big\}$$

$$= \exp\Big\{-\sum_{j=1}^{\infty}\frac{p^j}{j}\sum_{k=1}^{n-1}\frac{1}{k^{\gamma j}}\Big\} = \exp\Big\{-\sum_{j=1}^{[1/\gamma]+1}\frac{p^j}{j}\sum_{k=1}^{n-1}\frac{1}{k^{\gamma j}}\Big\}\exp\Big\{-\sum_{[1/\gamma]+1}^{\infty}\frac{p^j}{j}\sum_{k=1}^{n-1}\frac{1}{k^{\gamma j}}\Big\}. \tag{6}$$

Here $[1/\gamma]$ is an integer part of $1/\gamma$. It is not difficult to see that

$$\exp\{-\sum_{[1/\gamma]+1}^{\infty}(p^j/j)\sum_{k=1}^{n-1}k^{-\gamma j}\}$$

has a finite positive limit as $n \to \infty$. This limit may depend on $p$ and $\gamma$. Let us denote it by $C_1 = C_1(\gamma, p)$. Therefore,

$$\prod_{k=1}^{n-1}(1-\frac{p}{k^{\gamma}}) \sim C_1 \exp\Big\{-\sum_{j=1}^{[1/\gamma]+1}\frac{p^j}{j}\sum_{k=1}^{n-1}\frac{1}{k^{\gamma j}}\Big\} \quad \text{as} \quad n \to \infty. \tag{7}$$

Relations (5) and (7) give us

$$\mathbb{P}\{X=n\} \sim C_1 \cdot \frac{p}{n^{\gamma}} \cdot \exp\Big\{-\sum_{j=1}^{[1/\gamma]+1}\frac{p^j}{j}\sum_{k=1}^{n-1}\frac{1}{k^{\gamma j}}\Big\} \quad \text{as} \quad n \to \infty. \tag{8}$$

For $0 < \gamma j < 1$ the following asymptotic representation is known

$$\sum_{k=1}^{n-1}\frac{1}{k^{\gamma j}} = \frac{n^{1-\gamma j}}{1-\gamma j} + \zeta(\gamma j) + o(1) \quad \text{as} \quad n \to \infty, \tag{9}$$

where $\zeta(u)$ is Riemann zeta function. Further considerations depend on properties of the number $\gamma$.

(i) Suppose that $1/\gamma$ is not integer. Then $\gamma \cdot [1/\gamma] < 1$ and

$$\sum_{j=1}^{[1/\gamma]+1}\frac{p^j}{j}\sum_{k=1}^{n-1}\frac{1}{k^{\gamma j}} = \sum_{j=1}^{[1/\gamma]}\frac{n^{1-\gamma j}}{1-\gamma j}\frac{p^j}{j} + \sum_{j=1}^{[1/\gamma]}\zeta(\gamma j)\frac{p^j}{j} + \frac{p^{[1/\gamma]+1}}{[1/\gamma]+1}\sum_{k=1}^{n-1}\frac{1}{k^{\gamma([1/\gamma]+1)}} + o(1). \tag{10}$$

However, $\gamma([1/\gamma]+1) > 1$ and, therefore,

$$\lim_{n\to\infty}\sum_{k=1}^{n-1}\frac{1}{k^{\gamma([1/\gamma]+1)}} = \sum_{k=1}^{\infty}\frac{1}{k^{\gamma([1/\gamma]+1)}} < \infty.$$

From this and (10) it follows

$$\mathbb{P}\{X=n\} \sim C_2 \cdot \frac{p}{n^{\gamma}} \cdot \exp\Big\{\sum_{j=1}^{[1/\gamma]}\frac{n^{1-\gamma j}}{1-\gamma j}\cdot\frac{p^j}{j}\Big\}, \tag{11}$$

where $C_2$ depends on $p$ and $\gamma$ only.

(ii) Suppose that $1/\gamma$ is positive integer. Then $\gamma[1/\gamma] = 1$ and

$$\sum_{j=1}^{[1/\gamma]+1}\frac{p^j}{j}\sum_{k=1}^{n-1}\frac{1}{k^{\gamma j}} = \sum_{j=1}^{[1/\gamma]-1}\frac{n^{1-\gamma j}}{1-\gamma j}\frac{p^j}{j} + \sum_{j=1}^{[1/\gamma]-1}\zeta(\gamma j)\frac{p^j}{j} \tag{12}$$

$$+\frac{p^{[1/\gamma]}}{[1/\gamma]}\sum_{k=1}^{n-1}\frac{1}{k} + \frac{p^{[1/\gamma]+1}}{[1/\gamma]+1}\sum_{k=1}^{n-1}\frac{1}{k^2}.$$

It is known that

$$\lim_{n\to\infty}\sum_{k=1}^{n-1}\frac{1}{k^2}=\sum_{k=1}^{\infty}\frac{1}{k^2}<\infty$$

and

$$\sum_{k=1}^{n-1}\frac{1}{k}=\log(n)+\gamma_e+o(1),$$

where $\gamma_e$ is Euler's constant. Therefore,

$$\mathbb{P}\{X=n\}\sim C_3\cdot\frac{p}{n^{\gamma+p^{[1/\gamma]}/[1/\gamma]}}\cdot\exp\Big\{\sum_{j=1}^{[1/\gamma]-1}\frac{n^{1-\gamma j}}{1-\gamma j}\cdot\frac{p^j}{j}\Big\}\text{ as }n\to\infty. \tag{13}$$

Now we see that the asymptotic behavior of the probability $\mathbb{P}\{X=n\}$ in the case A) is given by (11) and (13). From the relations (11) and (13) it follows

$$\mathbb{P}\{X=\infty\}=\lim_{n\to\infty}\prod_{k=1}^{n-1}(1-p/k^\gamma)=0,$$

so that $X$ is a proper random variable.

Denote by

$$b_m=\prod_{k=1}^{m-1}(1-p/k^\gamma).$$

For the distribution tail $T_m$ we have

$$T_m=\sum_{n=m}^{\infty}\mathbb{P}\{X=n\}=(b_m-b_{m+1})+\ldots+(b_s-b_{s+1})+\ldots=b_m.$$

Particularly,

$$\sum_{n=1}^{\infty}\mathbb{P}\{X=n\}=1.$$

If $1/\gamma$ is not a positive integer, then

$$T_m=\prod_{k=1}^{m-1}(1-p/k^\gamma)\sim C_4\cdot\exp\Big\{\sum_{j=1}^{[1/\gamma]}\frac{n^{1-\gamma j}}{1-\gamma j}\cdot\frac{p^j}{j}\Big\},\quad\text{as}\quad n\to\infty, \tag{14}$$

where $C_4$ depends on $p$ and $\gamma$. Similarly, for the case of integer $1/\gamma$,

$$T_m\sim C_5\cdot\frac{p}{n^{p^{[1/\gamma]}/[1/\gamma]}}\cdot\exp\Big\{\sum_{j=1}^{[1/\gamma]-1}\frac{n^{1-\gamma j}}{1-\gamma j}\cdot\frac{p^j}{j}\Big\}\text{ as }n\to\infty. \tag{15}$$

Let us consider the case (B). We have

$$\mathbb{P}\{X=n\}=\frac{p}{n^\gamma}\cdot\prod_{k=1}^{n-1}(1-\frac{p}{k^\gamma}), \tag{16}$$

where $\gamma>1$. Transform the product in the right-hand-side:

$$b_n=\prod_{k=1}^{n-1}(1-\frac{p}{k^\gamma})=\exp\Big\{\sum_{k=1}^{n-1}\log(1-p/k^\gamma)\Big\}$$

$$= \exp\Big\{-\sum_{j=1}^{\infty}\sum_{k=1}^{n-1} p^j/(jk^{\gamma j})\Big\} = \exp\Big\{-\sum_{k=1}^{n-1}\sum_{j=1}^{\infty} p^j/(jk^{\gamma j})\Big\}$$

$$= \exp\Big\{-\sum_{k=1}^{n-1} p/(k^\gamma - p)\Big\} \overset{[}{n}\!\to \infty]\!\longrightarrow \exp\Big\{-\sum_{k=1}^{\infty} p/(k^\gamma - p)\Big\}.$$

The series under an exponential sign converges because $\gamma > 1$. From latest relation we see that

$$\mathbb{P}\{X = \infty\} = \exp\Big\{-\sum_{k=1}^{\infty} p/(k^\gamma - p)\Big\} > 0, \tag{17}$$

and $X$ is an improper random variable.

Therefore, for conditional probabilities we have

$$\mathbb{P}\{X = \eta | X < \infty\} \sim C_6 \frac{p}{n^\gamma} \quad \text{as} \quad n \to \infty, \tag{18}$$

where $C_6$ depends on $p$ and $\gamma$ only.

Summarizing, we obtain the following theorem

**Theorem 1.** *For the considered experiment scheme with probabilities given in (5) the following statements are true:*

- *If $\gamma = 0$ then $\mathbb{P}\{X = n\} = p(1-p)^{n-1}$, $n = 1, 2, \ldots$.*
- *If $0 < \gamma < 1$ and $1/\gamma$ is not a positive integer then*

$$\mathbb{P}\{X = n\} \sim C_2 \cdot \frac{p}{n^\gamma} \cdot \exp\Big\{-\sum_{j=1}^{[1/\gamma]} \frac{n^{1-\gamma j}}{1-\gamma j} \cdot \frac{p^j}{j}\Big\} \text{ as } n \to \infty. \tag{19}$$

*If $0 < \gamma < 1$ and $1/\gamma$ is a positive integer then*

$$\mathbb{P}\{X = n\} \sim C_3 \cdot \frac{p}{n^{\gamma + p^{[1/\gamma]}/[1/\gamma]}} \cdot \exp\Big\{-\sum_{j=1}^{[1/\gamma]-1} \frac{n^{1-\gamma j}}{1-\gamma j} \cdot \frac{p^j}{j}\Big\} \text{ as } n \to \infty. \tag{20}$$

- *If $\gamma = 1$ then*

$$\mathbb{P}\{X = n\} \sim p/(n^{p+1}\Gamma(1-p)), \quad n \to \infty. \tag{21}$$

- *If $\gamma > 1$ then*

$$\mathbb{P}\{X = n | X < \infty\} \sim C_4 \frac{p}{n^\gamma} \quad \text{as} \quad n \to \infty, \tag{22}$$

*and*

$$\mathbb{P}\{X = \infty\} = \exp\Big\{-\sum_{k=1}^{\infty} p/(k^\gamma - p)\Big\} > 0, \tag{23}$$

*All $C, C_1 - C_6$ depend on parameters $p$ and $\gamma$ only.*

One of the reviewers of the first version of the paper advised us to study the form of the constants for some particular cases. We are very grateful him for the advice. Below we consider the case $\gamma \in (1/2, 1)$. In this case $[1/\gamma] = 1$ so that the sum under exponential sign in (19) contains only one summand. The calculations similar to give above leads to the following expression

$$\mathbb{P}\{X = n\} = \frac{p}{n^\gamma} \exp\Big\{-\frac{p}{1-\gamma} n^{1-\gamma} - \sum_{k=1}^{\infty} \frac{p^k}{k} \zeta(k\gamma) + o(1)\Big\}.$$

In other words, the constant $C_2$ has form

$$C_2 = \exp\Big\{ - \sum_{k=1}^{\infty} \frac{p^k}{k} \zeta(k\gamma) \Big\} > 0.$$

However, precise calculation of all other constant is rather difficult. We do not these constants for the aims of this paper and omit any other calculations of constants.

## 5. Comments

Theorem 1 shows that for $0 \leq \gamma < 1$, the tail of the corresponding distribution is not heavy. Namely, the distribution has finite moments of all positive orders. However, the tail becomes heavier with growing $\gamma \in [0, 1)$. In the case of $\gamma \in [0, 1]$ the distribution is unimodal with mode equal to 1. For the values $\gamma \in [1, \infty)$, the distribution has a power-type tail, which is heavier than the ones occurring for $\gamma \in [0, 1)$. In the case $\gamma \in [1, 2)$ the conditional distribution under condition $X < \infty$ does not have the finite mean. However, for growing values of $\gamma \in [1, \infty)$ the tails of conditional distributions look to be less heavy. In the case of $\gamma \in [1, \infty)$ the conditional distribution has mode at 1.

## 6. The Case of Growing $p_n$

Above, we considered the case of the probability of event $A$ decreasing with increasing iment number. For completeness, consider the case of an increase of this probability.

Namely, suppose that in (1) $p_n = 1 - q/n^\gamma$ for $q \in (0, 1)$ and $\gamma > 0$. Then

$$\mathbb{P}\{X = n\} = (1 - q/n^\gamma) \prod_{k=1}^{n-1} \frac{q}{k^\gamma} = \frac{q^{n-1}}{((n-1)!)^\gamma} - \frac{q^n}{(n!)^\gamma}. \tag{24}$$

It is clear that $\mathbb{P}\{X = \infty\} = 0$, and the tail of the distribution

$$T_m = \frac{q^{m-1}}{(\Gamma(m))^\gamma}$$

is a quickly decreasing function of $m$. Of course, distribution of $X$ has finite moments of all orders and it may have a mode not only at 1.

## 7. Back to the Distribution of Citation Number of One Author

We suppose now that the distribution of citation number of one paper has the form (5):

$$\mathbb{P}\{X = n\} = \frac{p}{n^\gamma} \cdot \prod_{k=1}^{n-1} (1 - \frac{p}{k^\gamma}), \quad n = 1, 2, \ldots$$

with $\gamma > 0$. Corresponding probability generating function is

$$\mathcal{P}(z) = \sum_{n=1}^{\infty} z^n \mathbb{P}\{X = n\}. \tag{25}$$

As was mentioned above, the number of cited paper is distributed according to geometric law with probability generating function (1):

$$\mathcal{Q}(z) = \frac{q}{1 - (1-q)z}, \quad q \in (0, 1).$$

The probability generating function of citation number of one author equals to the composition of $\mathcal{P}$ and $\mathcal{Q}$, i.e., it is $\mathcal{P}(\mathcal{Q}(z))$. It is clear that the tail of corresponding distribution is not heavy for $\gamma \in [0, 1)$, it is heavy for $\gamma = 1$, and the distribution is improper for $\gamma > 1$.

Although the case of improper distribution seems to be not realistic, we discuss it for some particular cases below, after consideration of proper cases $\gamma \in [0,1]$.

Let us remind that the case $\gamma \in (0,1)$ leads to the light tailed distributions while $\gamma = 1$ leads to the laws with the heavy tail. The choice between models with light or heavy tails can only be made based on real data. Below we analyze some data of this kind.

### 7.1. Analyzing Data from Scholar Google "Mathematics"

Let us give the data for the part "Mathematics" on 16 February 2020 (see Table 1). The data given concern are the first 10 in the number of citations of authors. We do not give the names of these scientists. The table shows:

1. The serial number of the author;
2. The total number of citations by the author;
3. Hirsch Index;
4. The number of citations of the most popular work (By the most popular work we understand the work of this author having the largest number of citations among the works of this scientist);
5. Ratio of citations to squared Hirsch index;

**Table 1.** Citations "Mathematics".

| 1 | 2 | 3 | 4 | 5 |
|---|---|---|---|---|
| 1. | 448,557 | 270 | 28,303 | 6.15 |
| 2. | 162,457 | 98 | 44,406 | 16.92 |
| 3. | 159,123 | 147 | 26,929 | 7.36 |
| 4. | 138,820 | 64 | 110,393 | 33.89 |
| 5. | 101,662 | 59 | 35,640 | 29.20 |
| 6. | 99,206 | 78 | 41,647 | 16.31 |
| 7. | 85,288 | 59 | 55,293 | 24.50 |
| 8. | 84,918 | 48 | 18,901 | 36.86 |
| 9. | 77,319 | 98 | 11,715 | 8.05 |
| 10. | 73,989 | 72 | 17,153 | 14.27 |

Table 1 shows the first scientist has 2.76 times more citations than the second. In other words, the maximum of the observations is essentially greater than previous one. This observation leads us to think that the corresponding distribution has heavy tails (see [8,9]). As we have seen, it is possible for the case $\gamma = 1$ only.

### 7.2. Analyzing Data from Scholar Google "Biostatistics"

Let us give the data for the part "Biostatistics" on 16 February 2020 (see Table 2). The structure of Table 2 is the same as that of Table 1.

**Table 2.** Citations "Biostatistics".

| 1 | 2 | 3 | 4 | 5 |
|---|---|---|---|---|
| 1. | 478,691 | 227 | 66,611 | 9.29 |
| 2. | 301,786 | 132 | 59,613 | 17.32 |
| 3. | 253,221 | 208 | 26,127 | 5.85 |
| 4. | 223,038 | 218 | 10,184 | 4.69 |
| 5. | 199,143 | 169 | 23,447 | 6.97 |
| 6. | 178,855 | 117 | 39,271 | 13.07 |
| 7. | 150,695 | 105 | 42,485 | 13.67 |
| 8. | 119,199 | 111 | 20,666 | 9.67 |
| 9. | 108,648 | 140 | 20,842 | 5.54 |
| 10. | 100,491 | 111 | 30,315 | 8.16 |

Table 2 shows the first scientist has 1.59 times more citations than the second. Although it is it is less than the case of Table 1, the number is large enough to support our hypothesis on the presence of a heavy tail.

We do not give the data on the part "Statistics" but mention the situation is similar to that of the Tables 1 and 2.

### 7.3. Final Model for the Distribution of Citations

From the considerations of the two previous subsections, it follows that the most natural way to describe the distribution of citations is to choose $\gamma = 1$. This means

$$\mathcal{P}(z) = 1 - (1-z)^p, \quad \mathcal{Q}(z) = \frac{q}{1-(1-q)z}$$

and the probability generation function of citations distribution is given by

$$\mathcal{R}(z) = \mathcal{P}(\mathcal{Q}(z)) = 1 - \left(1 - \frac{q}{1-(1-q)z}\right)^p.$$

Denote by $Y$ the number of citations of a given scientist. It is clear that $\mathbb{P}\{Y = n\}$ may be found as the $n$-th coefficient of expansion $\mathcal{R}(z)$ in power series. We have

$$\begin{aligned}
\mathcal{R}(z) &= 1 - (1-q)^p(1-z)^p(1-(1-q)z)^{-p} \\
&= 1 - (1-q)^p \sum_{s=0}^{\infty}(-1)^s \left( \sum_{m=0}^{s} \binom{-p}{m}\binom{p}{s-m}(1-q)^m \right) z^s \\
&= 1 - (1-q)^p + \sum_{s=1}^{\infty}(-1)^{s+1}\binom{p}{s} {}_2F_1(p, -s, 1+p-s, 1-q) z^s,
\end{aligned}$$

where ${}_2F_1$ is a hypergeometric function. Therefore,

$$\begin{aligned}
&\mathbb{P}\{Y = 0\} = 1 - (1-q)^p; \\
&\mathbb{P}\{Y = s\} = (-1)^{s+1}\binom{p}{s} {}_2F_1(p, -s, 1+p-s, 1-q), \quad s = 1, 2, \ldots
\end{aligned} \tag{26}$$

It is possible to verify that $\mathbb{P}\{Y = 0\} > \mathbb{P}\{Y = 0\} > \mathbb{P}\{Y = s\}$ for all integers $s \geq 2$. Therefore, we meet a scientist without papers or with citing papers with maximal probability. If we limit ourselves by consideration of the scientists having at least one citation then the highest probability corresponds to authors with one citation.

The Laplace transform of the distribution of $Y$ has form

$$\mathcal{R}(e^{-t}) = 1 - \left(1 - \frac{q}{1-(1-q)e^{-t}}\right)^p, \quad t \in [0, \infty).$$

Its asymptotic as $t \to 0$ is

$$1 - \mathcal{R}(e^{-t}) \sim \left(\frac{1-q}{q}\right)^p \cdot t^p, \quad \text{as} \quad t \to +0. \tag{27}$$

This relation shows that the random variable $Y$ has moments of order less than $p$ and does not have moments of higher order. Because $p < 1$ the variable $Y$ has infinite mean. In practice, this means that some scholars have a very large number of citations. These citations refer to publications by a relatively small number of scholars. Of course, the data in Tables 1 and 2 are in agreement with these statements. It is important that the model is built on the assumption of the same capabilities of scientists. Even so, we must observe a greater variability in the number of citations of their publications.

Thus, the difference in the number of citations can be purely random and not say anything about the real contribution of the scientist into corresponding science field.

Of course, the proposed model is very idealistic, since it does not take into account the real difference in the capabilities of scientists, as well as in their equipping with the necessary tools and equipment. Taking into account the noted differences is likely to lead to the need to consider mixtures of the proposed distributions with different parameters $p$ and $q$. However, such a complication will not make it possible to distinguish scientists with a large contribution to science from those with a smaller impact.

Surely, the arguments presented for the choice of $\gamma = 1$ are rather crude, i.e., in reality, it may happen that $\gamma$ is close to unity. Although in this case, the distribution tail is not heavy, but over a very large (but finite) interval it is close to heavy. So, qualitatively, our conclusions will remain unchanged.

Based on the foregoing, we conclude that it is practically senseless to use the number of citations of a scientist's work to assess his contribution to science.

### 7.4. Remarks on the Model with $\gamma > 1$

In this subsection, we are trying to justify the possibility of using models with gamma greater than one. As already noted, in this model the probability $\mathbb{P}\{Y = \infty\}$ is not equal to zero. It is unlikely that this corresponds to the situation with the consideration of all scientists working in this field of science. However, a very long citation process (ideally, endless) is quite possible in the case of the most prominent scientists. For example, in the field of Mathematics, the works of Professor Andrei Nikolaevich Kolmogorov (1903–1987) continue to be cited. Over the past 15 years, they have been cited about 30,000 times, although more than 30 years have passed since the death of their author. It is highly probable that the citation process for these works will continue for a long time.

In addition, the concept of citation is somewhat arbitrary in our opinion. For example, in Mathematics, some theorems or other objects bear the names of scientists who were related to their preparation. Does the mention of these theorems and the corresponding names in some articles mean their citation? For example, many articles and books mention the Gaussian distribution without reference to the corresponding publication by Gauss. Is this mention a quotation? It seems to us that such kind of nominal results are not counted in determining the citation index. However, they certainly indicate the scientific significance of the result. It is very likely that for accounting for citations of this kind, models with a $\gamma$ greater than 1 may be required.

## 8. Hirsch Index

Recall that the definition of the Hirsch index was given on Page 1. Hirsch states that the proposed index $h$ is intended to rank authors of articles in the field of Physics. At the same time, it is noted that the index can be used in other fields of science. Since the number of citations is used in determining the index $h$, it seems plausible that $h$ is associated with this number. Hirsch notes that the number of citations is given by $N = \kappa h^2$. He wrote: "I find empirically that $\kappa$ ranges between 3 and 5" (We change notations of Hirsch. Namely, his $a$ is our $\kappa$.). Further, Hirsch wrote: "$\kappa > 5$ is very atypical value".

Below we show that the Hirsch statements presented here are doubtful. In addition, the use of this index seems unreasonable.

Let's start by analyzing the data in Tables 1 and 2. Remind that the column 5 gives corresponding values of $\kappa$. Table 1 does not contain any $\kappa \leq 6$ while Table 2 has only one such value $\kappa = 4.69$. Other values of $\kappa$ are "very atypical", especially for Table 1. Table 2 contains 2 values of $\kappa \in (5, 6)$. Therefore, at least for such fields as "Mathematics" and "Biostatistics", Hirsch's conclusion about the "typical" form of proportionality between the number of citations of an author and the square of corresponding Hirsch's index seems to be incorrect. However, was Hirsch right in the field of "Physics"?

### 8.1. Data in "Physics"

Now we give the data on field "Physics", arranging them into a table in the same way as for Table 1.

Again, Table 3 has only one $\kappa \le 5$, namely $\kappa = 4.88$. However, there are six values $\kappa \in (5,6)$. The kappa values for the "Physics" area look smaller than for the "Biostatistics" area and significantly smaller than for the "Mathematics" area. The value of the Hirsch index for Physics has much less variability than for Biostatistics and Mathematics. The differences in citation numbers are much greater for Mathematics than in the case of Physics.

So, we see that Hirsch's understanding of the situation in Physics is closer to reality than in the case of Biostatistics and, especially, Mathematics.

### *8.2. Data Comparison*

Continue the analysis of the data in Tables 1–3.

**Table 3.** Citations "Physics".

| 1 | 2 | 3 | 4 | 5 |
|---|---|---|---|---|
| 1. | 326,718 | 206 | 25,605 | 7.70 |
| 2. | 259,321 | 223 | 7275 | 5.21 |
| 3. | 240,376 | 200 | 15,651 | 6.01 |
| 4. | 232,057 | 206 | 26,535 | 5.47 |
| 5. | 231,746 | 218 | 15,589 | 4.88 |
| 6. | 227,530 | 206 | 15,684 | 5.36 |
| 7. | 217,495 | 144 | 35,746 | 10.49 |
| 8. | 200,565 | 191 | 11,807 | 5.50 |
| 9. | 198,735 | 190 | 7497 | 5.50 |
| 10. | 197,679 | 198 | 25,649 | 5.04 |

The average value of the Hirsch index in the case of Table 1 is 99.3 with a standard deviation of 66.45. The same indicators for Table 2 are 153.8 and 47.97, and for Table 3—198.2 and 21.73. We see that the standard deviation of the Hirsch index in the case of Mathematics is three times greater than in the case of Physics. On the contrary, the average value of the index is maximum in the case of Physics and minimum in the case of Mathematics. This shows that if Hirsch index is useful in the field of Physics, then its usefulness in the field of Mathematics is doubtful. Probably, it is true for Biostatistics too.

Authors with a higher Hirsch index are often inferior to others in the number of citations of the most popular works. For example, in Table 1, Author 1, having the highest Hirsch index, is inferior to Authors 2, 4, 5, 6 and 7 in the number of citations of the most popular work. In this case, Author 1 wrote his most cited work with co-authors, while author 2 did without co-authors.

It is clear that the Hirsch index does not exceed the number of cited publications of the author, which has an exponential distribution. Thus, the distribution of the Hirsch index has a light tail. Since the number of citations has a heavy tail, it is more variable than the Hirsch index. However, these two indicators are stochastically strongly related. Indeed, for the data in Table 1, the sample correlation coefficient between these indicators is $\rho 1 = 0.94$. On the other hand, the correlation coefficient between the Hirsch index and the number of citations of the most popular works is $\rho 2 = -0.23$. This coefficient indicates a small relationship between the indicators, and it is negative. In other words, a large Hirsch index is most likely not found among authors with highly cited individual articles. For Table 2, the values of the correlation coefficients equal to $\rho 1 = 0.702$, $\rho 2 = 0$, and for Table 3 $\rho 1 = 0.36$, $\rho 2 = -0.57$.

The increase in the Hirsch index with a decrease in the number of citations of the most popular work may result in the division of the work into a series of publications. However, when assessing the quality of a scientist's contribution, one should take into account that the publication of a series of articles instead of one may be caused not by a desire to increase the number of publications, but, for example, by a gradual insight into the essence of the problem under consideration. Such insight often requires a very long time, i.e., publication of a series of articles is justified. It should be noted that the publication of a series of articles naturally leads to an increase in the number of self-citations. This increase cannot be considered as a flaw of the author and does not mean attempts to artificially increase the number of citations. At the same time, the presence of a series of publications (which increases the Hirsch index) cannot be considered as preferable to one highly cited work.

The presence of higher values of the Hirsch index in Physics compared to Mathematics can be explained by the use in modern Physics of expensive equipment in experimental Physics and/or the results obtained on it in theoretical Physics. Often this equipment is used by some laboratory or scientific group, and then transferred to another or others. After some time, this equipment again becomes available to the first group. Thus, new experimental facts arrive intermittently, and during the break they are processed and published. A theoretical analysis of the observed facts is also taking place. Then comes new information related to new experiments. Therefore, the very flow of information (both experimental and theoretical) contributes to the publication of not a single article, but a series of articles. This circumstance leads to an increase in the Hirsch index with a relative decrease in the number of citations of popular works.

A similar situation is absent in Pure Mathematics. Therefore, there the appearance of the series has much fewer reasons. Separate works appear, which often cover a substantial part of the problem under consideration. They cause a stream of citation of this particular work, and in a series of works. Thus, the Hirsch index becomes smaller than it would be if a series of articles were published instead of this one, but the most popular work causes more citations than each individual work in the series.

So, the use of the Hirsch's index has some basis in the field of Physics, but it is not related to what is happening in Mathematics.

For some areas of Applied Mathematics, a situation may be observed that is intermediate between what is happening in Physics and in Pure Mathematics.

However, it is not clear to us why not replace the Hirsch index with two. The first of these could be the number of all citations, and the second - the number of citations of the most popular work. The Hirsch index is stochastically quite closely linked to the number of all citations, so it and this number are "interchangeable". However, after the termination of the work of a scientist in a given field of science, the number of his publications does not increase and, therefore, the Hirsch index remains limited, while the number of citations can continue to grow unlimitedly. This is exactly what happens with the works of the most outstanding scientists of the past.

## 9. Distribution of the Hirsch Index

In this section, we obtain the probability distribution of the Hirsch index.

We introduce some notation. It is clear that the Hirsch index is a random variable. Let us denote it by $H$. We will denote the values of this $H$ by $h$. Our aim here is to determine the probabilities that $H = h$, i.e., $\mathbb{P}\{H = h\}$. In order for the event $H = h$ to occur, it is necessary and sufficient that:

(a) no less than $h$ works were published;
(b) $h$ of the published works are cited at least $h$ times, and the rest - less than $h$ times.

Suppose that $l$ works are published, and $l \geq h$. The probability of this event is $q(1-q)^l$. Recall, the probability that a published work will be quoted $k$ times equals to $(p/k)\prod_{j=1}^{k-1}(1-p/j)$. Therefore, the probability that the published work will be cited at least $h$ times equals to

$$\sum_{k=h}^{\infty} \frac{p}{k} \cdot \prod_{j=1}^{k-1}(1-p/j) = \frac{\Gamma(h-p)}{\Gamma(h)\cdot\Gamma(1-p)},$$

where $\Gamma$ is Euler gamma function.

The probability that a published work will be cited less than $h$ times is defined as

$$1 - \frac{\Gamma(h-p)}{\Gamma(h)\cdot\Gamma(1-p)}.$$

Thus, the probability that $l$ papers are published, and the Hirsch index $H$ has taken the value $h$ is

$$q(1-q)^l \binom{l}{h} \cdot \left(\frac{\Gamma(h-p)}{\Gamma(h)\cdot\Gamma(1-p)}\right)^h \cdot \left(1-\frac{\Gamma(h-p)}{\Gamma(h)\cdot\Gamma(1-p)}\right)^{l-h}.$$

Now we see that

$$\mathbb{P}\{H=h\} = \sum_{l=h}^{\infty} q(1-q)^l \binom{l}{h} \cdot \left(\frac{\Gamma(h-p)}{\Gamma(h)\cdot\Gamma(1-p)}\right)^h \cdot \left(1-\frac{\Gamma(h-p)}{\Gamma(h)\cdot\Gamma(1-p)}\right)^{l-h}$$

$$= \left(\frac{\Gamma(h-p)}{\Gamma(h)\cdot\Gamma(1-p)-\Gamma(h-p)}\right)^h \cdot q \cdot \frac{\mu^h}{(1-\mu)^{h+1}},$$

where

$$\mu = \left(1-\frac{\Gamma(h-p)}{\Gamma(h)\cdot\Gamma(1-p)}\right)\cdot(1-q).$$

So, the random variable $H$ has the following distribution

$$\mathbb{P}\{H=h\} = (1-\nu)\cdot\nu^h,$$

where

$$\nu = \frac{(1-q)\Gamma(h-p)}{q\Gamma(h)\Gamma(1-p)+(1-q)\Gamma(h-p)}.$$

Note that this distribution is not geometric one because the value of $\nu$ depends on $h$.

Next, we are interested in estimating the tail of the distribution of $H$. To do this, we estimate the asymptotic behavior of the $\nu$. An application of the Stirling formula allows one to easily obtain that

$$\nu = \nu(h) \sim \frac{1-q}{q\Gamma(1-p)} \cdot \frac{1}{h^p}.$$

This formula immediately leads us to an asymptotic expression for the logarithm of probability $\mathbb{P}\{H=h\}$ for $h \to \infty$. Namely,

$$\log \mathbb{P}\{H=h\} \sim p\cdot h\cdot \log h, \quad h\to\infty.$$

It follows that the probability of the event $\{H=h\}$ decreases faster than the exponential function for $n \to \infty$. Of course, the tail of the distribution of $H$ also decreases faster than the exponential function. Therefore, there are moments of all orders of this distribution. Note that the distribution of the number of citations of articles by this author has an infinite mean value. So, if an author has a fairly large number of citations, then the ratio of the number of citations to the square of the Hirsch index can be arbitrarily large. This fact contradicts Hirsch's claim that $\kappa$ is bounded.

**Author Contributions:** Conceptualization, L.B.K.; investigation, L.B.K., Y.V.K. and Z.E.V. The authors have equally contributed to the writing, editing and style of the paper. All authors have read and agreed to the published version of the manuscript.

**Funding:** The study was partially supported by grant GAČR 19-04412S (Lev Klebanov).

**Conflicts of Interest:** The authors declare no conflict of interest.

## References

1. Garfield, E. Citation Indexes for Science. *Science* **1955**, *122*, 108–111. [CrossRef] [PubMed]
2. Garfield, E. Citation Index in Sociological and Historical research. *Curr. Contents* **1969**, *9*, 42–46.

3. Garfield, E. The evolution of the Science Citation Index. *Int. Microbiol.* **2007**, *10*, 65–69. [CrossRef] [PubMed]
4. Hirsch, J.E. An index to quantify an individual's scientific research output. *Proc. Natl. Acad. Sci. USA* **2005**, *102*, 16569–16572. [CrossRef] [PubMed]
5. Richter, M. Was misst der h-Index (nicht)?—Kritische Überlegungen zu einer populären Kennzahl für Forschungsleistungen. *WiSt Wirtsch. Stud.* **2018**, *47*, 64–68 [CrossRef]
6. Klebanov, L.B. One look at the rating of scientific publications and corresponding toy-models. *arXiv* **2017**, arXiv:1706.01238v1.
7. Sibuya, M. Generalized Hypergeometric, Digamma and Trigamma Distributions. *Ann. Inst. Statist. Math.* **1979**, *31*, 373–390. [CrossRef]
8. Klebanov, L.B.; Antoch, J.; Karlova, A.; Kakosyan, A.V. Outliers and related problems. *arXiv* **2017**, arXiv:1701.06642v1.
9. Volchenkova, I.V.; Klebanov, L.B. Characterization of the Pareto distribution by the properties of neighboring order statistics. *Zap. Nauchnih Semin. POMI* **2019**, *486*, 63–70. (In Russian)

mathematics

MDPI

Article

# Second Order Expansions for High-Dimension Low-Sample-Size Data Statistics in Random Setting

**Gerd Christoph** [1,2,†] **and Vladimir V. Ulyanov** [2,3,*,†]

1 Department of Mathematics, Otto-von-Guericke University Magdeburg, 39016 Magdeburg, Germany; gerd.christoph@ovgu.de
2 Moscow Center for Fundamental and Applied Mathematics, Lomonosov Moscow State University, 119991 Moscow, Russia
3 Faculty of Computer Science, National Research University Higher School of Economics, 167005 Moscow, Russia
* Correspondence: vulyanov@cs.msu.ru or vulyanov@hse.ru
† These authors contributed equally to this work.

Received: 12 May 2020; Accepted: 9 July 2020; Published: 14 July 2020

**Abstract:** We consider high-dimension low-sample-size data taken from the standard multivariate normal distribution under assumption that dimension is a random variable. The second order Chebyshev–Edgeworth expansions for distributions of an angle between two sample observations and corresponding sample correlation coefficient are constructed with error bounds. Depending on the type of normalization, we get three different limit distributions: Normal, Student's $t$-, or Laplace distributions. The paper continues studies of the authors on approximation of statistics for random size samples.

**Keywords:** second order expansions; high-dimensional; low sample size; random sample size; Laplace distribution; Student's $t$-distribution

**MSC:** 62E17 (Primary) 62H10; 60E05 (Secondary)

---

## 1. Introduction

Let $\vec{X}_1 = (X_{11}, ..., X_{1m})^T, \ldots, \vec{X}_k = (X_{k1}, ..., X_{km})^T$ be a random sample from $m$-dimensional population. The data set can be regarded as $k$ vectors or points in $m$-dimensional space. Recently, there has been significant interest in a high-dimensional datasets when the dimension is large. In a high-dimensional setting, it is assumed that either (i) $m$ tends to infinity and $k$ is fixed, or (ii) both $m$ and $k$ tend to infinity. Case (i) is related to high-dimensional low sample size (HDLSS) data. One of the first results for HDLSS data appeared in Hall et al. [1]. It became the basis of research in mathematical statistics for the analysis of high-dimensional data, see, e.g., Fujikoshi et al. [2], which are an important part of the current data analysis fashionable area called *Big data*. Scientific areas where these settings have proven to be very useful include genetics and other types of cancer research, neuroscience, and also image and shape analysis. See a recent survey on HDLSS asymptotics and its applications in Aoshima et al. [3].

For examining the features of the data set, it is necessary to study the asymptotic behavior of three functions: the length $\|\vec{X}_i\|$ of a $m$-dimensional observation vector, the distance $\|\vec{X}_i - \vec{X}_j\|$ between any two independent observation vectors, and the angle $\text{ang}(\vec{X}_i, \vec{X}_j)$ between these vectors at the population mean. Assuming that $\vec{X}_i$'s are a sample from $N(0, I_m)$, it was shown in Hall et al. [1] that for HDLSS data the three geometric statistics satisfy the following relations:

$$||\vec{X}_i|| = \sqrt{m} + \mathcal{O}_p(1), \quad i = 1, \ldots, k, \tag{1}$$

$$||\vec{X}_i - \vec{X}_j|| = \sqrt{2m} + \mathcal{O}_p(1), \ i, j = 1, \ldots k, \ i \neq j, \tag{2}$$

$$\text{ang}(\vec{X}_i, \vec{X}_j) = \frac{1}{2}\pi + \mathcal{O}_p(m^{-1/2}), \ i, j = 1, \ldots k, \ i \neq j, \tag{3}$$

where $|| \cdot ||$ is the Euclidean distance and $\mathcal{O}_p$ denotes the stochastic order. These interesting results imply that the data converge to the vertices of a deterministic regular simplex. These properties were extended for non-normal sample under some assumptions (see Hall et al. [1] and Aoshima et al. [3]). In Kawaguchi et al. [4], the relations (1)–(3) were refined by constructing second order asymptotic expansions for distributions of all three basic statistics. The refinements of (1) and (2) were achieved by using the idea of Ulyanov et al. [5] who obtained the computable error bounds of order $\mathcal{O}(m^{-1})$ for the chi-squared approximation of transformed chi-squared random variables with $m$ degrees of freedom.

The aim of the present paper is to study approximation for the third statistic $\text{ang}(\vec{X}_1, \vec{X}_2)$ under generalized assumption that $m$ is a realization of a random variable, say $N_n$, which represents the sample dimension and is independent of $\vec{X}_1$ and $\vec{X}_2$. This problem is closely related to approximations of statistics constructed from the random size samples, in particular, to this kind of problem for the sample correlation coefficient $R_m$.

The use of samples with random sample sizes has been steadily growing over the years. For an overview of statistical inferences with a random number of observations and some applications, see Esquível et al. [6] and the references cited therein. Gnedenko [7] considered the asymptotic properties of the distributions of sample quantiles for samples of random size. In Nunes et al. [8] and Nunes et al. [9], unknown sample sizes are assumed in medical research for analysis of one and more than one-way fixed effects ANOVA models to avoid false rejections, obtained when using the classical fixed size $F$-tests. Esquível et al. [6] considered inference for the mean with known and unknown variance and inference for the variance in the normal model. Prediction intervals for the future observations for generalized order statistics and confidence intervals for quantiles based on samples of random sizes are studied in Barakat et al. [10] and Al-Mutairi and Raqab [11], respectively. They illustrated their results with real biometric data set, the duration of remission of leukemia patients treated by one drug. The present paper continues studies of the authors on non-asymptotic analysis of approximations for statistics based on random size samples. In Christoph et al. [12], second order expansions for the normalized random sample sizes are proved, see below Propositions 1 and 2. These results allow for proving second order asymptotic expansions of random sample mean in Christoph et al. [12] and random sample median in Christoph et al. [13]. See also Chapters 1 and 9 in Fujikoshi and Ulyanov [14].

The structure of the paper is the following. In Section 2, we describe the relation between $\text{ang}(\vec{X}_1, \vec{X}_2)$ and $R_m$. We recall also previous approximation results proved for distributions of $\text{ang}(\vec{X}_1, \vec{X}_2)$ and $R_m$. Section 3 is on general transfer theorems, which allow us to construct asymptotic expansions for distributions of randomly normalized statistics on the base of approximation results for non-randomly normalized statistics and for the random size of the underlying sample, see Theorems 1 and 2. Section 4 contains the auxiliary lemmas. Some of them have independent interest. For example, Lemma 3 on the upper bounds for the negative order moments of a random variable having negative binomial distribution. We formulate and discuss main results in Sections 5 and 6. In Theorems 3–8, we construct the second order Chebyshev–Edgeworth expansions for distributions of $\text{ang}(\vec{X}_1, \vec{X}_2)$ and $R_m$ in random setting. Depending on the type of normalization, we get three different limit distributions: Normal, Laplace, or Student's $t$-distributions. All proofs are given in the Appendix A.

## 2. Sample Correlation Coefficient, Angle between Vectors and Their Normal Approximations

We slightly simplify notation. Let $\vec{X}_m = (X_1, ..., X_m)^T$ and $\vec{Y}_m = (Y_1, ..., Y_m)^T$ be two vectors from an $m$-dimensional normal distribution $N(0, \mathrm{I}_m)$ with zero mean, identity covariance matrix $\mathrm{I}_m$ and the sample correlation coefficient

$$R_m = R_m(\vec{X}_m, \vec{Y}_m) = \frac{\sum_{k=1}^m X_k Y_k}{\sqrt{\sum_{k=1}^m X_k^2 \ \sum_{k=1}^m Y_k^2}}. \tag{4}$$

Under the null hypothesis $H_0$: $\{\vec{X}_m$ *and* $\vec{Y}_m$ *are uncorrelated*$\}$, the so-called null density $p_{R_m}(y;n)$ of $R_m$ is given in Johnson, Kotz and Balakrishnan [15], Chapter 32, Formula (32.7):

$$p_{R_m}(y;m) = \frac{\Gamma((m-1)/2)}{\sqrt{\pi}\,\Gamma((m-2)/2)} \left(1 - y^2\right)^{(m-4)/2} \mathbb{I}_{(-1\ 1)}(y)$$

for $m \geq 3$, where $\mathbb{I}_A(.)$ denotes indicator function of a set $A$.

- Note $\mu = \mathbb{E}R_m = 0$ and $\sigma^2 = \mathrm{Var}(R_m) = 1/(m-1)$ for $m \geq 2$,
- $R_2$ is two point distributed with $P(R_2 = -1) = P(R_2 = 1) = 1/2$,
- $R_3$ is $U$-shaped with $p_{R_3}(y;3) = (1/\pi)\ (1-y^2)^{-1/2}\,\mathbb{I}_{(-1,1)}(y)$ and
- $R_4$ is uniform with density $p_{R_4}(y;4) = 1/2\,\mathbb{I}_{(-1,1)}(y)$.
- Moreover, for $m \geq 5$, the density function $p_{R_m}(y;m)$ is unimodal.

Consider now the standardized correlation coefficient

$$\overline{R}_m = \sqrt{m-c}\ R_m \tag{5}$$

with some correcting real constant $c < m$ having density

$$p_{\overline{R}_m}(y;m,c) \quad = \quad \frac{\Gamma((m-1)/2)}{\sqrt{m-c}\,\sqrt{\pi}\,\Gamma((m-2)/2)} \left(1 - \frac{y^2}{m-c}\right)^{(m-4)/2} \mathbb{I}_{\{|r|<\sqrt{m-c}\}}(y), \tag{6}$$

which converges with $c = \mathcal{O}(1)$ as $m \to \infty$ to the standard normal density

$$\varphi(y) = \frac{1}{\sqrt{2\,\pi}} e^{-\,y^2/2}, \quad y \in (-\infty\ \infty)$$

and by Konishi [16], Section 4, Formula (4.1) as $m \to \infty$:

$$F_m^*(x,c) := P\Big(\sqrt{m-c}\,R_m \leq x\Big) = \Phi(x) + \frac{x^3 + (2(c-1)-3)x}{4\,(m-c)}\,\varphi(x) + \mathcal{O}(m^{-3/2}), \tag{7}$$

where $\Phi(x) = \int_{-\infty}^x \varphi(y)dy$ is the standard normal distribution function. Note that in Konishi [16] the sample size (in our case the dimension of vectors) is $m+1$ and $c = 1 + 2\Delta$ with Konishi's correcting constant $\Delta$. Moreover, (7) follows from the more general Theorem 2.2 in the mentioned paper for independent components in the pairs $(X_k\ Y_k)$, $k = 1, ..., m$.

In Christoph et al. [17], computable error bounds of approximations in (7) with $c = 2$ and $c = 2.5$ of order $\mathcal{O}(m^{-2})$ for all $m \geq 7$ are proved:

$$\sup_x \left| P\Big(\sqrt{m-2.5}\ R_m \leq x\Big) - \Phi(x) - \frac{x^3\,\varphi(x)}{4\,(m-2.5)} \right| \leq \frac{B_m}{(m-2.5)^2} \leq \frac{B}{m^2} \tag{8}$$

and

$$\sup_x \left| P\Big(\sqrt{m-2}\ R_m \leq x\Big) - \Phi(x) - \frac{(x^3-x)\,\varphi(x)}{4\,(m-2)} \right| \leq \frac{B_m^*}{(m-2)^2} \leq \frac{B^*}{m^2} \tag{9}$$

where for some $m \geq 7$ constants $B_m$ and $B_m^*$ are calculated and presented in Table 1 in Christoph et al. [17]: i.e., $B_7 = 1.875$, $B_7^* = 2.083$ and $B_{50} = 0.720$, $B_{50}^* = 0.982$.

Usually, the asymptotic for $\overline{R}_m$ is (9), where $c = 2$ since it is related to the $t$-distributed statistic $\sqrt{m-2}\,R_m \,/\, \sqrt{1-R_m^2}$. With the correcting constant $c = 2.5$, one term in the asymptotic in (8) vanishes.

In order to use a transfer theorem from non-random to random dimension of the vectors, we prefer (7) with $c = 0$. In a similar manner as proving (8) and (9) in Christoph et al. [17], one can verify the following inequalities for $m \geq 3$:

$$\sup_x \left| P\left(\sqrt{m}R_m \leq x\right) - \Phi(x) - \frac{(x^3 - 5\,x)}{4\,m}\,\varphi(x)\right| \leq C_1 m^{-2}. \tag{10}$$

Let us consider now the connection between the correlation coefficient $R_m$ and the angle $\theta_m$ of the involved vectors $\vec{X}_m, \vec{Y}_m$:

$$\theta_m = \text{ang}(\vec{X}_m, \vec{Y}_m). \tag{11}$$

Hall et al. [1] showed that under the given conditions

$$\theta_m = \frac{1}{2}\pi + \mathcal{O}_p(m^{-1/2}) \quad \text{as} \quad m \to \infty,$$

where $\mathcal{O}_p$ denotes the stochastic order. Since

$$\cos\theta_m = \frac{\|\vec{X}_m\|^2 + \|\vec{Y}_m\|^2 - \|\vec{X}_m - \vec{Y}_m\|^2}{2\,\|\vec{X}_m\|\,\|\vec{Y}_m\|} = R_m(\vec{X}_m, \vec{Y}_m) = R_m,$$

the computable error bounds for $\theta_m$ follows from computable error bounds for $R_m$.

For any fixed constant $c < m$, and arbitrary $x$ with $|x| < \sqrt{m-c}\,\pi/2$, we obtain for the angle $\theta_m : 0 < \theta_m < \pi :$

$$\begin{aligned} P\left(\sqrt{m-c}(\theta_m - \pi/2) \leq x\right) &= P\left(\theta_m \leq \pi/2 + x/\sqrt{m-c}\right) \\ &= P\left(\cos\theta_m \geq \cos(\pi/2 + x/\sqrt{m-c})\right) \\ &= P\left(R_m \geq -\sin(x/\sqrt{m-c})\right) \\ &= P\left(\sqrt{m-c}\;R_m \leq \sqrt{m-c}\;\sin(x/\sqrt{m-c})\right) \end{aligned} \tag{12}$$

because $R_m$ is symmetric and $P(R_m \leq x) = P(-R_m \leq x)$.

Equation (12) shows the connection between the correlation coefficient $R_m$ and the angle $\theta_m$ among the vectors involved. In Christoph et al. [17], computable error bound of approximation in (8) are used to obtain similar bound for the approximation of the angle between two vectors, defined in (11). Here, the approximation (10) and (12) with $c = 0$ lead for any $m \geq 3$ and for $|x| \leq \pi\sqrt{m}\,/2$ to

$$\sup_x \left| P\left(\sqrt{m}(\theta_m - \frac{\pi}{2}) \leq x\right) - \Phi(x) - \frac{(1/3)x^3 - 5x}{4\,m}\,\varphi(x)\right| \leq C_1\, m^{-2}. \tag{13}$$

Many authors investigated limit theorems for the sums of random vectors when their dimension tends to infinity, see, e.g., Prokhorov [18]. In (6) and (7), the dimension $m$ of the vectors $\vec{X}_m$ and $\vec{Y}_m$ tends to infinity.

Now, we consider the correlation coefficient of vectors $\vec{X}_m$ and $\vec{Y}_m$, where the non-random dimension $m$ is replaced by a random dimension $N_n \in \mathbb{N}_+ = \{1, 2, ...\}$ depending on some natural parameter $n \in \mathbb{N}_+$ and $N_n$ is independent of $\vec{X}_m$ and $\vec{Y}_m$ for any $m, n \in \mathbb{N}_+$. Define

$$R_{N_n} = \frac{\sum_{k=1}^{N_n} X_k Y_k}{\sqrt{\sum_{k=1}^{N_n} X_k^2 \ \sum_{k=1}^{N_n} Y_k^2}}.$$

## 3. Statistical Models with a Random Number of Observations

Let $X_1, X_2, \ldots \in \mathbb{R} = (-\infty\ \infty)$ and $N_1, N_2, \ldots \in \mathbb{N}_+ = \{1, 2, ...\}$ be random variables on the same probability space $(\Omega, \mathbb{A}, \mathbb{P})$. Let $N_n$ be a random size of the underlying sample, i.e., the random number of observations, which depends on parameter $n \in \mathbb{N}_+$. We suppose for each $n \in \mathbb{N}_+$ that $N_n \in \mathbb{N}_+$ is independent of random variables $X_1, X_2, \ldots$ and $N_n \to \infty$ in probability as $n \to \infty$. Let $T_m := T_m(X_1, \ldots, X_m)$ be some statistic of a sample with *non-random sample size* $m \in \mathbb{N}_+$. Define the random variable $T_{N_n}$ for every $n \in \mathbb{N}_+$:

$$T_{N_n}(\omega) := T_{N_n(\omega)}\left(X_1(\omega), \ldots, X_{N_n(\omega)}(\omega)\right), \quad \omega \in \Omega,$$

i.e., $T_{N_n}$ is some statistic obtained from a random sample $X_1, X_2, \ldots, X_{N_n}$.

The randomness of the sample size may crucially change asymptotic properties of $T_{N_n}$, see, e.g., Gnedenko [7] or Gnedenko and Korolev [19].

### 3.1. Random Sums

Many models lead to random sums and random means

$$S_{N_n} = \sum_{k=1}^{N_n} X_k \quad \text{and} \quad M_{N_n} = \frac{1}{N_n} \sum_{k=1}^{N_n} X_k, . \tag{14}$$

A fundamental introduction to asymptotic distributions of random sums is given in Döbler [20].

It is worth mentioning that a suitable scaled factor by $S_{N_n}$ affects the type of limit distribution. In fact, consider random sum $S_{N_n}$ given in (14). For the sake of convenience, let $X_1, X_2, ...$ be independent standard normal random variables and $N_n \in \mathbb{N}_+$ be geometrically distributed with $E(N_n) = n$ and independent of $X_1, X_2, ....$ Then, one has

$$\mathbb{P}\left(\frac{1}{\sqrt{N_n}} S_{N_n} \leq x\right) = \int_{-\infty}^{x} \frac{1}{\sqrt{2\pi}} e^{-u^2/2} du \quad \text{for all } n \in \mathbb{N}, \tag{15}$$

$$\mathbb{P}\left(\frac{1}{\sqrt{\mathbb{E}(N_n)}} S_{N_n} \leq x\right) \to \int_{-\infty}^{x} \frac{1}{\sqrt{2}} e^{-\sqrt{2}|u|} du \quad \text{as } n \to \infty, \tag{16}$$

$$\mathbb{P}\left(\frac{\sqrt{\mathbb{E}(N_n)}}{N_n} S_{N_n} \leq x\right) \to \int_{-\infty}^{x} \left(2 + u^2\right)^{-3/2} du \quad \text{as } n \to \infty. \tag{17}$$

We have three different limit distributions. The suitable scaled geometric sum $S_{N_n}$ is standard normal distributed or tends to the Laplace distribution with variance 1 depending on whether we take the random scaling factor $1/\sqrt{N_n}$ or the non-random scaling factor $1/\sqrt{\mathbb{E}N_n}$, respectively. Moreover, we get the Student distribution with two degrees of freedom as the limit distribution if we use scaling with the mixed factor $\sqrt{\mathbb{E}(N_n)}/N_n$. Similar results also hold for the normalized random mean $M_{N_n} = \frac{1}{N_n} S_{N_n}$.

Assertion (15) is obtained by conditioning and the stability of the normal law. Moreover, using Stein's method, quantitative Berry–Esseen bounds in (15) and (16) for arbitrary centered random variables $X_1$ with $\mathbb{E}(|X_1|^3) < \infty$ were proved in (Chen et al. [21], Theorem 10.6), (Döbler [20] Theorems 2.5 and 2.7) and (Pike and Ren [22] Theorem 3), respectively. Statement (17) follows from (Bening and Korolev [23] Theorem 2.1).

First order asymptotic expansions are obtained for the distribution function of random sample mean and random sample median constructed from a sample with two different random sizes in Bening et al. [24] and in the conference paper Bening et al. [25]. The authors make use of the rate of convergence of $\mathbb{P}(N_n \leq g_n x)$ to the limit distribution $H(x)$ with some $g_n \uparrow \infty$. In Christoph et al. [12], second order expansions for the normalized random sample sizes are proved, see below Propositions 1 and 2. These results allow for proving second order asymptotic expansions of random sample mean in Christoph et al. [12] and random sample median in Christoph et al. [13].

### 3.2. Transfer Proposition from Non-Random to Random Sample Sizes

Consider now the statistic $T_{N_n} = T_{N_n}\left(\vec{X}_{N_n}, \vec{Y}_{N_n}\right)$ ,, where the dimension of the vectors $\vec{X}_{N_n}, \vec{Y}_{N_n}$ is a random number $N_n \in \mathbb{N}_+$.

In order to avoid too long expressions and at the same time to preserve a necessary accuracy, we limit ourselves to obtaining limit distributions and terms of order $m^{-1}$ in the following non-asymptotic approximations with a bounds of order $m^{-a}$ for some $a > 1$.

We suppose that the following condition on the statistic $T_m = T_m(\vec{X}_m, \vec{Y}_m)$ with $\mathbb{E}T_m = 0$ is met for a non-random sample size $m \in \mathbb{N}_+$:

**Condition 1.** *There exist differentiable bounded function $f_2(x)$ with $\sup_x |x f_2'(x)| < c_0$ and real numbers $a > 1$, $C_1 > 0$ such that for all integer $m \geq 1$*

$$\sup_x \left|\mathbb{P}(m^{\gamma} T_m \leq x) - \Phi(x) - m^{-1} f_2(x)\right| \leq C_1\, m^{-a}, \tag{18}$$

*where $\gamma \in \{-1/2, 0, 1/2\}$.*

**Remark 1.** *Relations (10) and (13) give the examples of statistics such that Condition 1 is met. For other examples of multivariate statistics of this kind, see Chapters 14–16 in Fujikoshi et al. [2].*

Suppose that the limiting behavior of distribution functions of the normalized random size $N_n \in \mathbb{N}_+$ is described by the following condition.

**Condition 2.** *There exist a distribution function $H(y)$ with $H(0+) = 0$, a function of bounded variation $h_2(y)$, a sequence $0 < g_n \uparrow \infty$ and real numbers $b > 0$ and $C_2 > 0$ such that for all integer $n \geq 1$*

$$\left.\begin{array}{ll} \sup_{y \geq 0} \left|\mathbb{P}(g_n^{-1} N_n \leq y) - H(y)\right| \leq C_2 n^{-b}, & 0 < b \leq 1 \\ \sup_{y \geq 0} \left|\mathbb{P}(g_n^{-1} N_n \leq y) - H(y) - n^{-1} h_2(y)\right| \leq C_2 n^{-b}, & b > 1 \end{array}\right\} \tag{19}$$

**Remark 2.** *In Propositions 1 and 2 below, we get the examples of discrete random variables $N_n$ such that Condition 2 is met.*

Conditions 1 and 2 allow us to construct asymptotic expansions for distributions of randomly normalized statistics on the base of approximation results for normalized fixed-size statistics (see relation (18)) and for the random size of the underlying sample (see relation (19)). As a result, we obtain the following transfer theorem.

**Theorem 1.** *Let $|\gamma| \le K < \infty$ and both Conditions 1 and 2 be satisfied. Then, the following inequality holds for all $n \in \mathbb{N}_+$ :*

$$\sup_{x\in\mathbb{R}} \left|\mathbb{P}\left(g_n^{\gamma} T_{N_n} \le x\right) - G_n(x, 1/g_n)\right| \le C_1\, \mathbb{E}\left(N_n^{-a}\right) + (C_3 D_n + C_4)\, n^{-b}, \tag{20}$$

$$G_n(x, 1/g_n) = \int_{1/g_n}^{\infty} \left(\Phi(x\, y^{\gamma}) + \frac{f_2(xy^{\gamma})}{g_n y}\right) d\left(H(y) + \frac{h_2(y)}{n}\right), \tag{21}$$

$$D_n = \sup_x \int_{1/g_n}^{\infty} \left|\frac{\partial}{\partial y}\left(\Phi(xy^{\gamma}) + \frac{f_2(xy^{\gamma})}{y g_n}\right)\right| dy, \tag{22}$$

*where $a > 1, b > 0, f_2(z), h_2(y)$ are given in (18) and (19). The constants $C_1, C_3, C_4$ do not depend on $n$.*

**Remark 3.** *Later, we use only the cases $\gamma \in \{-1/2, 0, 1/2\}$.*

**Remark 4.** *The domain $[1/g_n, \infty)$ of integration in (21) depends on $g_n$. Thus, it is not clear how $G_n(x, 1/g_n)$ is represented as a polynomial in $g_n^{-1}$ and $n^{-1}$. To overcome this problem (see (26)), we prove the following theorem.*

**Theorem 2.** *Under the conditions of Theorem 1 and the additional conditions on functions $H(.)$ and $h_2(.)$, depending on the convergence rate $b$ in (19):*

$$H(1/g_n) \le c_1\, g_n^{-b}, \quad b > 0, \tag{23}$$

$$\left.\begin{array}{ll} i: & \int_0^{1/g_n} y^{-1} dH(y) \le c_2\, g_n^{-b+1}, \\ ii: & h_2(0) = 0 \quad and \quad |h_2(1/g_n)| \le c_3\, n\, g_n^{-b}, \\ iii: & \int_0^{1/g_n} y^{-1} |h_2(y)| dy \le c_4\, n\, g_n^{-b}, \end{array}\right\} \quad for \quad b > 1, \tag{24}$$

*we obtain for the function $G_n(x, 1/g_n)$ defined in (21):*

$$\sup_x \left|G_n(x, 1/g_n) - G_{n,2}(x)\right| \le C\, g_n^{-b} + \sup_x \left(|I_1(x,n)|\, \mathbb{I}_{\{b<1\}}(b) + |I_2(x,n)|\right) \tag{25}$$

*with*

$$G_{n,2}(x) = \begin{cases} \int_0^{\infty} \Phi(x\, y^{\gamma}) dH(y), & 0 < b < 1, \\ \int_0^{\infty} \Phi(x\, y^{\gamma}) dH(y) + \frac{1}{g_n} \int_0^{\infty} \frac{f_2(x\, y^{\gamma})}{y} dH(y), & b = 1 \\ \int_0^{\infty} \Phi(x\, y^{\gamma}) dH(y) + \frac{1}{g_n} \int_0^{\infty} \frac{f_2(x\, y^{\gamma})}{y} dH(y)\, \mathbb{I}_{\{\gamma=0\}}(\gamma) + \frac{1}{n} \int_0^{\infty} \Phi(x\, y^{\gamma}) dh_2(y), & b > 1. \end{cases} \tag{26}$$

$$I_1(x,n) = \int_{1/g_n}^{\infty} \frac{f_2(xy^{\gamma})}{g_n\, y} dH(y) \;\; for \;\; b \le 1 \quad and \quad I_2(x,n) = \int_{1/g_n}^{\infty} \frac{f_2(xy^{\gamma})}{g_n\, n\, y} dh_2(y) \;\; for \;\; b > 1.$$

**Remark 5.** *The additional conditions (23) and (24) guarantee to extend the integration range from $[1/g_n, \infty)$ to $(0, \infty)$ of the integrals in (26).*

Theorems 1 and 2 are proved in Appendix A.1.

## 4. Auxiliary Propositions and Lemmas

Consider the standardized correlation coefficient (5) having density (6) with correcting real constant $c = 0$ and standardized angle $\sqrt{m}(\theta_m - \pi/2)$, see (12). By (10) and (13) for $m \ge 3$, we have

$$\sup_x \left|P\left(\sqrt{m}\, R_m \le x\right) - \Phi(x) - \frac{(x^3 - 5\,x)}{4\,m}\, \varphi(x)\right| \le C_1 m^{-2}, \qquad m \in \mathbb{N}_+, \tag{27}$$

and for the angle $\theta_m$ between the vectors for $|x| \leq \pi\sqrt{m}/2$

$$\sup_x \left| P\left(\sqrt{m}(\theta_m - \frac{\pi}{2}) \leq x\right) - \Phi(x) - \frac{(1/3)x^3 - 5x}{4m}\varphi(x)\right| \leq C_1 m^{-2}, \qquad m \in \mathbb{N}_+, \tag{28}$$

where (27) and (28) for $m = 1$ and $m = 2$ are trivial and $C_1$ does not depend on $m$.

Suppose $f_2(x;a) = (a\,x^3 - 5\,x)\,\varphi(x)/4$ with $a = 1$ or $a = 1/3$ when (27) or (28) are considered. Since a product of polynomials in $x$ with $\varphi(x)$ is always bounded, numerical calculus leads to

$$\sup_x |x\,f_2'(x;a)| = \sup_x |x\,(ax^4 - (3a+5)\,x^2 + 5)|\,\varphi(x)/4 \leq 0.4.$$

Condition 1 of the transfer Theorem 1 to the statistics $R_m$ and $\theta_m$ are satisfied with $c_0 = 0.4$ and $a = 2$.

Next, we estimate $D_n(x)$ defined in (22).

**Lemma 1.** *Let $g_n$ a sequence with $0 < g_n \uparrow \infty$ as $n \to \infty$. Then, with some $0 < c(\gamma, a) < \infty$, we obtain with $a = 1$ and $a = 1/3$:*

$$D_n = \sup_x \int_{1/g_n}^{\infty} \left| \frac{\partial}{\partial y}\left(\Phi(x\,y^\gamma) + \frac{f_2(xy^\gamma;a)}{yg_n}\right)\right| dy \leq \frac{1}{2} + \frac{c(\gamma,a)}{4}.$$

In the next subsection, we consider the cases when the random dimension $N_n$ is negative binomial distributed with success probability $1/n$.

### 4.1. Negative Binomial Distribution as Random Dimension of the Normal Vectors

Let the random dimension $N_n(r)$ of the underlying normal vectors be negative binomial distributed (shifted by 1) with parameters $1/n$ and $r > 0$, having probability mass function

$$\mathbb{P}(N_n(r) = j) = \frac{\Gamma(j + r - 1)}{\Gamma(j)\,\Gamma(r)}\left(\frac{1}{n}\right)^r\left(1 - \frac{1}{n}\right)^{j-1}, \; j = 1, 2, \ldots \tag{29}$$

with $\mathbb{E}(N_n(r)) = r\,(n-1) + 1$. Then, $\mathbb{P}(N_n(r)/g_n \leq x)$ tends to the Gamma distribution function $G_{r,r}(x)$ with the shape and rate parameters $r > 0$, having density

$$g_{r,r}(x) = \frac{r^r}{\Gamma(r)}x^{r-1}e^{-rx}\,\mathbb{I}_{(0\,\infty)}(x), \quad x \in \mathbb{R}. \tag{30}$$

If the statistic $T_m$ is asymptotically normal, the limit distribution of the standardized statistic $T_{N_n(r)}$ with random size $N_n(r)$ is Student's $t$-distribution $S_{2r}(x)$ having density

$$s_\nu(x) = \frac{\Gamma((\nu+1)/2)}{\sqrt{\nu\pi}\,\Gamma(\nu/2)}\left(1 + \frac{x^2}{\nu}\right)^{-(\nu+1)/2}, \quad \nu > 0, \quad x \in \mathbb{R}, \tag{31}$$

with $\nu = 2r$, see Bening and Korolev [23] or Schluter and Trede [26].

**Proposition 1.** *Let $r > 0$, discrete random variable $N_n(r)$ have probability mass function (29) and $g_n := \mathbb{E}N_n(r) = r(n-1) + 1$. For $x > 0$ and all $n \in \mathbb{N}$ there exists a real number $C_2(r) > 0$ such that*

$$\sup_{x \geq 0}\left| \mathbb{P}\left(\frac{N_n(r)}{g_n} \leq x\right) - G_{r,r}(x) - \frac{h_{2;r}(x)}{n}\right| \leq C_2(r)\,n^{-\min\{r,2\}}, \tag{32}$$

*where*

$$h_{2;r}(x) = \begin{cases} 0, & \text{for } r \leq 1, \\ \dfrac{g_{r,r}(x)\,\big((x-1)(2-r) + 2Q_1(g_n\,x)\big)}{2r}, & \text{for } r > 1. \end{cases}$$

$$Q_1(y) = 1/2 - (y - [y]) \text{ and } [.] \text{ denotes the integer part of a number.} \quad (33)$$

Figure 1 shows the approximation of $\mathbb{P}(N_n(r) \le (r(n-1)+1)x)$ by $G_{2,2}(x)$ and $G_{2,2}(x) + h_2(x)/n$.

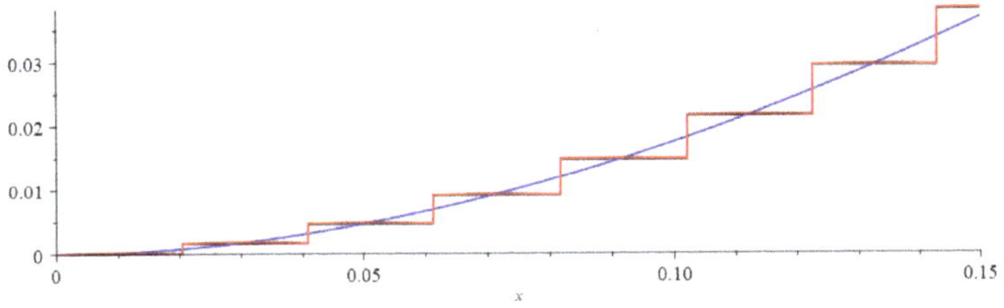


**Figure 1.** Distribution function $\mathbb{P}(N_n(r) \le (r(n-1)+1)x)$ (black line, almost covered by the red line), the limit law $G_{2,2}(x)$ (blue line) and the second approximation $G_{2,2}(x) + h_2(x)/n$ (red line) with $n = 25$ and $r = 2$.

**Remark 6.** *The convergence rate for $r \le 1$ is given in Bening et al. [24] or Gavrilenko et al. [27]. The Edgeworth expansion for $r > 1$ is proved in Christoph et al. [12], Theorem 1. The jumps of the sample size $N_n(r)$ have an effect only on the function $Q_1(.)$ in the term $h_{2;r}(.)$.*

The negative binomial random variable $N_n$ satisfies Condition 2 of the transfer Theorem 1 with $H(x) = G_{r,r}(x)$, $h_2(x) = h_{2;r}(x)$, $g_n = \mathbb{E}N_n(r) = r(n-1)+1$ and $b = \min\{r\ 2\}$.

**Lemma 2.** *In Theorem 2 the additional conditions (23) and (24) are satisfied with $H(x) = G_{r,r}(x)$, $h_2(x) = h_{2;r}(x)$, $g_n = \mathbb{E}N_n(r) = r(n-1)+1$ and $b = \min\{r\ 2\}$. Moreover, one has for $\gamma \in \{-1/2, 0, 1/2\}$ and $f_2(z;a) = (a\,z^3 - 5\,z)\,\varphi(z)/4$, with $a = 1$ or $a = 1/3$:*

$$|I_1(x,n)| = \begin{cases} \left| \int_{1/g_n}^{\infty} \frac{f_2(x\,y^\gamma;a)}{g_n\,y} dG_{r,r}(y) \right| \le c_5\,g_n^{-r} & r < 1, \\ \left| \int_{1/n}^{\infty} \frac{f_2(x\,y^\gamma;a)}{n\,y} dG_{1,1}(y) - n^{-1} f_2(x;a) \ln n\, \mathbb{I}_{\{\gamma=0\}}(\gamma) \right| \le c_6 n^{-1}, & r = 1, \end{cases} \quad (34)$$

$$|I_2(x,n)| = \left| \int_{1/g_n}^{\infty} \frac{f_2(x\,y^\gamma;a)}{g_n\,n\,y} dh_{2;r}(y) \right| \le \begin{cases} c_7 g_n^{-r}, & r > 1, r \ne 2, \\ (c_7 + c_8 \ln n\, \mathbb{I}_{\{\gamma=0\}}(\gamma))\,g_n^{-2}, & r = 2. \end{cases} \quad (35)$$

*Furthermore, we have*

$$0 \le g_n^{-1} - (r\,n)^{-1} \le (r-1)\,(r\,n)^{-2}\,e^{-1/2} \quad for \quad r > 1, \quad n \ge 2. \quad (36)$$

In addition to the expansion of $N_n(r)$ a bound of $\mathbb{E}(N_n(r))^{-a}$ is required, where $m^{-a}$ is rate of convergence of Edgeworth expansion for $T_m$, see (18).

**Lemma 3.** *Let $r > 0$, $\alpha > 0$ and the random variable $N_n(r)$ is defined by (29). Then,*

$$\mathbb{E}(N_n(r))^{-\alpha} \le C(r) \begin{cases} n^{-\min\{r,\alpha\}}, & r \ne \alpha \\ \ln(n)\,n^{-\alpha}, & r = \alpha \end{cases} \quad (37)$$

*and the convergence rate in case $r = \alpha$ cannot be improved.*

### 4.2. Maximum of n Independent Discrete Pareto Random Variables Is the Dimension of the Normal Vectors

Let $Y(s) \in \mathbb{N}$ be discrete Pareto II distributed with parameter $s > 0$, having probability mass and distribution functions

$$\mathbb{P}(Y(s) = k) = \frac{s}{s+k-1} - \frac{s}{s+k} \quad \text{and} \quad \mathbb{P}(Y(s) \le k) = \frac{k}{s+k}, \ k \in \mathbb{N}, \tag{38}$$

which is a particular class of a general model of discrete Pareto distributions, obtained by discretization continuous Pareto II (Lomax) distributions on integers, see Buddana and Kozubowski [28].

Now, let $Y_1(s), Y_2(s), ...,$ be independent random variables with the same distribution (38). Define for $n \in \mathbb{N}$ and $s > 0$ the random variable

$$N_n(s) = \max_{1 \le j \le n} Y_j(s) \quad \text{with} \quad \mathbb{P}(N_n(s) \le k) = \left(\frac{k}{s+k}\right)^n, \quad n \in \mathbb{N}. \tag{39}$$

It should be noted that the distribution of $N_n(s)$ is extremely spread out on the positive integers. In Christoph et al. [12], the following Edgeworth expansion was proved:

**Proposition 2.** *Let the discrete random variable $N_n(s)$ have distribution function (39). For $x > 0$, fixed $s > 0$ and all $n \in \mathbb{N}$, then there exists a real number $C_3(s) > 0$ such that*

$$\sup_{y>0} \left| \mathbb{P}\left(\frac{N_n(s)}{n} \le y\right) - H_s(y) - \frac{h_{2;s}(y)}{n} \right| \le \frac{C_3(s)}{n^2},$$

$$H_s(y) = e^{-s/y} \ \text{and} \ h_{2;s}(y) = s\, e^{-s/y} \left(s - 1 + 2Q_1(n\,y)\right) / (2\,y^2), \ y > 0 \tag{40}$$

*where $Q_1(y)$ is defined in (33).*

**Remark 7.** *The continuous function $H_s(y) = e^{-s/y}\mathbb{I}_{(0\ \infty)}(y)$ with parameter $s > 0$ is the distribution function of the inverse exponential random variable $W(s) = 1/V(s)$, where $V(s)$ is exponentially distributed with rate parameter $s > 0$. Both $H_s(y)$ and $\mathbb{P}(N_n(s) \le y)$ are heavy tailed with shape parameter 1.*

**Remark 8.** *Therefore, $\mathbb{E}(N_n(s)) = \infty$ for all $n \in \mathbb{N}$ and $\mathbb{E}(W(s)) = \infty$. Moreover:*

- *First absolute pseudo moment $\nu_1 = \int_0^\infty x \left| d\left(\mathbb{P}(N_n(s) \le n\,x) - e^{-s/x}\right)\right| = \infty$,*
- *Absolute difference moment $\chi_u = \int_0^\infty x^{u-1} \left|\mathbb{P}(N_n(s) \le n\,x) - e^{-s/x}\right| dx < \infty$ for $1 \le u < 2$, see Lemma 2 in Christoph et al. [12].*

*On pseudo moments and some of their generalizations, see Chapter 2 in Christoph and Wolf [29].*

**Lemma 4.** *In Transfer Theorem 2, the additional conditions (23) and (24) are satisfied with $H(y) = H_s(y) = e^{-s/y}$, $h_2(y) = h_{2;s}(y) = s\,e^{-s/y}\left(s - 1 + 2Q_1(n\,y)\right)/(2\,y^2)$, $y > 0$, $g_n = n$ and $b = 2$. Moreover, one has for $|\gamma| \le K < \infty$ and $f_2(z;a) = (a\,z^3 - 5\,z)\,\varphi(z)/4$, with $a = 1$ or $a = 1/3$:*

$$I_2(x,n) = \left| \int_{1/n}^\infty \frac{f_2(x\,y^\gamma;a)}{n^2\,y} dh_{2;s}(y) \right| \le C(s) n^{-2}.$$

**Lemma 5.** *For random size $N_n(s)$ with probabilities (39) with reals $s \ge s_0 > 0$ and arbitrary small $s_0 > 0$ and $n \ge 1$, we have*

$$\mathbb{E}\left(N_n(s)\right)^{-\alpha} \le C(s) n^{-\alpha}. \tag{41}$$

The Lemmas are proved in Appendix A.2.

## 5. Main Results

Consider the sample correlation coefficient $R_m = R_m(\vec{X}_m, \vec{Y}_m)$, given in (4) and the two statistics $R_m^* = \sqrt{m}\, R_m$ and $R_m^{**} = m\, R_m$ which differ from $R_m$ by scaling factors. Hence, by (10),

$$\mathbb{P}(\sqrt{m} R_m \le x) = \mathbb{P}(R_m^* \le x) = \mathbb{P}\left(\frac{1}{\sqrt{m}} R_m^{**} \le x\right) = \Phi(x) + \frac{(x^3 - 5\,x)}{4\,m}\, \varphi(x) + r(m) \tag{42}$$

with $|r(m)| \le C m^{-2}$.

Let $\theta_m$ be the angle between the vectors $\vec{X}_m$ and $\vec{Y}_m$. Contemplate the statistics $\Theta_m = \theta_m - \pi/2$, $\Theta_m^* = \sqrt{m}\,(\theta_m - \pi/2)$ and $\Theta_m^{**} = m\,(\theta_m - \pi/2)$ which differ only in scaling . Then, by (13),

$$\mathbb{P}(\sqrt{m}\,\Theta_m \le x) = \mathbb{P}(\Theta_m^* \le x) = \mathbb{P}\left(\frac{1}{\sqrt{m}} \Theta_m^{**} \le x\right) = \Phi(x) + \frac{((1/3)x^3 - 5\,x)}{4\,m}\, \varphi(x) + r^*(m)$$

with $|r^*(m)| \le C m^{-2}$.

Consider now the statistics $R_{N_n}$, $R_{N_n}^*$ and $R_{N_n}^{**}$ as well as $\Theta_{N_n}$, $\Theta_{N_n}^*$ and $\Theta_{N_n}^{**}$ when the vectors have random dimension $N_n$. The normalized statistics have different limit distributions as $n \to \infty$.

### 5.1. *The Random Dimension $N_n = N_n(r)$ Is Negative Binomial Distributed*

Let the random dimension $N_n(r)$ be negative binomial distributed with probability mass function (29) and $g_n = \mathbb{E} N_n(r) = r(n-1)+1$. "The negative binomial distribution is one of the two leading cases for count models, it accommodates the overdispersion typically observed in count data (which the Poisson model cannot)", see Schluter and Trede [26].

It follows from Theorems 1 and 2 and Proposition 1 that if limit distributions for $\mathbb{P}\left(g_n^{\gamma} N_n(r)^{1/2-\gamma} R_{N_n(r)} \le x\right)$ for $\gamma \in \{1/2, 0 - 1/2\}$ exist they are $\int_0^\infty \Phi(x\, y^\gamma) dG_{r,r}(y)$ with densities given bellow in the proof of the corresponding theorems:

$$\frac{r^r}{\sqrt{2\,\pi}\,\Gamma(r)} \int_0^\infty y^{r-1/2} e^{-(x\,y^\gamma + r\,y)} dy = \begin{cases} s_{2\,r}(x) = \dfrac{\Gamma(r+1/2)}{\sqrt{2\,r\pi}\,\Gamma(r)} \left(1 + \dfrac{x^2}{2\,r}\right)^{-(r+1/2)}, & \gamma = 1/2, \\ \varphi(x) = \dfrac{1}{\sqrt{2\,\pi}} e^{-x^2/2}, & \gamma = 0, \\ l_1(x) = \dfrac{1}{\sqrt{2}} e^{-\sqrt{2}|x|}, \quad \text{for} \quad r = 1, & \gamma = -1/2, \end{cases} \tag{43}$$

where in case $\gamma = -1/2$ for $r \ne 1$ generalized Laplace distributions occur.

#### 5.1.1. Student's $t$-Distribution

We start with the case $\gamma = 1/2$ in Theorems 1 and 2. Consider the statistic $\overline{R}_{N_n(r)} = \sqrt{g_n} R_{N_n(r)}$. The limit distribution is the Student's $t$-distribution $S_{2r}(x)$ with $2\,r$ degrees of freedom with density (31).

**Theorem 3.** *Let $r > 0$ and (29) be the probability mass function of the random dimension $N_n = N_n(r)$ of the vectors under consideration. If the representation (42) for the statistic $R_m$ and the inequality (32) with $g_n = \mathbb{E} N_n(r) = r(n-1)+1$ hold, then there exists a constant $C_r$ such that for all $n \in \mathbb{N}_+$*

$$\sup_x \left|\mathbb{P}\left(\sqrt{g_n}\, R_{N_n(r)} \le x\right) - S_{2r;n}(x;1)\right| \le C_r \begin{cases} n^{-\min\{r,2\}}, & r \ne 2, \\ \ln(n)\, n^{-2}, & r = 2, \end{cases} \tag{44}$$

*where*

$$S_{2r;n}(x;a) = S_{2r}(x) + \frac{s_{2r}(x)}{r\,n}\left(a\,x^3 - \frac{10\,r\,x + 5\,x^3}{2r-1} + \frac{(2-r)\,(x^3+x)}{4\,(2r-1)}\right). \tag{45}$$

*Moreover, the scaled angle $\theta_{N_n(r)}$ between the vectors $\vec{X}_{N_n(r)}$ and $\vec{Y}_{N_n(r)}$ allows the estimate*

$$\sup_x \left| \mathbb{P}\left(\sqrt{g_n}\,(\theta_{N_n(r)} - \pi/2) \le x\right) - S_{2r;n}(x;1/3) \right| \le C_r \begin{cases} n^{-\min\{r,2\}}, & r \neq 2, \\ \ln(n)\, n^{-2}, & r = 2, \end{cases}$$

*where $S_{2r;n}(x;1/3)$ is given in (45) with $a = 1/3$.*

Figure 2 shows the advantage of the Chebyshev–Edgeworth expansion versus the limit law in approximating the empirical distribution function.

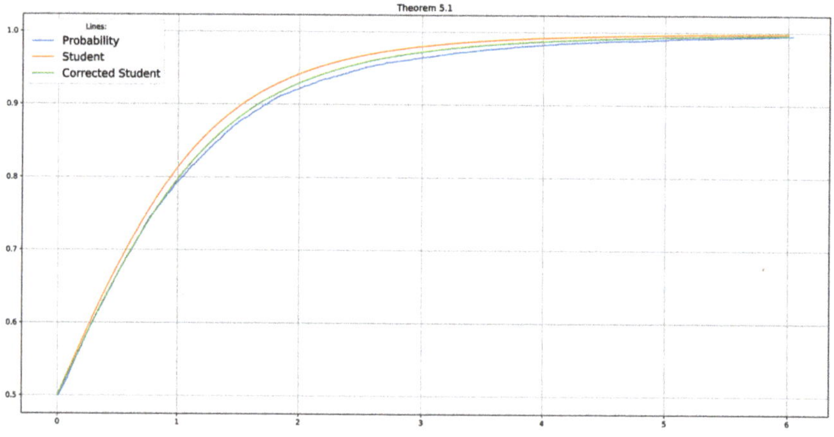


**Figure 2.** Empirical version of $\mathbb{P}\left(\sqrt{g_n}\,R_{N_n(r)} \le x\right)$ (blue line), limit Student law $S_{2r}(x)$ (orange line) and second approximation $S_{2r;n}(x;1)$ (green line) for the correlation coefficient for pairs of normal vectors with random dimension $N_{25}(2)$. Here, $x > 0, n = 25$ and $r = 2$.

**Remark 9.** *The limit Student's t-distribution $S_{2r}(x)$ is symmetric and a generalized hyperbolic distribution which can be written as a regularized incomplete beta function $I_z(a,b)$. For $x > 0$:*

$$S_{2r}(x) = \int_{-\infty}^{x} s_{2r}(u)\,du = \frac{1}{2}\left(1 + I_{2r/(x^2+2r)}(1/2, r)\right) \quad \text{and} \quad I_z(a,b) = \frac{\Gamma(a+b)}{\Gamma(a)\,\Gamma(b)} \int_0^z t^{a-1}(1-t)^{b-1}.$$

**Remark 10.** *For integer values $\nu = 2r \in \{1, 2, ...\}$ the Student's t-distribution $S_{2r}(x)$ is computable in closed form:*

$$\text{the Cauchy law} \quad S_1(x) = \frac{1}{2} + \frac{1}{\pi}\arctan(x), \qquad S_2(x) = \frac{1}{2} + \frac{x}{2\sqrt{2+x^2}},$$

$$S_3(x) = \frac{1}{2} + \frac{1}{\pi}\left(\frac{x}{\sqrt{3}(1+x^2/3)} + \arctan(x/\sqrt{3})\right) \quad \text{and} \quad S_4(x) = \frac{1}{2} + \frac{27\,(x^2+3)\,x\,(2\,x^2+9)}{8\,(3\,x^2+9)^{5/2}}.$$

**Remark 11.** *If the dimension of the vectors has the geometric distribution $N_n(1)$, then asymptotic distribution of the sample coefficient is the Student law $S_2(x)$ with two degrees of freedom.*

**Remark 12.** *The Cauchy limit distribution occurs when the dimension of the vectors has distribution $N_n(1/2)$.*

**Remark 13.** *The Student's t-distributions $S_{2r}(x)$ are heavy tailed and their moments of orders $\alpha \ge 2r$ do not exist.*

5.1.2. Standard Normal Distribution

Let $\gamma = 0$ in the Theorems 1 and 2 examining the statistics $R^*_{N_n(r)}$ and $\Theta^*_{N_n(r)} = \sqrt{N_n(r)}(\theta_{N_n(r)} - \pi/2)$.

**Theorem 4.** *Let $r > 0$ and $N_n = N_n(r)$ be the random vector dimension having probability mass function (29). If the representation (42) for the statistic $R_m$ and the inequality (32) with $g_n = \mathbb{E}N_n(r) = r(n-1)+1$ hold, then there exists a constant $C_r$ such that for all $n \in \mathbb{N}_+$*

$$\sup_x \left| \mathbb{P}\left(\sqrt{N_n(r)}\, R_{N_n(r)} \le x\right) - \Phi_{n;2}(x;1) \right| \le C_r \begin{cases} n^{-\min\{r,2\}}, & r \ne 2, \\ \ln(n)\, n^{-2}, & r = 2, \end{cases} \tag{46}$$

*where*

$$\Phi_{n;2}(x;a) = \Phi(x) + \frac{\varphi(x)}{n}\left(\frac{(a\,x^3 - 5\,x)\ln n}{4}\mathbb{I}_{\{r=1\}}(r) + \frac{\Gamma(r-1)\,(a\,x^3 - 5\,x)}{4\,\Gamma(r)}\mathbb{I}_{\{r>1\}}(r)\right). \tag{47}$$

*Moreover, the scaled angle $\theta^*_{N_n(r)}$ between the vectors $\vec{X}_{N_n(r)}$ and $\vec{Y}_{N_n(r)}$ allows the estimate*

$$\sup_x \left| \mathbb{P}\left(\sqrt{N_n(r)}\, (\theta_{N_n(r)} - \pi/2) \le x\right) - \Phi_{n;2}(x;1/3) \right| \le C_r \begin{cases} n^{-\min\{r,2\}}, & r \ne 2, \\ \ln(n)\, n^{-2}, & r = 2, \end{cases}$$

*where $\Phi_{n;2}(x;1/3)$ is given in (47) with $a = 1/3$.*

Figure 3 shows that the second order Chebyshev–Edgeworth expansion approximates the empirical distribution function better than the limit normal distribution.

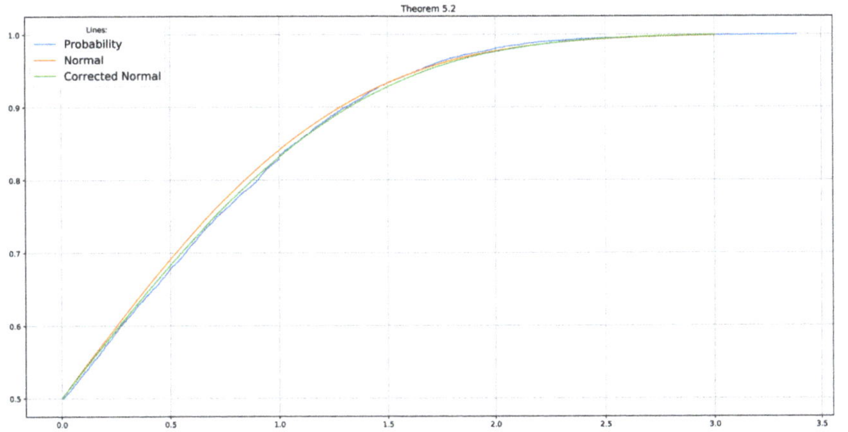


**Figure 3.** Empirical version of $\mathbb{P}\left(\sqrt{N_n(r)}\, R_{N_n(r)} \le x\right)$ (blue line), limit normal law $\Phi(x)$ (orange line) and second approximation $\Phi_{n;2}(x;1)$ (green line) for the correlation coefficient for pairs of normal vectors with random dimension $N_{25}(2)$. Here, $x > 0$, $n = 25$ and $r = 2$.

**Remark 14.** *When the distribution function of a statistic $T_m$ without standardization tends to the standard normal distribution $\Phi(x)$, i.e., $\mathbb{P}(T_m \le x) \to \Phi(x)$, then the limit law for $\mathbb{P}(T_{N_n} \le x)$ remains the standard normal distribution $\Phi(x)$.*

5.1.3. Generalized Laplace Distribution

Finally, we use $\gamma = -1/2$ in Theorems 1 and 2 examining the statistic $g_n^{-1/2} R^{**}_{N_n(r)}$. Theorems 1 and 2 state that if there exists a limit distribution of $\mathbb{P}\left(g_n^{-1/2} R^{**}_{N_n} \le x\right)$ as $n \to \infty$ then it has to be a scale mixture of normal distributions with zero mean and gamma distribution:

$$L_r(x) = \int_0^\infty \Phi(xy^{-1/2}) dG_{r,r}(y)$$

having density, see formula (A9) in the proof of Theorem 5:

$$l_r(x) = \frac{r^r}{\Gamma(r)} \int_0^\infty \varphi(xy^{-1/2})\, y^{r-3/2} e^{-ry} dy = \frac{2\, r^r}{\Gamma(r)\,\sqrt{2\,\pi}} \left(\frac{|x|}{\sqrt{2\,r}}\right)^{r-1/2} K_{r-1/2}(\sqrt{2\,r}\,|x|). \tag{48}$$

where $K_\alpha(u)$ is the *$\alpha$-order Macdonald function* or $\alpha$-order modified Bessel function of the third kind. See, e.g., Oldham et al. [30], Chapter 51, or Kotz et al. [31], Appendix, for properties of these functions.

For integer $r = 1, 2, 3, \ldots$ these densities $l_r(x)$, so-called *Sargan densities*, and their distribution functions are computable in closed forms:

$$\left.\begin{array}{lll} l_1(x) = \frac{1}{\sqrt{2}}\, e^{-\sqrt{2}\,|x|} & \text{and} & L_1(x) = 1 - \frac{1}{2} e^{-\sqrt{2}\,|x|}, \quad x > 0 \\ l_2(x) = \left(\frac{1}{2} + |x|\right) e^{-2\,|x|} & \text{and} & L_2(x) = 1 - \frac{1}{2}(1+x)\, e^{-2\,|x|}, \quad x > 0 \\ l_3(x) = \frac{3\sqrt{6}}{16}\left(1 + \sqrt{6}\,|x| + 2\,x^2\right) e^{-\sqrt{6}\,|x|)} & \text{and} & L_3(x) = 1 - \left(\frac{1}{2} + \frac{5\sqrt{6}\,x}{16} + \frac{3\,x^2}{8}\right) e^{-\sqrt{6}\,|x|}, \end{array}\right\} \tag{49}$$

where $L_r(-x) = 1 - L_r(x)$ for $x \ge 0$.

The standard Laplace distribution is $L_1(x)$ with variance 1 and density $l_1(x)$ given in (49). Therefore, Sargans distributions are a kind of generalizations of the standard Laplace distribution.

**Theorem 5.** *Let $r = 1, 2, 3$ and (29) be probability mass function of the random dimension $N_n = N_n(r)$ of the vectors under consideration. If the representation (42) for the statistic $R_m$ and the inequality (32) for $N_n(r)$ with $g_n = \mathbb{E}N_n(r) = r(n-1)+1$ hold, then there exists a constant $C_r$ such that for all $n \in \mathbb{N}_+$*

$$\sup_x \left|\mathbb{P}\left(g_n^{-1/2} N_n(r)\, R_{N_n(r)} \le x\right) - L_{n;2}(x;1)\right| \le C_r \begin{cases} n^{-\min\{r,2\}}, & r \ne 2, \\ \ln(n)\, n^{-2}, & r = 2, \end{cases} \tag{50}$$

*where*

$$L_{n;2}(x;a) = \begin{cases} L_1(x), & r = 1, \\ L_2(x) + \dfrac{a\,x\,|x| - 5\,x\,\sqrt{2}}{2(n-1)+1}\, e^{-2\,|x|}, & r = 2, \\ L_3(x) + \dfrac{27}{24(n-1)+8}\left(\dfrac{a\,x^3}{\sqrt{2}} - \dfrac{5\,x\,|x|}{6} - \dfrac{5\,x}{6\sqrt{6}}\right) e^{-\sqrt{6}\,|x|} & \\ \quad + \dfrac{9\,x}{2\,n}\left(\dfrac{1}{12\sqrt{6}} + \dfrac{|x|}{12} - \dfrac{x^2}{6\,\sqrt{6}}\right) e^{-\sqrt{6}\,|x|}, & r = 3. \end{cases} \tag{51}$$

*For arbitrary $r > 0$, the approximation rate is given by:*

$$\sup_x \left|\mathbb{P}\left(g_n^{-1/2} N_n(r)\, R_{N_n(r)} \le x\right) - L_r(x)\right| \le C_r n^{-\min\{r\ 1\}}.$$

*Moreover, the scaled angle $N_n(r)\,\theta_{N_n(r)}$ between the vectors $\vec{X}_{N_n(r)}$ and $\vec{Y}_{N_n(r)}$ allows the estimate*

$$\sup_x \left|\mathbb{P}\left(g_n^{-1/2} N_n(r)\, \theta_{N_n(r)} \le x\right) - L_{n;2}(x;1/3)\right| \le C_r \begin{cases} n^{-\min\{r,2\}}, & r \ne 2, \\ \ln(n)\, n^{-2}, & r = 2, \end{cases}$$

*where* $L_{n;2}(x;1/3)$ *is given in* (51) *with* $a = 1/3$.

Figure 4 shows that the Chebyshev–Edgeworth expansion approaches the empirical distribution function better than the limit Laplace law.

**Remark 15.** *One can find the distribution functions* $L_r(x)$ *for arbitrary* $r > 0$ *with formula 1.12.1.3 in Prudnikov et al.* [32]:

$$
\begin{aligned}
L_r(x) &= \frac{1}{2} + \frac{2\,r^r}{\sqrt{2\,\pi}\,\Gamma(r)} \int_0^x \left(\frac{|x|}{2\,r}\right)^{r-1/2} K_{r-1/2}(\sqrt{2\,r}\,|x|)dx \\
&= \frac{1}{2} + \frac{x}{(2\,r)^{(r-1/2)/2}} \Big( K_{r-1/2}(\sqrt{2\,r x}) \mathbf{L}_{r-3/2}(\sqrt{2\,r x}) + K_{r-3/2}(\sqrt{2\,r x}) \mathbf{L}_{r-1/2}(\sqrt{2\,r x}) \Big).
\end{aligned}
$$

*where* $\mathbf{L}_\alpha(x)$ *are the modified Struve functions of order* $\alpha$, *for properties of modified Struve functions see, e.g., Oldham et al.* [30], *Section 57:13.*

**Remark 16.** *The function* (48) *as density of a mixture of normal distributions with zero mean and random variance* $W_r$ *having gamma distribution* $\mathbb{P}(W_r \le y) = G_{r,r}(y)$ *is given also in Kotz et al.* [31], *Formula (4.1.32) with* $\tau = r$, $\sigma = 1/\sqrt{r}$, *using Formula (A.0.4) with* $\lambda = -r + 3/2$ *and the order-reflection formula* $K_{-\alpha}(x) = K_\alpha(x)$. *Such a variance gamma model is studied in Madan and Senata* [33] *for share market returns.*

**Remark 17.** *A systematic exposition about the Laplace distribution and its numerous generalization and diverse applications one finds in the useful and interesting monography by Kotz et al.* [31]. *Here, these generalized Laplace distributions* $L_1(x)$, $L_2(x)$ *and* $L_3(x)$ *are the leading terms in the approximations of the sample correlation coefficient* $R^{**}_{N_n(r)}$ *of two Gaussian vectors with negative binomial distributed random dimension* $N_n(r)$ *and the angle* $\theta^{**}_{N_n(r)}$ *between these vectors.*

**Remark 18.** *In Goldfeld and Quandt* [34] *and Missiakoulis* [35] *Sargans densities* $l_r(x)$ *and distribution functions* $L_r(x)$ *for arbitrary integer* $r = 1, 2, 3, \ldots$ *have been studied as an alternative to normal law in econometric models because they are computable in closed form, see also Kotz et al.* [31], *Section 4.4.3 and the references therein.*

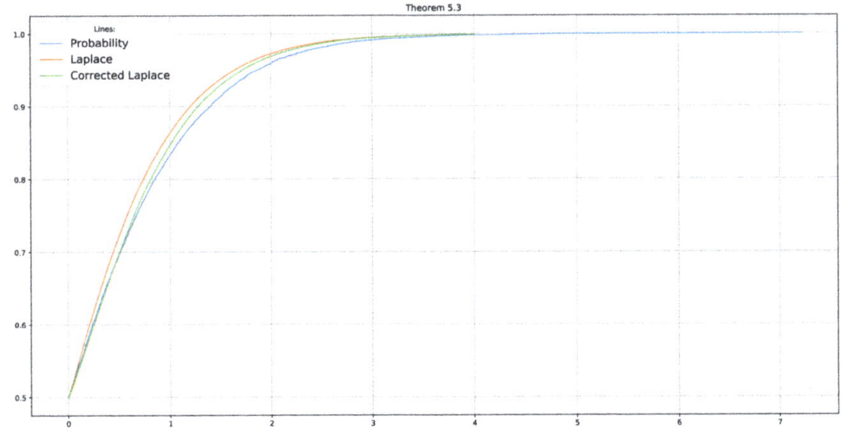


**Figure 4.** Empirical version of $\mathbb{P}\left(g_n^{-1/2}\, N_n(r)\, R_{N_n(r)} \le x\right)$ (blue line), limit Laplace law $L_r(x)$ (orange line) and second approximation $L_{n;2}(x;1)$ (green line) for the correlation coefficient for pairs of normal vectors with random dimension $N_{25}(2)$. Here, $x > 0, n = 25$ and $r = 2$.

### 5.2. The Random Dimension $N_n = N_n(s)$ Is the Maximum of $n$ Independent Discrete Pareto Random Variables

The random dimension $N_n(s)$ has probability mass function (39): Since $\mathbb{E}N_n(s) = \infty$ we choose $g_n = n$ and consider again the cases $\gamma = 1/2$, $\gamma = 0$ and $\gamma = -1/2$.

It follows from Theorems 1 and 2 and Proposition 2 that if limit distributions for $\mathbb{P}\left(g_n^\gamma R_{N_n(s)} \le x\right)$ for $\gamma \in \{1/2, 0 - 1/2\}$ exist, they are $\int_0^\infty \Phi(x\,y^\gamma) dH_s(y)$ with densities given below in the proof of the corresponding theorems

$$\frac{s}{\sqrt{2\pi}} \int_0^\infty y^{-3/2} e^{-(x^2 y^{2\gamma}/2 + s/y)} dy = \begin{cases} l_{1/\sqrt{s}}(x) = \frac{\sqrt{2s}}{2} e^{-\sqrt{2s}|x|}, & \gamma = 1/2, \\ \varphi(x) = \frac{1}{\sqrt{2\pi}} e^{-x^2/2}, & \gamma = 0, \\ s_2^*(x; \sqrt{s}) = \frac{1}{2\sqrt{2s}} \left(1 + \frac{x^2}{2s}\right)^{-3/2}, & \gamma = -1/2, \end{cases} \tag{52}$$

where $s_2^*(x; \sqrt{s})$ is the density of the *scaled Student's t-distribution* $S_2^*(x; \sqrt{s})$ *with 2 degrees of freedom*, see Definition B37 in Jackman [36], p.507. If $Z$ has density $s_2^*(x; \sqrt{s})$ then $Z/\sqrt{s}$ has a classic Student's $t$-distribution with two degrees of freedom.

#### 5.2.1. Laplace Distribution

We start with the case $\gamma = 1/2$ in Theorems 1 and 2. Consider the statistics $\sqrt{n}\,R_{N_n(s)}$ and $\sqrt{n}(\theta_{N_n(s)} - \pi/2)$. The limit distribution is now the Laplace distribution

$$L_{1/\sqrt{s}}(x) = \frac{1}{2} + \frac{1}{2}\text{sign}(x)\left(1 - e^{-\sqrt{2s}|x|}\right) \quad \text{with density} \quad l_{1/\sqrt{s}}(x) = \frac{\sqrt{2s}}{2} e^{-\sqrt{2s}|x|}.$$

**Theorem 6.** *Let $s > 0$ and (39) be the probability mass function of the random dimension $N_n = N_n(s)$ of the vectors under consideration. If the representation (42) for the statistic $R_m$ and the inequality (32) with $g_n = n$ hold, then there exists a constant $C_s$ such that for all $n \in \mathbb{N}_+$*

$$\sup_x \left| \mathbb{P}\left(\sqrt{n}\,R_{N_n(s)} \le x\right) - L_{1/\sqrt{s};n}(x; a)\right| \le C_s n^{-2},$$

*where*

$$L_{1/\sqrt{s};n}(x; a) = L_{1/\sqrt{s}}(x) + \frac{l_{1/\sqrt{s}}(x)}{8\,s\,n}\left(a\,2\,s\,x^3 - (4 - s)x\left(1 + \sqrt{2s}\,|x|\right)\right). \tag{53}$$

*Moreover, the scaled angle $\theta_{N_n(s)}$ between the vectors $\vec{X}_{N_n(s)}$ and $\vec{Y}_{N_n(s)}$ allows the estimate*

$$\sup_x \left| \mathbb{P}\left(\sqrt{n}\,(\theta_{N_n(s)} - \pi/2) \le x\right) - L_{1/\sqrt{s};n}(x; 1/3)\right| \le C_s\, n^{-2},$$

*where $L_{1/\sqrt{s};n}(x; 1/3)$ is given in (53) with $a = 1/3$.*

#### 5.2.2. Standard Normal Distribution

Let $\gamma = 0$ in the Theorems 1 and 2 examine the statistics $R^*_{N_n(s)}$ and $\Theta^*_{N_n(s)} = \sqrt{N_n(s)}(\theta_{N_n(s)} - \pi/2)$.

**Theorem 7.** *Let $s > 0$ and $N_n = N_n(s)$ be the random vector dimension having probability mass function (39). If the representation (42) for the statistic $R_m$ and the inequality (32) with $g_n = n$ hold, then there exists a constant $C_s$ such that, for all $n \in \mathbb{N}_+$*

$$\sup_x \left| \mathbb{P}\left(\sqrt{N_n(s)}\,R_{N_n(s)} \le x\right) - \Phi(x) - \frac{1}{4n}\varphi(x)\,s^2\,(x^3 - 5\,x)\right| \le C_s\, n^{-2},$$

*Moreover, the scaled angle $\theta^*_{N_n(s)}$ between the vectors $\vec{X}_{N_n(s)}$ and $\vec{Y}_{N_n(s)}$ allows the estimate*

$$\sup_x \left| \mathbb{P}\left( \sqrt{N_n(s)}\,(\theta_{N_n(s)} - \pi/2) \le x \right) - \Phi(x) - \frac{1}{4n}\varphi(x)\, s^2\,(\frac{1}{3}x^3 - 5\,x) \right| \le C_s n^{-2}.$$

### 5.2.3. Scaled Student's $t$-Distribution

Finally, we use $\gamma = -1/2$ in Theorems 1 and 2 examining the statistics $n^{-1/2}\,N_n(s)\,R_{N_n(s)}$ and $n^{-1/2}\,N_n(s)\,(\theta_{N_n(s)} - \pi/2)$. The limit *Scaled Student's t-Distribution $S_2^*(x;\sqrt{s})$ with two degrees of freedom* is a scale mixture of the normal distribution with zero mean and mixing exponential distribution $1 - e^{-sy}$, $y \ge 0$, and it is representable in a closed form, see (A15) below in the proof of Theorem 8:

$$\begin{aligned} \int_0^\infty \Phi(x/\sqrt{y})de^{-s/y} &= \int_0^\infty \Phi(x/\sqrt{y})sy^{-2}e^{-s/y}dy = \int_0^\infty \Phi(x\sqrt{z})se^{-sz}dz \\ &= \int_0^\infty \Phi(x\sqrt{z})d(1-e^{-sz})dz = \frac{1}{2} + \frac{x/\sqrt{s}}{2\sqrt{2}\sqrt{1+x^2/(2s)}} = S_2^*(x) \end{aligned}$$

**Theorem 8.** *Let $s > 0$ and $N_n = N_n(s)$ be the random vector dimension having probability mass function (39). If the representation (42) for the statistic $R_m$ and the inequality (32) with $g_n = n$ hold, then there exists a constant $C_s$ such that for all $n \in \mathbb{N}_+$*

$$\sup_x \left| \mathbb{P}\left( n^{-1/2}\,N_n(s)\,R_{N_n(s)} \le x \right) - S^*_{n;2}(x;1) \right| \le C_r\, n^{-2}, \tag{54}$$

*where*

$$S^*_{n;2}(x;\sqrt{s};a) = S_2^*(x;\sqrt{s}) + \frac{(15a + 3s - 18)x^3 - 6\,x\,s(6-s)}{4\,n\,(x^2 + 2\,s)^2}\, s_2^*(x;\sqrt{s}) \tag{55}$$

*Moreover, the scaled angle $\theta^*_{N_n(s)}$ between the vectors $\vec{X}_{N_n(s)}$ and $\vec{Y}_{N_n(s)}$ allows the estimate*

$$\sup_x \left| \mathbb{P}\left( n^{-1/2}\,N_n(s)\,\theta_{N_n(s)} \le x \right) - S_{n;2}(x;\sqrt{s};1/3) \right| \le C_s\, n^{-2},$$

*where $S_{n;2}(x;\sqrt{s};1/3)$ is given in (55) with $a = 1/3$.*

Theorems 3 to 8 are proved in Appendix A.3.

## 6. Conclusions

The asymptotic distributions of the sample correlation coefficient of vectors with random dimensions are normal scale mixtures. From (43) and (52), one can conclude that random dimension and corresponding scaling have significant influence on limit distributions A scale mixture of a normal distribution change the tail behavior of the distribution. Students $t$-Distributions have polynomial tails, as one class of heavy-tailed distributions, they can be used to model heavy-tail returns data in finance. The Laplace distributions have heavier tails than normal distributions.

**Author Contributions:** Conceptualization, G.C. and V.V.U.; methodology, V.V.U. and G.C.; writing–original draft, G.C. and V.V.U.; writing–review and editing, V.V.U. and G.C. All authors have read and agreed to the published version of the manuscript.

**Funding:** This research received no external funding. It was done within the framework of the Moscow Center for Fundamental and Applied Mathematics, Lomonosov Moscow State University, and HSE University Basic Research Programs.

**Acknowledgments:** The authors would like to thank the Managing Editor and the Reviewers for the careful reading of the manuscript and pertinent comments. Their constructive feedback helped to improve the quality of this work and shape its final form.

**Conflicts of Interest:** The authors declare no conflict of interest.

## Appendix A. Proofs of the Theorems and Lemmas

### *Appendix A.1. Proofs of Theorems 1 and 2*

**Proof of Theorem 1.** The proof follows along the similar arguments of the more general transfer Theorem 3.1 in Bening et al. [24]. Since in Theorem 3.1 in Bening et al. [24] the constant $\gamma$ has to be non-negative and in our Theorem 1, we also need $\gamma = -1/2$, therefore we repeat the proof. Conditioning on $N_n$, we have

$$\mathbb{P}\Big(g_n^\gamma T_{N_n} \le x\Big) = \mathbb{P}\Big(N_n^\gamma T_{N_n} \le x\,(N_n/g_n)^\gamma\Big) = \textstyle\sum_{m=1}^\infty \mathbb{P}\Big(m^\gamma T_m \le x(m/g_n)^\gamma\Big)\,\mathbb{P}(N_n = m).$$

Using now (18) with $\Phi_m(z) := \Phi(z) + m^{-1} f_2(z)$, we find

$$\begin{aligned}\textstyle\sum_{m=1}^\infty &\Big|\mathbb{P}\Big(m^\gamma T_m \le x(m/g_n)^\gamma\Big) - \Phi_m(x(m/g_n)^\gamma)\Big|\;\mathbb{P}(N_n = m)\\ &\overset{(18)}{\le} C_1 \textstyle\sum_{m=1}^\infty m^{-a}\,\mathbb{P}(N_n = m) = C_1\,\mathbb{E}(N_n^{-a}).\end{aligned} \tag{A1}$$

Taking into account $\mathbb{P}\Big(N_n/g_n < 1/g_n\Big) = \mathbb{P}\Big(N_n < 1\Big) = 0$, we obtain

$$\begin{aligned}\textstyle\sum_{m=1}^\infty \Phi_m(x\,(m/g_n)^\gamma)\mathbb{P}(N_n = m) &= \mathbb{E}\left(\Phi_{N_n}(x\,(N_n/g_n)^\gamma)\right)\\ &= \int_{1/g_n}^\infty \Delta_n(x,y)d\mathbb{P}\Big(\frac{N_n}{g_n} \le y\Big) = G_n(x, 1/g_n) + I_1,\end{aligned}$$

where $\Delta_n(x,y) := \Phi(xy^\gamma) + f_2(xy^\gamma)/(g_n y)$, $G_n(x,1/g_n)$ is defined in (21) and

$$I_1 = \int_{1/g_n}^\infty \Delta_n(x,y)d\Big(\mathbb{P}\Big(\frac{N_n}{g_n} \le y\Big) - H(y) - \frac{h_2(y)\,\mathbb{I}_{\{b>1\}}(b)}{n}\Big).$$

Estimating integral $I_1$, we use the integration by parts for Lebesgue–Stieltjes integrals, the boundedness of $f_2(z)$, say $\sup_z |f_2(z)| \le c_1^*$, and estimates (19)

$$\begin{aligned}|I_1| &\le \sup_x \lim_{L\to\infty} |\Delta_n(x,y)|\,\left|\mathbb{P}(N_n/g_n \le y) - H(y) - n^{-1}h_2(y)\,\mathbb{I}\{b > 1\}(b)\right|\Big|_{y=1/g_n}^{y=L}\\ &\quad + \sup_x \int_{1/g_n}^\infty \left|\frac{\partial}{\partial y}\Delta_n(x,y)\right|\,\left|\mathbb{P}(N_n/g_n \le y) - H(y) - n^{-1}h_2(y)\,\mathbb{I}_{\{b>1\}}(b)\right| dy\\ &\le (1 + c_1^*)\,C_2\,n^{-b} + C_2\,D_n\,n^{-b},\end{aligned}$$

where $D_n$ is defined in (22). Together with (A1), we obtain (20) and Theorem 1 is proved. □

**Proof of Theorem 2.** Using (23), we find for $b > 0$

$$\int_0^{1/g_n} \Phi(x\,y^\gamma)dH(y) \le \int_0^{1/g_n} dH(y) = H(1/\sqrt{g_n}) - H(0) \overset{(23)}{\le} c_1 g_n^{-b}.$$

Let now $b > 1$. Since $f_2(z)$ is supposed to be bounded, it follows from $|f_2(z)| \le c_1^* < \infty$ and (24i) that

$$\int_0^{1/g_n} |f_2(x\,y^\gamma)|y^{-1}dH(y) \le c_1^* \int_0^{1/g_n} y^{-1}dH(y) \overset{(24i)}{\le} c_1^*\,c_2 g_n^{-b+1}.$$

Integration by parts, $|z|\varphi(z) \le c^* = (2\,\pi\,e)^{-1/2}$, (24ii) and (24iii) lead to

$$\left|\int_0^{1/g_n} \Phi(x\,y^\gamma)dh_2(y)\right| \le |h_2(1/g_n)| + \gamma c^* \int_0^{1/g_n} y^{-1}|h_2(y)|dy \le (c_3 + \gamma c^* c_4)n\,g_n^{-b}.$$

Theorem 2 is proved. □

*Appendix A.2. Proofs of Lemmas 1 to 5*

**Proof of Lemma 1.** To estimate $D_n$, we consider three cases:

$$D_n = \sup_x |D_n(x)| = \max\{\sup_{x>0}|D_n(x)|, \sup_{x<0}|D_n(x)|, |D_n(0)|\}.$$

Let $x > 0$. Since $\frac{\partial}{\partial y}\Phi(x\,y^\gamma) = \gamma\,x\,y^{\gamma-1}\varphi(x\,y^\gamma) \geq 0$, we find

$$\int_{1/g_n}^{\infty}\left|\frac{\partial}{\partial y}\Phi(x\,y^\gamma)\right|dy = \int_{1/g_n}^{\infty}\gamma\,x\,y^{\gamma-1}\varphi(x\,y^\gamma)dy = \int_{x\,g_n^{-\gamma}}^{\infty}\varphi(u)du = \Phi(\infty) - \Phi(x\,g_n^{-\gamma}) \leq 1/2.$$

Consider now $f_2(x\,y^\gamma; a) = (a(x\,y^\gamma)^3 - 5\,x\,y^\gamma)\,\varphi(x\,y^\gamma)/4$ with $a = 1$ or $a = 1/3$. Then,

$$\frac{\partial}{\partial y}\left(\frac{f_2(x\,y^\gamma;a)}{y}\right) = \frac{Q_5(x\,y^\gamma;a)}{4\,y^2}, \quad Q_5(z;a) = -(\gamma\,a\,z^5 - ((3a+5)\,\gamma - a)\,z^3 + 5\,(\gamma-1)z)\,\varphi(z). \quad \text{(A2)}$$

Since $\sup_z |Q_5(z;a)| \leq c(\gamma;a) < \infty$ and $g_n^{-1}\int_{1/g_n}^{\infty} y^{-2}dy = 1$, inequality (29) holds for $x > 0$. Taking into account $|D_n(x)| = |D_n(-x)|$ and $D_n(0) = 0$, Lemma 1 is proved. □

**Proof of Lemma 2.** Using (30), we find $G_{r,r}(1/g_n) \leq c_1 g_n^{-r}$ with $c_1 = r^{r-1}/\Gamma(r)$. For $r > 1$, then $\int_0^{1/g_n} y^{-1}dG_{r,r}(y) \leq c_2 g_n^{-r+1}$ with $c_2 = r^r/((r-1)\Gamma(r))$. Since $g_{r,r}(0) = 0$, $h_{2;r}(0) = 0$ and $g_n \leq r\,n$ for $r > 1$, then (24ii) and (24iii) hold with $c_3 = c_r^*$ and $c_4 = c_r^*/(r-1)$, where $c_r^* = \frac{r^r}{2r\,\Gamma(r)}\sup_y\{e^{-r\,y}\,(|y-1||2-r|+1)\} < \infty$.

It remains to prove the bounds in (34) and (35). Let first $r < 1$. With $c_1^* = \sup_z |f_2(z;a)|$, we find

$$|I_1(x,n)| \leq \frac{c_1^*\,r^r}{g_n\,\Gamma(r)}\int_{1/g_n}^{\infty} y^{r-2}dy \leq \frac{c_1^*\,r^r}{(r-1)\,\Gamma(r)}\,g_n^{-r} \quad \text{with} \quad c_5 = \frac{c_1^*\,r^r}{(r-1)\,\Gamma(r)}.$$

If $r = 1$ with $c_1^{**} = \sup_z\{|a\,z^2 - 5|\varphi(z/\sqrt{2})\}$, we find $|f_2(z;a)| \leq c_1^{**}|z|\,\varphi(z/\sqrt{2})$ and

$$|I_1(x,n)| \leq \frac{c_1^{**}\,|x|}{\sqrt{2\,\pi}\,n}\int_{1/n}^{\infty} y^{\gamma-1}e^{-\,(y+x^2\,y^{2\gamma}/4)}\,dy \quad \text{with} \quad \gamma \in \{-1/2, 0, 1/2\}.$$

For $\gamma = 1/2$ using $|x|\,(1+x^2/4)^{-1/2} \leq 2$, we obtain

$$|I_1(x,n)| \leq \frac{c_1^{**}\,|x|}{\sqrt{2\,\pi}\,n}\int_{1/n}^{\infty} y^{1/2-1}e^{-\,(1+x^2/4)\,y}\,dy \leq \frac{c_1^{**}\,|x|\Gamma(1/2)}{\sqrt{2\,\pi}\,(1+x^2/4)^{1/2}}n^{-1} \leq c_6\,n^{-1}, \quad c_6 = \sqrt{2}\,c_1^{**}.$$

If $\gamma = -1/2$, then Prudnikov et al. [37], formula 2.3.16.3, for $x \neq 0$ leads to

$$I_1(x,n) \leq \frac{c_1^{**}\,|x|}{\sqrt{2\,\pi}\,n}\int_{1/n}^{\infty} y^{-1-1/2}\,e^{-(2\,y+x^2/(4\,y))}dy \leq \frac{c_1^{**}\,|x|}{\sqrt{2\,\pi}\,n}\,\frac{2\sqrt{\pi}}{|x|}e^{-\sqrt{2}\,|x|} \leq \frac{\sqrt{2}\,c_1^{**}}{n}, \quad c_6 = \sqrt{2}\,c_1^{**}.$$

Finally, if $\gamma = 0$, then $f_2(x\,y^\gamma;a) = f_2(x;a)$ does not depend on $y$. Using now

$$0 \leq \ln n - \int_{1/n}^{1} y^{-1}dG_{1,1}(y) = \int_{1/n}^{1}\frac{1-e^{-y}}{y}dy \leq 1 \text{ and } \int_{1}^{\infty} y^{-1}dG_{1,1}(y) \leq e^{-1},$$

then (34) for $r = 1$ holds with $c_6 = c_1^*(1+e^{-1})$.

Let $r > 1$. Integration by parts for Lebesgue–Stieltjes integrals in $I_2(x,n)$ in (35) and (A2) lead to

$$I_2(x,n) \leq \frac{1}{n\,g_n}\left(c_1^*\,g_n\,|h_{2;r}(1/g_n)| + \int_{1/g_n}^{\infty}\frac{|Q_5(xy^\gamma;a)|}{4\,y^2}\,|h_{2;r}(y)|dy\right). \quad \text{(A3)}$$

Since $c(\gamma;a) = \sup_z |Q_5(z;a)| < \infty$ and with above defined $c_r^*$, we find

$$\int_{1/g_n}^{\infty} \frac{|h_{2;r}(y)|}{y^2} dy \le c_r^* \int_{1/g_n}^{\infty} y^{r-3} dy = \frac{c_r^*}{(2-r)}\, g_n^{-r+2} \quad \text{for} \quad 1 < r < 2$$

and with $c_r^{**} = \frac{r^{r-1}}{2\,\Gamma(r)} \sup_y\{(e^{-ry/2}\,(|y-1|\,|2-r|+1)\} < \infty$, we obtain

$$\int_{1/g_n}^{\infty} \frac{|h_{2;r}(y)|}{y^2} dy \le c_r^{**} \int_{1/g_n}^{\infty} y^{r-3} e^{-ry/2} dy \le \frac{c_r^{**}\Gamma(r-2)}{(r/2)^{r-2}} \quad \text{for} \quad r > 2.$$

Hence, we obtain (35) for $r > 1$, $r \neq 2$ with some constant $0 < c_7 < \infty$.

For $r = 2$, the second integral in line above is an exponential integral. Therefore, we estimate the integral $I_2(x,n)$ in (35) more precisely like in estimating $I_1(x,n)$ above, taking into account the given function $f_2(z;a)$.

Using $|h_{2;2}(y)| \le 4\,y\,e^{-2y}$ and consider (A2), define $P_4(z;a)$ by $Q_5(z;a) = -z\,P_4(z;a)\varphi(z/\sqrt{2})$ with $c_2^* = \sup_z |P_4(z;a)|\varphi(z/\sqrt{2}) < \infty$, we obtain in (A3)

$$\int_{1/g_n}^{\infty} \frac{|Q_5(x\,y^{\gamma})|}{4\,y^2}\,|h_{2;2}(y)|dy \le \frac{c_2^*\,|x|}{\sqrt{2\,\pi}} \int_{1/g_n}^{\infty} y^{\gamma-1}\, e^{-(2y+x^2\,y^{2\gamma}/4)} dy.$$

We estimate the latter integral in the same way as $I_1(x,n)$ for the two cases $\gamma = 1/2$ $\gamma = -1/2$ and find (35) for $r = 2$ with some constants $0 < c_7 < \infty$.

In order to prove (35) for $r = 2$ and $\gamma = 0$, we consider for $\alpha > 0$ the following inequalities:

$$\int_{1/g_n}^{\infty} y^{-1} e^{-\alpha y} dy \begin{cases} \le \int_{1/g_n}^{1} y^{-1} dy + \int_1^{\infty} e^{-\alpha y} dy \le \ln g_n + \alpha^{-1} e^{-\alpha}, \\ \ge \int_{1/g_n}^{1} y^{-1} e^{-\alpha y} dy \ge e^{-\alpha} \int_{1/g_n}^{1} y^{-1} dy \ge e^{-\alpha} \ln g_n \end{cases}. \tag{A4}$$

The upper bound in (A4) leads to (35) for $r = 2$, $\gamma = 0$, too. The lower bound in (A4) shows that the $\ln n$-term cannot be improved.

Bound (36) for $n \ge 2$, $r > 1$ results from $0 \le \frac{1}{g_n} - \frac{1}{r\,n} = \frac{r-1}{r^2\,n^2(1-(r-1)/(r\,n))} \le \frac{2(r-1)}{r^2\,n^2}$. □

**Proof of Lemma 3.** Let $r > 0$. If $n = 1$, then $\mathbb{P}(N_1(r) = 1) = 1$ and (37) holds with $C(r) = 1$. Let $n \ge 2$ and $\alpha > 0$

$$\mathbb{E}(N_n(r))^{-\alpha} = \frac{1}{n^r}\left(1 + \sum_{k=2}^{\infty} \frac{\Gamma(k+r-1)}{k^{\alpha}\,\Gamma(r)\,\Gamma(k)}\left(1-\frac{1}{n}\right)^{k-1}\right).$$

It follows from the relations (49) and (50) with their corresponding bounds in the proof of Theorem 1 in Christoph et al. [12] that

$$\frac{\Gamma(k+r-1)}{\Gamma(r)\Gamma(k)} = \frac{1}{(k+r-1)\,B(r\;k)} = \frac{k^{r-1}}{\Gamma(r)}\Big(1+R_1(k)\Big), \quad |R_1(k)| \le \frac{c_1(r)}{k}. \tag{A5}$$

For $x \ge k \ge 2$ using $(1-1/n)^x \le e^{-x/n}$, we find

$$\frac{k^{r-1}(1-1/n)^{k-1}}{k^{\alpha}} \le \int_k^{k+1} \frac{x^r\,(1-1/n)^{x-2}}{(x-1)^{1+\alpha}} dx \le 2^{3+\alpha} \int_k^{k+1} x^{r-3}\, e^{-x/n} dx.$$

Then, with $c_2(r) = 2^{3+\alpha}(1+c_1(r))/\Gamma(r)$, we obtain

$$\mathbb{E}(N_n(r))^{-\alpha} \le c_2(r) n^{-r} J_r(n), \quad \text{where} \quad J_r(n) = \int_1^{\infty} x^{r-\alpha-1}\, e^{-x/n} dx = n^{r-\alpha} \int_{1/n}^{\infty} y^{r-\alpha-1}\, e^{-y} dy.$$

Since $J_r(n) \le (\alpha - r)^{-1}$ for $0 < r < \alpha$, $J_r(n) \le n^{r-\alpha}\,\Gamma(r-\alpha)$ for $r > \alpha$ and using (A4) with $r = \alpha$ $J_r(n) \le \ln n + e^{-1}$, the upper bound (37) is proved.

Let $r = \alpha > 0$. Considering the formula (A5), $0 \le \sum_{k=2}^{\infty} k^{-1}|R_1(k)| \le c_1(r)\pi^2/(6\Gamma(r)) < \infty$, $\sum_{(n)} := \sum_{k=2}^{n-1} k^{-1} \ge \ln n - \ln 2$ and $\sum_{k=2}^{n-1} \dfrac{1-(1-1/n)^{k-1}}{k} \le \sum_{k=2}^{n-1} \dfrac{k-1}{k\,n} \le 1$, we find:

$$\mathbb{E}(N_n(r))^{-r} \ge \frac{1}{n^r\,\Gamma(r)}\left(\sum_{k=2}^{n-1}\frac{1}{k}\Big(1-\frac{1}{n}\Big)^{k-1} - c_3\right) \ge \frac{1}{n^r\,\Gamma(r)}\left(\sum_{k=2}^{n-1}\frac{1}{k} - c_4\right) \ge \frac{1}{n^r\,\Gamma(r)}\,(\ln n - c_5)\,,$$

where $c_3 = c_1(r)\pi^2/6$, $c_4 = 1 + c_3$ and $c_5 = c_4 - \ln 2$. Hence, the $\ln n$-term cannot be dropped. □

**Proof of Lemma 4.** The upper bounds in the estimates (23) and (24) with $H_s(y)$, $h_{2;s}(y)$ and $I_2(x,n)$ given in (40) are $C(s)e^{-sn/2}$. For example, (24ii):
$\int_0^{1/n} y^{-1}|h_{2;s}(y)|dy \le s(s+1)/2\int_0^{1/n} y^{-3}\,e^{-s/y}dy \le (s+1)/(2s)\int_{sn}^{\infty} z\,e^{-z}dz \le (s+1)/(2s)e^{-sn/2}$. □

**Proof of Lemma 5.** Proceeding as in Bening et al. [24] using

$$\mathbb{P}(N_n(s)=k) = \left(\frac{k}{s+k}\right)^n - \left(\frac{k-1}{s+k-1}\right)^n = s\,n\int_{k-1}^{k}\frac{x^{n-1}}{(s+x)^{n+1}}dx$$

and Formula 2.2.4.24 in Prudnikov et al. [37], p. 298, then

$$\mathbb{E}(N_n^{-\alpha}) = s\,n\sum_{k=1}^{\infty}\frac{1}{k^\alpha}\int_{k-1}^{k}\frac{x^{n-1}}{(s+x)^{n+1}}dx \le s\,n\int_0^{\infty}\frac{x^{n-1-\alpha}}{(s+x)^{n+1}}dx = s\,n\,B(n-\alpha,1+\alpha).$$

Using $B(n-\alpha,1+\alpha) = \Gamma(1+\alpha)\,(n+1)^{-1+\alpha}(1+R_1/n)$ with $|R_1| \le c < \infty$, we obtain (41). □

*Appendix A.3. Proofs of Theorems 3 to 8*

**Proof of Theorem 3.** Since the additional assumptions (23) and (24) in the transfer Theorem 2 for the limit Gamma distribution $H(x) = G_{r,r}(x)$ of the normalized sample size $N_n(r)$ are satisfied by Lemma 2 with $b = r > 0$ and by Lemma 3 for $\alpha = 2$, it remains to calculate the integrals in (26). Define

$$J_1^*(x) = \int_0^{\infty}\Phi(x\sqrt{y})dG_{r,r}(y), \qquad J_2^*(x) = \int_0^{\infty}\frac{a\,(x\sqrt{y})^3 - 5x\sqrt{y}\,\varphi(x\sqrt{y})}{4y}dG_{r,r}(y), \qquad \text{and}$$

$$J_3^*(x) = \int_0^{\infty}\Phi(x\sqrt{y})dh_{2;r}(y) \quad \text{with} \quad h_{2;r}(y) = \Big((y-1)\,(2-r) + 2Q_1((r(n-1)+1)y)\Big)\,\frac{g_{r,r}(y)}{2r},$$

and $Q_1(y) = 1/2 - (y - [y])$. Then,

$$G_{2;n}(x;0) = J_1^*(x) + \frac{J_2^*(x)}{g_n} + \frac{J_3^*(x)}{n} \quad \text{with} \quad g_n = \mathbb{E}N_n(r) = r(n-1)+1\,. \tag{A6}$$

Using formula 2.3.3.1 in Prudnikov et al. [37], p. 322, with $\alpha = r - 1/2,\ r + 1/2,\ \ p = 1 + x^2/(2\,r)$ and $q = 1$:

$$M_\alpha(x) = \frac{r^r}{\Gamma(r)\,\sqrt{2\pi}}\int_0^{\infty} y^{\alpha-1}e^{-(r+x^2/2)y}dy = \frac{\Gamma(\alpha)\,r^{r-\alpha}}{\Gamma(r)\,\sqrt{2\,\pi}}\left(1 + x^2/(2r)\right)^{-\alpha} \tag{A7}$$

we calculate the integrals occurring in (A6). Consider

$$\frac{\partial}{\partial x}J_1^*(x) = \int_0^{\infty} y^{1/2}\varphi(x\sqrt{y})g_{r,r}(y)dy = \frac{r^r}{\Gamma(r)\,\sqrt{2\,\pi}}\int_0^{\infty} y^{r-1/2}e^{-(r+x^2/2)y}dy$$
$$= M_{r+1/2}(x) = s_{2r}(x) \quad \text{and} \qquad J_1^*(x) = S_{2r}(x)\,.$$

The integral $J_2^*(x)$ in (A6) we calculate again with (A7) using $M_{r-1/2}(x) = s_{2r}(x)\,(2r + x^2)/(2r - 1)$ and $M_{r+1/2}(x) = s_{2r}(x)$

$$\begin{aligned} J_2^*(x) &:= \frac{r^r}{\sqrt{2\pi}\,\Gamma(r)} \int_0^\infty \frac{1}{y}\left(a\,x^3 y^{3/2} - 5\,x\,y^{1/2}\right) y^{r-1}\, e^{-(r+x^2/2)y} dy \\ &= \left(a\,x^3\, M_{r+1/2}(x) - 5\,x\, M_{r-1/2}(x)\right) = \left(a\,x^3 - \frac{10\,r\,x + 5\,x^3}{2r-1}\right) s_{2r}(x). \end{aligned}$$

The integral $J_3^*(x)$ in (A6) is the same as the integral $J_4(x)$ in the proof of Theorem 2 in Christoph et al. [12] with the estimate

$$\sup_x \left| J_3^*(x) - \frac{(2-r)x(x^2+1)}{4r(2r-1)}\, s_{2r}(x) \right| \le c(r)\, n^{-r+1}.$$

With (36), we proved (44). □

**Proof of Theorem 4.** By Lemma 2, the additional assumptions (23) and (24) in the transfer Theorem 2 are satisfied with the limit Gamma distribution $H(x) = G_{r,r}(x)$ of the normalized sample size $N_n(r)$ with $b = r > 0$. In Transfer Theorem 1 for $T_{N_n} = \sqrt{N_n}\, R_{N_n}$, the right-hand side of (20) is estimated by Lemma 1 and Lemma 3 for $\alpha = 2$ for the case $\gamma = 0$. Then, we have by (21) with (35)

$$G_n(x, 1/g_n) = \Phi(x)\left(1 - G_{r,r}(1/g_n) - n^{-1} h_{2;r}(1/g_n)\mathbb{I}_{\{r>1\}}(r)\right) + \frac{f_2(x;a)}{g_n} \int_{1/g_n}^\infty \frac{1}{y} dG_{r,r}(y)\mathbb{I}_{\{r>1\}}(r).$$

The estimates (23), (24i), (24ii), (34) and $\int_0^\infty y^{-1} dG_{r,r}(y) = r\Gamma(r-1)/\Gamma(r)$ for $r > 1$ lead to (46) with $\Phi_{n;2}(x;1)$ defined in (47). Thus, Theorem 4 is proved. □

**Proof of Theorem 5.** By Lemma 2, the additional assumptions (23) and (24) in the transfer Theorem 2 are satisfied with the limit Gamma distribution $H(x) = G_{r,r}(x)$ of the normalized sample size $N_n(r)$ with $b = r > 0$. In Transfer Theorem 1 for $T_{N_n} = g_n^{-1/2}\, N_n\, R_{N_n}$, the right-hand side of (20) is estimated by Lemma 1 and Lemma 3 for $\alpha = 2$ for the case $\gamma = -1/2$. Then, we have in (25)

$$G_{2;n}(x;0) = J_{1;r}^*(x) + \left(\frac{J_{2;r}^*(x)}{g_n} + \frac{J_{3;r}^*(x)}{n}\right)\mathbb{I}_{\{r>1\}}(r) \quad \text{with} \quad g_n = \mathbb{E}N_n(r) = r(n-1)+1 \qquad \text{(A8)}$$

$$\begin{aligned} J_{1;r}^*(x) &= \int_0^\infty \Phi(x/\sqrt{y}) dG_{r,r}(y), \quad J_{2;r}^*(x) = \int_0^\infty \frac{(a\,x^3\,y^{-3/2} - 5x\,y^{-1/2})\varphi(x/\sqrt{y})}{4\,y} dG_{r,r}(y), \text{ and} \\ J_{3;r}^*(x) &= \int_0^\infty \Phi(x/\sqrt{y}) dh_{2;r}(y) \text{ with } h_{2;r}(y) = \left((y-1)\,(2-r) + 2Q_1\left((r(n-1)+1)y\right)\right)\frac{g_{r,r}(y)}{2r}. \end{aligned}$$

Consider formula 2.3.16.1 in Prudnikov et al. [37], p. 444:

$$I_\alpha := \int_0^\infty y^{\alpha-1}\, e^{-py-q/y} dy = 2\left(\frac{q}{p}\right)^{\alpha/2} K_\alpha(2\sqrt{pq}) \quad p > 0, \quad q > 0,$$

where $K_\alpha(u)$ is the $\alpha$-order Macdonald function (or $\alpha$-order modified Bessel function of the second kind), see, e.g., Oldham et al. [30], Chapter 51, for properties of these functions.

Let us calculate the integral $J_{1;r}^*(x)$ occurring in (A8). Consider

$$\begin{aligned} \frac{d}{dx} J_{1;r}^*(x) &= \frac{r^r}{\sqrt{2\,\pi}\,\Gamma(r)} \int_0^\infty y^{r-3/2} e^{-ry-(x^2/(2y))} dy \\ &= \frac{2\,r^r}{\sqrt{2\,\pi}\,\Gamma(r)} \left(\frac{|x|}{2\,r}\right)^{r-1/2} K_{r-1/2}(\sqrt{2\,r}\,|x|) =: l_r(x). \end{aligned} \qquad \text{(A9)}$$

If $\alpha = \pm 1/2, \pm 3/2, \pm 5/2, \ldots$ the integral $I_\alpha$ and consequently $K_\alpha(x)$ are computable in closed-form expressions with formula 2.3.16.2 in Prudnikov et al. [37], p. 444:

$$I_m^* = \int_0^\infty y^{m-1/2} e^{-py-q/y} dy = (-1)^m \sqrt{\pi} \frac{\partial^m}{\partial p^m} \left( p^{-1/2} e^{-2\sqrt{pq}} \right), \; p > 0, \, q > 0, \, m = 0, 1, 2, \ldots \quad \text{(A10)}$$

and with formula 2.3.16.3 in Prudnikov et al. [37], p. 444:

$$I_{-m}^* = \int_0^\infty y^{-m-1/2} e^{-py-q/y} dy = (-1)^m \sqrt{\frac{\pi}{p}} \frac{\partial^m}{\partial q^m} e^{-2\sqrt{pq}}, \;\; p > 0, \;\; q > 0, \;\; m = 0, 1, 2, \ldots$$

For $r = 1, 2, 3$ using (A10) with $m = r - 1$, we find

$$l_r(x) = \frac{d}{dx} J_{1,r}^*(x) = \frac{r^r}{\Gamma(r)\sqrt{2\pi}} \int_0^\infty y^{r-3/2} e^{-ry-x^2/(2y)} dy = \frac{r^r}{\Gamma(r)\sqrt{2\pi}} I_{r-1}^*$$

and we obtain the densities $l_r(x)$ in (49) with

$$I_m^*(x) == \begin{cases} \sqrt{2\pi} \, \frac{1}{|x|} \, e^{-\sqrt{2r}|x|}, \quad x \neq 0 & m = -1, \\ \sqrt{\pi} \; e^{-\sqrt{2r}|x|}, & m = 0, \\ \sqrt{\pi} \left( \frac{1}{2 r^{3/2}} + \frac{|x|\sqrt{2}}{2r} \right) e^{-\sqrt{2r}|x|}, & m = 1, \\ \sqrt{\pi} \left( \frac{3}{4 r^{5/2}} + \frac{3|x|\sqrt{2}}{4 r^2} + \frac{|x|^2}{2 r^{3/2}} \right) e^{-\sqrt{2r}|x|}, & m = 2. \end{cases}$$

Consider now $J_{2;r}(x)$ for $r = 2$ and $r = 3$:

$$J_{2;r}^*(x) = \int_0^\infty \frac{(a\, x^3 y^{-3/2} - 5x\, y^{-1/2})\, r^r\, y^{r-1}\, e^{-ry-x^2/(2y)}}{4y\sqrt{2\pi}\,\Gamma(r)} dy = \frac{r^r}{4\sqrt{2\pi}\,\Gamma(r)} \left( a\, x^3\, I_{r-3}^*(x) - 5\, x\, I_{r-2}^*(x) \right).$$

Hence,

$$J_{2;2}^*(x) = (a\, x\, |x| - 5\, x/\sqrt{2})\, e^{-2|x|} \quad \text{and} \quad J_{2;3}^*(x) = \frac{27}{8} \left( \frac{a\, x^3}{\sqrt{2}} - 5\, x \Big( \frac{1}{6\sqrt{6}} + \frac{|x|}{6} \Big) \right) e^{-\sqrt{6}|x|}.$$

Integration by parts in the integral $J_{3;r}^*$ in (A8) leads to

$$\begin{aligned} J_{3;r}^*(x) \;\; &:= \;\; \int_0^\infty \Phi(x\, y^{-1/2}) d(h_{2;r}(y)) = \frac{x}{2} \int_0^\infty y^{-3/2} \varphi(x\, y^{-1/2})\, h_{2;r}(y) dy \\ &= \;\; \frac{r^r\, x}{2\, r\, \Gamma(r) \sqrt{2\pi}} \int_0^\infty y^{r-5/2} e^{-r\,y-x^2/(2\,y)} \Big( (y-1)\,(2-r) + 2Q_1(g_n\, y) \Big) dy, \\ &= \;\; \frac{r^{r-1}\, x}{2\,\Gamma(r)\,\sqrt{2\pi}} \int_0^\infty y^{r-5/2} (y-1)\,(2-r) e^{-r\,y-x^2/(2\,y)} dy \\ &+ \;\; \frac{r^{r-1}\, x}{\Gamma(r)\sqrt{2\pi}} \int_0^\infty y^{r-5/2} Q_1(g_n\, y)\, e^{-r\,y-x^2/(2\,y)} dy = J_{3;r,1}(x) + J_{3;r,2}(x). \end{aligned}$$

Since $J_{3;2,1}(x)$ vanishes, we calculate $J_{3;3,1}(x)$:

$$J_{3;3,1}(x) = \frac{9\, x}{2\sqrt{2\pi}} (I_1^*(x) - I_2^*(x)) = \frac{9\, x}{2} \left( \frac{1}{12\sqrt{6}} + \frac{|x|}{12} - \frac{|x|^2}{6\sqrt{6}} \right) e^{-\sqrt{6}|x|}.$$

It remains to estimate $J_{3;2,2}(x)$ and $J_{3;3,2}(x)$. The function $Q_1(y)$ is periodic with period 1:

$$Q_1(y) = Q_1(y+1) \;\; \text{for all} \;\; y \in \mathbb{R} \;\; \text{and} \;\; Q_1(y) := 1/2 - y \;\; \text{for} \;\; 0 \le y < 1$$

It is right-continuous and has the jump 1 at every integer point $y$. The Fourier series expansion of $Q_1(y)$ at all non-integer points $y$ is

$$Q_1(y) = 1/2 - (y - [y]) = \sum_{k=1}^{\infty} \frac{\sin(2\pi k y)}{k\pi} \qquad y \neq [y], \tag{A11}$$

see formula 5.4.2.9 in Prudnikov et al. [37], p. 726, with $a = 0$.

Using the Fourier series expansion (A11) of the periodic function $Q_1(y)$ and interchange sum and integral, we find

$$J^*_{3;r,2} = \frac{x}{\sqrt{2\pi}} \sum_{k=1}^{\infty} \frac{1}{k} \int_0^{\infty} y^{r-5/2} e^{-ry-x^2/(2y)} \sin(2\pi k g_n y) dy. \tag{A12}$$

First, we consider $r = 2$. Let $p > 0$, $q > 0$ and $b > 0$ be some real constants. Formula 2.5.37.4 in Prudnikov et al. [37], p. 453 reads

$$\int_0^{\infty} y^{-1/2} e^{-py-q/y} \sin(by) dy = \sqrt{\frac{\pi}{p^2+b^2}}\, e^{-2\sqrt{q} z_+} \left(z_+ \sin(2\sqrt{q} z_-) + z_- \cos(2\sqrt{q} z_-)\right). \tag{A13}$$

with $2z_\pm^2 = \sqrt{p^2+b^2} \pm p$. Consider $z_\pm$ with $p = r,\ q = x^2/2,\ b = 2\pi k g_n$, $k \geq 1$ and $n \geq 1$: Then,

$$\sqrt{\frac{\pi}{p^2+b^2}} = \sqrt{\frac{\pi}{r^2+(2\pi k g_n)^2}} \leq \frac{\sqrt{\pi}}{2\pi k g_n}, \quad \sqrt{2}|x| z_+ e^{-\sqrt{2}|x| z_+} \leq e^{-1} \quad \text{and} \quad 0 < z_- \leq z_+$$

leads to

$$\begin{aligned} |J^*_{3;2,2}(x)| &\leq \frac{2|x|}{\sqrt{2\pi}} \sum_{k=1}^{\infty} \frac{1}{k} \sqrt{\frac{\pi}{p^2+b^2}}\, e^{-2\sqrt{q} z_+} \left(z_+ \sin(2\sqrt{q} z_-) + z_- \cos(2\sqrt{q} z_-)\right) \\ &\leq \frac{2}{\sqrt{2\pi}} \sum_{k=1}^{\infty} \frac{1}{k} \frac{\sqrt{\pi}\sqrt{2} e^{-1}}{2\pi k g_n} = \frac{1}{2\pi e g_n} \sum_{k=1}^{\infty} \frac{1}{k^2} = \frac{\pi}{12 e g_n}. \end{aligned}$$

Together with $g_n \geq n$, we find $n^{-1}|J^*_{3;2,2}(x)| \leq C n^{-2}$.

Consider finally $J^*_{3;3,2}$ given in (A12). In order to estimate $J^*_{3;3,2}(x)$, we apply Leibniz's rule for differentiation under the integral sign with respect to $p$ in (A13) and obtain

$$\int_0^{\infty} y^{1/2} e^{-py-q/y} \sin(by) dy = \frac{\partial}{\partial p} \left\{ \sqrt{\frac{\pi}{p^2+b^2}}\, e^{-2\sqrt{q} z_+} \left(z_+ \sin(2\sqrt{q} z_-) + z_- \cos(2\sqrt{q} z_-)\right) \right\}.$$

Simple calculation considering $\sqrt{q} = |x|/\sqrt{2}$ and $|x|^m e^{-\sqrt{2}|x| z_+} \leq \frac{m}{2^{m/2} z_+} \leq \frac{m}{2^{m/2} b^m}$ for $m = 1, 2$, leads to

$$|x| \frac{\partial}{\partial p} \left\{ \sqrt{\frac{\pi}{p^2+b^2}}\, e^{-2\sqrt{q} z_+} \left(z_+ \sin(2\sqrt{q} z_-) + z_- \cos(2\sqrt{q} z_-)\right) \right\} \leq \frac{C}{b^2} = \frac{C}{(2\pi k g_n)^2}$$

and we find equation (A12) with $r = 3$ that $n^{-1}|J^*_{3;3,2}| \leq C n^{-3}$ and (50) is proved. The approximation (52) holds since Lemmas 1, 2, and 3 are valid for arbitrary $r > 0$. Theorem 5 is proved. □

**Proof of Theorem 6.** By Lemma 4, the additional assumptions (23) and (24) in the transfer Theorem 2 are satisfied with the limit inverse exponential distribution $H_s(y)$ and $h_{2;s}(y)$ given in (40), $g_n = n$ and $b = 2$. In Transfer Theorem 1, the right-hand side of (20) is estimated by Lemma 1 and Lemma 5 for $\alpha = 2$ for the case $\gamma = 1/2$. Then, we have in (25) with (35)

$$G_{2;n}(x;0) = J^*_{1;s}(x) + n^{-1} J^*_{2;s}(x) + n^{-1} J^*_{3;s}(x),$$

where $J^*_{1;s}(x) = \int_0^\infty \Phi(x\sqrt{y})de^{-s/y}$, $\qquad J^*_{2;s}(x) = \int_0^\infty \frac{(ax^3y^{3/2} - 5x\sqrt{y})\varphi(x\sqrt{y})}{4y}de^{-s/y}$,

and $J^*_{3;s}(x) = \int_0^\infty \Phi(x\sqrt{y})dh_{2;s}(y)$ with $h_{2;s}(y) = s\,e^{-s/y}\left(s - 1 + 2Q_1(n\,y)\right)/(2y^2)$.

To obtain (53), we calculate the above integrals as in the proof of Theorem 5 in Christoph et al. [12]. Here, we use Formula 2.3.16.3 in Prudnikov et al. [37], p. 344 with $p = x^2/2 > 0, s > 0, m = 1, 2$:

$$\int_0^\infty \frac{e^{-x^2y/2}}{\sqrt{2\pi}\,y^{m-3/2}}dH_s(y) = \int_0^\infty \frac{s\,e^{-x^2y/2-s/y}}{\sqrt{2\pi}\,y^{m+1/2}}dy = (-1)^m\frac{s}{|x|}\frac{\partial^m}{\partial s^m}e^{-\sqrt{2s}|x|}. \tag{A14}$$

In the mentioned proof we obtained with (A14) for $m = 1$

$$\int_0^\infty \Phi(x\sqrt{y})dH_s(y) = L_{1/\sqrt{s}}(x)$$

and with (A14) for $m = 2$

$$n^{-1}\sup_x\left|J^*_{3;s}(x) - \frac{(1-s)x(1+\sqrt{2s}|x|)}{8\,s}l_{1/\sqrt{s}}(x)\right| \le n^{-1}c(s)e^{-\sqrt{\pi s n}/2} \le C(s)n^{-2}.$$

Again using (A14) with $p = x^2/2 > 0, s > 0, m = 1, 2$ we find

$$\begin{aligned} J_{2;s}(x) &= \frac{s}{\sqrt{2\pi}}\int_0^\infty (ax^3y^{-1-1/2} - 5xy^{-2-1/2})e^{-(x^2\,y/2+s/y)}dy \\ &= \frac{2sax^3 - 5x(\sqrt{2s}\,|x| + 1)}{8\,s}\,l_{1/\sqrt{s}}(x). \end{aligned}$$

□

**Proof of Theorem 7.** By Lemma 4, the additional assumptions (23) and (24) in Transfer Theorem 2 are satisfied with the limit inverse exponential distribution $H_s(y)$ and $h_{2;s}(y)$ given in (40), $g_n = n$ and $b = 2$. In Transfer Theorem 1, the right-hand side of (20) is estimated by Lemma 1 and Lemma 5 for $\alpha = 2$ in the case $\gamma = 0$. Then, we have in (21) with (35)

$$G_n(x, 1/n) = \Phi(x)\left(1 - e^{-sn} - n^{-1}h_{2;s}(1/n)\right) + \frac{f_2(x;a)}{n}\int_{1/n}^\infty \frac{1}{y}de^{-s/y}.$$

The estimates (24i), (24ii) for $b = 2$ and $\int_0^\infty y^{-1}de^{-s/y} = s\int_0^\infty y^{-3}e^{-s/y}dy = s^2\int_0^\infty ze^{-z}dz = s^2$ lead to

$$\left|G_n(x, 1/g_n) - \Phi(x) - n^{-1}s^2f_2(x;a)\right| \le C_s\,n^{-2}$$

and Theorem 7 is proved. □

**Proof of Theorem 8.** By Lemma 4, the additional assumptions (23) and (24) in Transfer Theorem 2 are satisfied with the limit inverse exponential distribution $H_s(y)$ and $h_{2;s}(y)$ given in (40), $g_n = n$ and $b = 2$. In Transfer Theorem 1, the right-hand side of (20) is estimated by Lemma 1 and Lemma 5 for $\alpha = 2$ in the case $\gamma = -1/2$. Then, we have in (21) with (35)

$$G_{2;n}(x;0) = J^*_{1;s}(x) + n^{-1}\,J^*_{2;s}(x) + n^{-1}\,J^*_{3;s}(x),$$

where $J^*_{1;s}(x) = \int_0^\infty \Phi(xy^{-1/2})de^{-s/y}$, $\qquad J^*_{2;s}(x) = \int_0^\infty \frac{(ax^3y^{-3/2} - 5xy^{-1/2})\varphi(xy^{-1/2})}{4y}de^{-s/y}$,

and $J^*_{3;s}(x) = \int_0^\infty \Phi(xy^{-1/2})dh_{2;s}(y)$ with $h_{2;s}(y) = s\,e^{-s/y}\left(s - 1 + 2Q_1(n\,y)\right)/(2y^2)$.

To obtain (54), we calculate the above integrals:

$$\begin{aligned}\frac{\partial}{\partial x}\int_0^\infty \Phi(x\sqrt{y})de^{-s/y} &= \frac{s}{\sqrt{2\pi}}\int_0^\infty y^{-3/2}e^{-(x^2/2+s)/y)}dy = \frac{s}{\sqrt{2\pi}}\int_0^\infty z^{1/2-1}e^{-(x^2/2+s)z)}dz \\ = \frac{1}{2\sqrt{2s}}\left(1+\frac{x^2}{2s}\right)^{-3/2} &= s_2^*(x;\sqrt{s}), \quad \text{and} \quad \int_0^\infty \Phi(x\sqrt{y})de^{-s/y} = S_2^*(x;\sqrt{s}). \qquad \text{(A15)}\end{aligned}$$

Define $K = (s + x^2/2)$. With $z = K/y$ and $\Gamma(\alpha) = \int_0^\infty z^{\alpha-1}e^{-z}dz$, $\alpha > 0$, we obtain

$$\begin{aligned}J_{2;s}^*(x) &= \frac{s}{4\sqrt{2\pi}}\int_0^\infty (ax^3y^{-9/2} - 5xy^{-7/2})e^{-K/y}dy = \frac{s\,K^{-7/2}}{4\sqrt{2\pi}}\int_0^\infty (ax^3z^{5/2} - 5xz^{3/2}K)e^{-z}dz \\ &= \frac{s\,K^{-7/2}}{4\sqrt{2\pi}}\Big(ax^3\Gamma(7/2) - 5x\,K\,\Gamma(5/2)\Big) = \frac{1}{4\,(x^2+2s)^2}\Big(15(a-1)x^3 - 30xs\Big)\,s_2^*(x;\sqrt{s}).\end{aligned}$$

Integration by parts in $J_{3;s}^*(x)$ leads to

$$J_{3;s}^*(x) = \frac{x}{2\sqrt{2\pi}}\int_0^\infty y^{-3/2}e^{-x^2/(2y)}s\,y^{-2}e^{-s/y}\,((s-1)/2 + Q_1(n\,y))dy = J_{4;s}^*(x) + J_{5;s}^*(x),$$

where

$$J_{4;s}^*(x) = \frac{s\,(s-1)\,x}{4\sqrt{2\pi}}\int_0^\infty y^{-7/2}e^{-K/y}dy = \frac{s\,(s-1)\,x\,\Gamma(5/2)}{4\sqrt{2\pi}\,K^{5/2}} = \frac{3\,(s-1)\,x}{4\,(x^2+2s)}\,s_2^*(x;\sqrt{s})$$

and using the Fourier series expansion (A11) of the periodic function $Q_1(y)$ and interchange sum and integral, we find

$$\begin{aligned}J_{5;s}^*(x) &= \frac{s\,x}{2\sqrt{2\pi}}\int_0^\infty y^{-7/2}e^{-K/y}\,Q_1(n\,y)dy = \frac{s\,x}{2\sqrt{2\,\pi}}\sum_{k=1}^\infty \frac{1}{k}\int_0^\infty y^{-7/2}\,e^{-K/y}\,\sin(2\pi\,k\,n\,y)dy \\ &= \frac{s\,x}{2\sqrt{2\,\pi}}\sum_{k=1}^\infty \frac{1}{k}\int_0^\infty y^{-7/2}\,e^{-K/y}\,\sin(2\pi\,k\,n\,y)dy.\end{aligned}$$

Integration by parts in the latter integral and $|x|/\sqrt{K} \le \sqrt{2}$ leads now to

$$\sup_x |J_{5;s}^*(x)| \le \sup_x \frac{s\,|x|}{(2\,\pi)^{3/2}\,n}\sum_{k=1}^\infty \frac{1}{k^2}\int_0^\infty \left(\frac{7}{2}y^{-9/2} + K\,y^{11/2}\right)e^{-K/y}dy \le c_s n^{-1}$$

with $c_s = \dfrac{s\sqrt{2}}{(2\,\pi)^{3/2}\,n}\left(\dfrac{7\,\Gamma(11/2)}{2\,s^3} + \dfrac{\Gamma(13/2)}{s^4}\right)\dfrac{\pi^2}{6}$ and Theorem 8 is proved. □

## References


1. Hall, P.; Marron, J.S.; Neeman, A. Geometric representation of high dimension, low sample size data. *J. R. Stat. Soc. Ser.* **2005**, *67*, 427–444. [CrossRef]
2. Fujikoshi, Y.; Ulyanov, V.V.; Shimizu, R. *Multivariate Statistics. High-Dimensional and Large-Sample Approximations*; Wiley Series in Probability and Statistics; John Wiley & Sons, Inc.: Hoboken, NJ, USA, 2010.
3. Aoshima, M.; Shen, D.; Shen, H.; Yata, K.; Zhou, Y.-H.; Marron, J.S. A survey of high dimension low sample size asymptotics. *Aust. N. Z. J. Stat.* **2018**, *60*, 4–19. [CrossRef] [PubMed]
4. Kawaguchi, Y.; Ulyanov, V.V.; Fujikoshi, Y. Asymptotic distributions of basic statistics in geometric representation for high-dimensional data and their error bounds (Russian). *Inform. Appl.* **2010**, *4*, 12–17.
5. Ulyanov, V.V.; Christoph, G.; Fujikoshi, Y. On approximations of transformed chi-squared distributions in statistical applications. *Sib. Math. J.* **2006**, *47*, 1154–1166. [CrossRef]
6. Esquível, M.L.; Mota, P.P.; Mexia, J.T. On some statistical models with a random number of observations. *J. Stat. Theory Pract.* **2016**, *10*, 805–823. [CrossRef]

7. Gnedenko, B.V. Estimating the unknown parameters of a distribution with a random number of independent observations. (Probability theory and mathematical statistics (Russian)). *Tr. Tbil. Mat. Instituta* **1989**, *92*, 146–150.
8. Nunes, C.; Capristrano, G.; Ferreira, D.; Ferreira, S.S.; Mexia, J.T. Fixed effects ANOVA: an extension to samples with random size. *J. Stat. Comput. Simul.* **2014**, *84*, 2316–2328. [CrossRef]
9. Nunes, C.; Capristrano, G.; Ferreira, D.; Ferreira, S.S.; Mexia, J.T. Exact critical values for one-way fixed effects models with random sample sizes. *J. Comput. Appl. Math.* **2019**, *354*, 112–122. [CrossRef]
10. Barakat, H.M.; Nigm, E.M.; El-Adll, M.E.; Yusuf, M. Prediction of future generalized order statistics based on exponential distribution with random sample size. *Stat. Pap.* **2018**, *59*, 605–631. [CrossRef]
11. Al-Mutairi, J.S.; Raqab, M.Z. Confidence intervals for quantiles based on samples of random sizes. *Stat. Pap.* **2020**, *61*, 261–277. [CrossRef]
12. Christoph, G.; Monakhov, M.M.; Ulyanov, V.V. Second-order Chebyshev–Edgeworth and Cornish-Fisher expansions for distributions of statistics constructed with respect to samples of random size. *J. Math. Sci.* **2020**, *244*, 811–839. [CrossRef]
13. Christoph, G.; Ulyanov, V.V.; Bening, V.E. Second Order Expansions for Sample Median with Random Sample Size. *arXiv* **2019**, arXiv:1905.07765v2.
14. Fujikoshi, Y.; Ulyanov, V.V. *Non-Asymptotic Analysis of Approximations for Multivariate Statistics*; Springer: Singapore, 2020.
15. Johnson, N.L.; Kotz, S.; Balakrishnan, N. *Continuous Univariate Distributions*, 2nd ed.; Wiley: New York, NY, USA, 1995; Volume 2.
16. Konishi, S. Asymptotic expansions for the distributions of functions of a correlation matrix. *J. Multivar. Anal.* **1979**, *9*, 259–266. [CrossRef]
17. Christoph, G.; Ulyanov, V.V.; Fujikoshi, Y. Accurate approximation of correlation coefficients by short Edgeworth-Chebyshev expansion and its statistical applications. In *Prokhorov and Contemporary Probability Theory*; Proceedings in Mathematics & Statistics 33; Shiryaev, A.N., Varadhan, S.R.S., Presman, E.L., Eds.; Springer: Heidelberg, Germany, 2013; pp. 239–260.
18. Prokhorov, Y.V. Limit theorems for the sums of random vectors whose dimension tends to infinity. *Teor. Veroyatnostei Primen.* **1990**, *35*, 751–753. English translation: *Theory Probab. Appl.* **1991**, *35*, 755–757. [CrossRef]
19. Gnedenko, B.V.; Korolev, V.Y. *Random Summation. Limit Theorems and Applications*; CRC Press: Boca Raton, FL, USA, 1996.
20. Döbler, C. New Berry-Esseen and Wasserstein bounds in the CLT for non-randomly centered random sums by probabilistic methods. *ALEA Lat. Am. J. Probab. Math. Stat.* **2015**, *12*, 863–902.
21. Chen, L.H.Y.; Goldstein, L.; Shao, Q.-M. *Normal Approximation by Stein's Method. Probability and its Applications*; Springer: Heidelberg, Germany, 2011.
22. Pike, J.; Ren, H. Stein's method and the Laplace distribution. *ALEA Lat. Am. J. Probab. Math. Stat.* **2014**, *11*, 571–587.
23. Bening, V.E.; Korolev, V.Y. On the use of Student's distribution in problems of probability theory and mathematical statistics. *Theory Probab. Appl.* **2005**, *49*, 377–391. [CrossRef]
24. Bening, V.E.; Galieva, N.K.; Korolev, V.Y. Asymptotic expansions for the distribution functions of statistics constructed from samples with random sizes (Russian). *Inform. Appl.* **2013**, *7*, 75–83.
25. Bening, V.E.; Korolev, V.Y.; Zeifman, A.I. Asymptotic expansions for the distribution function of the sample median constructed from a sample with random size. In *Proceedings 30th ECMS 2016 Regensburg*; Claus, T., Herrmann, F., Manitz, M., Rose, O., Eds.; European Council for Modeling and Simulation: Regensburg, Germany, 2016; pp. 669–675.
26. Schluter, C.; Trede, M. Weak convergence to the Student and Laplace distributions. *J. Appl. Probab.* **2016**, *53*, 121–129. [CrossRef]
27. Gavrilenko, S.V.; Zubov, V.N.; Korolev, V.Y. The rate of convergence of the distributions of regular statistics constructed from samples with negatively binomially distributed random sizes to the Student distribution. *J. Math. Sci.* **2017**, *220*, 701–713. [CrossRef]
28. Buddana, A.; Kozubowski, T.J. Discrete Pareto distributions. *Econ. Qual. Control.* **2014**, *29*, 143–156. [CrossRef]
29. Christoph, G.; Wolf, W. *Convergence Theorems with a Stable Limit Law, Mathematical Research 70*; Akademie-Verlag: Berlin, Germany, 1992.

30. Oldham, K.B.; Myl, J.C.; Spanier, J. *An Atlas of Functions*, 2nd ed.; Springer Science + Business Media: New York, NY, USA, 2009.
31. Kotz, S.; Kozubowski, T.J.; Podgórski, K. *The Laplace distribution and Generalizations: A Revisit with Applications to Communications, Economics, Engineering, and Finance*; Birkhäuser: Boston, MA, USA, 2001.
32. Prudnikov, A.P.; Brychkov, Y.A.; Marichev, O.I. *Integrals and Series, Volume 2: Special Functions*, 3rd ed.; Gordon & Breach Science Publishers: New York, NY, USA, 1992.
33. Madan, D.B.; Senata, E. The variance gamma (V.G.) model for share markets returns. *J. Bus.* **1990** *63*, 511–524. [CrossRef]
34. Goldfeld, S.M.; Quandt, R.E. Econometric modelling with non-normal disturbances. *J. Econom.* **1981**, *17*, 141–155. [CrossRef]
35. Missiakeles, S. Sargan densities which one? *J. Econom.* **1983**, *23*, 223–233. [CrossRef]
36. Jackman, S. *Bayesian Analysis for the Social Sciences*; Wiley Series in Probability and Statistics; John Wiley & Sons, Ltd.: Chichester, UK, 2009.
37. Prudnikov, A.P.; Brychkov, Y.A.; Marichev, O.I. *Integrals and Series, Volume 1: Elementary Functions*, 3rd ed.; Gordon & Breach Science Publishers: New York, NY, USA, 1992.

mathematics

MDPI

Review

# Two Approaches to the Construction of Perturbation Bounds for Continuous-Time Markov Chains

**Alexander Zeifman [1,2,3,*], Victor Korolev [2,4,5] and Yacov Satin [1]**

1 Department of Applied Mathematics, Vologda State University, 160000 Vologda, Russia; yacovi@mail.ru
2 Institute of Informatics Problems of the Federal Research Center "Computer Science and Control", Russian Academy of Sciences, 119333 Moscow, Russia; vkorolev@cs.msu.ru
3 Vologda Research Center of the Russian Academy of Sciences, 160014 Vologda, Russia
4 Faculty of Computational Mathematics and Cybernetics, Moscow State University, 119991 Moscow, Russia
5 Department of Mathematics, School of Science, Hangzhou Dianzi University, Hangzhou 310018, China
* Correspondence: a_zeifman@mail.ru

Received: 20 January 2020; Accepted: 11 February 2020; Published: 14 February 2020

**Abstract:** This paper is largely a review. It considers two main methods used to study stability and to obtain appropriate quantitative estimates of perturbations of (inhomogeneous) Markov chains with continuous time and a finite or countable state space. An approach is described to the construction of perturbation estimates for the main five classes of such chains associated with queuing models. Several specific models are considered for which the limit characteristics and perturbation bounds for admissible "perturbed" processes are calculated.

**Keywords:** continuous-time Markov chains; non-stationary Markovian queueing model; stability; perturbation bounds; forward Kolmogorov system

## 1. Introduction

In this paper, some topics are considered that are related to the stability of both homogeneous and non-homogeneous continuous-time Markov chains with respect to the perturbation of their intensities (infinitesimal characteristics). It is assumed that the evolution of the system under consideration is described by a Markov chain with the known state space, and it is the infinitesimal matrix that is given inexactly. Different classes of admissible perturbations can be considered. The "perturbed" infinitesimal matrix can be arbitrary, and the small deviation of its norm from that of the original matrix is assumed or it can be assumed that the structure of the infinitesimal matrix is known and only its elements are "perturbed" within the same structure. Below we will give a detailed description of these cases. In some papers it is assumed that the perturbations have a special form and, for example, are expanded in a power series of a small parameter. This assumption seems to be too restrictive and unrealistic.

The study of stability of characteristics of stochastic models has been actively developing since the 1970s [1–3]. At that time, Zolotarev proposed to treat limit theorems of probability theory as special stability theorems. Zolotarev created the theoretical foundation of the key method used within this approach, namely, the theory of probability metrics [4]. This approach assumes that statements establishing the convergence must be accompanied by statements establishing the convergence rate. Zolotarev called the conditions of convergence that simultaneously serve as convergence rate estimates "natural." This approach was developed in the works of Zolotarev, Kalashnikov, Kruglov, Senatov, Yu, Korolev, Yu, Khokhlov, and their colleagues in the framework of international seminars on stability problems for stochastic models. This seminar was founded by Zolotarev in the early 1970s and still continues to hold its regular (as a rule, annual) international sessions (see the series of the proceedings of the seminar published as Springer Lecture Notes starting from [5] or as issues of the Journal of

Mathematical Sciences). In particular, this approach proved to be very productive for the study of random sums in queueing theory, renewal theory, and the theory of branching processes [6].

Since the 1980s, the problems related to the estimation of stability of Markov chains with respect to perturbations of their characteristics have been thoroughly studied by Kartashov for homogeneous discrete-time chains with general state space and, in parallel, by Zeifman for inhomogeneous continuous-time chains within the seminar mentioned above (see [7–9]). In particular, a general approach for inhomogeneous continuous-time chains was developed in [9]. That paper was published in the proceedings of the seminar "Stability Problems for Stochastic Models" and dealt with both uniform and strong cases.

Later birth-death processes were considered in [10], and general properties and estimates for inhomogeneous finite chains were considered in [11]. The paper [12] was specially devoted to estimates for general birth-death processes, with the queueing system $M_t|M_t|N$ considered as an example. It should be mentioned that these papers were not noticed by Western authors. For example, in [13], it was stated that there were no papers on the stability of the (simplest stationary!!) system $M|M|1$. For the first time, we used the term "perturbation bounds" instead of "stability" in the paper [14] on the referee's prompt. The same situation takes place with Kartashov's papers cited above. The methods proposed in those papers seem to be used by most authors of subsequent studies in estimations of perturbations of discrete-time chains. Possibly, poor acquaintance with the early papers of Kartashov and Zeifman can be explained by the differences in terminology mentioned above: in the original (and foundational) papers, the term "stability" was used (in the proceedings of the seminar with the consonant appellation "Stability Problems for Stochastic Models").

The present paper deals only with continuous-time chains, so the subsequent remarks mainly regard such a case.

Note that, to obtain explicit and exact estimates of the perturbation bounds of a chain, it is required to have estimates of the rate of convergence of the chain to its limit characteristics in the form of explicit inequalities. Moreover, the sharper the convergence rate estimates are, the more accurate the perturbation bounds are. These bounds can be more easily obtained for finite homogeneous Markov chains. Therefore, most publications concern this situation only (see, e.g., [15–20]). Thus, two main approaches can be highlighted.

The first of them can be used for the case of weak ergodicity of a chain in the uniform operator topology. The first bounds in this direction were obtained in [9]. The principal progress related to the replacement of the constant $S$ with $\log S$ in the bound was implemented in [17] and continued in Mitrophanov's papers [18–20] for the case of homogeneous chains and then in [14,21] and in the subsequent papers of these authors for the inhomogeneous chains. The contemporary state of affairs in this field and new applied problems related to the link between convergence rate and perturbation bounds in the "uniform" case were described in [22]. In some recent papers, uniform perturbation bounds of homogeneous Markov chains were studied by the techniques of stochastic differential equations (see, for instance, [23] and the references therein).

The second approach is used in the case where the uniform ergodicity is not assured, which is typical for the processes most interesting from a practical viewpoint. For example, birth-death processes used for modeling queueing systems, and real processes in biology, chemistry, and physics, as a rule, are not uniformly ergodic.

Following the ideas of Kartashov (see a detailed description in [24]), most authors use the probability methods to study ergodicity and perturbation bounds of stationary chains (with a finite, countable, or general state space) in various norms [13,25,26]. For a wide class of (mainly) stationary discrete-time chains, a close approach was considered in [27] and more recent papers [28–38].

In the works of the authors of the present paper, perturbation bounds for non-stationary finite or infinite continuous-time chains were studied by other methods.

The first papers dealing with non-stationary queueing models appeared in the 1970s (see [39,40], and the more recent paper [41]). Moreover, as far back as the year in which [42] was published, it was noted that it is principally possible to use the logarithmic matrix norm for the study of the convergence rate of continuous-time Markov chains. The corresponding general approach employing the theory of differential equations in Banach spaces was developed in a series of papers by the authors of the present paper(see a detailed description in [43,44]). In [9] (see also [10,11]), a method for the study of perturbation bounds for the vector of state probabilities of a continuous-time Markov chain with respect to the perturbations of infinitesimal characteristics of the chain in the total variation norm ($l_1$-norm) was proposed. The paper [12] contained a detailed study of the stability of essentially non-stationary birth-death processes with respect to conditionally small perturbations. Convergence rate estimates in terms of weight norms, and hence the corresponding bounds for new classes of Markov chains, were considered in [45–48].

In the present paper, both approaches are considered along with the classes of inhomogeneous Markov chains, for which at least one of these approaches yields reasonable perturbation bounds for basic probability characteristics.

The paper is organized as follows. In Section 2, basic notions and preliminary results are introduced. In Section 3, general theorems on perturbation bounds are considered. Section 4 contains convergence rate estimates and perturbation bounds for basic classes of the chains under consideration. Finally, in Section 5, some special queueing models are studied.

## 2. Basic Notions and Preliminaries

Let $X = X(t)$, $t \geq 0$, be, in general, an inhomogeneous continuous-time Markov chain with a finite or countable state space $E_S = 0, 1, \ldots, S$, $S \leq \infty$. The transition probabilities for $X = X(t)$ will be denoted $p_{ij}(s,t) = \Pr\{X(t) = j\,|X(s) = i\}$, $i, j \geq 0$, $0 \leq s \leq t$. Let $p_i(t) = \Pr\{X(t) = i\}$ be the state probabilities of the chain and $\mathbf{p}(t) = (p_0(t), p_1(t), \ldots)^T$ be the corresponding vector of state probabilities. In what follows, it is assumed that

$$\Pr\{X(t+h) = j|X(t) = i\} =$$

$$= \begin{cases} q_{ij}(t)\,h + \alpha_{ij}(t,h), & \text{if } j \neq i \\ 1 - \sum\limits_{k \neq i} q_{ik}(t)\,h + \alpha_i(t,h), & \text{if } j = i \end{cases} \tag{1}$$

where all $\alpha_i(t,h)$ are $o(h)$ uniformly in $i$, that is, $\sup_i |\alpha_i(t,h)| = o(h)$.

As usual, we assume that, if a chain is inhomogeneous, then all the infinitesimal characteristics (intensity functions) $q_{ij}(t)$ are integrable in $t$ on any interval $[a,b]$, $0 \leq a \leq b$.

Let $a_{ij}(t) = q_{ji}(t)$ for $j \neq i$ and $a_{ii}(t) = -\sum_{j \neq i} a_{ji}(t) = -\sum_{j \neq i} q_{ij}(t)$.

Further, to provide the possibility to obtain more evident estimates, we will assume that

$$|a_{ii}(t)| \leq L < \infty \tag{2}$$

for almost all $t \geq 0$.

The state probabilities then satisfy the forward Kolmogorov system

$$\frac{d\mathbf{p}}{dt} = A(t)\mathbf{p}(t) \tag{3}$$

where $A(t) = Q^T(t)$, and $Q(t)$ is the infinitesimal matrix of the process.

Let $\|\cdot\|$ be the usual $l_1$-norm, i.e. $\|\mathbf{x}\| = \sum |x_i|$, and $\|B\| = \sup_j \sum_i |b_{ij}|$ for $B = (b_{ij})_{i,j=0}^{\infty}$. Denote $\Omega = \{\mathbf{x} : \mathbf{x} \in l_1^+ \;\&\; \|\mathbf{x}\| = 1\}$. Therefore,

$$\|A(t)\| = 2 \sup_k |a_{kk}(t)| \leq 2L$$

for almost all $t \geq 0$, and we can apply the results of [49] to Equation (3) in the space $l_1$. Namely, in [49] it was shown that the Cauchy problem for Equation (3) has a unique solution for an arbitrary initial condition. Moreover, if $\mathbf{p}(s) \in \Omega$, then $\mathbf{p}(t) \in \Omega$, for any $0 \leq s \leq t$ and any initial condition $\mathbf{p}(s)$.

Let $p_0(t) = 1 - \sum_{i \geq 1} p_i(t)$.

Put $\mathbf{z} = (p_1, p_2, \ldots)^T$.

Therefore, from Equation (3), we obtain the equation

$$\frac{d\mathbf{z}}{dt} = B(t)\mathbf{z}(t) + \mathbf{f}(t), \tag{4}$$

where $\mathbf{f} = (a_{10}, a_{20}, \cdots)^T$,

$$B = \begin{pmatrix} a_{11} - a_{10} & a_{12} - a_{10} & \cdots & a_{1r} - a_{10} & \cdots \\ a_{21} - a_{20} & a_{22} - a_{20} & \cdots & a_{2r} - a_{20} & \cdots \\ a_{31} - a_{30} & a_{32} - a_{30} & \cdots & a_{3r} - a_{30} & \cdots \\ \cdots & & & & \\ a_{r1} - a_{r0} & a_{r2} - a_{r0} & \cdots & a_{rr} - a_{r0} & \cdots \\ \cdots\cdots\cdots & & & & \end{pmatrix}, \tag{5}$$

where all expressions depend on $t$.

By $\bar{X} = \bar{X}(t)$, we will denote the "perturbed" Markov chain with the same state space, state probabilities $\bar{p}_i(t)$, transposed infinitesimal matrix $\bar{A}(t) = (\bar{a}_{ij}(t))_{i,j=0}^{\infty}$, and so on, and the "perturbations" themselves, that is, the differences between the corresponding "perturbed" and original characteristics will be denoted by $\hat{a}_{ij}(t)$, $\hat{A}(t)$.

Let $E(t,k) = E\{X(t)\,|X(0) = k\}$. Recall that a Markov chain $X(t)$ is weakly ergodic if $\|\mathbf{p}^*(t) - \mathbf{p}^{**}(t)\| \to 0$ as $t \to \infty$ for any initial condition, and it has the limiting mean $\phi(t)$ if $|E(t,k) - \phi(t)| \to 0$ as $t \to \infty$ for any $k$.

Now we briefly describe the main classes of the chains under consideration. The details concerning the first four classes can be found in [47,50].

**Case 1.** *Let $a_{ij}(t) = 0$ for all $t \geq 0$ if $|i-j| > 1$, and $a_{i,i+1}(t) = \mu_{i+1}(t)$, $a_{i+1,i}(t) = \lambda_i(t)$. This is an inhomogeneous birth-death process (BDP) with the intensities $\lambda_i(t)$ (of birth) and $\mu_{i+1}(t)$ (of death) correspondingly.*

**Case 2.** *Now let $a_{ij}(t) = 0$ for $i < j-1$, $a_{i+k,i}(t) = a_k(t)$ for $k \geq 1$, and $a_{i,i+1}(t) = \mu_{i+1}(t)$. This chain describes, for instance, the number of customers in a queueing system in which the customers arrive in groups, but are served one by one (in this case, $a_k(t)$ is the arrival intensity of a group of k customers, and $\mu_i(t)$ is the service intensity of the ith customer). The simplest models of this type were considered in [51] (see also [47,50]).*

**Case 3.** *Let $a_{ij}(t) = 0$ for $i > j+1$, $a_{i,i+k}(t) = b_k(t)$, $k \geq 1$, and $a_{i+1,i}(t) = \lambda_i(t)$. This situation occurs in modeling queueing systems with the arrivals of single customers and group service.*

**Case 4.** *Let $a_{i+k,i}(t) = a_k(t)$, $a_{i,i+k}(t) = b_k(t)$ for $k \geq 1$. This process appears in the description of a system with group arrival and group service, for earlier studies see [46,52,53].*

**Case 5.** *Consider a Markov chain with "catastrophes" used for modeling of some queueing systems (see, e.g., [14,54–58]). Here the intensities have a general form, whereas a single (although substantial) restriction consists in that the zero state is attainable from any other state, and the corresponding intensities $q_{k,0}(t) = a_{0,k}(t)$ for $k \geq 1$ are called the intensities of catastrophes.*

Now consider the following example illustrating some specific features of the problem under consideration.

**Example 1** ([14]). *Consider a homogeneous BDP (Class I) with the intensities* $\lambda_k(t) = 1$, $\mu_k(t) = 4$ *for all* $t$ *and* $k$ *and denote by* $A$ *the corresponding transposed intensity matrix. Therefore, as is known (see, e.g., [59]), the BDP is strongly ergodic and stable in the corresponding norm. On the other hand, take a perturbed process with the transposed infinitesimal matrix* $\bar{A} = A + \hat{A}$, *where* $\hat{a}_{00} = -\varepsilon$, $\hat{a}_{k0} = \frac{\varepsilon}{k(k+1)}$ *for* $k \geq 1$, *and* $\hat{a}_{ij} = 0$ *for the other* $i, j$. *The perturbed Markov chain* $\bar{X}(t)$ *(describing the "M* | *M* | *c queue with mass arrivals when empty" (see [54,58,60])) is then not ergodic, since, from the condition* $\bar{A}\bar{\mathbf{p}} = 0$, *it follows that the coordinates of the stationary distribution (if it exists) must satisfy the condition* $4\bar{p}_{k+1} = \bar{p}_k + \bar{p}_0\frac{\varepsilon}{k+1} \geq \bar{p}_0\frac{\varepsilon}{k+1}$, *which is impossible.*

As has already been noted, the (upper) bounds of perturbations are closely connected with the (correspondingly, upper) estimates for the convergence rate (see also the two next sections). On the other side, it is also possible to construct important lower estimates of the rate of convergence provided that the influence of the initial conditions cannot fade too rapidly (see [61]). It turns out that it is principally impossible to construct lower bounds for perturbations. Indeed, if we consider the same BDP and, as a perturbed BDP, choose a BDP with the intensities $\bar{\lambda}_k(t) = 1 + \varepsilon$, $\bar{\mu}_k(t) = 4(1+\varepsilon)$, then the stationary distribution for the perturbed process will be the same as that for the original BDP for any positive $\varepsilon$.

### 3. General Theorems Concerning Perturbation Bounds

First consider *uniform* bounds that provide the first approach to perturbation estimation. This approach is applied to uniformly ergodic Markov chains and the study of stability of the state probability vector. The most important class of such processes is that of Markov chains on finite state space, both homogeneous and inhomogeneous.

**Theorem 1.** *Let the Markov chain* $X(t)$ *be exponentially weakly ergodic; that is, for any initial conditions* $\mathbf{p}^*(s) \in \Omega$, $\mathbf{p}^{**}(s) \in \Omega$ *and any* $s \geq 0$, $t \geq s$, *there holds the inequality*

$$\|\mathbf{p}^*(t) - \mathbf{p}^{**}(t)\| \leq 2ce^{-b(t-s)}. \tag{6}$$

*Therefore, for the perturbations small enough* ($\hat{A}(t) \leq \varepsilon$ *for almost all* $t \geq 0$), *the perturbed chain* $\bar{X}(t)$ *is also exponentially weakly ergodic, and the following perturbation bound takes place:*

$$\limsup_{t\to\infty} \|\mathbf{p}(t) - \bar{\mathbf{p}}(t)\| \leq \frac{(1+\log(c/2))\,\varepsilon}{b}. \tag{7}$$

For the proof, we will use the approach proposed in [17] and modified in [21] for the inhomogeneous case, see also [14]. Let

$$\beta(t,s) = \sup_{\|\mathbf{v}\|=1, \sum v_i = 0} \|U(t)\mathbf{v}\| = \tfrac{1}{2}\sup_{i,j}\sum_k |p_{ik}(t,s) - p_{jk}(t,s)|. \tag{8}$$

Therefore,

$$\|\mathbf{p}(t) - \bar{\mathbf{p}}(t)\| \leq \beta(t,s)\|\mathbf{p}(s) - \bar{\mathbf{p}}(s)\| + \int_s^t \|\hat{A}(u)\|\beta(u,s)du. \tag{9}$$

Moreover,

$$\beta(t,s) \leq 1, \quad \beta(t,s) \leq \tfrac{c}{2}e^{-b(t-s)}, \ 0 \leq s \leq t. \tag{10}$$

Hence,

$$\|\mathbf{p}(t) - \bar{\mathbf{p}}(t)\| \leq$$

$$\begin{cases} \|\mathbf{p}(s)-\bar{\mathbf{p}}(s)\|+(t-s)\varepsilon, & if \quad 0<t-s<\frac{1}{b}\ln\frac{c}{2}, \\ \frac{c}{2}e^{-b(t-s)}\|\mathbf{p}(s)-\bar{\mathbf{p}}(s)\|+\frac{1}{b}(\ln\frac{c}{2}+1-ce^{-b(t-s)})\varepsilon, & if \quad t-s\geq\frac{1}{b}\ln\frac{c}{2} \end{cases} \tag{11}$$

whence, as $t \to \infty$, we obtain Equation (7).

**Corollary 1.** *If under the conditions of Theorem 1 the Markov chain $X(t)$ has a finite state space, then both Markov chains $X(t)$ and $\bar{X}(t)$ have limit expectations and*

$$|\phi(t)-\bar{\phi}(t)| \leq \tfrac{1}{b}S\,(1+\log(c/2))\,\varepsilon. \tag{12}$$

Now consider the second approach. Namely, we turn to weighted bounds. Such estimates can be applied to a wide class of Markov chains which are exponentially ergodic in some weighted norms. Moreover, as a rule, these estimates also allow one to study stability characteristics of the mathematical expectation for countable Markov chains, both homogeneous and inhomogeneous. Here we use the approach proposed in [9] (see also the detailed description in [43,44]).

Let $1 \leq d_1 \leq d_2 \leq \ldots$,

$$D = \begin{pmatrix} d_1 & d_1 & d_1 & \cdots \\ 0 & d_2 & d_2 & \cdots \\ 0 & 0 & d_3 & \cdots \\ & \ddots & \ddots & \ddots \end{pmatrix}. \tag{13}$$

Let $l_{1D} = \{\mathbf{z} = (p_1, p_2, \cdots)^T : \|\mathbf{z}\|_{1D} \equiv \|D\mathbf{z}\| < \infty\}$. Therefore, $\|B\|_{1D} = \|DBD^{-1}\|$. In addition, let $\|\mathbf{p}\|_{1D} = \|\mathbf{z}\|_{1D}$.

Below we will assume that the following conditions hold:

$$\|B(t)\|_{1D} \leq \mathfrak{B} < \infty, \quad \|\mathbf{f}(t)\|_{1D} \leq \mathfrak{f} < \infty \tag{14}$$

for almost all $t \geq 0$.

Recall that $X(t)$ is a $1D$-exponentially weakly ergodic Markov chain if

$$\|\mathbf{p}^*(t)-\mathbf{p}^{**}(t)\|_{1D} \leq Me^{-a(t-s)}\|\mathbf{p}^*(s)-\mathbf{p}^{**}(s)\|_{1D}. \tag{15}$$

for some $M > 0$, $a > 0$ and any $s, t$: $t \geq s \geq 0$, any initial conditions $\mathbf{p}^*(s) \in l_{1D}$, $\mathbf{p}^{**}(s) \in l_{1D}$.

If one can choose $\mathbf{p}^{**}(t) = \mathfrak{ß}$, then the chain is $1D$-exponentially strongly ergodic.

Let

$$\|B(t)-\bar{B}(t)\|_{1D} \leq |\mathfrak{B}-\bar{\mathfrak{B}}|, \quad \|\mathbf{f}(t)-\bar{\mathbf{f}}(t)\|_{1D} \leq |\mathfrak{f}-\bar{\mathfrak{f}}|. \tag{16}$$

for almost all $t \geq 0$.

**Theorem 2.** *If a Markov chain $X(t)$ is $1D$-exponentially weakly ergodic, then $\bar{X}(t)$ is also $1D$-exponentially weakly ergodic and the following perturbation estimate in the $1D$-norm holds:*

$$\limsup_{t\to\infty} \|\mathbf{p}(t)-\bar{\mathbf{p}}(t)\|_{1D} \leq \frac{M\,(M\,|\mathfrak{B}-\bar{\mathfrak{B}}|\,\mathfrak{f}+a\,|\mathfrak{f}-\bar{\mathfrak{f}}|)}{a\,(a-M\,|\mathfrak{B}-\bar{\mathfrak{B}}|)}. \tag{17}$$

*If $W = \inf_{i\geq 1}\frac{d_i}{i} > 0$, then both chains $X(t)$ and $\bar{X}(t)$ have limiting means and*

$$\limsup_{t\to\infty} |\phi(t)-\bar{\phi}(t)| \leq \frac{M\,(M\,|\mathfrak{B}-\bar{\mathfrak{B}}|\,\mathfrak{f}+a\,|\mathfrak{f}-\bar{\mathfrak{f}}|)}{Wa\,(a-M\,|\mathfrak{B}-\bar{\mathfrak{B}}|)}. \tag{18}$$

**Proof.** The detailed consideration can be found in [44]. Here we only outline the scheme of reasoning. Let $V(t,s)$ and $\bar{V}(t,s)$ be the Cauchy operators for Equation (4) and for the corresponding "perturbed" equation, respectively. Therefore,

$$\|V(t,s)\|_{1D} \le Me^{-a(t-s)}, \quad \|\bar{V}(t,s)\|_{1D} \le Me^{-(a-M|\mathfrak{B}-\bar{\mathfrak{B}}|)(t-s)} \tag{19}$$

for all $t \ge s \ge 0$. Therefore, rewriting Equation (4) as

$$\frac{d\mathbf{z}}{dt} = \bar{B}(t)\mathbf{z}(t) + \mathbf{f}(t) + \left(B(t) - \bar{B}(t)\right)\mathbf{z}(t), \tag{20}$$

after some algebra, we obtain the following inequality in the $1D$-norm:

$$\|\mathbf{z}(t) - \bar{\mathbf{z}}(t)\| \le \int_0^t \|\bar{V}(t,\tau)\| \left(\|B(\tau) - \bar{B}(\tau)\|\|\mathbf{z}(\tau)\| + \|\mathbf{f}(\tau) - \bar{\mathbf{f}}(\tau)\|\right) d\tau \le$$

$$\le \int_0^t Me^{-(a-M|\mathfrak{B}-\bar{\mathfrak{B}}|)(t-\tau)} \left(|\mathfrak{B} - \bar{\mathfrak{B}}| \, \|\mathbf{z}(\tau)\| + |\mathfrak{f} - \bar{\mathfrak{f}}|\right) d\tau. \tag{21}$$

On the other hand, $\|\mathbf{z}(t)\|_{1D} \le Me^{-at}\|\mathbf{z}(0)\|_{1D} + \frac{M}{a}\mathfrak{f}$, for any $0 \le s \le t$. Hence, under any initial condition $\mathbf{z}(0) \in l_{1D}$, we obtain the following inequalities for the $1D$-norm:

$$\|\mathbf{z}(t) - \bar{\mathbf{z}}(t)\| \le M\left(|\mathfrak{B} - \bar{\mathfrak{B}}| \frac{M}{a}\mathfrak{f} + |\mathfrak{f} - \bar{\mathfrak{f}}|\right) \int_0^t e^{-(a-M|\mathfrak{B}-\bar{\mathfrak{B}}|)(t-\tau)} d\tau +$$

$$+M \int_0^t e^{-(a-M|\mathfrak{B}-\bar{\mathfrak{B}}|)(t-\tau)} |\mathfrak{B} - \bar{\mathfrak{B}}| \, Me^{-a\tau}\|\mathbf{z}(0)\| \, d\tau \le$$

$$\le \frac{M\left(M|\mathfrak{B} - \bar{\mathfrak{B}}|\,\mathfrak{f} + a\,|\mathfrak{f} - \bar{\mathfrak{f}}|\right)}{a\left(a - M|\mathfrak{B} - \bar{\mathfrak{B}}|\right)} + o(1). \tag{22}$$

Therefore, the first assertion of the theorem is proved.

Therefore, the second assertion follows from the inequality $\|\mathbf{z}\|_{1E} \le W^{-1}\|\mathbf{z}\|_{1D}$ (see, e.g., [62]) and the estimate expressed by Equation (22), where $l_{1E} = \left\{z = (p_1, p_2, \ldots)^T \,:\, \|z\|_{1E} \equiv \sum n|p_n| < \infty\right\}$. □

**Remark 1.** *A number of consequences of this statement can be formulated, for example,*

$$\limsup_{t\to\infty} \|\mathbf{p}(t) - \bar{\mathbf{p}}(t)\| \le \frac{4M\left(M|\mathfrak{B} - \bar{\mathfrak{B}}|\,\mathfrak{f} + a\,|\mathfrak{f} - \bar{\mathfrak{f}}|\right)}{ad\left(a - M|\mathfrak{B} - \bar{\mathfrak{B}}|\right)}, \tag{23}$$

*which follows from*

$$\|\mathbf{p}^* - \mathbf{p}^{**}\| \le 2\|\mathbf{z}^* - \mathbf{z}^{**}\| \le \frac{4}{d}\|\mathbf{z}^* - \mathbf{z}^{**}\|_{1D}. \tag{24}$$

*The respective perturbation bounds can be formulated for strongly ergodic (for instance, homogeneous) Markov chains (see [44]).*

**Remark 2.** *As shown in [44], the bounds presented in Theorem 2 and its corollaries are sufficiently sharp. Namely, in [44], we considered the queue-length process for the simplest ordinary $M/M/1$ queue and proved that the bounds established in Theorem 2 have the proper order.*

## 4. Convergence Rate Estimates and Perturbation Bounds for Main Classes

For Markov chains of Classes 1–4, an important role is played by the matrix $B^{**}(t) = DB(t)D^{-1}$. To begin with, write out this matrix for each of these classes.

For Class 1, this matrix has the form

$$B^{**}(t) =$$

$$= \begin{pmatrix} -(\lambda_0+\mu_1) & \frac{d_1}{d_2}\mu_1 & 0 & \cdots & 0 & \cdots & \cdots \\ \frac{d_2}{d_1}\lambda_1 & -(\lambda_1+\mu_2) & \frac{d_2}{d_3}\mu_2 & \cdots & 0 & \cdots & \cdots \\ \ddots & \ddots & \ddots & \ddots & \ddots & \cdots & \\ 0 & \cdots & \cdots & \frac{d_r}{d_{r-1}}\lambda_{r-1} & -(\lambda_{r-1}+\mu_r) & \frac{d_r}{d_{r+1}}\mu_r & \cdots \\ \cdots & \cdots & \cdots & \cdots & \cdots & \cdots & \cdots \end{pmatrix} \tag{25}$$

in the case of a countable state space ($S = \infty$);

$$B^{**}(t) =$$

$$= \begin{pmatrix} -(\lambda_0+\mu_1) & \frac{d_1}{d_2}\mu_1 & 0 & \cdots & 0 \\ \frac{d_2}{d_1}\lambda_1 & -(\lambda_1+\mu_2) & \frac{d_2}{d_3}\mu_2 & \cdots & 0 \\ \ddots & \ddots & \ddots & \ddots & \ddots \\ 0 & \cdots & \cdots & \frac{d_S}{d_{S-1}}\lambda_{S-1} & -(\lambda_{S-1}+\mu_S) \end{pmatrix} \tag{26}$$

in the case of a finite state space ($S < \infty$).

For Class 2, this matrix has the form

$$B^{**}(t) = \begin{pmatrix} a_{11} & \frac{d_1}{d_2}\mu_1 & 0 & \cdots & 0 \\ \frac{d_2}{d_1}a_1 & a_{22} & \frac{d_2}{d_3}\mu_2 & \cdots & 0 \\ \frac{d_3}{d_1}a_2 & \frac{d_3}{d_2}a_1 & a_{33} & \frac{d_3}{d_4}\mu_3 & \cdots \\ \ddots & \ddots & \ddots & \ddots & \ddots \\ \ddots & \ddots & \ddots & \ddots & \ddots \end{pmatrix} \tag{27}$$

in the case of a countable state space ($S = \infty$);

$$B^{**}(t) =$$

$$= \begin{pmatrix} a_{11}-a_S & \frac{d_1}{d_2}\mu_1 & 0 & \cdots & 0 \\ \frac{d_2}{d_1}(a_1-a_S) & a_{22}-a_{S-1} & \frac{d_2}{d_3}\mu_2 & \cdots & 0 \\ \ddots & \ddots & \ddots & \ddots & \ddots \\ \frac{d_S}{d_1}(a_{S-1}-a_S) & \cdots & \cdots & \frac{d_S}{d_{S-1}}(a_1-a_2) & a_{SS}-a_1 \end{pmatrix} \tag{28}$$

in the case of a finite state space ($S < \infty$).

For Class 3, this matrix has the form

$$B^{**}(t) =$$

$$= \begin{pmatrix} -(\lambda_0+b_1) & \frac{d_1}{d_2}(b_1-b_2) & \frac{d_1}{d_3}(b_2-b_3) & \cdots & \cdots \\ \frac{d_2}{d_1}\lambda_1 & -(\lambda_1+\sum\limits_{i\le 2} b_i) & \frac{d_2}{d_3}(b_1-b_3) & \cdots & \cdots \\ \ddots & \ddots & \ddots & \ddots & \ddots \\ 0 & \cdots & \cdots & \frac{d_r}{d_{r-1}}\lambda_{r-1} & -(\lambda_{r-1}+\sum\limits_{i\le r} b_i)\cdots \\ \ddots & \ddots & \ddots & \ddots & \ddots \end{pmatrix} \quad (29)$$

in the case of a countable state space ($S = \infty$);

$$B^{**}(t) =$$

$$= \begin{pmatrix} -(\lambda_0+b_1) & \frac{d_1}{d_2}(b_1-b_2) & \frac{d_1}{d_3}(b_2-b_3) & \cdots & \frac{d_1}{d_S}(b_{S-1}-b_S) \\ \frac{d_2}{d_1}\lambda_1 & -(\lambda_1+\sum\limits_{i\le 2} b_i) & \frac{d_2}{d_3}(b_1-b_3) & \cdots & \frac{d_2}{d_S}(b_{S-2}-b_S) \\ \ddots & \ddots & \ddots & \ddots & \ddots \\ 0 & \cdots & \cdots & \frac{d_S}{d_{S-1}}\lambda_{S-1} & -(\lambda_{S-1}+\sum\limits_{i\le S} b_i) \end{pmatrix} \quad (30)$$

in the case of a finite state space ($S < \infty$).

Finally, for Class 4, this matrix has the form

$$B^{**} = \begin{pmatrix} a_{11} & \frac{d_1}{d_2}(b_1-b_2) & \frac{d_1}{d_3}(b_2-b_3) & \cdots & \cdots & \\ \frac{d_2}{d_1}a_1 & a_{22} & \frac{d_2}{d_3}(b_1-b_3) & \cdots & \cdots & \\ \ddots & \ddots & \ddots & \ddots & \ddots & \\ \frac{d_r}{d_1}a_{r-1} & \cdots & \cdots & \frac{d_r}{d_{r-1}}a_1 & a_{rr} & \cdots \\ \cdots & \cdots & \cdots & \cdots & \cdots & \cdots \end{pmatrix} \quad (31)$$

in the case of a countable state space ($S = \infty$);

$$B^{**}(t) =$$

$$= \begin{pmatrix} a_{11}-a_S & \frac{d_1}{d_2}(b_1-b_2) & \frac{d_1}{d_3}(b_2-b_3) & \cdots & \frac{d_1}{d_S}(b_{S-1}-b_S) \\ \frac{d_2}{d_1}(a_1-a_S) & a_{22}-a_{S-1} & \frac{d_2}{d_3}(b_1-b_3) & \cdots & \frac{d_2}{d_S}(b_{S-2}-b_S) \\ \ddots & \ddots & \ddots & \ddots & \ddots \\ \frac{d_S}{d_1}(a_{S-1}-a_S) & \cdots & \cdots & \frac{d_S}{d_{S-1}}(a_1-a_2) & a_{SS}-a_1 \end{pmatrix} \quad (32)$$

in the case of a finite state space ($S < \infty$).

In the proofs of the following theorems, we use the notion of the logarithmic norm of a linear operator function and the related estimates of the norm of the Cauchy operator of a linear differential equation. The corresponding results are described in detail in our preceding works (see [47,62,63]). Here we restrict ourselves only to the necessary minimum.

Recall that the logarithmic norm of an operator function $B^{**}(t)$ is defined as the number

$$\gamma(B^{**}(t)) = \lim_{h\to+0} h^{-1}\left(\|I + hB^{**}(t)\| - 1\right).$$

Let $V(t,s) = V(t)V^{-1}(s)$ be the Cauchy operator of the differential equation

$$\frac{d\mathbf{w}}{dt} = B^{**}(t)\mathbf{w}.$$

Therefore, the estimate

$$\|V(t,s)\| \le e^{\int_s^t \gamma(B^{**}(u))\,du}$$

holds. Moreover, if for each $t \ge 0$ $B^{**}(t)$ maps $l_1$ into itself, then the logarithmic norm can be calculated by the formula

$$\gamma(B^{**}(t)) = \sup_{1\le j\le S} \left( b_{jj}^{**}(t) + \sum_{i\neq j} |b_{ij}^{**}(t)| \right). \tag{33}$$

Now let

$$\alpha_i\left(t\right) = -\left( b_{jj}^{**}\left(t\right) + \sum_{i\neq j} |b_{ij}^{**}\left(t\right)| \right),\ \alpha\left(t\right) = \inf_{i\ge 1} \alpha_i\left(t\right). \tag{34}$$

Also note that, if in Classes 2–4 the intensities $a_k(t)$ and $b_k(t)$ do not increase in $k$ for each $t$, then in all the cases the matrix $B^{**}(t)$ is essentially nonnegative (that is, its non-diagonal elements are nonnegative); therefore, in Equations (33) and (34), the signs of the absolute value can be omitted.

The following statement ([47]) Theorem 1 is given here for convenience.

**Theorem 3.** *Let, for some sequence $\{d_i,\ i \ge 1\}$ of positive numbers, the conditions $d_1 = 1$, $d = \inf_{i\ge 1} d_i > 0$ and*

$$\int_0^\infty \alpha(t)\,dt = +\infty \tag{35}$$

*hold. Therefore, the Markov chain $X(t)$ is weakly ergodic and for any initial condition $s \ge 0$, $\mathbf{w}(s)$, and for all $t \ge s$ the following estimate holds:*

$$\|\mathbf{w}\left(t\right)\| \le e^{-\int_s^t \alpha(u)du}\|\mathbf{w}(s)\|. \tag{36}$$

Now let, instead of Equation (35), for all $0 \le s \le t$, a stronger condition

$$e^{-\int_s^t \alpha(\tau)\,d\tau} \le M^* e^{-a^*(t-s)} \tag{37}$$

hold.

**Theorem 4.** *Let, under the conditions of Theorem 3, Inequality (37) hold. Therefore, the Markov chain $X(t)$ is $1D$-exponentially weakly ergodic, and for all $t \ge s \ge 0$ and $\mathbf{p}^*(s) \in l_{1D}$, $\mathbf{p}^{**}(s) \in l_{1D}$, Inequality (15) holds with $M = M^*$ and $a = a^*$.*

**Remark 3.** *In the case of a homogeneous Markov chain, or if all intensities are periodic with one and the same period, conditions expressed by Equations (35) and (37) are equivalent.*

**Theorem 5.** *Let the conditions of Theorem 4 hold. Therefore, the Markov chain $X(t)$ is $1D$-exponentially weakly ergodic. Under perturbations small enough (see Equation (16)), the perturbed chain $\bar{X}(t)$ is also $1D$-exponentially weakly ergodic, and the perturbation bound expressed by Equation (38) in the $1D$-norm holds. If, moreover, $W = \inf_{i\ge 1} \frac{d_i}{i} > 0$, then both chains $X(t)$ and $\bar{X}(t)$ have limit expectations and the estimate expressed by Equation (18) holds for the perturbation of the mathematical expectation.*

To obtain perturbation bounds in the natural norm, it suffices to use Inequality (24) mentioned above.

**Corollary 2.** *Under the conditions of Theorem 5, the following perturbation bound in the natural $l_1$- (total variation) norm holds:*

$$\limsup_{t\to\infty} \|\mathbf{p}(t) - \bar{\mathbf{p}}(t)\| \le \frac{4M\left(M\,|\mathfrak{B} - \bar{\mathfrak{B}}|\,\mathfrak{f} + a\,|\mathfrak{f} - \bar{\mathfrak{f}}|\right)}{ad\left(a - M\,|\mathfrak{B} - \bar{\mathfrak{B}}|\right)}. \tag{38}$$

Note that it is convenient to use the results formulated above for the construction of perturbation bounds for Markov chains of the first four classes (see, e.g., [12,44,46,47]).

For chains of the fifth class, as a rule, it is convenient to use the approach based on uniform bounds as shown below. These models were considered, e.g., in [14,64,65].

Let

$$\beta_*(t) = \inf_k a_{0k}(t). \tag{39}$$

**Theorem 6.** *Let the intensities of catastrophes be essential, that is*

$$\int_0^\infty \beta_*(t)\,dt = +\infty. \tag{40}$$

*Therefore, the chain $X(t)$ is weakly ergodic in the uniform operator topology and for any initial conditions $\mathbf{p}^*(0), \mathbf{p}^{**}(0)$, and any $0 \le s \le t$, the following convergence rate estimate holds:*

$$\|\mathbf{p}^*(t) - \mathbf{p}^{**}(t)\| \le 2e^{-\int_s^t \beta_*(\tau)\,d\tau}. \tag{41}$$

To prove this theorem, we will use the same technique as in [14]. Rewrite the forward Kolmogorov system expressed by Equation (3) in the form

$$\frac{d\mathbf{p}}{dt} = A^*(t)\,\mathbf{p} + \mathbf{g}(t), \quad t \ge 0. \tag{42}$$

Here $\mathbf{g}(t) = (\beta_*(t), 0, 0, \dots)^T$, $A^*(t) = \left(a^*_{ij}(t)\right)_{i,j=0}^{\infty}$, and

$$a^*_{ij}(t) = \begin{cases} a_{0j}(t) - \beta_*(t), & \text{if } i = 0, \\ a_{ij}(t), & \text{otherwise}. \end{cases} \tag{43}$$

The solution to this equation can be written as

$$\mathbf{p}(t) = U^*(t,0)\,\mathbf{p}(0) + \int_0^t U^*(t,\tau)\,\mathbf{g}(\tau)\,d\tau \tag{44}$$

where $U^*(t,s)$ is the Cauchy operator of the differential equation

$$\frac{d\mathbf{z}}{dt} = A^*(t)\,\mathbf{z}. \tag{45}$$

Note that the matrix $A^*(t)$ is essentially nonnegative for all $t \ge 0$. Its logarithmic norm is equal to

$$\gamma(A^*(t)) = \sup_i \left( a^*_{ii}(t) + \sum_{j \ne i} a^*_{ji}(t) \right) = -\beta_*(t). \tag{46}$$

Hence,

$$\|\mathbf{p}^*(t) - \mathbf{p}^{**}(t)\| \le e^{-\int_s^t \beta_*(\tau)\,d\tau} \|\mathbf{p}^*(s) - \mathbf{p}^{**}(s)\| \le 2e^{-\int_s^t \beta_*(\tau)\,d\tau}. \tag{47}$$

**Theorem 7.** *Let, instead of Equation (40), the stronger condition*

$$e^{-\int_s^t \beta_*(\tau)\,d\tau} \le c^* e^{-b^*(t-s)} \tag{48}$$

*hold. Therefore, the chain $X(t)$ is weakly exponentially ergodic in the uniform operator topology, and if the perturbations are small enough, that is, $\|\hat{A}(t)\| \le \varepsilon$ for almost all $t \ge 0$, then the perturbed chain $\bar{X}(t)$ is also exponentially weakly ergodic, and the perturbation bound expressed by Equation (7) holds with $c = c^*$ and $b = b^*$.*

## 5. Examples

First note that many examples of perturbation bounds for queueing systems have been considered in [11,12,14,44,46,66,67].

Here, to compare both approaches, we will mostly deal with the queueing system $M_t|M_t|N|N$ with losses and 1-periodic intensities. In the preceding papers on this model, other problems were considered. For example, in [68], the asymptotics of the rate of convergence to the stationary mode as $N \to \infty$, was studied, whereas the paper [69] dealt with the asymptotics of the convergence parameter under various limit relations between the intensities and the dimensionality of the model. In [66,67], perturbation bounds were considered under additional assumptions.

Let $N \ge 1$ be the number of servers in the system. Assume that the customers arrival intensity $\lambda(t)$ and the service intensity of a server $\mu(t)$ are 1-periodic nonnegative functions integrable on the interval $[0,1]$. Therefore, the number of customers in the system (queue length) $X(t)$ is a finite Markov chain of Class 1, that is, a BDP with the intensities $\lambda_{k-1}(t) = \lambda(t)$, $\mu_k(t) = k\mu(t)$ for $k = 1, \dots, N$.

It should be especially noted that the process $X(t)$ is weakly ergodic (obviously exponentially and uniformly ergodic, since the intensities are periodic and the state space is finite) if and only if

$$\int_0^1 (\lambda(t) + \mu(t))\,dt > 0 \tag{49}$$

(see, e.g., [70]).

For definiteness, assume that $\int_0^1 \mu(t)\,dt > 0$.

Apply the approach described in Theorems 3 and 4.

Let all $d_k = 1$. Therefore,

$$B^{**}(t) = \begin{pmatrix} -(\lambda+\mu) & \mu & 0 & \cdots & 0 \\ \lambda & -(\lambda+2\mu) & 2\mu & \cdots & 0 \\ \ddots & \ddots & \ddots & \ddots & \ddots \\ 0 & \cdots & \cdots & \lambda & -(\lambda+N\mu) \end{pmatrix}, \tag{50}$$

and in Equation (34) we have $\alpha_i(t) = \mu(t)$ for all $i$; hence, $\alpha(t) = \mu(t)$.

Therefore, Theorem 3 yields the estimate

$$\|\mathbf{p}^*(t) - \mathbf{p}^{**}(t)\|_{1D} \le e^{-\int_s^t \mu(\tau)\,d\tau} \|\mathbf{p}^*(s) - \mathbf{p}^{**}(s)\|_{1D}. \tag{51}$$

To find the constants in the estimates, let $\mu^* = \int_0^1 \mu(\tau)\,d\tau$ and consider

$$\int_0^t \mu(\tau)\,d\tau = \mu^* t + \int_0^{\{t\}} (\mu(\tau) - \mu^*)\,d\tau. \tag{52}$$

Find the bound for the second summand in Equation (52). Assuming $u = \{t\}$, we obtain

$$\left| \int_0^u (\mu(\tau) - \mu^*)\, d\tau \right| \leq K^* = \sup_{u \in [0,1]} \int_0^u (\mu(\tau) - \mu^*)\, d\tau. \tag{53}$$

Therefore,

$$e^{-\int_s^t \mu(\tau)\, d\tau} \leq e^{K^*} e^{-\mu^*(t-s)}. \tag{54}$$

Therefore, for the queueing system $M_t|M_t|N|N$, the conditions of Theorem 5 and Corollary 2

$$d = 1, \quad M = M^* = e^{K^*}, \quad a = a^* = \mu^*, \quad W = \frac{1}{N}. \tag{55}$$

These statements imply the following perturbation bounds:

$$\limsup_{t\to\infty} \|\mathbf{p}(t) - \bar{\mathbf{p}}(t)\| \leq \frac{4e^{K^*}\left(e^{K^*}|\mathfrak{B} - \bar{\mathfrak{B}}|\,\mathfrak{f} + \mu^*|\mathfrak{f} - \bar{\mathfrak{f}}|\right)}{\mu^*\left(\mu^* - e^{K^*}|\mathfrak{B} - \bar{\mathfrak{B}}|\right)} \tag{56}$$

for the vector od=f state probabilities, and

$$\limsup_{t\to\infty} |\phi(t) - \bar{\phi}(t)| \leq \frac{Ne^{K^*}\left(e^{K^*}|\mathfrak{B} - \bar{\mathfrak{B}}|\,\mathfrak{f} + \mu^*|\mathfrak{f} - \bar{\mathfrak{f}}|\right)}{\mu^*\left(\mu^* - e^{K^*}|\mathfrak{B} - \bar{\mathfrak{B}}|\right)}, \tag{57}$$

for limit expectations.

Moreover, for these bounds to be consistent, additional information is required concerning the form of the perturbed intensity matrix. The simplest bounds can be obtained, if it is assumed that the perturbed Markov chain is also a BDP with the same state space and the birth and death intensities $\lambda_{k-1}(t)$ and $\mu_k(t)$, respectively. Therefore, if the birth and death intensities themselves do not exceed $\varepsilon$ for almost all $t \geq 0$, then $|\mathfrak{f} - \bar{\mathfrak{f}}| \leq \varepsilon$ and $|\mathfrak{B} - \bar{\mathfrak{B}}| \leq 5\varepsilon$, so that the bounds expressed by Equations (56) and (57) have the form

$$\limsup_{t\to\infty} \|\mathbf{p}(t) - \bar{\mathbf{p}}(t)\| \leq \frac{4e^{K^*}\left(5Le^{K^*} + \mu^*\right)\varepsilon}{\mu^*\left(\mu^* - 5\varepsilon e^{K^*}\right)} \tag{58}$$

for the vectors of state probabilities, and

$$\limsup_{t\to\infty} |\phi(t) - \bar{\phi}(t)| \leq \frac{4Ne^{K^*}\left(5Le^{K^*} + \mu^*\right)\varepsilon}{\mu^*\left(\mu^* - 5\varepsilon e^{K^*}\right)} \tag{59}$$

for the limit expectations.

On the other hand, Theorem 7 can be applied as well. To construct the bounds for the corresponding parameters, Equation (24) and the fact that $\|D\|_1 = N$ is exploited. Therefore, Theorem 7 is valid for the queueing system $M_t|M_t|N|N$ with the following values of the parameters:

$$c = c^* = 4Ne^{K^*}, \quad b = b^* = \mu^*. \tag{60}$$

According to this theorem, we obtain the estimate

$$\limsup_{t\to\infty} \|\mathbf{p}(t) - \bar{\mathbf{p}}(t)\| \leq \frac{(1 + K^* + \log(2N))\,\varepsilon}{\mu^*}. \tag{61}$$

Moreover, the Markov chains $X(t)$ and $\bar{X}(t)$ have limit expectations and

$$|\phi(t) - \bar{\phi}(t)| \leq \frac{N\,(1 + K^* + \log(2N))\,\varepsilon}{\mu^*}. \tag{62}$$

It is worth noting that, for the estimates expressed by Equations (61) and (62) to hold, only the condition of the smallness of perturbations is required, and *no* additional information concerning the structure of the intensity matrix is required.

Thus, in the example with the finite state space under consideration, uniform bounds turn out to be more exact.

Now consider a more special example. Let $N = 299$, $\lambda(t) = 200(1 + \sin 2\pi\omega t)$, $\mu(t) = 1$.

In Figures 1–5, there are plots of the expected number of customers in the system for some of most probable states with $\omega = 1$; in Figures 6 and 7, there are plots of the expected number of customers with $\omega = 0.5$.

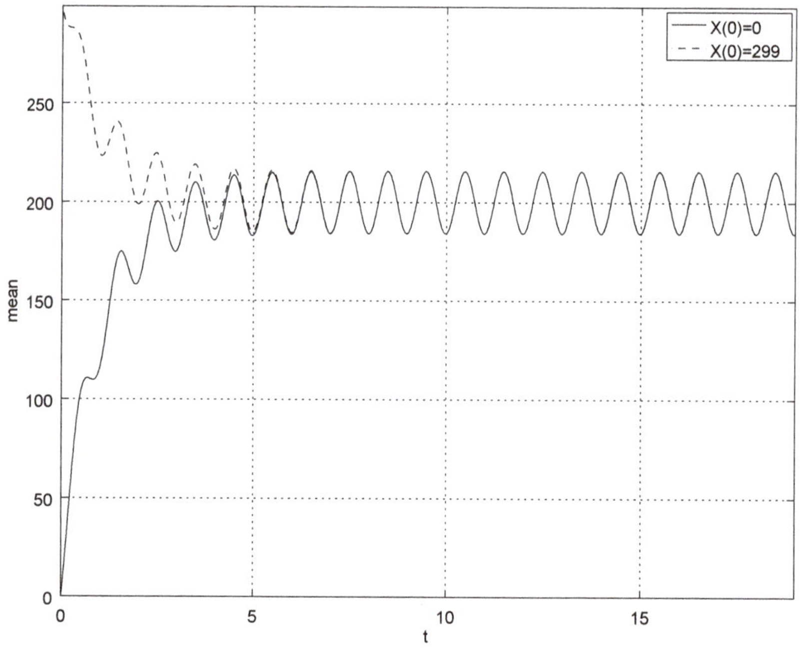


**Figure 1.** Example 1. The mean $E(t, 0)$ and $E(t, N)$ for the original process $t \in [0, 19]$, $\omega = 1$.

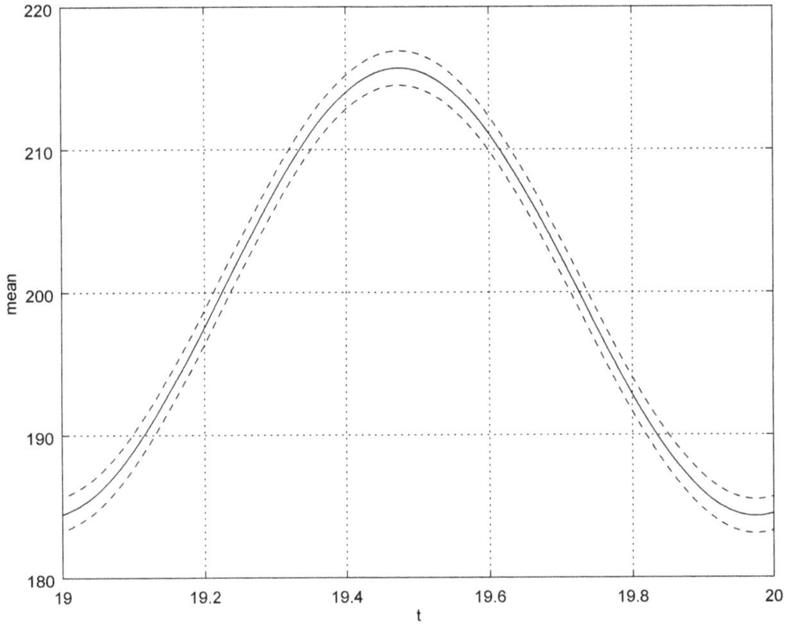


**Figure 2.** Example 1. The perturbation bounds for the limit expectation $E(t,0)$, $t \in [19,20]$, $\omega = 1$.

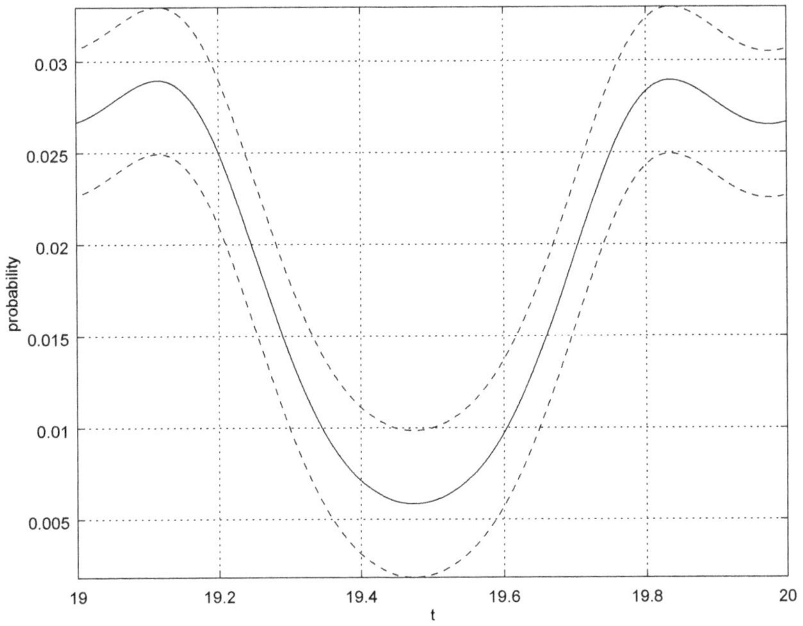


**Figure 3.** Example 1. The perturbation bounds for the "limit" probability $\Pr(X(t) = 190)$, $t \in [19,20]$, $\omega = 1$.

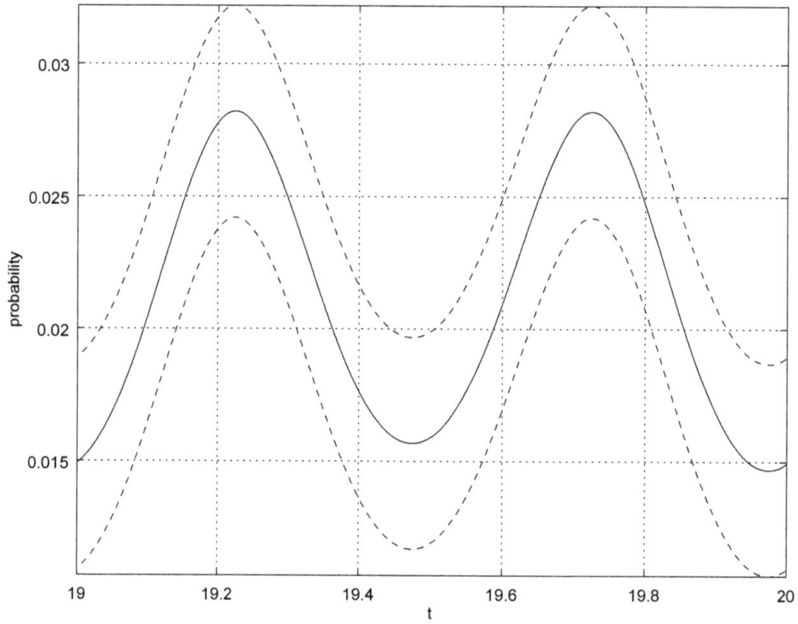


**Figure 4.** Example 1. The perturbation bounds for the "limit" probability $\Pr(X(t) = 200)$, $t \in [19, 20]$, $\omega = 1$.

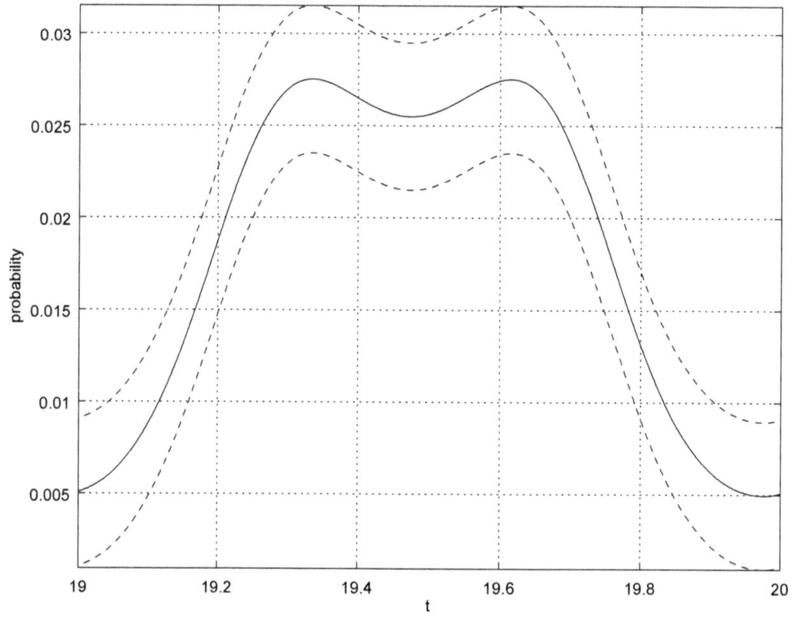


**Figure 5.** Example 1. The perturbation bounds for the "limit" probability $\Pr(X(t) = 210)$, $t \in [19, 20]$, $\omega = 1$.

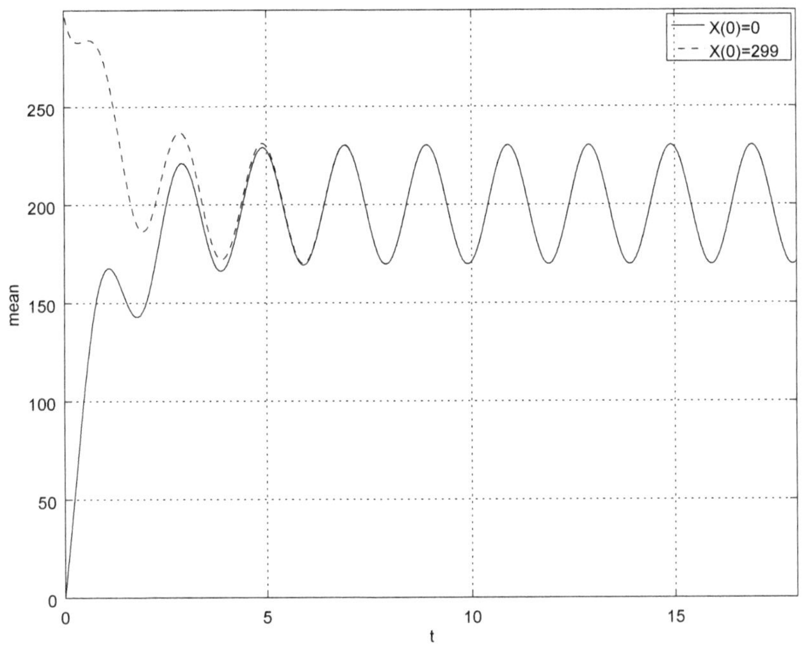


**Figure 6.** Example 1. The expectations $E(t,0)$ and $E(t,N)$ for the original process $t \in [0,18]$, $\omega = 0.5$.

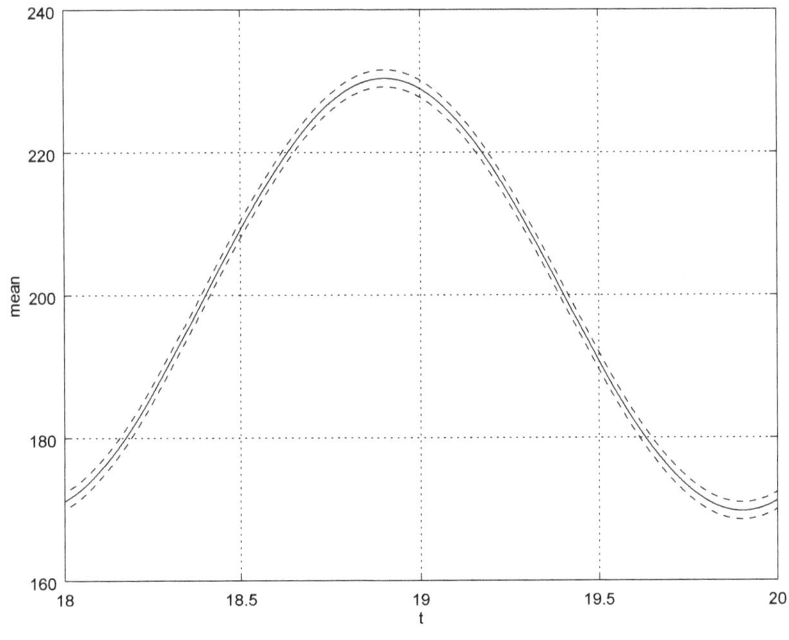


**Figure 7.** Example 1. The perturbation bounds for the limit expectation $E(t,0)$, $t \in [18,20]$, $\omega = 0.5$.

On the other hand, as has already been noted, for the Markov chains of Classes 1–4 with countable state space, no uniform bounds could be constructed.

Consider the construction of bounds on the example of a rather simple model, which, however, does not belong to the most well-studied Class 1 (that is, which is not a BDP).

Let a queueing system be given in which the customers can appear separately or in pairs with the corresponding intensities $a_1(t) = \lambda(t)$ and $a_2(t) = 0.5\lambda(t)$, but are served one by one on one of two servers with constant intensities $\mu_k(t) = \min(k, 2)\mu$, where $\lambda(t)$ is a 1-periodic function integrable on the interval $[0, 1]$. Therefore, the number of customers in this system belongs to Class 2, and the corresponding matrix $B^{**}(t)$ has the form

$$B^{**}(t) = \begin{pmatrix} a_{11} & \frac{d_1}{d_2}\mu & 0 & \cdots & 0 \\ \frac{d_2}{d_1}\lambda & a_{22} & \frac{d_2}{d_3}2\mu & \cdots & 0 \\ \frac{d_3}{d_1}0.5\lambda & \frac{d_3}{d_2}\lambda & a_{33} & \frac{d_3}{d_4}2\mu & \cdots \\ 0 & \ddots & \ddots & \ddots & \ddots \\ \ddots & \ddots & \ddots & \ddots & \ddots \end{pmatrix} \quad (63)$$

where $a_{11}(t) = -(1.5\lambda(t) + \mu)$, $a_{kk}(t) = -(1.5\lambda(t) + 2\mu)$, if $k \geq 2$. This matrix is essentially nonnegative, such that, in the expression for the logarithmic norm, the signs of the absolute value can be omitted. Let $d_1 = 1$, $d_{k+1} = \delta d_k$, and $\delta > 1$. For this purpose, consider the expressions from Equation (34). We have

$$\alpha_1(t) = \mu - \lambda(t)\left(0.5\delta^2 + \delta - 1.5\right),$$

$$\alpha_2(t) = \mu\left(2 - \delta^{-1}\right) - \lambda(t)\left(0.5\delta^2 + \delta - 1.5\right),$$

$$\alpha_k(t) = 2\mu\left(1 - \delta^{-1}\right) - \lambda(t)\left(0.5\delta^2 + \delta - 1.5\right), \ k \geq 3.$$

Therefore, for $\delta \leq 2$, we obtain

$$\alpha(t) = \inf_{i \geq 1} \alpha_i(t) = 2\mu\left(1 - \delta^{-1}\right) - \lambda(t)\left(0.5\delta^2 + \delta - 1.5\right) = \\ = (\delta - 1)\left(\frac{2\mu}{\delta} - 0.5\lambda(t)(\delta + 3)\right), \quad (64)$$

and the condition

$$\alpha^* = \int_0^1 (\delta - 1)\left(\frac{2\mu}{\delta} - 0.5\lambda(t)(\delta + 3)\right) dt = \frac{\delta - 1}{2}\left(\frac{4\mu}{\delta} - \lambda^*(\delta + 3)\right) > 0 \quad (65)$$

will a fortiori hold if $\mu > \lambda^*$ with a corresponding choice of $\delta \in (1, 2]$.

The further reasoning is almost the same as in the preceding example: instead of Equation (54), we obtain

$$e^{-\int_s^t \alpha(\tau)\, d\tau} \leq e^{K^*} e^{-\alpha^*(t-s)} \quad (66)$$

where now

$$K^* = \sup_{u \in [0,1]} \int_0^u (\alpha(\tau) - \alpha^*)\, d\tau. \quad (67)$$

Hence, the conditions of Theorem 5 and Corollary 2 for the number of customers in the system under consideration hold for

$$d = 1, \quad M = M^* = e^{K^*}, \quad a = a^* = \alpha^*, \quad W = \inf_{k \geq 1} \frac{\delta^{k-1}}{k}. \quad (68)$$

To construct meaningful perturbation bounds, it is necessarily required to have additional information concerning the form of the perturbed intensity matrix. Therefore, Example 1 in Section 2 shows that, if a possibility of the arrival of an arbitrary number of customers ("mass arrival" in the terminology of [58]) to an empty queue is assumed, then an arbitrarily small (in the uniform norm) perturbation of the intensity matrix can "spoil" all the characteristics of the process. For example, satisfactory bounds can be constructed if we know that the intensity matrix of the perturbed system has the same form; that is, the customers can appear either separately or in pairs and are served one by one. Therefore, if the perturbations of the intensities themselves do not exceed $\varepsilon$ for almost all $t \geq 0$, then $|\mathfrak{f} - \bar{\mathfrak{f}}| \leq 5\varepsilon$ and $|\mathfrak{B} - \bar{\mathfrak{B}}| \leq 5\varepsilon$, such that, instead of Equations (56) and (57), we obtain

$$\limsup_{t \to \infty} \|\mathbf{p}(t) - \bar{\mathbf{p}}(t)\| \leq \frac{20e^{K^*}\varepsilon\left(Le^{K^*} + \alpha^*\right)}{\alpha^*\left(\alpha^* - 20\varepsilon e^{K^*}\right)} \tag{69}$$

for the vectors of state probabilities and

$$\limsup_{t \to \infty} |\phi(t) - \bar{\phi}(t)| \leq \frac{20e^{K^*}\varepsilon\left(Le^{K^*} + \alpha^*\right)}{\alpha^* W\left(\alpha^* - 20\varepsilon e^{K^*}\right)} \tag{70}$$

for the limit expectations.

For example, let $\lambda(t) = 1 + \sin 2\pi t$, $\mu(t) = 3$, and $\delta = 2$. Therefore, we have

$$\alpha(t) = \mu - 2.5\lambda(t), \quad \alpha^* = 0.5, \quad W = 1. \tag{71}$$

Furthermore, we follow the method described in [71,72] in detail. Namely, we choose the dimensionality of the truncated process (300 in our case), the interval on which the desired accuracy is achieved ([0, 100]) in the example under consideration) and the limit interval itself (here it is [100, 101]).

Figures 8–13 expose the plots of the expected number of customers in the system and some of the most probable states.

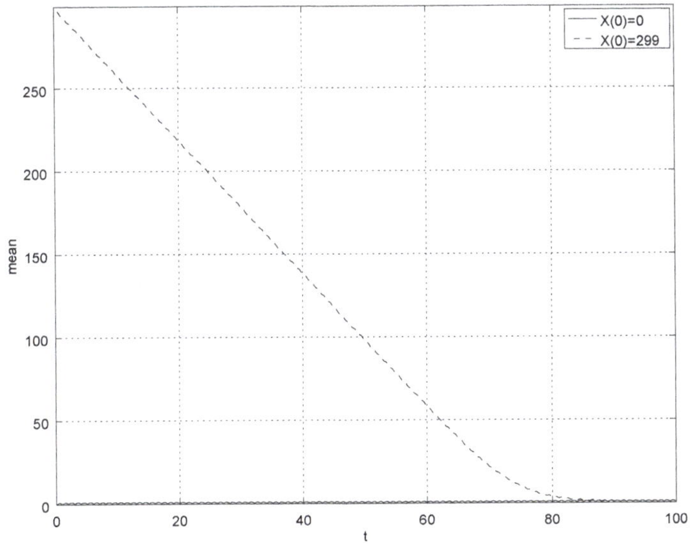


**Figure 8.** Example 2. The expectations $E(t, 0)$ and $E(t, 299)$ for the original process $t \in [0, 100]$.

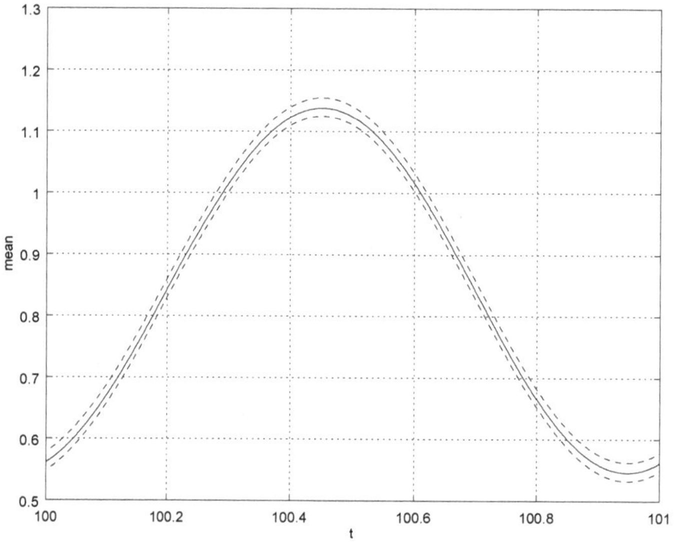


**Figure 9.** Example 2. The perturbation bounds for the limit expectation $E(t, 0)$, $t \in [100, 101]$.

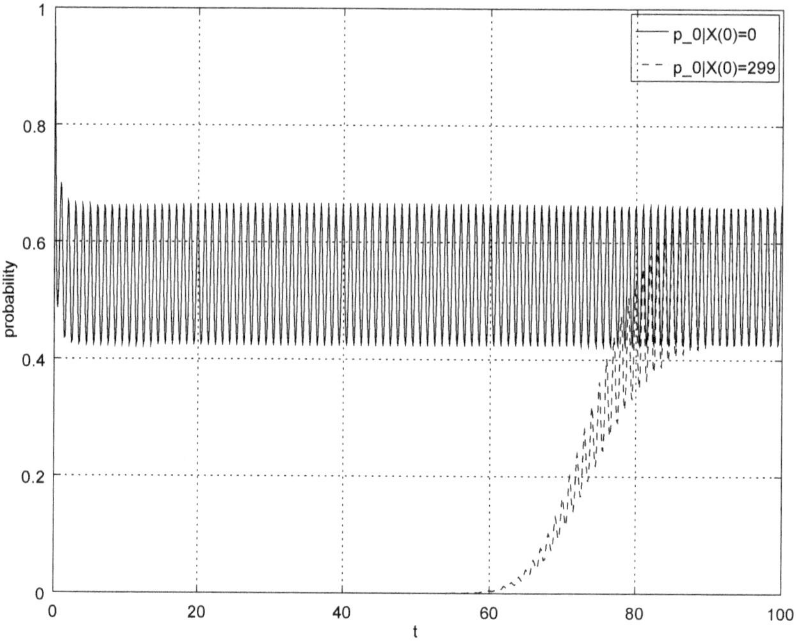


**Figure 10.** Example 2. The probabilities of the empty queue for $X(0) = 0$ and $X(0) = 299$ for the original process $t \in [0, 100]$.

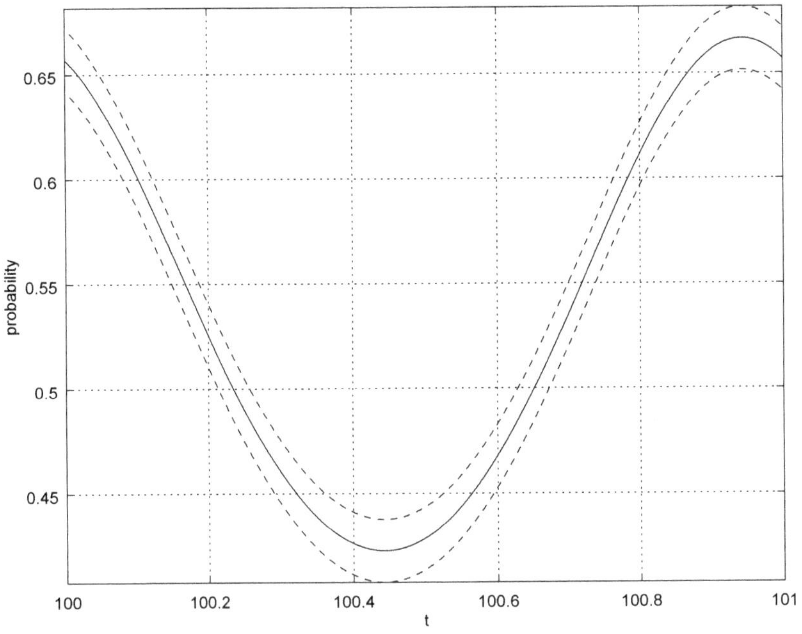


**Figure 11.** Example 2. The perturbation bounds for the "limit" probability $\Pr(X(t) = 0)$, $t \in [100, 101]$.

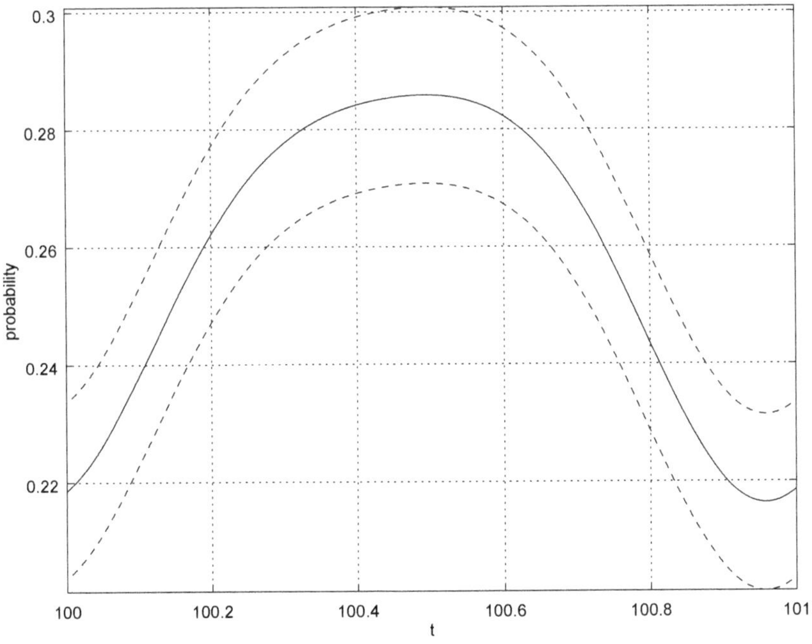


**Figure 12.** Example 2. The perturbation bounds for the "limit" probability $\Pr(X(t) = 1)$, $t \in [100, 101]$.

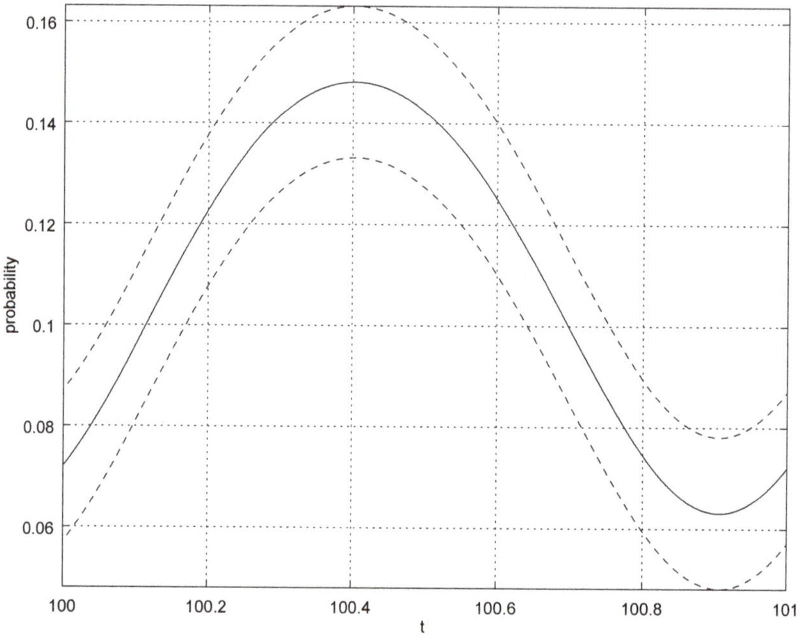


**Figure 13.** Example 2. The perturbation bounds for the "limit" probability $\Pr(X(t) = 2)$, $t \in [100, 101]$.

**Author Contributions:** Conceptualization, A.Z and V.K.; Software, Y.S.; Investigation, A.Z.(Sections 1–5), V.K. (Sections 1–3) and Y.S. (Sections 4–5); Writing—original draft preparation, A.Z., V.K. and Y.S.; Writing—review and editing, A.Z., V.K. and Y.S. All authors have read and agreed to the published version of the manuscript.

**Funding:** Sections 1–3 were written by Korolev and Zeifman under the support of the Russian Science Foundation, project 18-11-00155. Sections 4 and 5 were written by Satin and Zeifman under the support of the Russian Science Foundation, project 19-11-00020.

**Conflicts of Interest:** The authors declare that there is no conflict of interest.

## References

1. Kalashnikov, V.V. Qualitative analysis of the behavior of complex systems by the method of test functions. *M. Nauka* **1978**, 247. (In Russian)
2. Stoyan, D. *Qualitative Eigenschaften Und Abschätzungen Stochastischer Modelle*; Akademie-Verlag: Berlin, Germany, 1977. (In Germany)
3. Zolotarev, V.M. Quantitative estimates for the continuity property of queueing systems of type $G/G/\infty$. *Theory Prob. Its Appl.* **1977**, *22*, 679–691. [CrossRef]
4. Zolotarev, V.M. Probability metrics. *Theory Prob. Its Appl.* **1983**, *28*, 264–287. [CrossRef]
5. Kalashnikov, V.V.; Zolotarev, V.M. *Stability Problems for Stochastic Models*; Lecture Notes in Mathematics; VSP: Perm, Russia, 1983; Volume 982.
6. Gnedenko, B.V.; Korolev, V.Y. *Random Summation: Limit Theorems and Applications*; CRC Press: Boca, OH, USA, 1996.
7. Kartashov, N.V. Strongly stable Markov chains. *J. Soviet Math.* **1986**, *34*, 1493–1498. [CrossRef]
8. Kartashov, N.V. Criteria for uniform ergodicity and strong stability of Markov chains with a common phase space. *Theory Prob. Its Appl.* **1985**, *30*, 71–89.
9. Zeifman, A.I. Stability for contionuous-time nonhomogeneous Markov chains. *Lect. Notes Math.* **1985**, *1155*, 401–414.

10. Zeifman, A. Qualitative properties of inhomogeneous birth and death processes. *J. Sov. Math.* **1991**, *57*, 3217–3224. [CrossRef]
11. Zeifman, A.I.; Isaacson, D. On strong ergodicity for nonhomogeneous continuous-time Markov chains. *Stochast. Process. Their Appl.* **1994**, *50*, 263–273. [CrossRef]
12. Zeifman, A.I. Stability of birth and death processes. *J. Math. Sci.* **1998**, *91*, 3023–3031. [CrossRef]
13. Altman, E.; Avrachenkov, K.; Núnez-Queija, R. Perturbation analysis for denumerable Markov chains with application to queueing models. *Adv. Appl. Prob.* **2004**, *36*, 839–853. [CrossRef]
14. Zeifman, A.I.; Korotysheva, A. Perturbation bounds for $M_t/M_t/N$ queue with catastrophes. *Stochast. Models* **2012**, *28*, 49–62. [CrossRef]
15. Aldous, D.J.; Fill, J. Reversible Markov Chains and Random Walks on Graphs. Chapter 8. Available online: http://www.stat.berkeley.edu/users/aldous/RWG/book.html (accessed on 13 Fabruary 2020).
16. Diaconis, P.; Stroock, D. Geometric Bounds for Eigenvalues of Markov Chains. *Ann. Appl. Prob.* **1991**, *1*, 36–61. [CrossRef]
17. Mitrophanov, A.Y. Stability and exponential convergence of continuous-time Markov chains. *J. Appl. Prob.* **2003**, *40*, 970–979. [CrossRef]
18. Mitrophanov, A.Y. The spectral gap and perturbation bounds for reversible continuous-time Markov chains. *J. Appl. Prob.* **2004**, *41*, 1219–1222. [CrossRef]
19. Mitrophanov, A.Y. Ergodicity coefficient and perturbation bounds for continuous-time Markov chains. *Math. Inequal. Appl.* **2005**, *8*, 159–168. [CrossRef]
20. Mitrophanov, A.Y. Estimates of sensitivity to perturbations for finite homogeneous continuous-time Markov chains. *Theory Probab. Its Appl.* **2006**, *50*, 319–326. [CrossRef]
21. Zeifman, A.I.; Korotysheva, A.V.; Panfilova, T.Y.L.; Shorgin, S.Y. Stability bounds for some queueing systems with catastrophes. *Inf. Primeneniya* **2011**, *5*, 27–33.
22. Mitrophanov, A.Y. Connection between the Rate of Convergence to Stationarity and Stability to Perturbations for Stochastic and Deterministic Systems. In Proceedings of the 38th International Conference Dynamics Days Europe (DDE 2018), Loughborough, UK, 3–7 September 2018.
23. Shao, J.; Yuan, C. Stability of regime-switching processes under perturbation of transition rate matrices. *Nonlinear Anal. Hybrid Syst.* **2019**, *33*, 211–226. [CrossRef]
24. Kartashov, N.V. *Strong Stable Markov Chains*; Utrecht VSP: Perm, Russia, 1996.
25. Ferre, D.; Herve, L.; Ledoux, J. Regular perturbation of V-geometrically ergodic Markov chains. *J. Appl. Prob.* **2013**, *50*, 84–194. [CrossRef]
26. Mouhoubi, Z.; Aïssani D. New perturbation bounds for denumerable Markov chains. *Linear Algebra Its Appl.* **2010**, *432*, 1627–1649. [CrossRef]
27. Meyn, S.P.; Tweedie, R.L. Computable bounds for geometric convergence rates of Markov chains. *Ann. Appl. Prob.* **1994**, *4*, 981–1012. [CrossRef]
28. Abbas, K.; Berkhout, J.; Heidergott, B. A critical account of perturbation analysis of Markov chains. *arXiv* **2016**, arXiv:1609.04138.
29. Jiang, S.; Liu, Y.; Tang, Y. A unified perturbation analysis framework for countable Markov chains. *Linear Algebra Its Appl.* **2017**, *529*, 413–440. [CrossRef]
30. Liu, Y.; Li, W. Error bounds for augmented truncation approximations of Markov chains via the perturbation method. *Adv. Appl. Prob.* **2018**, *50*, 645–669. [CrossRef]
31. Medina-Aguayo, F.; Rudolf, D.; Schweizer, N. Perturbation bounds for Monte Carlo within Metropolis via restricted approximations. *Stochast. Process. Appl.* **2019**. [CrossRef]
32. Negrea, J.; Rosenthal, J.S. Error bounds for approximations of geometrically ergodic Markov chains. *arXiv* **2017**, arXiv:1702.07441.
33. Rudolf, D.; Schweizer, N. Perturbation theory for Markov chains via Wasserstein distance. *Bernoulli* **2018**, *24*, 2610–2639. [CrossRef]
34. Liu, Y. Perturbation bounds for the stationary distributions of Markov chains. *SIAM J. Matrix Anal. Appl.* **2012**, *33*, 1057–1074. [CrossRef]
35. Thiede, E.; Van Koten, B.; Weare, J. Sharp entrywise perturbation bounds for Markov chains. *SIAM J. Matrix Anal. Appl.* **2015**, *36*, 917–941. [CrossRef]
36. Truquet, L. A perturbation analysis of some Markov chains models with time-varying parameters. *arXiv* **2017**, arXiv:1706.03214.

37. Vial, D.; Subramanian, V. Restart perturbations for lazy, reversible Markov chains: Trichotomy and pre-cutoff equivalence. *arXiv* **2019**, arXiv:1907.02926.
38. Zheng, Z.; Honnappa, H.; Glynn, P.W. Approximating Performance Measures for Slowly Changing Non-stationary Markov Chains. *arXiv* **2018**, arXiv:1805.01662.
39. Gnedenko, B.; Soloviev, A. On the conditions of the existence of final probabilities for a Markov process. *Math. Operationsforsch. Statist.* **1973**, *4*, 379–390.
40. Gnedenko, D.B. On a generalization of Erlang formulae. *Zastosow. Mater.* **1972**, *12*, 239–242.
41. Massey, W.A.; Whitt, W. Uniform acceleration expansions for Markov chains with time-varying rates. *Ann. Appl. Prob.* **1998**, *8*, 1130–1155. [CrossRef]
42. Gnedenko, B.V.; Makarov, I.P. Properties of a problem with losses in the case of periodic intensities. *Differ. Eq.* **1971**, *7*, 1696–1698. (In Russian)
43. Zeifman, A.I.; Korolev, V.Y.E.; Korotysheva, A.V.; Shorgin, S.Y. General bounds for nonstationary continuous-time Markov chains. *Inf. Primeneniya* **2014**, *8*, 106–117.
44. Zeifman, A.I.; Korolev, V.Y. On perturbation bounds for continuous-time Markov chains. *Stat. Probab. Lett.* **2014**, *88*, 66–72. [CrossRef]
45. Satin, Y.; Zeifman, A.; Kryukova, A. On the Rate of Convergence and Limiting Characteristics for a Nonstationary Queueing Model. *Mathematics* **2019**, *7*, 678. [CrossRef]
46. Zeifman, A.; Korotysheva, A.; Korolev, V.; Satin, Y.; Bening, V. Perturbation bounds and truncations for a class of Markovian queues. *Queueing Syst.* **2014**, *76*, 205–221. [CrossRef]
47. Zeifman, A.; Razumchik, R.; Satin, Y.; Kiseleva, K.; Korotysheva, A.; Korolev, V. Bounds on the Rate of Convergence for One Class of Inhomogeneous Markovian Queueing Models with Possible Batch Arrivals and Services. *Int. J. Appl. Math. Comput. Sci.* **2018**, *28*, 141–154. [CrossRef]
48. Zeifman, A.; Satin, Y.; Kiseleva, K.; Korolev, V.; Panfilova, T. On limiting characteristics for a non-stationary two-processor heterogeneous system. *Appl. Math. Comput.* **2019**, *351*, 48–65. [CrossRef]
49. Daleckii, J.L.; Krein, M.G. Stability of solutions of differential equations in Banach space. *Am. Math. Soc.* **2002**, *43*, 1024–1102.
50. Zeifman, A.; Sipin, A.; Korolev, V.; Shilova, G.; Kiseleva, K.; Korotysheva, A.; Satin, Y. On Sharp Bounds on the Rate of Convergence for Finite Continuous-Time Markovian Queueing Models. In *Computer Aided Systems Theory EUROCAST 2017*; Lecture Notes in Computer Science; Moreno-Diaz, R., Pichler, F., Quesada-Arencibia, A., Eds.; Springer: Cham, Switzerland, 2018; Volume 10672, pp. 20–28.
51. Nelson, R.; Towsley, D.; Tantawi, A.N. Performance analysis of parallel processing systems. *IEEE Trans. Softw. Eng.* **1988**, *14*, 532–540. [CrossRef]
52. Satin, Y.A.; Zeifman, A.I.; Korotysheva, A.V.; Shorgin, S.Y. On a class of Markovian queues. *Inf. Its Appl.* **2011**, *5*, 6–12. (In Russian)
53. Satin, Y.A.; Zeifman, A.I.; Korotysheva, A.V. On the rate of convergence and truncations for a class of Markovian queueing systems. *Theory Prob. Its Appl.* **2013**, *57*, 529–539. [CrossRef]
54. Li, J.; Zhang, L. $M^X/M/c$ Queue with catastrophes and state-dependent control at idle time. *Front. Math. China* **2017**, *12*, 1427–1439. [CrossRef]
55. Di Crescenzo, A.; Giorno, V.; Nobile, A.G.; Ricciardi, L.M. A note on birthdeath processes with catastrophes. *Stat. Prob. Lett.* **2008**, *78*, 2248–2257. [CrossRef]
56. Dudin, A.; Nishimura, S. A BMAP/SM/1 queueing system with Markovian arrival input of disasters. *J. Appl. Prob.* **1999**, *36*, 868–881. [CrossRef]
57. Dudin, A.; Karolik, A. BMAP/SM/1 queue with Markovian input of disasters and non-instantaneous recovery. *Perform. Eval.* **2001**, *45*, 19–32. [CrossRef]
58. Zhang, L.; Li, J. The M/M/c queue with mass exodus and mass arrivals when empty. *J. Appl. Prob.* **2015**, *52*, 990–1002. [CrossRef]
59. Zeifman, A.I. Upper and lower bounds on the rate of convergence for nonhomogeneous birth and death processes. *Stochast. Process. Their Appl.* **1995**, *59*, 157–173. [CrossRef]
60. Chen, A.; Renshaw, E. The $M/M/1$ queue with mass exodus and mass arrivals when empty. *J. Appl. Prob.* **1997**, *34*, 192–207. [CrossRef]
61. Zeifman, A.I.; Korolev, V.Y.; Satin, Y.A.; Kiseleva, K.M. Lower bounds for the rate of convergence for continuous-time inhomogeneous Markov chains with a finite state space. *Statist. Prob. Lett.* **2018**, *137*, 84–90. [CrossRef]

62. Zeifman, A.I.; Leorato, S.; Orsingher, E.; Satin, Y.; Shilova, G. Some universal limits for nonhomogeneous birth and death processes. *Queueing Syst.* **2006**, *52*, 139–151. [CrossRef]
63. Granovsky, B.L.; Zeifman, A.I. Nonstationary Queues: Estimation of the Rate of Convergence. *Queueing Syst.* **2004**, *46*, 363–388. [CrossRef]
64. Zeifman, A.; Korotysheva, A.; Satin, Y.; Razumchik, R.; Korolev, V.; Shorgin, S. Ergodicity and truncation bounds for inhomogeneous birth and death processes with additional transitions from and to origin. *Stochast. Models* **2017**, *33*, 598–616. [CrossRef]
65. Zeifman, A.I.; Korotysheva, A.; Satin, Y.; Kiseleva, K.; Korolev, V.; Shorgin, S. Bounds For Markovian Queues With Possible Catastrophes. In ECMS. May 2017; pp. 628–634. Available online: http://www.scs-europe.net/dlib/2017/2017-0628.htm (accessed on 13 February 2020).
66. Zeifman, A.; Korotysheva, A.; Satin, Y. On stability for Mt/Mt/N/N queue. *Int. Congr. Ultra Modern Telecommun. Control Syst. Moscow* **2010**, 1102–1105. [CrossRef]
67. Zeifman, A.; Shorgin, S.; Korotysheva, A.; Bening, V. Stability bounds for Mt/Mt/N/N + R queue. In Proceedings of the 5th International ICST Conference on Performance Evaluation Methodologies and Tools (VALUETOOLS '11), ICST (Institute for Computer Sciences, Social-Informatics and Telecommunications Engineering), ICST, Brussels, Belgium, 16–20 May 2011; pp. 434–438.
68. Van Doorn, E.A.; Zeifman, A.I.; Panfilova, T.L. Bounds and asymptotics for the rate of convergence of birth-death processes. *Theory Prob. Its Appl.* **2010**, *54*, 97–113. [CrossRef]
69. Van Doorn, E.A. Rate of convergence to stationarity of the system M/M/N/N+R. *Top* **2011**, *19*, 336–350. [CrossRef]
70. Zeifman, A.I. On the nonstationary Erlang loss model. *Autom. Remote Control* **2009**, *70*, 2003–2012. [CrossRef]
71. Zeifman, A.; Satin, Y.; Korolev, V.; Shorgin, S. On truncations for weakly ergodic inhomogeneous birth and death processes. *Int. J. Appl. Math. Comput. Sci.* **2014**, *24*, 503–518 [CrossRef]
72. Zeifman, A.I.; Korotysheva, A.V.; Korolev, V.Y.; Satin, Y.A. Truncation bounds for approximations of inhomogeneous continuous-time Markov chains. *Theory Prob. Appl.* **2017**, *61*, 513–520. [CrossRef]

mathematics

MDPI

Article

# The Calculation of the Density and Distribution Functions of Strictly Stable Laws

**Viacheslav Saenko**

Department of Theoretical Physics, Ulyanovsk State University, 432970 Ulyanovsk, Russia; saenkovv@gmail.com

Received: 2 April 2020; Accepted: 6 May 2020; Published: 12 May 2020

**Abstract:** Integral representations for the probability density and distribution function of a strictly stable law with the characteristic function in the Zolotarev's "C" parametrization were obtained in the paper. The obtained integral representations express the probability density and distribution function of standard strictly stable laws through a definite integral. Using the methods of numerical integration, the obtained integral representations allow us to calculate the probability density and distribution function of a strictly stable law for a wide range of admissible values of parameters $(\alpha, \theta)$. A number of cases were given when numerical algorithms had difficulty in calculating the density. Formulas were given to calculate the density and distribution function with an arbitrary value of the scale parameter $\lambda$.

**Keywords:** stable distribution; probability density function; distribution function

**MSC:** 60E07

## 1. Introduction

The problem of calculating the density of a strictly stable law with the characteristic function

$$\hat{g}(t, \alpha, \theta, \lambda) = \exp\{-\lambda|t|^{\alpha} \exp\{-i\tfrac{\pi}{2}\alpha\theta \operatorname{sign} t\}\}, \quad t \in \mathbf{R}, \tag{1}$$

where $0 < \alpha \leqslant 2$, $|\theta| \leqslant \min(1, 2/\alpha - 1)$, $\lambda > 0$ is considered in the paper. One of the reasons why it became necessary to calculate the density of a strictly stable law with this characteristic function is the need to calculate the density of a fractionally stable law which is defined by the expression

$$q(x, \alpha, \nu, \theta) = \int_0^{\infty} g(xy^{\nu/\alpha}, \alpha, \theta) g(y, \nu, 1) y^{\nu/\alpha} dy. \tag{2}$$

Here, $g(x, \alpha, \theta)$ and $g(y, \nu, 1)$ are the densities of strictly stable and one-sided strictly stable laws with the characteristic function in Equation (1) and parameter $\lambda = 1$. For the first time, the density in Equation (2) was obtained in the article [1]. The density in Equation (2) got its name in the work [2], since the random variable $Z(\alpha, \nu, \theta)$ distributed by this law is defined by the ratio $Z(\alpha, \nu, \theta) = Y(\alpha, \theta)[V(\nu, 1)]^{-\nu/\alpha}$. Here the random variables $Y(\alpha, \theta)$ and $V(\nu, 1)$ are distributed by the laws $g(x, \alpha, \theta)$ and $g(y, \nu, 1)$, respectively.

The density Equation (2) appears as a limit distribution with the following random walk scheme. Let the particle be at the origin $x = 0$ at the initial time $t = 0$ and it stays at this point during random time $T_1$. Then, it instantly moves with an equal probability to the right or left at random distance $X_1$ and it stays at rest again random time $T_2$. Then, the whole process is repeated in the same way. Values $X_i, i = 1, 2, \ldots$ are independent identically distributed random variables belonging to the domain of normal attraction of a strictly stable law $g(t, \alpha, \theta)$ with the characteristic function in Equation (1).

Values $T_i, i = 1, 2, \ldots$ are independent both between themselves and of the sequence $\{X_i\}$ by identically distributed random variables belonging to the domain of normal attraction of a strictly stable law $g(t, \nu, 1)$ with the characteristic function in Equation (1) and $0 < \nu \leqslant 1$. We will form the sum of these random variables $S(\tau) = \sum_{i=1}^{N(\tau)} X_i, \tau > 0$, where $N(\tau)$ is the counting process: $N(\tau) = \max\{n : \sum_{i=1}^{n} T_i \leqslant \tau\}$. The physical interpretation of the sum $S(\tau)$ is the coordinate of particle $x$ at time $\tau$. In the works [1,2] it has been shown that the asymptotic (at $\tau \to \infty$) distribution of the sum $S(\tau)$ is described by the distribution Equation (2).

The described random walk scheme is called Continuous Time Random Walk (CTRW). For the first time it was considered in the work [3]. Later, it was described in the works [4–6]. For more detailed familiarity with this model one can look through the overviews [7,8]. In the work [1] it has been shown that the asymptotic (at $t \to \infty$) distribution of the CTRW process is described with the distribution (2). In the work [9], it has been shown that the CTRW process in large time asymptotics is described with the fractional-differential equation of diffusion. The solution of this equation is expressed through fractional-stable distributions in Equation (2). This is one of the factors determining the interest in studying the class of the fractional-stable laws.

Another factor is the appearance of these distributions in various processes occurring in a plasma [10,11] or in biology processes [12–14]. In particular, the fractional-stable distributions were used to describe a distribution of the gene expression in cells of tissues of various organisms in the following papers [12–14]. It is known that the distribution of the gene expression is described by laws with the power decrease in the density [15–17]. Since the density in Equation (2) decreases as $x^{-\alpha-1}$ at $x \to \infty$, therefore this class of distributions was used to describe the gene expression distribution. In the articles [12,13] the fractional-stable distributions were used to describe the gene expression obtained with the microarray technology. In the paper [14], these distributions were used to describe the results obtained with the Next Generation Sequence technology. In the papers [12–14] the Monte Carlo method was used to calculate the density $q(x, \alpha, \nu, \theta)$. To estimate the parameters $(\alpha, \nu, \theta)$ of the fractional-stable law a method described in [18] was used which is also based on the Monte Carlo method. However, to construct more effective estimators of the parameters $(\alpha, \nu, \theta)$, for example, the maximum likelihood estimation, one should be able to calculate the density $q(x, \alpha, \nu, \theta)$. As a result we come again to the necessity of calculating the integral of Equation (2).

As we can see from Equation (2), the density of a fractional-stable law is defined using the Mellin convolution of two strictly stable densities with the characteristic function in Equation (1). Hence, to calculate the density in Equation (2) it is necessary to be able to calculate densities $g(x, \alpha, \theta)$ for any admissible set of parameters $(\alpha, \theta)$. It should be pointed out that the problem of calculating densities of stable laws at present is well studied. The solution to this problem is based on the inverse Fourier transform of the characteristic function of a stable law. There are several methods for performing the inverse Fourier transform: a direct calculation of the inverse Fourier transform [19–26], the use of the fast Fourier transform algorithm [27,28], the use of the inversion formula followed by the numerical calculation of the integral [29,30], and the use of the inversion method by V. Zolotarev [31–34].

Direct implementation of the inverse Fourier transform of the characteristic function of the stable law leads to the appearance of special functions. As a rule, such a transformation can be implemented if the shift parameter of the stable law $\gamma = 0$. Therefore, practically all cases when it is possible to express the density of a stable law through special functions are referred to strictly stable laws. In addition, the density of a strictly stable law can be obtained only for rational values of characteristic exponent $\alpha$ and parameter of skewness. For instance, in the works [19,20], representations were obtained for the densities of stable laws through the Fox H-function. Representations for the densities of stable laws through an incomplete hypergeometric function were obtained in the article [21]. Later, the results of the article [21] were generalized in [22] in which representations were obtained through Meijer's G-function. In the work [23] the expressions for density were obtained through the Fox $H$-function and hypergeometric function. In the work [24] expressions for the density of a one-sided stable law were obtained at $0 < \alpha < 1$ through the hypergeometric function. In the subsequent work [25] using the law of duality and the Mellin transform, the authors generalized the result to the case of two-sided distributions ($-\infty < x < \infty$) and to the range of values $0 < \alpha \leqslant 2$, where $\alpha$ is a rational number. It has been mentioned earlier that it is possible to directly implement the inverse Fourier transform of the characteristic function if the shift parameter $\gamma = 0$. An exception may be represented by the work [26]. In this work, the author was able to invert the characteristic function of the stable law for arbitrary values of the shift parameter and scale. As a result, for $\alpha \in (1,2]$ it was possible to express the density of stable laws through a generalization of the Srivastava-Daoust of Kampé de Fériet two-variable hypergeometric function.

In the work [27], to invert the characteristic function of the stable law, the fast Fourier transform (FFT) algorithm is used. Using the FFT algorithm allows one to quickly reverse the characteristic function of the stable law and obtain numerical values of the density. However, the FFT algorithm allows one to calculate density values only on a grid of equally spaced coordinate values. This is not always convenient, since one should use interpolation methods to calculate density values at intermediate points. In the paper [28], standard quadrature numerical integration algorithms are redefined to invert the characteristic function. In the proposed approach, the FFT algorithm is used to calculate the value of the integrand at the nodes of the grid. This approach makes it possible to reduce the approximation error in the central part of the distribution. To calculate the density in the tails of the distribution, the Bergström expansion of the density of a stable law in a series is used [35] (see also § 2.4 in [32]). However, the accuracy of the proposed method depends on the values $\alpha$ and $\beta$ and turns out to be effective only with values $\alpha \in (1,2]$.

In the papers [29,30], the inversion formula is used to calculate the density of a stable law

$$g(x,\alpha,\beta) = \frac{1}{\pi}\Re\int_0^\infty e^{itx}\hat{g}(t,\alpha,-\beta)dt = \frac{1}{\pi}\int_0^\infty \cos(h(t,x,\alpha,\beta))e^{-t^\alpha}dt, \tag{3}$$

where $\hat{g}(t,\alpha,\beta)$ is the characteristic function of a stable law with the scale parameter $\lambda = 1$ and shift parameter $\gamma = 0$. In this case, the density $g(x,\alpha,\beta)$ is expressed through the improper integral of real variables. To calculate it, one can use standard algorithms of numerical integration. This approach was used in the work [29] where the characteristic function was chosen as

$$\hat{g}(t,\alpha,\beta,\lambda,\gamma) = \begin{cases} \exp\left\{it\lambda\gamma - \lambda|t|^\alpha + it\lambda\left(|t|^{\alpha-1}-1\right)\beta\tan(\pi\alpha/2)\right\}, & \alpha \neq 1, \\ \exp\left\{it\lambda\gamma - \lambda|t|^\alpha - it\lambda\beta(2/\pi)\ln|t|\right\}, & \alpha = 1. \end{cases} \tag{4}$$

Here the parameters vary within $0 < \alpha \leqslant 2$, $-1 \leqslant \beta \leqslant 1$, $-\infty < \gamma < \infty$, $\lambda > 0$. As it was pointed out in the paper, this approach does not have difficulty with the values $\alpha > 1.1$. Difficulties with calculation arise at $\alpha < 0.75$, $\alpha \approx 1$, and $\beta \neq 0$. In addition to it, at greater values of $x$ the integrand begins to oscillate fast which leads to difficulties in numerical integration. In the paper [30], it is proposed to use an optimized generalized Gaussian scheme of numerical integration to calculate the integral of Equation (3) with the characteristic function in Equation (4). In this work the constants $B_{80}^\infty$ and $B_{40}^\infty$ were introduced (more detailed information about the definition of these constants see [30]). If $\beta = 0$ the proposed integration scheme is effective at $0.5 \leqslant \alpha \leqslant 2$ and $|x| \leqslant B_{40}^\infty$. If $\beta \neq 0$ the scheme is effective for values $\alpha \in [0.5, 0.9] \cup [1.1, 2.0]$ and $|x-\zeta| \leqslant B_{80}^\infty$. With the values of $|x| > B_{40}^\infty$ and $|x-\zeta| > B_{80}^\infty$ an asymptotic expansion of the density is used in a series. With the values $\alpha \in (0.9, 1.1), \beta \neq 0$ and $\alpha \in (0, 0.5), \beta \in [-1, 1]$ the scheme is not applicable.

The use of Equation (3) leads to the appearance of fast oscillating functions under the sign of the integral. To get around this problem, in the paper [31], Zolotarev V.M. developed a method of inverting the characteristic function of a stable law. Using this method, in the paper [31] (see also § 2.2 in [32], § 4.4 in [36]) an integral representation was obtained for the density of a stable law with the characteristic function

$$\hat{g}(t,\alpha,\beta,\lambda,\gamma) = \begin{cases} \exp\left\{it\lambda\gamma - \lambda|t|^\alpha \exp\left\{-i\frac{\pi}{2}\beta K(\alpha)\,\mathrm{sign}\,t\right\}\right\}, & \alpha \neq 1, \\ \exp\left\{it\lambda\gamma - \lambda|t|^\alpha\left(\frac{\pi}{2} + i\beta\log|t|\,\mathrm{sign}\,t\right)\right\}, & \alpha = 1, \end{cases} \tag{5}$$

where $0 < \alpha \leqslant 2, -1 \leqslant \beta \leqslant 1, \lambda > 0, -\infty < \gamma < \infty$, $K(\alpha) = \alpha - 1 + \mathrm{sign}\,(1-\alpha)$. The obtained integral representation of the density of the stable law is expressed through a definite integral. It is not possible to calculate this integral analytically. However, using the methods of numerical integration, it is possible to calculate and obtain the probability density and the distribution function of a stable law. Using the specified method of inverting the characteristic function in the work [33] integral representations for probability density and distribution functions of a stable law with the characteristic function in Equation (4) were obtained. In the paper [37], a slight modification of the characteristic function in Equation (4) is considered and it is noted that for calculation purposes it is more convenient to use this particular modification. Subsequently, this integral representation formed the basis of various software packages for calculating the probability density and distribution function of stable laws [38–42]. In the paper [43], it is indicated that difficulties in calculating the integral in the integral representation obtained in [33] arise with (1) small values of $\alpha$ and $x \to 0$, (2) $x \to \infty$ and (3) $\alpha$ close either to 1 or 2. In this paper, the authors proposed a method of solving the last two problems for symmetric stable laws and note that using the proposed approach, it is possible to calculate the densities of stable laws for values $\alpha$ close to either 1 or 2 as well as at $x \to 0$ and $x \to \infty$.

Having slightly modified Zolotarev's method [31,32] of inverting the characteristic function in the paper [34] expansions were obtained for the density of stable laws in power series. Investigating trans-stable distributions, the authors obtained expansions in the power series of densities of stable laws for the cases $0 < \alpha < 1$ and $1 < \alpha < 2$. In each of the ranges $0 < \alpha < 1$ and $1 < \alpha < 2$ expansions are represented in the form of "internal" $(x \to 0)$ and "external" $(x \to \infty)$ expansions. To describe the behavior of the density of a stable law in the whole range of values $0 < x < \infty$ these two expansions are put together.

Thus, all the results related to obtaining expressions for the probability density of stable laws were obtained for laws with characteristic functions in Equations (4) and (5). However, to calculate the density in Equation (2), it is necessary to have an expression for the probability density $g(x,\alpha,\theta)$ with a characteristic function Equation (1). It should be emphasized that an integral representation for the density of a stable law with the characteristic function in Equation (1) is presented in the paper [44]. However, the expression cited is valid only for $x > 0$ and $\alpha \neq 1$. In this paper, we will obtain an integral representation for the density and distribution function of a stable law with a characteristic function Equation (1) for arbitrary $x$ and any admissible values of parameters $\alpha$ and $\theta$.

## 2. Auxiliary Results

Thus, the objective is to obtain an integral representation of the density of a strictly stable law with a characteristic function Equation (1). Without losing generality we will further assume everywhere that $\lambda = 1$. A strictly stable law with a parameter $\lambda = 1$ is commonly called the standard strictly stable law. An abbreviated notation of the characteristic function is accepted for standard strictly stable laws $\hat{g}(t,\alpha,\theta,1) \equiv \hat{g}(t,\alpha,\theta)$, for density $g(x,\alpha,\theta,1) \equiv g(x,\alpha,\theta)$, for the distribution function $G(x,\alpha,\theta,1) \equiv G(x,\alpha,\theta)$, and random variable $Y(\alpha,\theta,1) \equiv Y(\alpha,\theta)$. Everywhere below, for standard strictly stable laws, we will use this notation. To obtain an integral representation, we use the method of inverting the characteristic function of a stable law for the first time proposed by V. Zolotarev in the work [31] and described in detail in his monograph [32] (see also § 4.4 in [36]). To prove the main theorem, we will need some auxiliary results.

**Property 1** (Property of inversion). *For any admissible set of values of parameters* $(\alpha, \theta)$

$$Y(\alpha, -\theta) \stackrel{d}{=} -Y(\alpha, \theta). \tag{6}$$

**Proof.** The proof of this property is simple enough. Applying the definition of the characteristic function in Equation (1) and by making the substitution of a variable $t \to -\tau$, we obtain

$$\mathsf{E}\exp\{itY(\alpha,-\theta)\} = \exp\{-|t|^\alpha \exp\{i\tfrac{\pi}{2}\alpha\theta\,\mathrm{sign}\,t\}\} = \\ \exp\{-|t|^\alpha \exp\{-i\tfrac{\pi}{2}\alpha\theta\,\mathrm{sign}\,\tau\}\} = \mathsf{E}\exp\{-i\tau Y(\alpha,\theta)\}.$$

It follows directly from here Equation (6). □

In terms of the characteristic function $\hat{g}(t,\alpha,\theta)$, probability density functions $g(x,\alpha,\theta)$ and distribution functions $G(x,\alpha,\theta)$ of a strictly stable law the property of inversion is written in the form

$$\hat{g}(-t,\alpha,\theta) = \hat{g}(t,\alpha,-\theta), \quad g(-x,\alpha,\theta) = g(x,\alpha,-\theta), \quad G(-x,\alpha,\theta) = 1 - G(x,\alpha,-\theta). \tag{7}$$

The utility of this property consists in the fact that owing to this property it is sufficient to consider the issue of the density representation $g(x,\alpha,\theta)$ or the distribution function $G(x,\alpha,\theta)$ only for $x \geqslant 0$ or for $\theta \geqslant 0$. For negative values of the argument $x$ or the parameter $\theta$ expressions can be obtained according to the expressions given earlier.

The following property will be useful further.

**Property 2.** *For any two admissible sets of parameters* $(\alpha,\theta,\lambda)$ *and* $(\alpha,\theta,\lambda')$*, there is such a unambiguously defined real* $a > 0$*, that* $Y(\alpha,\theta,\lambda) \stackrel{d}{=} aY(\alpha,\theta,\lambda')$*. For the characteristic function in Equation (1), the value* $a$ *is connected with parameters in the following way* $a = (\lambda/\lambda')^{1/\alpha}$*.*

This property is a full analog of property 2.1 in [32] (see also § 3.7 in [36]) formulated for strictly stable random variables with the characteristic function in Equation (1). This property is proved in the same way to the one which is performed in [32]. In the particular case that is of interest to us $\lambda' = 1$, we obtain

$$Y(\alpha,\theta,\lambda) = \lambda^{1/\alpha} Y(\alpha,\theta,1). \tag{8}$$

We now formulate a lemma which makes it possible to perform the inverse Fourier transform of the characteristic function and obtain the density of a strictly stable law.

**Lemma 1.** *The probability density function* $g(x,\alpha,\theta)$ *for any admissible values of parameters* $(\alpha,\theta)$ *and any* $x$ *can be obtained with the help of the inversion formulas*

$$g(x,\alpha,\theta) = \frac{1}{2\pi}\int_{-\infty}^{\infty} e^{-itx}\hat{g}(t,\alpha,\theta)dt = \begin{cases} \dfrac{1}{\pi}\Re\displaystyle\int_0^\infty e^{itx}\hat{g}(t,\alpha,-\theta)dt, \\ \dfrac{1}{\pi}\Re\displaystyle\int_0^\infty e^{-itx}\hat{g}(t,\alpha,\theta)dt. \end{cases} \tag{9}$$

**Proof.** Performing the inverse of the characteristic function $\hat{g}(t,\alpha,\theta)$ we obtain

$$g(x,\alpha,\theta) = \frac{1}{2\pi}\int_{-\infty}^{\infty} e^{-itx}\exp\left\{-|t|^{\alpha}\exp\{-i\alpha(\pi/2)\theta\operatorname{sign}t\}\right\}dt = \frac{1}{2\pi}\int_{-\infty}^{0}\exp\left\{-itx-|t|^{\alpha}\exp\{-i\alpha(\pi/2)\theta\operatorname{sign}t\}\right\}dt + \frac{1}{2\pi}\int_{0}^{\infty}\exp\left\{-itx-|t|^{\alpha}\exp\{-i\alpha(\pi/2)\theta\operatorname{sign}t\}\right\}dt = I_1 + I_2.$$

Let us consider the integral $I_1$. By substituting the integration variable in this integral $-t\to\tau$, we obtain

$$I_1 = \frac{1}{2\pi}\int_{-\infty}^{0}\exp\left\{-itx-|t|^{\alpha}\exp\{-i\alpha(\pi/2)\theta\operatorname{sign}t\}\right\}dt = \frac{1}{2\pi}\int_{0}^{\infty}\exp\left\{i\tau x-|\tau|^{\alpha}\exp\{i\alpha(\pi/2)\theta\operatorname{sign}\tau\}\right\}d\tau.$$

Now having calculated the sum $I_1+I_2$, we will obtain

$$I_1+I_2 = \frac{1}{2\pi}\int_{0}^{\infty}\exp\left\{i\tau x-|\tau|^{\alpha}\exp\{i\alpha\tfrac{\pi}{2}\theta\operatorname{sign}\tau\}\right\}d\tau + \frac{1}{2\pi}\int_{0}^{\infty}\exp\left\{-itx-|t|^{\alpha}\exp\{-i\alpha\tfrac{\pi}{2}\theta\operatorname{sign}t\}\right\}dt = \frac{1}{\pi}\int_{0}^{\infty}\exp\{-t^{\alpha}\cos(\tfrac{\pi}{2}\alpha\theta)\}\cos(tx-t^{\alpha}\sin(\tfrac{\pi}{2}\alpha\theta))dt = \frac{1}{\pi}\Re\int_{0}^{\infty}\exp\{-t^{\alpha}\cos(\tfrac{\pi}{2}\alpha\theta)\}\left(\cos(tx-t^{\alpha}\sin(\tfrac{\pi}{2}\alpha\theta))+i\sin(tx-t^{\alpha}\sin(\tfrac{\pi}{2}\alpha\theta))\right)dt = \frac{1}{\pi}\Re\int_{0}^{\infty}e^{itx}\exp\left\{-t^{\alpha}\exp\{i\tfrac{\pi}{2}\alpha\theta\}\right\}dt = \frac{1}{\pi}\Re\int_{0}^{\infty}e^{itx}\hat{g}(t,\alpha,-\theta)dt. \quad (10)$$

As a result, we obtained the first formula in Equation (9). In order to obtain the second formula in (9) it is necessary to subtract in the penultimate manipulation the imaginary component in Equation (10). □

Later, we need analytic continuation of the characteristic function $\hat{g}(t,\alpha,\theta)$ in the complex plane $z$. We will carry out this analytic continuation in the complex plane $z$ with a semiaxis $\Re z = t > 0$ with a cut along the negative part of the real axis $\arg z = -\pi$. The resulting analytic continuation of the function $\hat{g}(t,\alpha,\theta)$ with a half-line $t>0$ will be designated as $g^{+}(z,\alpha,\theta)$. Using the characteristic function in Equation (1), we obtain

$$g^{+}(z,\alpha,\theta) = \exp\left\{-z^{\alpha}\exp\left\{-i\tfrac{\pi}{2}\alpha\theta\right\}\right\}. \quad (11)$$

The idea of analytic continuation of the integrand in the formula of inversion in Equation (9) in the complex plane $z$ and subsequent calculation of the resulting integral $\int_{\Gamma}\exp\{izx\}g^{+}(z,\alpha,\theta)dz$ underlies the method of inverting a characteristic function developed by Zolotarev V.M. in the work [31]. This integral is calculated due to such a change in the integration contour $\Gamma$ at which its real part does not change (for more details see [32]). To substantiate the change in the integration contour, we need the following lemma.

**Lemma 2.** *For any arbitrarily small $\varepsilon > 0$ of any admissible values of parameters $\alpha$ and $\theta$ and any $x \geqslant 0$ the integral is*

$$I(C_R) = \int_{C_R} e^{izx}g^{+}(z,\alpha,-\theta)dz \to 0,$$

*if:*

1. $\alpha < 1$, $-1 \leqslant \theta \leqslant 1$, $x > 0$, *the contour* $C_R$ *has the form* $C_R = \{z : \ |z| = R, \ 0 \leqslant \arg z \leqslant \pi - \varepsilon\}$ *and* $R \to \infty$;
2. $\alpha > 1$, $|\theta| \leqslant (2/\alpha - 1)$, $x > 0$, *the contour* $C_R$ *has the form* $C_R = \{z : \ |z| = R, \ -\frac{\pi}{2\alpha} - \frac{\pi\theta}{2} + \varepsilon \leqslant \arg z \leqslant \frac{\pi}{2\alpha} - \frac{\pi\theta}{2}\}$ *and* $R \to \infty$;
3. $0 < \alpha \leqslant 2$, $|\theta| \leqslant \min(1, 2/\alpha - 1)$ *the contour* $C_R$ *has the form* $C_R = \{z : \ |z| = R, \ -\pi \leqslant \arg z \leqslant \pi\}$ *and* $R \to 0$;
4. $0 < \alpha \leqslant 2$, $x = 0$, $|\theta| \leqslant \min(1, 2/\alpha - 1)$ *the contour* $C_R$ *has the form* $C_R = \{z : \ |z| = R, \ -\frac{\pi}{2\alpha} - \frac{\pi\theta}{2} + \varepsilon \leqslant \arg z \leqslant \frac{\pi}{2\alpha} - \frac{\pi\theta}{2} - \varepsilon\}$ *and* $R \to \infty$;
5. $\alpha = 1$, $x > 1$, $-1 \leqslant \theta \leqslant 1$, *the contour* $C_R$ *has the form* $C_R = \{z : \ |z| = R, \ \varphi_0(\theta, x) + \varepsilon \leqslant \arg z \leqslant \frac{\pi}{2}\}$ *and* $R \to \infty$;
6. $\alpha = 1$, $0 \leqslant x < 1$, $-1 \leqslant \theta < \theta_0$, *the contour* $C_R$ *has the form* $C_R = \{z : \ |z| = R, \ \varphi_0(\theta, x) + \varepsilon \leqslant \arg z \leqslant \frac{\pi}{2}\}$ *and* $R \to \infty$;
7. $\alpha = 1$, $0 \leqslant x < 1$, $\theta = \theta_0$, *the contour* $C_R$ *has the form* $C_R = \{z : \ |z| = R, \ -\frac{\pi}{2} + \varepsilon \leqslant \arg z \leqslant \frac{\pi}{2} - \varepsilon\}$ *and* $R \to \infty$;
8. $\alpha = 1$, $0 \leqslant x < 1$, $\theta_0 < \theta \leqslant 1$, *the contour* $C_R$ *has the form* $C_R = \{z : \ |z| = R, \ -\frac{\pi}{2} \leqslant \arg z \leqslant \varphi_0(\theta, x) - \varepsilon\}$ *and* $R \to \infty$;
9. $\alpha = 1$, $x = 1$, $-1 \leqslant \theta < 1$, *the contour* $C_R$ *has the form* $C_R = \{z : \ |z| = R, \ \varphi_0(\theta, x) + \varepsilon \leqslant \arg z \leqslant \frac{\pi}{2}\}$ *and* $R \to \infty$.

*Here,* $\theta_0 = \frac{2}{\pi} \arcsin x$ *and*

$$\varphi_0(\theta, x) = \arctan\left(\frac{\cos(\pi\theta/2)}{\sin(\pi\theta/2) - x}\right). \tag{12}$$

**Proof.** Let us consider the integral $I(C_R) = \int_{C_R} e^{izx} g^+(z, \alpha, -\theta) dz$. As an integration contour $C_R$ we will consider contour lines that represent an arc of a circle of radius $R$ which has $\varphi_1 \leqslant \arg z \leqslant \varphi_2$ or

$$C_R = \{z : \ |z| = R, \ \varphi_1 \leqslant \arg z \leqslant \varphi_2\}. \tag{13}$$

The task is to determine under what conditions imposed on the contour $C_R$ and the parameters $\alpha$ and $\theta$ limits $\lim_{R\to 0} I(C_R) = 0$ and $\lim_{R\to\infty} I(C_R) = 0$.

For any contour $C_R$ the inequality

$$|I(C_R)| \leqslant \int_{C_R} |e^{izx} g^+(z, \alpha, -\theta)||dz|,$$

is true. Assuming in this expression $z = re^{i\varphi}$ and taking into account that $C_R$ is an arc of a circle of radius $R$ , we obtain

$$\begin{aligned} |I(C_R)| &\leqslant R \int_{\varphi_1}^{\varphi_2} |e^{ixR\exp\{i\varphi\}} g^+(Re^{i\varphi}, \alpha, -\theta)| d\varphi \\ &= R \int_{\varphi_1}^{\varphi_2} \left| \exp\left\{ ixRe^{i\varphi} - R^\alpha \exp\left\{ i\alpha\left(\varphi + \pi\theta/2\right) \right\} \right\} \right| d\varphi = \int_{\varphi_1}^{\varphi_2} U(R, \varphi, \alpha, \theta) d\varphi, \end{aligned} \tag{14}$$

where

$$U(r, \varphi, \alpha, \theta) = \exp\left\{-rx \sin\varphi - r^\alpha \cos(\alpha(\varphi + \pi\theta/2)) + \ln r\right\}. \tag{15}$$

(1) Let us consider the behavior of this integral at $R \to \infty$ and $\alpha < 1$ and $x > 0$. Let us consider the integrand $U(R, \varphi, \alpha, \theta)$. Assuming that $R \to \infty$ and taking into account that $\alpha < 1$ we obtain

$$\lim_{R\to\infty} U(R, \varphi, \alpha, \theta) = \lim_{R\to\infty} \exp\{-R(x \sin\varphi + R^{\alpha-1}\cos(\alpha(\varphi + \pi\theta/2))) + \ln R\}$$
$$= \lim_{R\to\infty} \exp\{-Rx \sin\varphi + \ln R\}.$$

In view of the fact that $x > 0$ and the fact that the linear function grows faster than $\ln R$, we obtain $\lim_{R\to\infty} \exp\{-Rx \sin\varphi + \ln R\} = 0$, if $0 < \varphi < \pi$ and any $\alpha < 1$ and $-1 \leqslant \theta \leqslant 1$. If $\varphi = 0$, then

$$U(R, 0, \alpha, \theta) = \exp\{-R^\alpha \cos(\alpha\pi\theta/2) + \ln R\}.$$

Taking into account that $0 < \alpha < 1$ and $-1 \leqslant \theta \leqslant 1$, we deduce that $-\pi/2 < \alpha\theta\pi/2 < \pi/2$ and, therefore, $\cos(\alpha\theta\pi/2) > 0$. Assuming that $R \to \infty$ in this expression and in view of the fact that $R^\alpha$ grows faster than $\ln R$, we deduce $\lim_{R\to\infty} U(R, 0, \alpha, \theta) = 0$. Without loss of generality the case $\varphi = \pi$ can be excluded from consideration. Hence, $\varphi_1 = 0$, and $\varphi_2$ we represent in the form $\varphi_2 = \pi - \varepsilon$, where $0 < \varepsilon \leqslant \pi/6$ is an arbitrary fixed number. As a result, we deduce

$$\lim_{R\to\infty} U(R, \varphi, \alpha, \theta) = 0 \quad \text{if} \quad 0 \leqslant \varphi < \pi - \varepsilon, \tag{16}$$

and integration contour in Equation (13) takes the form

$$C_R = \{z : |z| = R, \ 0 \leqslant \arg z \leqslant \pi - \varepsilon\}. \tag{17}$$

Now assuming that $R \to \infty$ in Equation (14) and using Equation (16) we deduce

$$\lim_{R\to\infty} |I(C_R)| \leqslant \lim_{R\to\infty} \int_0^{\pi-\varepsilon} U(R, \varphi, \alpha, \theta) d\varphi = 0.$$

From here it follows that in the case $\alpha < 1$ and $x > 0$ the integral $I(C_R) = 0$, at $R \to \infty$ where contour integration has the form (17). The first item of the lemma is proved.

(2) Let us consider the case $\alpha > 1$. As it is known that at $1 < \alpha \leqslant 2$ the parameter $\theta$ can vary within the range $-(2/\alpha - 1) \leqslant \theta \leqslant (2/\alpha - 1)$. We are interested in the conditions under which the integral in Equation (14) will tend to zero at $R \to \infty$. Assuming $R \to \infty$ in Equation (14), we deduce

$$\lim_{R\to\infty} |I(C_R)| \leqslant \lim_{R\to\infty} \int_{\varphi_1}^{\varphi_2} U(R, \varphi, \alpha, \theta) d\varphi. \tag{18}$$

From here it follows that the behavior of this integral at $R \to \infty$ is defined by the behavior $U(R, \varphi, \alpha, \theta)$. Applying Equation (15), we deduce

$$\lim_{R\to\infty} U(R, \varphi, \alpha, \theta) = \lim_{R\to\infty} \exp\{-R^\alpha(R^{1-\alpha} x \sin\varphi + \cos(\alpha(\varphi + \pi\theta/2))) + \ln R\}. \tag{19}$$

Taking into account that $1 < \alpha \leqslant 2$, then $1 - \alpha < 0$. Consequently, at $R \to \infty$ we can ignore the summand $R^{1-\alpha} \sin\varphi$ in comparison with the second summand in these brackets. As a result, we have $\lim_{R\to\infty} U(R, \varphi, \alpha, \theta) = \lim_{R\to\infty} \exp\{-R^\alpha \cos(\alpha(\varphi + \pi\theta/2)) + \ln R\}$. Taking into account that $R^\alpha$ grows faster than $\ln R$ we deduce that $\lim_{R\to\infty} U(R, \varphi, \alpha, \theta) = 0$ if $\cos(\alpha(\varphi + \pi\theta/2)) > 0$, or $-\pi/2 + 2k\pi < \alpha(\varphi + \pi\theta/2) < \pi/2 + 2k\pi, \quad k = 0, \pm1, \pm2, \ldots$. In this problem we will be interested in the case $k = 0$.

It should be noted that here there is a strict inequality, that is why the cases $\alpha(\varphi + \pi\theta/2) = \pm\pi/2$ will be considered separately. As a result, we have $-\pi/2 < \alpha(\varphi + \pi\theta/2) < \pi/2$. From here it follows that

$$\lim_{R\to\infty} U(R,\varphi,\alpha,\theta) = 0, \quad \text{if} \quad -\frac{\pi}{2\alpha} - \frac{\pi\theta}{2} < \varphi < \frac{\pi}{2\alpha} - \frac{\pi\theta}{2}. \tag{20}$$

Now we consider the cases $\alpha(\varphi + \pi\theta/2) = \pm\pi/2$. With this value of the argument $\cos(\alpha(\varphi + \pi\theta/2)) = 0$ and, therefore, it is impossible to ignore the summand $R^{1-\alpha}x\sin\varphi$ in Equation (19) now. We will assume that $x > 0$. In view of the aforesaid Equation (19) takes the form

$$\lim_{R\to\infty} U(R,\varphi,\alpha,\theta) = \lim_{R\to\infty} \exp\{-Rx\sin\varphi + \ln R\}, \quad \text{at} \quad \alpha(\varphi + \pi\theta/2) = \pm\pi/2. \tag{21}$$

Since $R$ grows faster than $\ln R$, then we will take interest in the constraints imposed on $\varphi$ at which $\sin\varphi > 0$. For the case $\alpha(\varphi + \pi\theta/2) = \pi/2$ we deduce that $\varphi = \frac{\pi}{2\alpha} - \frac{\pi\theta}{2}$. Now we will take into account that the parameter $\theta$ can take values in a range $-(2/\alpha - 1) \leqslant \theta \leqslant (2/\alpha - 1)$. Thus, if $\theta = -(2/\alpha - 1)$, then $\varphi = \frac{3\pi}{2\alpha} - \frac{\pi}{2}$. If $\theta = 2/\alpha - 1$, then $\varphi = -\frac{\pi}{2\alpha} + \frac{\pi}{2}$. As a result, we deduce

$$-\frac{\pi}{2\alpha} + \frac{\pi}{2} \leqslant \varphi \leqslant \frac{3\pi}{2\alpha} - \frac{\pi}{2} \tag{22}$$

We will take into account now that $1 < \alpha \leqslant 2$. Substituting the values $\alpha = 1$ and $\alpha = 2$ in Equation (22), alternately we deduce $0 < \varphi < \pi$, if $\alpha \to 1$ and $\pi/4 \leqslant \varphi \leqslant \pi/4$, if $\alpha = 2$. It should be pointed out that since in the considered case $\alpha$ cannot take the value $\alpha = 1$, then in the corresponding inequality there is a strict inequality. As a result we deduce $\sin\varphi > 0$, if $\alpha(\varphi + \pi\theta/2) = \pi/2$ for any $1 < \alpha \leqslant 2$ and any $-(2/\alpha - 1) \leqslant \theta \leqslant 2/\alpha - 1$. Applying this result in Equation (21), we deduce

$$\lim_{R\to\infty} U(R,\varphi,\alpha,\theta) = 0, \quad \text{if} \quad \alpha(\varphi + \pi\theta/2) = \pi/2 \quad \text{and} \quad x > 0. \tag{23}$$

The case $\alpha(\varphi + \pi\theta/2) = -\pi/2$ is considered in the similar way as the previous case. As a result, we obtain $-\pi < \varphi < 0$ if $\alpha \to 1$ and $-\pi/4 \leqslant \varphi \leqslant -\pi/4$, if $\alpha = 2$. From here it follows that $\sin\varphi < 0$ if $\alpha(\varphi + \pi\theta/2) = -\pi/2$ for any $1 < \alpha \leqslant 2$ and any $-(2/\alpha - 1) \leqslant \theta \leqslant 2/\alpha - 1$. Making use of this result in Equation (21), we obtain

$$\lim_{R\to\infty} U(R,\varphi,\alpha,\theta) = \infty, \quad \text{if} \quad \alpha(\varphi + \pi\theta/2) = -\pi/2. \tag{24}$$

and, consequently, it is necessary to exclude this case from our consideration. Putting together Equations (20) and (23), and taking account of Equation (24) we obtain $\lim_{R\to\infty} U(R,\varphi,\alpha,\theta) = 0$ if $-\frac{\pi}{2\alpha} - \frac{\pi\theta}{2} < \varphi \leqslant \frac{\pi}{2\alpha} - \frac{\pi\theta}{2}$ and $x > 0$. In view of the obtained result, the expression in Equation (18) takes the form

$$\lim_{R\to\infty} |I(C_R)| \leqslant \lim_{R\to\infty} \int_{-\frac{\pi}{2\alpha}-\frac{\pi\theta}{2}+\varepsilon}^{\frac{\pi}{2\alpha}-\frac{\pi\theta}{2}} U(R,\varphi,\alpha,\theta)d\varphi = 0,$$

where $\varepsilon > 0$ is an arbitrary small number. Thus, in the considered case contour integration Equation (13) takes the form $C_R = \left\{z : |z| = R, -\frac{\pi}{2\alpha} - \frac{\pi\theta}{2} + \varepsilon \leqslant \varphi \leqslant \frac{\pi}{2\alpha} - \frac{\pi\theta}{2}\right\}$, and for this contour $\lim_{R\to\infty} I(C_R) = 0$. This proves the second item of the lemma.

(3) Now let us consider the case $0 < \alpha \leqslant 2$ and $R \to 0$. It is known that for a specified range of values of the parameter $\alpha$ the parameter $\theta$ can take the values $|\theta| \leqslant \min(1, 2/\alpha - 1)$. Assuming that $R \to 0$ in Equation (14) we obtain

$$\lim_{R\to 0} |I(C_R)| \leqslant \lim_{R\to 0} \int_{\varphi_1}^{\varphi_2} U(R, \varphi, \alpha, \theta) d\varphi. \tag{25}$$

From here we can see that the behavior of this integral at $R \to 0$ will be determined by the behavior $U(R, \varphi, \alpha, \theta)$ at $R \to 0$. Using Equation (15), we obtain

$$\lim_{R\to 0} U(R, \varphi, \alpha, \theta) = \lim_{R\to 0} \exp\left\{-Rx \sin\varphi - R^{\alpha} \cos(\alpha(\varphi + \pi\theta/2)) + \ln R\right\} = 0 \tag{26}$$

For any values of $\varphi$ and any $x$. Choosing $\varphi_1 = -\pi$ and $\varphi_2 = \pi$ a contour $C_R$ will take the form

$$C_R = \{z : \ |z| = R, \ -\pi \leqslant \arg z \leqslant \pi\}. \tag{27}$$

Now using Equations (26) and (27) in Equation (25) we will finally obtain $\lim_{R\to 0} |I(C_R)| \leqslant \lim_{R\to 0} \int_{-\pi}^{\pi} U(R, \varphi, \alpha, \theta) d\varphi = 0$, for any admissible values of parameters $\alpha$ and $\theta$ and for any $x$. It proves the third item of the lemma.

(4) We will consider the case $x = 0$ and $R \to \infty$. In this case, the expression in Equation (15) will take the form

$$U(R, \varphi, \alpha, \theta) = \exp\left\{-R^{\alpha} \cos(\alpha(\varphi + \pi\theta/2)) + \ln R\right\}. \tag{28}$$

Thus, the integral in Equation (14) will tend to zero at $R \to \infty$ if $\cos(\alpha(\varphi + \pi\theta/2)) > 0$. Consequently, $\varphi$ must meet the conditions

$$-\frac{\pi}{2\alpha} - \frac{\pi\theta}{2} < \varphi < \frac{\pi}{2\alpha} - \frac{\pi\theta}{2}. \tag{29}$$

Since no additional limitations for parameters $\alpha$ and $\theta$ were introduced here, then this result is true for any admissible values of these parameters. It should be pointed out that there are strict inequalities here. In fact, if $\varphi = \pm\frac{\pi}{2\alpha} - \frac{\pi\theta}{2}$, then $\cos(\alpha(\varphi + \pi\theta/2)) = 0$ and in this case $U(R, \varphi, \alpha, \theta) \to \infty$ at $R \to \infty$. Choosing in Equation (14) the values $\varphi_1 = -\pi/(2\alpha) - \pi\theta/2 + \varepsilon$ and $\varphi_2 = \pi/(2\alpha) - \pi\theta/2 - \varepsilon$ as the limits of integration, contour integration $C_R$ at $x = 0$ takes the form

$$C_R = \left\{z : \ |z| = R, \ -\frac{\pi}{2\alpha} - \frac{\pi\theta}{2} + \varepsilon \leqslant \arg z \leqslant \frac{\pi}{2\alpha} - \frac{\pi\theta}{2} - \varepsilon\right\}, \tag{30}$$

where $\varepsilon$ is an arbitrary small positive number. Now using Equations (28) and (30) in Equation (14), we obtain

$$\lim_{R\to\infty} |I(C_R)| \leqslant \lim_{R\to\infty} \int_{-\frac{\pi}{2\alpha} - \frac{\pi\theta}{2} + \varepsilon}^{\frac{\pi}{2\alpha} - \frac{\pi\theta}{2} - \varepsilon} U(R, \varphi, \alpha, \theta) d\varphi = 0$$

at $x = 0$ and any admissible values of parameters $\alpha$ and $\theta$. It proves the fourth item of the lemma.

(5) Let us consider the case $\alpha = 1$. In this case the parameter $\theta$ can vary within the limits $-1 \leqslant \theta \leqslant 1$. It is necessary to determine under which conditions the integral in Equation (14) tends to zero at $R \to \infty$. As in previous cases, assuming that $R \to \infty$ in Equation (14), we obtain

$$\lim_{R\to\infty} |I(C_R)| \leqslant \lim_{R\to\infty} \int_{\varphi_1}^{\varphi_2} U(R, \varphi, 1, \theta) d\varphi. \tag{31}$$

From this it follows that these conditions are a consequence of the behavior $U(R, \varphi, 1, \theta)$ at $R \to \infty$. Applying Equation (15), we obtain

$$\lim_{R\to\infty} U(R, \varphi, 1, \theta) = \lim_{R\to\infty} \exp\{-R(x \sin\varphi + \cos(\varphi + \pi\theta/2)) + \ln R\}. \tag{32}$$

In view of the fact that $R$ grows faster than $\ln R$ at $R \to \infty$ we obtain that

$$\exp\{-R(x \sin\varphi + \cos(\varphi + \pi\theta/2)) + \ln R\} \underset{R\to\infty}{\to} 0$$

if

$$x \sin\varphi + \cos(\varphi + \pi\theta/2) > 0. \tag{33}$$

This inequality allows us to determine the conditions imposed on $\varphi$. It should be noted that one should exclude the case $x = 1, \theta = 1$ from consideration. In fact, substituting these values in Equation (32), we obtain $\lim_{R\to\infty} U(R, \varphi, 1, 1) = \infty$ for any $\varphi$. Thus, from Equation (31) we obtain

$$I(C_R) \underset{R\to\infty}{\to} \infty, \quad \text{if} \quad \alpha = 1, \quad x = 1, \quad \theta = 1. \tag{34}$$

Making some transformations the inequality in Equation (33) can be written in the form

$$\tan\varphi > \frac{\cos(\pi\theta/2)}{\sin(\pi\theta/2) - x} \tag{35}$$

From this inequality we can see that it is necessary to consider three cases: $x > 1$, $0 \leqslant x < 1$, and $x = 1$.

In the case $x > 1$ for any $\theta$ the difference $\sin(\pi\theta/2) - x$ is negative. Taking into account that the function $\arctan x$ is a multi-valued function, then choosing the principal branch $\arctan x$ we deduce $\pi/2 \geqslant \varphi > \max(-\pi/2, \varphi_0(\theta, x))$ or

$$\varphi_0(\theta, x) < \varphi \leqslant \pi/2, \quad x > 1, \quad -1 \leqslant \theta \leqslant 1, \tag{36}$$

where

$$\varphi_0(\theta, x) = \arctan\left(\frac{\cos(\pi\theta/2)}{\sin(\pi\theta/2) - x}\right). \tag{37}$$

It should be noted here that this expression is the solution of the equation

$$x \sin\varphi + \cos(\varphi + \pi\theta/2) = 0. \tag{38}$$

We need to consider the case $0 \leqslant x < 1$. Here we need to consider three possible situations: (1) $\sin(\pi\theta/2) - x < 0$, if $\theta < \theta_0$, (2) $\sin(\pi\theta/2) - x = 0$, if $\theta = \theta_0$, (3) $\sin(\pi\theta/2) - x > 0$, if $\theta > \theta_0$. Here $\theta_0 = (2/\pi) \arcsin x$. We will introduce the notation as follows $f(\varphi, \theta, x) = x \sin\varphi + \cos(\varphi + \pi\theta/2)$. In view of this notation, the condition in Equation (33) will take the form $f(\varphi, \theta, x) > 0$. In Figure 1 the function graph $f(\varphi, \theta, x)$ is plotted for $0 \leqslant x < 1$ with different values of the parameter $\theta$ that correspond to three possible situations: $\theta < \theta_0$, $\theta = \theta_0$, $\theta > \theta_0$.

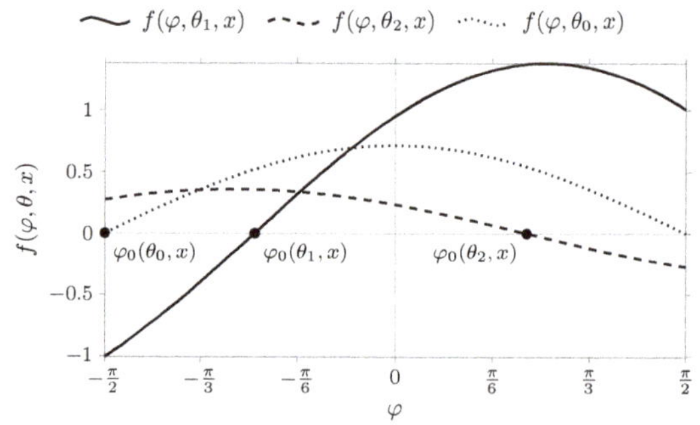


**Figure 1.** The graph of the function $f(\varphi,\theta,x)$ at $x<1$ (the curve is plotted at $x=0.7$) and different values of the parameter $\theta$: $\theta_0=(2/\pi)\arcsin x$, $\theta_1<\theta_0$, $\theta_2>\theta_0$. Heavy dots demonstrate the value of the solution of the Equation $f(\varphi,\theta,x)=0$ for values $\theta=\theta_0,\theta_1,\theta_2$.

Let us consider the case $\theta<\theta_0$. In Figure 1, a curve $f(\varphi,\theta_1,x)$ corresponds to this case. From this figure one can see that the condition $f(\varphi,\theta,x)>0$ is met for values $\varphi>\varphi_0(\theta,x)$. We need to remind that $\varphi_0(\theta,x)$ is the solution of the Equation (38). Thus, choosing the principal branch of the function $\arctan x$ we get $\pi/2\geqslant\varphi>\max(-\pi/2,\varphi_0(\theta,x))$ or

$$\varphi_0(\theta,x)<\varphi\leqslant\pi/2,\quad 0\leqslant x<1,\quad \theta<\theta_0. \tag{39}$$

This condition is easy to determine from Figure 2. The values of $\varphi$ lying above the curve $\varphi_0(\theta,x)$ and not exceeding the value $\pi/2$ correspond to the condition $f(\varphi,\theta,x)>0$ at $\theta<\theta_0$.

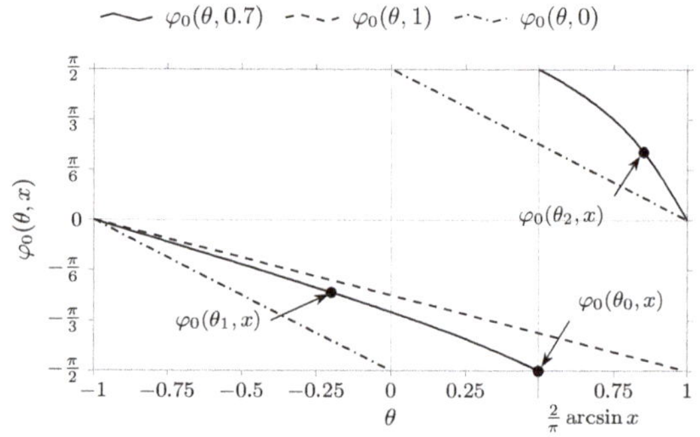


**Figure 2.** The graph of the function $\varphi_0(\theta,x)$ at $x\leqslant 1$. The figure shows the graphs for values $x=0,0.7,1$. Heavy dots designate the values of this function with the value of the parameter $\theta=\theta_0,\theta_1,\theta_2$ and $x=0.7$ corresponding to Figure 1.

In the case of $x=0$, we get $\theta_0=0$ and $\varphi_0(\theta,0)=\arctan(\cot(\pi\theta/2))$. Therefore, the condition $\theta<\theta_0$ will take the form $\theta<0$. One should pay attention that the argument $\pi\theta/2$ at $-1\leqslant\theta\leqslant 1$ takes the values in the range from $-\pi/2$ to $\pi/2$. In this range of values $\cot(\pi\theta/2)$ in the point $\theta=0$ has the point of discontinuity. We will write $\varphi_0(\theta,0)$ in the form

$$\varphi_0(\theta,0)=\arctan(1/\tan(\pi\theta/2)). \tag{40}$$

and will make use of the following trigonometric identities

$$\arctan y = \begin{cases} \operatorname{arccot}(1/y) - \pi, & y \leqslant 0, \\ \operatorname{arccot}(1/y), & y > 0 \end{cases} \tag{41}$$

and

$$\arctan y + \operatorname{arccot} y = \pi/2. \tag{42}$$

Now using these two trigonometric identities in Equation (40), we find

$$\varphi_0(\theta, x) = -\frac{\pi}{2} - \frac{\pi\theta}{2}, \quad \theta < 0 \tag{43}$$

Thus, at $x = 0$ the condition in Equation (39) will take the form $-\frac{\pi}{2} - \frac{\pi\theta}{2} < \varphi \leqslant \frac{\pi}{2}$, if $\alpha = 1$ and $\theta < 0$. The condition obtained is the same as the condition in Equation (29), if to limit the latter above with the value $\pi/2$ at $\theta < 0$.

Now we need to consider the case $\theta = \theta_0$. In Figure 2 we can see that when increasing the parameter $\theta$ the point $\varphi_0(\theta, x)$ will approach the value $-\pi/2$. With such a change of the parameter $\theta$ the function graph $f(\varphi, \theta, x)$ in Figure 1 will shift to the left and with the value $\theta = \theta_0$ will take the form that is given in Figure 1. As we can see from Figure 2, in this case the function $\varphi_0(\theta, x)$ has a discontinuity in the point $\theta = \theta_0$. It is connected with the fact that the principal branch of the function $\arctan x$ is investigated. In the vicinity of this point we have $\lim_{\theta \to \theta_0 - 0} \varphi_0(\theta, x) = \lim_{\theta \to \theta_0 - 0} \arctan\left(\frac{\cos(\pi\theta/2)}{\sin(\pi\theta/2) - x}\right) = -\frac{\pi}{2}$ and $\lim_{\theta \to \theta_0 + 0} \varphi_0(\theta, x) = \lim_{\theta \to \theta_0 + 0} \arctan\left(\frac{\cos(\pi\theta/2)}{\sin(\pi\theta/2) - x}\right) = \frac{\pi}{2}$. From this we can see that the point $\theta_0$ is the point of discontinuity of the first kind. It is possible to eliminate the discontinuity of the function $\varphi_0(\theta, x)$ in the point $\theta_0$ by defining the value of this function in the given point. We will select

$$\varphi_0(\theta_0, x) = -\pi/2, \quad 0 \leqslant x < 1. \tag{44}$$

Thus, the condition in Equation (33) is met if

$$-\pi/2 < \varphi < \pi/2, \quad 0 \leqslant x < 1, \quad \theta = \theta_0. \tag{45}$$

It should be noted that there are precisely strict inequalities here. In fact, if to take $\varphi = \pm\pi/2$, then in the case considered these values will be the solution of the Equation (38) which will lead to divergence of the integral in Equation (31).

Now we will consider the case $\theta > \theta_0$. From Figure 1 one can see that by increasing the parameter $\theta$ from $\theta_0$ to 1 the half-period of the function $f(\varphi, \theta, x)$ satisfying the condition $f(\varphi, \theta, x) > 0$ will keep moving to the left. At the same time, the left point which is the solution of an equation $f(\varphi, \theta, x) = 0$ will become smaller than $-\pi/2$. Since we take interest in the interval from $-\pi/2$ to $\pi/2$, then the left bound of the interval will be the value $-\pi/2$. Keeping in mind that the half-period of the function $f(\varphi, \theta, x)$ is equal to $\pi$, the right bound of the interval $f(\varphi, \theta, x) > 0$ will be the second solution of the equation $f(\varphi, \theta, x) = 0$. As a result, the case considered the condition in Equation (33) will be met with values $\varphi$ satisfying the inequality

$$-\pi/2 \leqslant \varphi < \varphi_0(\theta, x), \quad 0 \leqslant x < 1, \quad \theta > \theta_0. \tag{46}$$

The graph of the function $\varphi_0(\theta, x)$ is given in Figure 2.

If $x = 0$, we have $\theta_0 = 0$ and condition $\theta > \theta_0$ turns into a condition $\theta > 0$. As it was pointed out earlier, in this case it is convenient to represent $\varphi_0(\theta, x)$ in the form of Equation (40). Now using the trigonometric identities in Equations (41) and (42) in Equation (40) we obtain

$$\varphi_0(\theta, x) = \frac{\pi}{2} - \frac{\pi\theta}{2}, \quad \theta > 0. \tag{47}$$

Thus, at $x = 0$ the condition in Equation (46) takes the form $-\frac{\pi}{2} \leqslant \varphi < \frac{\pi}{2} - \frac{\pi\theta}{2}$, if $x = 0$ and $\theta > 0$. The condition obtained is the same as the condition in Equation (29), if to limit the latter below with a value $-\pi/2$ at $\theta > 0$.

If now we put together the expressions in Equations (43), (44), and (47), then $\varphi_0(\theta, 0)$ takes the form

$$\varphi_0(\theta, 0) = \begin{cases} -\pi/2 - \pi\theta/2, & \theta \leqslant 0 \\ \pi/2 - \pi\theta/2, & \theta > 0. \end{cases}$$

The graph of the function $\varphi_0(\theta, 0)$ is given in Figure 2.

Consider now the case $x = 1$. In this case the inequality in Equation (35) has the form

$$\tan\varphi(1 - \sin(\pi\theta/2)) > -\cos(\pi\theta/2). \tag{48}$$

From this it follows that $1 - \sin(\pi\theta/2) > 0$, for any $\theta < \theta_0$, where $\theta_0 = (2/\pi)\arcsin 1 = 1$. As a result, selecting the principal branch of the function $\arctan y$, the inequality in Equation (48) takes the form

$$\varphi_0(\theta, x) < \varphi \leqslant \pi/2, \quad x = 1, \quad -1 \leqslant \theta < 1. \tag{49}$$

This inequality gives a condition under which the inequality will be satisfied in Equation (35) in case $x = 1$. The graph of the function $\varphi_0(\theta, 1)$ is given in Figure 2.

Thus, the inequalities in Equations (36), (39), (45), (46), and (49) completely define the condition under which the inequality is satisfied in Equation (33). Combining these conditions for Equation (32), we obtain

$$\lim_{R\to\infty} U(R, \varphi, 1, \theta) = 0, \quad \text{if} \quad \begin{cases} \varphi_0(\theta, x) < \varphi \leqslant \pi/2, & x > 1, \quad -1 \leqslant \theta \leqslant 1, \\ \varphi_0(\theta, x) < \varphi \leqslant \pi/2, & 0 \leqslant x < 1, \quad \theta < \theta_0, \\ -\pi/2 < \varphi < \pi/2, & 0 \leqslant x < 1, \quad \theta = \theta_0, \\ -\pi/2 \leqslant \varphi < \varphi_0(\theta, x), & 0 \leqslant x < 1, \quad \theta > \theta_0, \\ \varphi_0(\theta, x) < \varphi \leqslant \pi/2, & x = 1, \quad -1 \leqslant \theta < 1, \end{cases} \tag{50}$$

where $\theta_0 = (2/\pi)\arcsin x$. It should be noted that if $\varphi = \varphi_0(\theta, x)$, then in this case $x\sin\varphi + \cos(\varphi + \pi\theta/2) = 0$ and, as a consequence, $\lim_{R\to\infty} U(R, \varphi_0(\theta, x), 1, \theta) = \infty$. Therefore, $\varphi \neq \varphi_0(\theta, x)$. In addition, in terms of Equation (34), one should exclude the case $x = 1, \theta = 1$ from consideration.

Now using Equation (50) in Equation (31), we ultimately obtain

$$I(C_R) \underset{R\to\infty}{\to} 0 \text{ if } \begin{cases} C_R = \{z: \ |z| = R, \ \varphi_0(\theta, x) + \varepsilon \leqslant \arg z \leqslant \pi/2\}, & x > 1, \ -1 \leqslant \theta \leqslant 1, \\ C_R = \{z: \ |z| = R, \ \varphi_0(\theta, x) + \varepsilon \leqslant \arg z \leqslant \pi/2\}, & 0 \leqslant x < 1, \ -1 \leqslant \theta < \theta_0, \\ C_R = \{z: \ |z| = R, \ -\pi/2 + \varepsilon \leqslant \arg z \leqslant \pi/2 - \varepsilon\}, & 0 \leqslant x < 1, \ \theta = \theta_0, \\ C_R = \{z: \ |z| = R, \ -\pi/2 \leqslant \arg z \leqslant \varphi_0(\theta, x) - \varepsilon\}, & 0 \leqslant x < 1, \ \theta_0 < \theta \leqslant 1, \\ C_R = \{z: \ |z| = R, \ \varphi_0(\theta, x) + \varepsilon \leqslant \arg z \leqslant \pi/2\}, & x = 1, \ -1 \leqslant \theta < 1, \end{cases}$$

where $\varepsilon$ is any arbitrary small positive number. This proves items 5, 6, 7, 8, and 9 of the lemma and proves the lemma completely. □

**Remark 1.** *The statements of the Lemma 2 are preserved if to substitute contours $C_R$ with their parts.*

Using the lemma which was proved one can substantiate the validity of transition from an integral $\int_0^\infty \exp\{itx\}\hat{g}(t,\alpha,\theta)dt$ along the real variable $t$ to an integral $\int_\Gamma \exp\{izx\}g^+(z,\alpha,\theta)dz$ in the complex variable $z$ along some contour $\Gamma$ for inversion formulas in Equation (9). We will state this result in the form of a lemma.

**Lemma 3.** *Let us consider the family of contours $\{\Gamma\}$ in the complex plane $z$ with a cut along a half-line $\arg z = -\pi$ satisfying the following conditions:*

1. *Every contour starts in the point $z = 0$.*
2. *None of the contours $\Gamma$ intersects the lines of the cut.*
3. *Moving from the point $z = 0$ along the contour $\Gamma$ we let it tend to infinity but in such a way that starting from some place all points $z \in \Gamma$ have values of arguments within the limits:*

$$0 \leqslant \arg z \leqslant \pi - \varepsilon, \quad \text{if} \quad 0 < \alpha < 1,\ -1 \leqslant \theta \leqslant 1,\ x > 0, \tag{51}$$

$$-\frac{\pi}{2\alpha} - \frac{\pi\theta}{2} + \varepsilon \leqslant \arg z \leqslant \frac{\pi}{2\alpha} - \frac{\pi\theta}{2}, \quad \text{if} \quad 1 < \alpha \leqslant 2,\ |\theta| \leqslant (2/\alpha - 1),\ x > 0, \tag{52}$$

$$-\frac{\pi}{2\alpha} - \frac{\pi\theta}{2} + \varepsilon \leqslant \arg z \leqslant \frac{\pi}{2\alpha} - \frac{\pi\theta}{2} - \varepsilon, \quad \text{if} \quad 0 < \alpha \leqslant 2,\ x = 0,\ |\theta| \leqslant \min(1, 2/\alpha - 1), \tag{53}$$

$$\varphi_0(\theta, x) + \varepsilon \leqslant \arg z \leqslant \frac{\pi}{2}, \quad \text{if} \quad \alpha = 1,\ x > 1,\ -1 \leqslant \theta \leqslant 1, \tag{54}$$

$$-\frac{\pi}{2} \leqslant \arg z \leqslant \varphi_0(\theta, x) - \varepsilon, \quad \text{if} \quad \alpha = 1,\ 0 \leqslant x < 1,\ \theta_0 < \theta \leqslant 1, \tag{55}$$

$$-\frac{\pi}{2} + \varepsilon \leqslant \arg z \leqslant \frac{\pi}{2} - \varepsilon, \quad \text{if} \quad \alpha = 1,\ 0 \leqslant x < 1,\ \theta = \theta_0, \tag{56}$$

$$\varphi_0(\theta, x) + \varepsilon \leqslant \arg z \leqslant \frac{\pi}{2}, \quad \text{if} \quad \alpha = 1,\ 0 \leqslant x < 1,\ -1 \leqslant \theta < \theta_0, \tag{57}$$

$$\varphi_0(\theta, x) + \varepsilon \leqslant \arg z \leqslant \frac{\pi}{2}, \quad \text{if} \quad \alpha = 1,\ x = 1,\ -1 \leqslant \theta < 1, \tag{58}$$

*where $\theta_0 = (2/\pi)\arcsin x$ and $\varphi_0(\theta, x)$ have the form in Equation (12) and $\varepsilon > 0$ is any arbitrary small number. Then for any contour of the specified type and any pair of admissible parameters $(\alpha, \theta)$ and any $x \geqslant 0$ (with the exception of the point $x = 1, \alpha = 1, \theta = 1$)*

$$\int_0^\infty e^{itx} g(t, \alpha, -\theta) dt = \int_\Gamma e^{izx} g^+(z, \alpha, -\theta) dz, \tag{59}$$

*where $t$ is real.*

**Proof.** From constraints imposed on contours $\Gamma$ one can see that it is possible to divide the whole family of contours $\{\Gamma\}$ into two kinds. The contours which start in the point $z = 0$ and tend to infinity without intersecting the positive part of a real semiaxis are referred to the contours of the first kind. The contours intersecting the positive part of a real semiaxis are referred to the contours of the second kind.

We will consider, at first, contours of the first kind. We will introduce the following notation: $z_r$ is the intersection point of the contour $\Gamma$ with a circle $|z| = r$, $C_r$ is an arc of a circle of radius $r$ (not crossing the cut) which is formed when moving from the point $z = r$ to the point $z_r$ and $\Gamma_{r,R}$ is a part of a contour $\Gamma$ which is formed when moving from the point $z_r$ to the point $z_R$. We form a closed contour $G_{r,R} = [r, R] \cup C_R \cup \bar{\Gamma}_{r,R} \cup C_r$ (see Figure 3). The line means that we go along the contour in the opposite direction. Since the function $h(z) = e^{izx}g^+(z,\alpha,-\theta)$ is analytic in the region restricted by the contour $G_{r,R}$, then by using the Cauchy theorem

$$\int_{G_{r,R}} h(z)dz = \int_r^R h(z)dz + \int_{C_R} h(z)dz - \int_{\Gamma_{r,R}} h(z)dz + \int_{C_r} h(z)dz = 0.$$

We will assume in this expression that $r \to 0$ and $R \to \infty$. Using Lemma 2 and Remark 1 we find that $\int_{C_R} h(z)dz + \int_{C_r} h(z)dz \to 0$, at $R \to \infty$ and $r \to 0$. Therefore, the equality in Equation (59) is true.

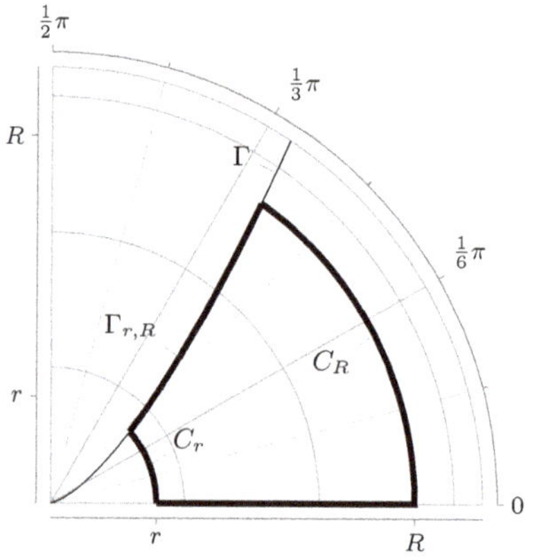


**Figure 3.** Auxiliary contour $G_{r,R}$ (heavy curve).

Now consider the contours of the second kind. These contours are characterized by the feature that they intersect the real axis. Therefore, to prove the lemma, we consider two closed auxiliary contours: the contour $G_{r,x} = [r,x] \cup \bar{\Gamma}_{r,x} \cup C_r$ and contour $G_{x,R} = \overline{[x,R]} \cup C_R \cup \Gamma_{x,R}$ (see Figure 4). Here $x$ is the intersection point of a contour $\Gamma$ with a real axis. Since the function $h(z)$ is analytic within the regions restricted with the contours $\Gamma_{r,x}$ and $\Gamma_{x,R}$, then by using the Cauchy theorem

$$\int_{G_{x,r}} h(z)dz = \int_r^x h(z)dz - \int_{\Gamma_{r,x}} h(z)dz + \int_{C_r} h(z)dz = 0, \tag{60}$$

$$\int_{G_{x,R}} h(z)dz = -\int_x^R h(z)dz + \int_{\Gamma_{x,R}} h(z)dz + \int_{C_R} h(z)dz = 0. \tag{61}$$

Now we will assume in these expressions that $r \to 0$ and $R \to \infty$. Using Lemma 2 and Remark 1, we find that $\int_{C_r} h(z)dz \to 0$ at $r \to 0$ and $\int_{C_R} h(z)dz \to 0$ at $R \to \infty$. Now summing up Equations (60) and (61) in view of this result we obtain

$$\int_0^x h(z)dz + \int_x^\infty h(z)dz = \int_{\Gamma_{0,x}} h(z)dz + \int_{\Gamma_{x,\infty}} h(z)dz,$$

hence, the equality in Equation (59) is true. □

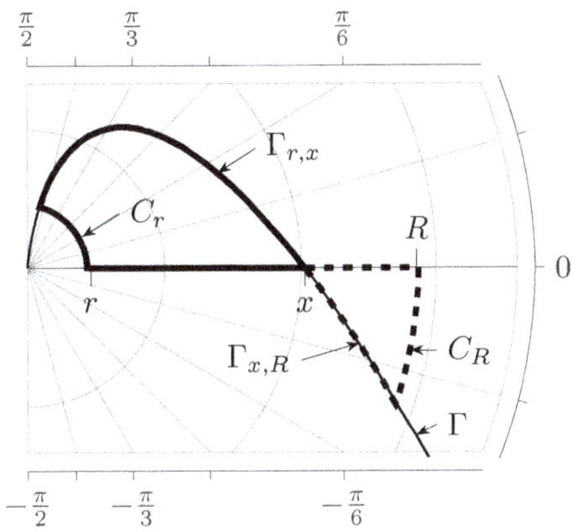


**Figure 4.** Auxiliary contours $G_{r,x}$ (heavy curve) and $G_{x,R}$ (dashed heavy curve).

The following lemma will be useful further

**Lemma 4.** *For any $x \geqslant 0$ and any $-1 \leqslant \theta \leqslant 1$ the function*

$$\varphi(\theta, x) = \arctan\left(\frac{x - \sin\left(\frac{\pi}{2}\theta\right)}{\cos\left(\frac{\pi}{2}\theta\right)}\right) \tag{62}$$

*and function $\varphi_0(\theta, x)$, defined by (12), are connected with relations between each other:*

1. *if $x \geqslant 1$, then*
$$\varphi_0(\theta, x) = \varphi(\theta, x) - \pi/2; \tag{63}$$
2. *if $0 \leqslant x < 1$, then*
$$\varphi_0(\theta, x) = \begin{cases} \varphi(\theta, x) - \pi/2, & -1 \leqslant \theta \leqslant \theta_0, \\ \varphi(\theta, x) + \pi/2, & \theta_0 < \theta \leqslant 1, \end{cases} \tag{64}$$

*where $\theta_0 = (2/\pi) \arcsin x$.*

**Proof.** Let us consider the case $x > 1$. Using the identity in Equation (42) as well as the identity

$$\operatorname{arccot}(-y) = \pi - \operatorname{arccot}(y), \tag{65}$$

we obtain

$$\varphi_0(\theta, x) = \arctan\left(\frac{\cos(\pi\theta/2)}{\sin(\pi\theta/2) - x}\right) = \frac{\pi}{2} - \operatorname{arccot}\left(\frac{\cos(\pi\theta/2)}{\sin(\pi\theta/2) - x}\right) = -\frac{\pi}{2} + \operatorname{arccot}\left(\frac{\cos(\pi\theta/2)}{x - \sin(\pi\theta/2)}\right).$$

Since $x > 1$, then the argument $\frac{\cos(\pi\theta/2)}{x - \sin(\pi\theta/2)} > 0$ for any $-1 \leqslant \theta \leqslant 1$. Now using the identity in Equation (41) for $y > 0$ we get

$$\varphi_0(\theta, x) = \varphi(\theta, x) - \pi/2, \quad x > 1, \quad -1 \leqslant \theta \leqslant 1. \tag{66}$$

Consider the case $x = 1$. In the same way as in our previous case, using the identities in Equations (42) and (65), we find

$$\varphi_0(\theta, 1) = \arctan\left(\frac{\cos(\pi\theta/2)}{\sin(\pi\theta/2) - 1}\right) = -\frac{\pi}{2} + \operatorname{arccot}\left(\frac{\cos(\pi\theta/2)}{1 - \sin(\pi\theta/2)}\right). \tag{67}$$

From this expression one can see that the argument $\frac{\cos(\pi\theta/2)}{1-\sin(\pi\theta/2)} \geqslant 0$ for any $-1 \leqslant \theta < 1$ and it has an indeterminate form $0/0$ at $\theta = 1$. Evaluating indeterminate forms according to L'Hôpital's rule we get $\lim_{\theta\to 1-0} \frac{\cos(\pi\theta/2)}{1-\sin(\pi\theta/2)} = \infty$. Consequently, $\cos(\pi\theta/2)/(1 - \sin(\pi\theta/2)) \geqslant 0$, if $-1 \leqslant \theta \leqslant 1$. Now using in Equation (67) the identity of Equation (41) for $y > 0$, we obtain

$$\varphi_0(\theta, 1) = \varphi(\theta, 1) - \pi/2, \quad -1 \leqslant \theta \leqslant 1.$$

Combining now this expression and the expression of Equation (66) we come to Equation (63). Thus, the first item of the lemma is proved.

Now we will consider the case $0 \leqslant x < 1$. Using the identities in Equations (42) and (65) we get

$$\varphi_0(\theta, x) = \arctan\left(\frac{\cos(\pi\theta/2)}{\sin(\pi\theta/2) - x}\right) = -\frac{\pi}{2} + \operatorname{arccot}\left(\frac{\cos(\pi\theta/2)}{x - \sin(\pi\theta/2)}\right). \tag{68}$$

Taking into consideration that $\cos(\pi\theta/2) \geqslant 0$ for any $\theta \in [-1, 1]$, we find that the sign of $\frac{\cos(\pi\theta/2)}{x-\sin(\pi\theta/2)}$ is defined by the denominator. We have three possible situations: (1) $x - \sin(\pi\theta/2) < 0$, if $\theta > \theta_0$, (2) $x - \sin(\pi\theta/2) = 0$, if $\theta = \theta_0$, (3) $x - \sin(\pi\theta/2) > 0$, if $\theta < \theta_0$. Here $\theta_0 = (2/\pi) \arcsin x$. Taking into consideration Equation (44), we obtain $\frac{\cos(\pi\theta/2)}{x-\sin(\pi\theta/2)} = +\infty$, if $\theta = \theta_0$. Thus, $\frac{\cos(\pi\theta/2)}{x-\sin(\pi\theta/2)} \leqslant 0$, if $\theta > \theta_0$, and $\frac{\cos(\pi\theta/2)}{x-\sin(\pi\theta/2)} \geqslant 0$, if $\theta \leqslant \theta_0$. Now in Equation (68) applying the identity in Equation (41) we get

$$\varphi_0(\theta, x) = \begin{cases} -\frac{\pi}{2} + \arctan\left(\frac{x-\sin(\pi\theta/2)}{\cos(\pi\theta/2)}\right), & \theta \leqslant \theta_0, \\ -\frac{\pi}{2} + \pi + \arctan\left(\frac{x-\sin(\pi\theta/2)}{\cos(\pi\theta/2)}\right), & \theta > \theta_0. \end{cases}$$

From here it follows Equation (64). The lemma is completely proved. □

In Figures 5 and 6, the graphs of the functions $\varphi_0(\theta, x)$ and $\varphi(\theta, x)$ are given for the cases $x \geqslant 1$ and $0 \leqslant x < 1$ which clearly illustrate the validity of the lemma that has just been proved.

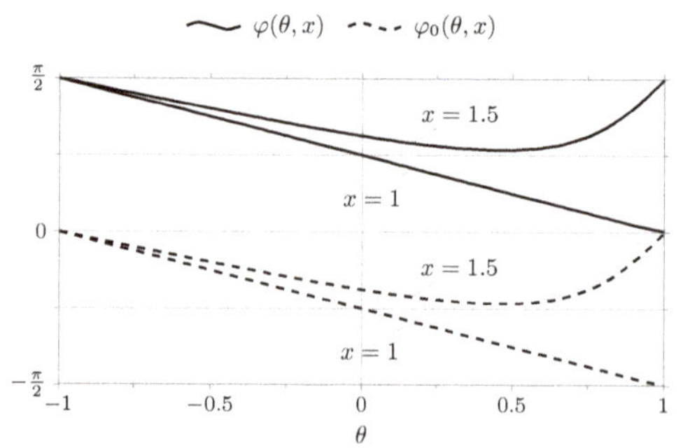


**Figure 5.** The graph of the function $\varphi(\theta, x)$ and $\varphi_0(\theta, x)$ depending on the parameter $\theta$ in the case of $x \geqslant 1$ (The graphs are plotted for $x = 1.5$ and $x = 1$).

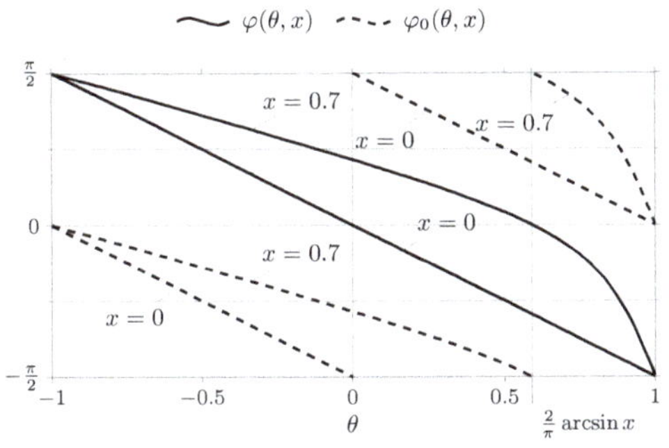


**Figure 6.** The graph of the function $\varphi(\theta, x)$ and $\varphi_0(\theta, x)$ depending on the parameter $\theta$ in the case $x < 1$ (The graphs are plotted for $x = 0.7$ and $x = 0$).

## 3. Main Results

Now we can formulate the main theorem which gives an integral representation for the probability density of a stable law $g(x, \alpha, \theta)$ with the characteristic function in Equation (1).

**Theorem 1.** *The distribution density $g(x, \alpha, \theta)$ of a strictly stable law with a characteristic function as in Equation (1) can be represented in the form*

1. *If $\alpha \neq 1$ and $x \neq 0$ for any values $|\theta| \leqslant \min(1, 2/\alpha - 1)$*

$$g(x, \alpha, \theta) = \frac{\alpha}{\pi|\alpha - 1|} \int_{-\pi\theta^*/2}^{\pi/2} \exp\left\{-|x|^{\alpha/(\alpha-1)} U(\varphi, \alpha, \theta^*)\right\} U(\varphi, \alpha, \theta^*)|x|^{1/(\alpha-1)} d\varphi, \tag{69}$$

*where $\theta^* = \theta \operatorname{sign} x$ and*

$$U(\varphi, \alpha, \theta) = \left(\frac{\sin\left(\alpha\left(\varphi + \frac{\pi}{2}\theta\right)\right)}{\cos\varphi}\right)^{\alpha/(1-\alpha)} \frac{\cos\left(\varphi(1-\alpha) - \frac{\pi}{2}\alpha\theta\right)}{\cos\varphi}. \tag{70}$$

2. *If $x = 0$, then for any $0 < \alpha \leqslant 2$ and $|\theta| \leqslant \min(1, 2/\alpha - 1)$*

$$g(0, \alpha, \theta) = \tfrac{1}{\pi} \cos\left(\pi\theta/2\right) \Gamma\left(1/\alpha + 1\right) \tag{71}$$

3. *If $\alpha = 1$, then for any $|\theta| \leqslant 1$ and any values $x$*

$$g(x, 1, \theta) = \frac{\cos(\pi\theta/2)}{\pi(x^2 - 2x\sin(\pi\theta/2) + 1)}. \tag{72}$$

**Proof.** To obtain the expression for the probability density we use the inversion formulas in Equation (9). In principle, it makes no difference which formula to use. The result will differ then only by the sign of the parameter $\theta$. We will use the first formula in Equation (9), we have

$$g(x, \alpha, \theta) = \frac{1}{\pi} \Re \int_0^\infty e^{itx} \hat{g}(t, \alpha, -\theta) dt. \tag{73}$$

Without loss of generality, we assume that $x > 0$. The density $g(x, \alpha, \theta)$ for $x < 0$ can be obtained with the inversion property in Equation (7). Next, let us make the substitution of integration variable

$t \to z$, where $z$ is complex-valued. Such a substitution means that we analytically extend the integral to the complex plane. With this analytical continuation is carried out starting with the positive part of a real semiaxis. As a result, we have that $\hat{g}(t,\alpha,-\theta) \to g^{+}(z,\alpha,-\theta)$, where $g^{+}(z,\alpha,\theta)$ is defined by Equation (11). The improper integral becomes the integral along the contour $\Gamma$. We will define the contour $\Gamma$ in such a way that $\Im e^{izx} g^{+}(z,\alpha,\theta) = 0$, and the contour itself $\Gamma$ should start in the point $z = 0$ tend to infinity. Since in the inversion formula of Equation (73) the variable is $t > 0$ then from this it follows that $\arg z$ must lie within the limits $-\pi/2 \leqslant \arg z \leqslant \pi/2$. As result, the contour $\Gamma$ will take the form

$$\Gamma = \{z : \Im e^{izx} g^{+}(z,\alpha,\theta) = 0,\ -\tfrac{\pi}{2} \leqslant \arg z \leqslant \tfrac{\pi}{2},\ |z| \geqslant 0\}, \tag{74}$$

However, the specific type of the contour has to be defined. In view of the foregoing, the expression in Equation (73) will be written in the form

$$g(x,\alpha,\theta) = \frac{1}{\pi}\Re \int_{\Gamma} \exp\{izx - z^{\alpha}\exp\{i\tfrac{\pi}{2}\alpha\theta\}\}dz. \tag{75}$$

As a result, the problem consists in determining the contour form $\Gamma$, in proving the validity of transition from Equation (73) to Equation (75) and in calculating this integral.

Let us consider the case $\alpha \neq 1$. Representing $z = re^{i\varphi}$ and using this representation in Equation (75) for the intergrand we obtain

$$\begin{aligned}\exp\{izx - z^{\alpha}\exp\{i\tfrac{\pi}{2}\alpha\theta\}\} &= \exp\left\{ixre^{i\varphi} - r^{\alpha}e^{i\alpha\varphi}\exp\{i\tfrac{\pi}{2}\alpha\theta\}\right\} = \\ &\exp\{-rx\sin\varphi - r^{\alpha}\cos(\alpha(\varphi + \tfrac{\pi}{2}\theta))\}\exp\left\{i[rx\cos\varphi - r^{\alpha}\sin(\alpha(\varphi+\tfrac{\pi}{2}\theta))]\right\}.\end{aligned} \tag{76}$$

From the condition $\Im e^{izx} g^{+}(z,\alpha,\theta) = 0$ and $-\frac{\pi}{2} \leqslant \arg z \leqslant \frac{\pi}{2}$ follows that

$$rx\cos\varphi - r^{\alpha}\sin(\alpha(\varphi + \tfrac{\pi}{2}\theta)) = 0. \tag{77}$$

The solution of this equation is in an explicit form

$$r(\varphi) = \left(\frac{\sin(\alpha(\varphi + \frac{\pi}{2}\theta))}{x\cos(\varphi)}\right)^{\frac{1}{1-\alpha}}. \tag{78}$$

This expression determines the form of contour integration $\Gamma$.

We will determine the admissible region $\arg z \equiv \varphi$. From the condition $|z| \geqslant 0$ follows that $(\sin(\alpha(\varphi + \frac{\pi}{2}\theta)))/(x\cos(\varphi)) \geqslant 0$. Taking into account that $x > 0$ we obtain that the condition $r(\varphi) > 0$ is met if $\sin(\alpha(\varphi + \frac{\pi}{2}\theta)) \geqslant 0$ and $\cos\varphi \geqslant 0$. From the condition $\sin(\alpha(\varphi + \frac{\pi}{2}\theta)) \geqslant 0$ we obtain that $-\frac{\pi}{2}\theta \leqslant \varphi \leqslant \frac{\pi}{\alpha} - \frac{\pi}{2}\theta$, and from the condition $\cos\varphi \geqslant 0$ we get that $-\frac{\pi}{2} \leqslant \varphi \leqslant \frac{\pi}{2}$. Combining these two inequalities we obtain

$$\max\left(-\frac{\pi}{2}, -\frac{\pi}{2}\theta\right) \leqslant \varphi \leqslant \min\left(\frac{\pi}{2}, \frac{\pi}{\alpha} - \frac{\pi}{2}\theta\right). \tag{79}$$

Taking into consideration that at $0 < \alpha < 1$ the parameter $\theta$ takes values from the range $-1 \leqslant \theta \leqslant 1$ and at $1 < \alpha \leqslant 2$ from the range $-(\frac{2}{\alpha} - 1) \leqslant \theta \leqslant \frac{2}{\alpha} - 1$ we obtain that $\max\left(-\frac{\pi}{2}, -\frac{\pi}{2}\theta\right) = -\frac{\pi}{2}\theta$ and $\min\left(\frac{\pi}{2}, \frac{\pi}{\alpha} - \frac{\pi}{2}\theta\right) = \frac{\pi}{2}$. Using this result in Equation (79) we get $-\pi\theta/2 \leqslant \varphi \leqslant \pi/2$. Taking account of this condition and the expression in Equation (78) we obtain that in the case $\alpha \neq 1$ the contour $\Gamma$ takes the form

$$\Gamma = \left\{z : r(\varphi) = \left(\frac{\sin(\alpha(\varphi + \frac{\pi}{2}\theta))}{x\cos(\varphi)}\right)^{\frac{1}{1-\alpha}},\ -\frac{\pi}{2}\theta \leqslant \varphi \leqslant \frac{\pi}{2},\ \alpha \neq 1\right\}. \tag{80}$$

We consider now the contour form $\Gamma$ with different values of parameters. Replacing the boundary values $\varphi$ in the expression for $r(\varphi)$, we obtain

$$\begin{array}{lll} r\left(-\frac{\pi}{2}\theta\right)=0, & r\left(\frac{\pi}{2}\right)=\infty, & \text{at} \quad 0<\alpha<1 \\ r\left(-\frac{\pi}{2}\theta\right)=\infty, & r\left(\frac{\pi}{2}\right)=0, & \text{at} \quad 1<\alpha\leqslant 2. \end{array} \tag{81}$$

Thus, at $0<\alpha<1$ the contour $\Gamma$ starts in the point $z=0$ at $\varphi=-\frac{\pi}{2}\theta$ and tends to infinity at $\varphi=\frac{\pi}{2}$. In the case $1<\alpha\leqslant 2$ the situation is opposite: the contour $\Gamma$ starts in the point $z=0$ at $\varphi=\frac{\pi}{2}$ and tends to infinity at $\varphi=-\frac{\pi}{2}\theta$.

As we can see the contours $\Gamma$ described in Equation (80) with different values of the parameter $\theta$ differ in its type (see Figures 7 and 8). They can be divided into 7 main groups.

1. The contours of the first group are made up of the contours with values $0<\alpha<1$ and $\theta=-1$. In Figure 7 the contour $\Gamma_1$ corresponds to this case. From the definition of the contour in Equation (80) one can see that in this case the admissible region of an angle $\varphi$ takes the form $\frac{\pi}{2}\leqslant\varphi\leqslant\frac{\pi}{2}$. This means that the contour goes along the positive part of the imaginary axis: $\Gamma\equiv\Gamma_1=I^+$.
2. The contours of the second group include the contours with values $0<\alpha<1$ and $-1<\theta\leqslant 0$. In Figure 7 the contours $\Gamma_2,\Gamma_2',\Gamma_2'',\Gamma_2'''$ are referred to this case. The contours of this group start in the point $z=0$ at $\varphi=-\frac{\pi}{2}\theta$ and tend to infinity at $\varphi\to\frac{\pi}{2}$. As one can see, in this case $-\frac{\pi}{2}\theta\geqslant 0$, and contours of this group do not cross the real semiaxis.
3. The third group is made up of the contours with values of parameters $0<\alpha<1$ and $0<\theta<1$. In Figure 7 this group consists of the contours $\Gamma_3,\Gamma_3',\Gamma_3'',\Gamma_3'''$. The contours of this group start in the point $z=0$ coming out at an angle $\varphi=-\frac{\pi}{2}\theta$, and tend to infinity at $\varphi\to\frac{\pi}{2}$. As we can see in this case $-\frac{\pi}{2}\theta<0$. Therefore, the contours of this group approach the point $z=0$ at values of $\varphi<0$ which, in its turn, means that these contours intersect the positive part of the real axis.
4. The fourth group is composed of the contours $0<\alpha<1$, $\theta=1$. In Figure 7 the contour $\Gamma_4$. corresponds to this case. From the expression in Equation (78) we can see that in this case at $\varphi\to-\frac{\pi}{2}$ in this expression there is an indeterminate form $0/0$. Evaluating this indeterminate form according to L'Hôpital's rule we get $\lim_{\varphi\to-\frac{\pi}{2}} r(\varphi)=(\alpha/x)^{1/(1-\alpha)}$. Thus, the contours of this group start in the point $z=-i(\alpha/x)^{1/(1-\alpha)}$ at $\varphi=-\frac{\pi}{2}$ and tend to infinity at $\varphi\to\pi/2$. As we can see the contours of this group also cross the positive part of the real axis.
5. The fifth group includes the contours with parameters $1<\alpha\leqslant 2$ and $-(2/\alpha-1)\leqslant\theta\leqslant 0$. In Figure 8 the contours $\Gamma_5,\Gamma_5',\Gamma_5'',\Gamma_5'''$ are referred to this case. The contours of this group start in the point $z=0$ at $\varphi=\pi/2$ and tend to infinity at $\varphi\to-\frac{\pi}{2}\theta$. Since in this case $\theta\leqslant 0$, then the condition $\varphi>0$ is met for all points of the contour. Therefore, the contours of this group do not cross the positive part of the real axis.
6. The sixth group consists of the contours with parameters $1<\alpha\leqslant 2$, $0<\theta<2/\alpha-1$. In Figure 8 the contours $\Gamma_6,\Gamma_6'$ correspond to this case. The contours of this groups start in the point $z=0$ at $\varphi=\pi/2$ and tend to infinity at $\varphi\to-\frac{\pi}{2}\theta$. Since in this case $\theta>0$, then $-\frac{\pi}{2}\theta<0$ and, consequently, the contours of this group cross the positive part of the real semiaxis.
7. The seventh group comprises the contours with parameters $1<\alpha\leqslant 2$ and $\theta=2/\alpha-1$. In Figure 8 the contour $\Gamma_7$ corresponds to this case. One should pay attention that in this case at $\varphi=\pi/2$ the function $r(\varphi)$ defined by Equation (78) has an indeterminate form $0/0$. Evaluating this indeterminate form according to L'Hôpital's rule we obtain $\lim_{\varphi\to\pi/2} r(\varphi)=(\alpha/x)^{1/(1-\alpha)}$. Thus, the contours of this group start in the point $z=i(\alpha/x)^{1/(1-\alpha)}$ at $\varphi=\pi/2$ and tend to infinity at $\varphi\to-\frac{\pi}{2}\theta$. The contours of this group also cross the positive part of the real axis.

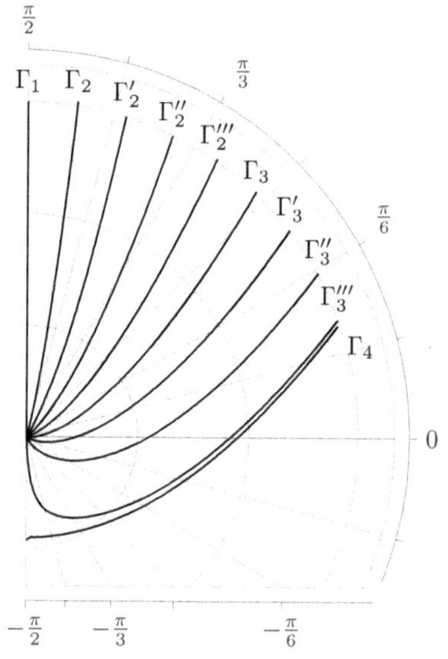


**Figure 7.** The type of a contour $\Gamma$ at $\alpha \in (0,1)$ and various values of parameter $\theta$. The contours are given for the case $\alpha = 0.6$ and $\Gamma_1$ - $\theta = -1$, $\Gamma_2$ - $\theta = -0.75$, $\Gamma_2'$ - $\theta = -0.5$, $\Gamma_2''$ - $\theta = -0.25$, $\Gamma_2'''$ - $\theta = 0$, $\Gamma_3$ - $\theta = 0.25$, $\Gamma_3'$ - $\theta = 0.5$, $\Gamma_3''$ - $\theta = 0.75$, $\Gamma_3'''$ - $\theta = 0.98$, $\Gamma_4$ - $\theta = 1$.

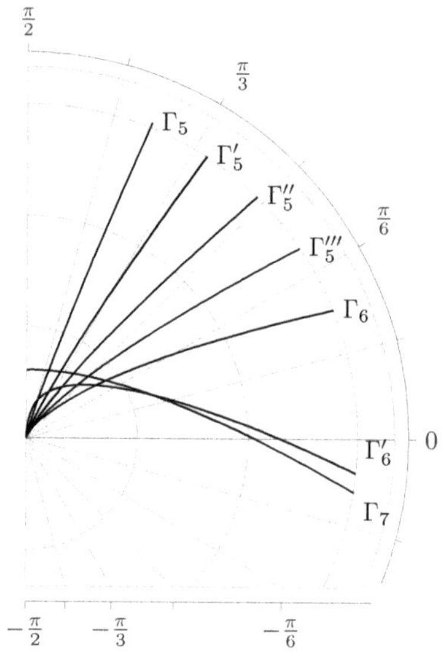


**Figure 8.** The type of a contour $\Gamma$ at $\alpha \in (1,2]$ and with different values of the parameter $\theta$. The contours are given for the case $\alpha = 1.25$ and $\Gamma_5$ - $\theta = -0.6$, $\Gamma_5'$ - $\theta = -0.4$, $\Gamma_5''$ - $\theta = -0.2$, $\Gamma_5'''$ - $\theta = 0$, $\Gamma_6$ - $\theta = 0.2$, $\Gamma_6'$ - $\theta = 0.57$, $\Gamma_7$ - $\theta = 0.6$.

This indicates that the contours of the first, second and third groups ($0 < \alpha < 1, -1 \leqslant \theta < 1$) and contours of the fifth and sixth groups ($1 < \alpha \leqslant 2, -(2/\alpha - 1) \leqslant \theta < 2/\alpha - 1$) satisfy the conditions of the Lemma 3. In fact, the contours of these groups start in the point $z = 0$, do not cross the line of the cut that goes through a half-line $\arg z = -\pi$ and tend to infinity. The contours of the first, second, third groups tend to infinity at $\varphi = \frac{\pi}{2}$. This means that the contours of these groups satisfy the condition in Equation (51). The contours of the fifth and sixth groups tend to infinity at $\varphi = -\frac{\pi}{2}\theta$. Thus, these contours satisfy the condition in Equation (52).

Consider the contours of the fourth group $\Gamma_4$. The contours of this group start in the point $z = -i(\alpha/x)^{1/(1-\alpha)}$ and tend to infinity at $\varphi \to \frac{\pi}{2}$. Consider an auxiliary contour $\Gamma^* = \{z : \arg z = -\pi/2, 0 \leqslant |z| \leqslant (\alpha/x)^{1/(1-\alpha)}\}$. With the help of a contour $\Gamma^*$ we form a new contour $S = \Gamma^* \bigcup \Gamma_4$. The specific feature of the contour $\Gamma^*$ is then that $\Im \exp\{izx - z^\alpha \exp\{i\frac{\pi}{2}\alpha\theta\}\} = 0$ for $z \in \Gamma^*$. From this it follows that

$$\frac{1}{\pi}\Re \int_{\Gamma^*} \exp\{izx - z^\alpha \exp\{i\tfrac{\pi}{2}\alpha\theta\}\}dz = 0.$$

Therefore, for the case $0 < \alpha < 1$, $\theta = 1$ we get

$$\frac{1}{\pi}\Re \int_{S} \exp\{izx - z^\alpha \exp\{i\tfrac{\pi}{2}\alpha\theta\}\}dz = \frac{1}{\pi}\Re \int_{\Gamma_4} \exp\{izx - z^\alpha \exp\{i\tfrac{\pi}{2}\alpha\theta\}\}dz. \tag{82}$$

However, now the contour $S = \Gamma^* \bigcup \Gamma_4$ completely satisfies the conditions of the Lemma 3: it starts in the point $z = 0$ without crossing the line of the cut and tends to infinity at $\varphi = \pi/2$.

For the contours of the seventh froup $\Gamma_7$ we do the same. Consider an auxiliary contour $\Gamma^{**} = \{z : \arg z = \pi/2,\ 0 \leqslant |z| \leqslant (x/\alpha)^{1/(\alpha-1)}\}$. With the help of this contour we form the contour $S^* = \Gamma^{**} \bigcup \Gamma_7$. Now, the contour $S^*$ completely satisfies the conditions of the Lemma 3. As in the previous case $\Im \exp\{izx - z^\alpha \exp\{i\frac{\pi}{2}\alpha\theta\}\} = 0$ for $z \in \Gamma^{**}$. From this it follows

$$\frac{1}{\pi}\Re \int_{\Gamma^{**}} \exp\{izx - z^\alpha \exp\{i\tfrac{\pi}{2}\alpha\theta\}\}dz = 0.$$

As a result, for the case $1 < \alpha \leqslant 2, \theta = 2/\alpha - 1$ we get

$$\frac{1}{\pi}\Re \int_{S^*} \exp\{izx - z^\alpha \exp\{i\tfrac{\pi}{2}\alpha\theta\}\}dz = \frac{1}{\pi}\Re \int_{\Gamma_7} \exp\{izx - z^\alpha \exp\{i\tfrac{\pi}{2}\alpha\theta\}\}dz. \tag{83}$$

Now, applying Lemma 3 and taking account of the equalities in Equations (82) and (83) we find that

$$\frac{1}{\pi}\Re \int_0^\infty e^{itx}\hat{g}(t,\alpha,-\theta)dt = \frac{1}{\pi}\int_\Gamma e^{izx} g^+(z,\alpha,-\theta)dz, \tag{84}$$

where the contour $\Gamma$ is defined by the expression in Equation (80). As a result, an improper integral along the positive part of the real axis in the expression of Equation (73) can be replaced with an integral along the contour $\Gamma$. Thus, in the case considered ($\alpha \neq 1$) we showed the validity of transition from Equation (73) to Equation (75).

Returning to Equation (75), taking into consideration Equation (76) and representing complex $z$ in the form $z = re^{i\varphi}$, we obtain

$$g(z,\alpha,\theta) = \frac{1}{\pi}\Re \int_\Gamma \exp\{izx - z^\alpha \exp\{i\tfrac{\pi}{2}\alpha\theta\}\}dz = \frac{1}{\pi}\int_\Gamma \Re \exp\{izx - z^\alpha \exp\{i\tfrac{\pi}{2}\alpha\theta\}\}d(\Re z) =$$
$$\frac{1}{\pi}\int_\Gamma \exp\{-rx\sin\varphi - r^\alpha \cos(\alpha(\varphi + \tfrac{\pi}{2}\theta))\} \cos\left(rx\cos\varphi - r^\alpha \sin(\alpha(\varphi + \tfrac{\pi}{2}\theta))\right) d[r\cos\varphi].$$

We will place here the expression in Equation (78). One should pay attention that in this case the Equation (77) is valid, consequently, we obtain

$$g(x,\alpha,\theta) = \frac{1}{\pi}\int_{\Gamma} \exp\{A(\varphi)\} d[r(\varphi)\cos\varphi], \tag{85}$$

where $A(\varphi) = -r(\varphi)x\sin\varphi - r(\varphi)^{\alpha}\cos(\alpha(\varphi + \frac{\pi}{2}\theta))$ and $r(\varphi)$ is defined by Equation (78). We transform the function $A(\varphi)$

$$\begin{aligned} A(\varphi) &= -x\left(\frac{\sin(\alpha(\varphi+\frac{\pi}{2}\theta))}{x\cos(\varphi)}\right)^{1/(1-\alpha)}\sin\varphi - \left(\frac{\sin(\alpha(\varphi+\frac{\pi}{2}\theta))}{x\cos(\varphi)}\right)^{\alpha/(1-\alpha)}\cos(\alpha(\varphi+\tfrac{\pi}{2}\theta)) \\ &= -x^{\alpha/(\alpha-1)}\left(\frac{\sin(\alpha(\varphi+\frac{\pi}{2}\theta))}{\cos(\varphi)}\right)^{\alpha/(1-\alpha)}\frac{\cos\left(\varphi(1-\alpha)-\frac{\pi}{2}\alpha\theta\right)}{\cos\varphi} = -x^{\alpha/(\alpha-1)}U(\varphi,\alpha,\theta), \end{aligned} \tag{86}$$

where $U(\varphi,\alpha,\theta)$ has the form of Equation (70).

Now we consider the differential, we have $d[r(\varphi)\cos\varphi] = \cos\varphi d[r(\varphi)] - r(\varphi)\sin\varphi d\varphi$. For the differential $d[r(\varphi)]$ we get

$$d[r(\varphi)] = d\left[\left(\frac{\sin(\alpha(\varphi+\frac{\pi}{2}\theta))}{x\cos(\varphi)}\right)^{1/(1-\alpha)}\right] = \frac{r(\varphi)}{1-\alpha}\left(\alpha\cot\left(\alpha\left(\varphi+\tfrac{\pi}{2}\theta\right)\right)+\tan\varphi\right)d\varphi.$$

Using now this result in the expression for $d[r(\varphi)\cos\varphi]$, we have

$$\begin{aligned} d[r(\varphi)\cos\varphi] &= \frac{r(\varphi)}{1-\alpha}\cos\varphi\left(\alpha\cot\left(\alpha\left(\varphi+\tfrac{\pi}{2}\theta\right)\right)+\tan\varphi\right)d\varphi - r(\varphi)\sin\varphi d\varphi \\ &= \frac{\alpha r(\varphi)}{1-\alpha}\left(\cos\varphi\cot\left(\alpha\left(\varphi+\tfrac{\pi}{2}\theta\right)\right)+\sin\varphi\right)d\varphi \\ &= \frac{\alpha}{1-\alpha}\left(\frac{\sin\left(\alpha\left(\varphi+\frac{\pi}{2}\theta\right)\right)}{x\cos\varphi}\right)^{1/(1-\alpha)}\frac{\cos\left(\alpha\left(\varphi+\frac{\pi}{2}\theta\right)-\varphi\right)}{\sin\left(\alpha\left(\varphi+\frac{\pi}{2}\theta\right)\right)}d\varphi = \frac{\alpha}{1-\alpha}x^{1/(\alpha-1)}U(\varphi,\alpha,\theta)d\varphi, \end{aligned} \tag{87}$$

where $U(\varphi,\alpha,\theta)$ has the form of Equation (70).

Now using Equations (86) and (87) in Equation (85) and also taking into consideration that the motion along the contour $\Gamma$, having the form of Equation (80) now described by the parameter change $\varphi$, we obtain

$$g(x,\alpha,\theta) = \frac{\alpha}{\pi(1-\alpha)}\int_{\varphi_{min}}^{\varphi_{max}} \exp\left\{-x^{\alpha/(\alpha-1)}U(\varphi,\alpha,\theta)\right\} x^{1/(\alpha-1)}U(\varphi_{\alpha},\theta)d\varphi \tag{88}$$

From the expression in Equation (84) one can see that the integration limits $\varphi_{min}$ and $\varphi_{max}$ should be selected in such a way that the motion along the contour $\Gamma$ could correspond to a change from $r=0$ to $r=\infty$. From the expression in Equation (81) it is clear that in the case $0<\alpha<1$ a change in the angle $\varphi$ from $-\frac{\pi}{2}\theta$ to $\frac{\pi}{2}$ corresponds to the motion along the contour $\Gamma$ from $r=0$ to $r=\infty$. Therefore, in this case $\varphi_{min} = -\frac{\pi}{2}\theta$, $\varphi_{max} = \frac{\pi}{2}$. In the case $1<\alpha\leqslant 2$ a change in the angle $\varphi$ from $\frac{\pi}{2}$ to $-\frac{\pi}{2}\theta$ corresponds to the motion from $r=0$ to $r=\infty$. Therefore, in this case $\varphi_{min} = \frac{\pi}{2}$, and $\varphi_{max} = -\frac{\pi}{2}\theta$. Combining these two cases the expression in Equation (88) takes the form

$$g(x,\alpha,\theta) = \frac{\alpha}{\pi|1-\alpha|}\int_{-\frac{\pi}{2}\theta}^{\frac{\pi}{2}} \exp\left\{-x^{\alpha/(\alpha-1)}U(\varphi,\alpha,\theta)\right\} x^{1/(\alpha-1)}U(\varphi,\alpha,\theta)d\varphi, \quad \alpha\neq 1. \tag{89}$$

It should be pointed out that this formula was obtained on the assumption $x > 0$. The case $x < 0$ is easy to obtain using the property of inversion for the density of probabilities in Equation (7). For this it is enough to replace the parameter $\theta$ in the formula in Equation (89) with $-\theta$. It is possible to combine these two cases if in the formula of Equation (89) to substitute the parameter $\theta$ for the parameter $\theta^* = \theta \operatorname{sign} x$ and the value $x$ is taken in absolute value. As a result, we obtain the expression in Equation (69) valid for any $x \neq 0$ and any admissible $\alpha \neq 1$.

Now we consider the case $0 < \alpha \leqslant 2$ and $x = 0$. In this case the inversion formula of Equation (75) takes the form

$$g(0, \alpha, \theta) = \frac{1}{\pi} \Re \int_\Gamma \exp\left\{-z^\alpha \exp\left\{\tfrac{\pi}{2}\alpha\theta\right\}\right\} dz, \tag{90}$$

where the contour $\Gamma$ is determined by the expression in Equation (74), but it is necessary to determine the specific type of a contour in the case under consideration.

We represent the complex number $z$ in the form $z = re^{i\varphi}$. As a result, the integrand in Equation (90) takes the form

$$\exp\left\{-z^\alpha \exp\left\{\tfrac{\pi}{2}\alpha\theta\right\}\right\} = \exp\left\{-r^\alpha \cos\left(\alpha\left(\varphi + \tfrac{\pi}{2}\theta\right)\right) - ir^\alpha \sin\left(\alpha\left(\varphi + \tfrac{\pi}{2}\theta\right)\right)\right\}. \tag{91}$$

Using the condition $\Im g^+(0, \alpha, \theta) = 0$, we obtain

$$r^\alpha \sin\left(\alpha\left(\varphi + \tfrac{\pi}{2}\theta\right)\right) = 0. \tag{92}$$

This equation has two solutions: $r = 0$, for any $\varphi$, and $\varphi = -\pi\theta/2$, $r \geqslant 0$. It is clear that if $r = 0$, then $\Im g^+(0, \alpha, \theta) = 0$ for any value of $\varphi$, for definiteness we will select $\varphi = -\pi\theta/2$ if $r = 0$. As a result, the contour of integration of Equation (74) takes the form $\Gamma = \{z : |z| \geqslant 0, \arg z = -\pi\theta/2\}$, where $|\theta| \leqslant \min(1, 2/\alpha - 1)$, $0 < \alpha \leqslant 2$. From this it is clear that in the case under consideration the family of contours $\Gamma$ are represented by half-lines starting from the point $z = 0$ at an angle $\arg z = -\pi\theta/2$. Consequently, the family of contours $\Gamma$ satisfy items 1 and 2 of Lemma 3 and also the condition in Equation (53) in item 3. Thus, we substantiate the transition from the improper integral of Equation (73) to the contour integral in Equation (90).

Now we put the expression in Equation (91) in Equation (90) and taking into account Equation (92), we obtain

$$g(0, \alpha, \theta) = \frac{1}{\pi} \int_\Gamma \exp\left\{-r^\alpha \cos\left(\alpha\left(\varphi + \tfrac{\pi}{2}\theta\right)\right)\right\} \Re[dz] = \frac{\cos(\pi\theta/2)}{\pi} \int_0^\infty \exp\left\{-r^\alpha\right\} dr.$$

Here it was taken into account that $\Re[dz] = \cos(\pi\theta/2)dr$ on the contour $\Gamma$. The limits of integration were selected in such a way that when moving along the contour $r$ could change from 0 to $\infty$. Since the contour $\Gamma$ in the case under consideration is a half-line coming out of the point $z = 0$ at an angle $\arg z = -\pi\theta/2$, then the motion along the contour $\Gamma$ from 0 to $\infty$ corresponds to a change $r$ from 0 to $\infty$. Making in this integral a substitution of a variable $r^\alpha = y$ and using the definition of the gamma-function $\Gamma(n) = \int_0^\infty t^{n-1}e^{-t}dt$, $n > 0$ we obtain the expression Equation (71). Thus, the second item of the theorem is proved.

Now we consider the case $\alpha = 1$. We will make an assumption that $x \geqslant 0$. In this case, the inversion formula in Equation (75) takes the form

$$g(x, 1, \theta) = \frac{1}{\pi} \Re \int_\Gamma \exp\left\{izx - z\exp\left\{i\tfrac{\pi}{2}\theta\right\}\right\} dz, \tag{93}$$

where the contour of integration defined by Equation (74) will be written in the form

$$\Gamma = \{z:\ \Im e^{izx} g^{+}(z,1,\theta) = 0,\ \tfrac{\pi}{2} \leqslant \arg z \leqslant \tfrac{\pi}{2},\ |z| \geqslant 0\}, \tag{94}$$

Here it should be noted that analytic continuation of the function $\hat{g}(t,1,-\theta)$ from the positive part of the real axis $t$ to the complex plane $z$ at $\alpha = 1$ is an analytic function and it has the form $g^{+}(z,1,-\theta) = \exp\left\{-z\exp\left\{i\frac{\pi}{2}\theta\right\}\right\}$.

As in the previous case, we begin by defining the form of the integration contour $\Gamma$. Consider the integrand in Equation (93) and we represent the complex number $z$ in the form $z = re^{i\varphi}$. As a result, we obtain

$$\exp\{izx - z\exp\{i\tfrac{\pi}{2}\theta\}\} = \exp\left\{ixre^{i\varphi} - re^{i\varphi}\exp\{i\tfrac{\pi}{2}\theta\}\right\} = \\ \exp\{-r\left(x\sin\varphi + \cos((\varphi + \tfrac{\pi}{2}\theta))\right)\}\exp\left\{ir\left(x\cos\varphi - \sin((\varphi + \tfrac{\pi}{2}\theta))\right)\right\}. \tag{95}$$

From this expression we get that the condition $\Im e^{izx} g^{+}(z,1,-\theta) = 0$ leads to an equation

$$r\left(x\cos\varphi - \sin((\varphi + \tfrac{\pi}{2}\theta))\right) = 0. \tag{96}$$

This equation has two solutions. The first solution is $r = 0$. The second solution we obtain from the equation $x\cos\varphi - \sin((\varphi + \frac{\pi}{2}\theta)) = 0$, $r \geqslant 0$. Solving it with respect to $\varphi$ we get

$$\varphi(\theta,x) = \arctan\left(\frac{x - \sin(\pi\theta/2)}{\cos(\pi\theta/2)}\right). \tag{97}$$

As a result, the contour of integration in Equation (94) takes the form

$$\Gamma = \{z:\ \arg z = \varphi(\theta,x),\ |z| \geqslant 0, -1 \leqslant \theta \leqslant 1\} \tag{98}$$

Thus, the contours $\Gamma$ with different values $\theta$ are half-lines coming out of the origin at an angle $\varphi(\theta,x)$ and tending to infinity.

Next, it is necessary to substantiate the transition from the integral in Equation (73) to the integral along the contour in Equation (93). For this we will use Lemmas 3 and 4. From the definition in Equation (98), it is clear that for all admissible values $\theta$ and $x \geqslant 0$ the contour $\Gamma$ satisfies item 1 and 2 of the Lemma 3. There is only one thing left, to find out if the contour in Equation (98) satisfies item 3 of this lemma.

Consider the case $x > 1$ at first. According to item 3 of the Lemma 3 for the equality in Equation (59) to be valid the contour of Equation (98) must satisfy the condition in Equation (54). Now using the Lemma 4. According to this lemma in the case $x \geqslant 1$ the functions $\varphi(\theta,x)$ and $\varphi_0(\theta,x)$ are connected between each other with a ratio (63) from which it directly follows that $\varphi_0(\theta,x) < \varphi(\theta,x)$ for all $\theta \in [-1,1]$. Therefore, at $x > 1$ the condition in Equation (54) is met and we can move from the integral in Equation (73) to the integral in Equation (93). In the case $x = 1$ the contour $\Gamma$ must meet the condition in Equation (58). Now using the Lemma 4 we obtain that in this case for all $\theta \in [-1,1]$ the inequality $\varphi_0(\theta,x) < \varphi(\theta,x)$ is true. Therefore, in this case the condition in Equation (58) is satisfied. Now we consider the case $0 \leqslant x < 1$. Similar to previous case, applying the Lemma 4 namely, the formula in Equation (64) we get that in this case the contour in Equation (98) satisfies the conditions in Equations (55)–(57) Lemma 3. Thus, the contour in Equation (98) completely satisfies the conditions of the Lemma 3 and therefore, the equality

$$\frac{1}{\pi}\Re\int_0^{\infty} e^{itx}\hat{g}(t,1,-\theta)dt = \frac{1}{\pi}\int_{\Gamma} e^{izx}g^{+}(z,1,-\theta)dz$$

is true. This makes it possible to replace the improper integral in the expression of Equation (73) over the real variable with the integral over the contour $\Gamma$. Thus, the possibility for the transition from the formula in Equation (73) to the formula in Equation (75) is substantiated. We need to note that in the case under consideration $\alpha = 1$ the expression in Equation (75) takes the form of Equation (93). It should also be pointed out that that due to Equation (34) the case $\alpha = 1, \theta = 1, x = 1$ is excluded in the Lemma 3. That is why, here, this case should also be excluded from consideration.

Now taking into account Equation (95), the expression of Equation (93) will be written in the following form

$$g(x,1,\theta) = \frac{1}{\pi}\Re\int_{\Gamma}\exp\{-r\left(x\sin\varphi + \cos(\varphi + \tfrac{\pi}{2}\theta)\right)\}\exp\left\{ir\left(x\cos\varphi - \sin(\varphi + \tfrac{\pi}{2}\theta)\right)\right\}dz =$$
$$\frac{1}{\pi}\int_{\Gamma}\exp\{-r\left(x\sin\varphi + \cos(\varphi + \tfrac{\pi}{2}\theta)\right)\}\cos\left\{r\left(x\cos\varphi - \sin(\varphi + \tfrac{\pi}{2}\theta)\right)\right\}\Re[dz]$$

Now using here the definition of Equation (98) we obtain that $\Re[dz] = \cos(\varphi(\theta,x))dr$. The motion along the contour $\Gamma$ should take place in such a way that it would start in the point $z = 0$ in the process of moving it would tend to infinity. Therefore, the motion from $r = 0$ to $r = \infty$ corresponds to such motion. Taking into consideration that on the contour $\Gamma$ the condition of Equation (96) is met, we obtain

$$g(x,1,\theta) = \frac{1}{\pi}\int_0^{\infty}\exp\{-r\left(x\sin\varphi(\theta,x) + \cos(\varphi(\theta,x) + \tfrac{\pi}{2}\theta)\right)\}\cos(\varphi(\theta,x))dr$$

As we can see, the integral obtained is easy to calculate. As a result, we obtain

$$g(x,1,\theta) = \frac{1}{\pi}\frac{\cos(\varphi(\theta,x))}{x\sin\varphi(\theta,x) + \cos(\varphi(\theta,x) + \pi\theta/2)}.$$

Using now the definition of the function $\varphi(\theta,x)$ (97) after simple transformations we get

$$g(x,1,\theta) = \frac{\cos(\pi\theta/2)}{\pi\left(x^2 - 2x\sin(\pi\theta/2) + 1\right)}. \quad (99)$$

Recall that consideration was carried out for the case $x \geqslant 0$. The case $x < 0$ can be obtained using the inversion property for the density of probability of Equation (7). For this it is enough to replace $\theta$ with $-\theta$. It is possible to combine these two cases if to introduce a parameter $\theta^* = \theta\,\mathrm{sign}\,x$. However, we want to note that if to perform this replacement in the expression of Equation (99), then the expression itself will not change $g(x,1,\theta) \to g(x,1,\theta^*) \equiv g(x,1,\theta)$. Consequently, the formula in Equation (99) is true for any $x$. Thus, the theorem is completely proved. □

We will make some remarks on the proved theorem.

**Remark 2.** *The proof of the case $\alpha = 1$ was carried out under the assumption $x \geqslant 0$. Therefore, the formula in Equation (99) was obtained for the case $x \geqslant 0$. The generalization of this formula for the case $x < 0$ was carried out using the inversion property $g(-x,1,\theta) = g(x,1,-\theta)$. It should be noted here that in the process of proving this case the point $\theta = 1$, $x = 1$ was excluded from consideration. Therefore, in view of the inversion property,* *the point $\theta = -1$, $x = -1$ should also be excluded. In these two points the density $g(x, 1, \theta)$ has a peculiarity. In fact, substituting the value $\theta = \pm 1$, we obtain*

$$g(x, 1, \pm 1) = \frac{\cos(\pi/2)}{\pi(x \mp 1)^2} = \begin{cases} 0, & x \neq \pm 1, \\ 0/0, & x = \pm 1. \end{cases}$$

*However, an indeterminate form $0/0$ can be evaluated in the following remark.*

**Remark 3.** *Equations (71) and (72) can be obtained directly from the inversion formulas in Equation (9) without resorting to analytic continuation of the characteristic function $\hat{g}(t, \alpha, \theta)$ to the complex plane with the subsequent transition from the improper integral over the real variable (73) to the integral along the contour in Equation (75). We first consider the case $\alpha = 1$. Using the inversion formula (the first formula in Equation (9)) we obtain*

$$g(x, 1, \theta) = \frac{1}{\pi} \Re \int_0^\infty e^{itx} \hat{g}(t, 1, -\theta) dt = \frac{1}{\pi} \Re \int_0^\infty \exp\left\{t\left(ix - \exp\left\{i\tfrac{\pi}{2}\theta\right\}\right)\right\} dt = \frac{1}{\pi} \Re I, \tag{100}$$

*where $I = \int_0^\infty \exp\left\{t\left(ix - \exp\left\{i\frac{\pi}{2}\theta\right\}\right)\right\} dt$. We determine under which conditions this integral will converge. For the integral $I$ the inequality is valid*

$$|I| \leqslant \int_0^\infty \left|\exp\left\{t\left(ix - \exp\left\{i\tfrac{\pi}{2}\theta\right\}\right)\right\}\right| dt \tag{101}$$

*For the integrand we have*

$$\left|\exp\left\{t\left(ix - \exp\left\{i\tfrac{\pi}{2}\theta\right\}\right)\right\}\right| = \exp\{-t\cos(\pi\theta/2)\} \cos(t(x - \sin(\pi\theta/2))). \tag{102}$$

*Since in the case considered ($\alpha = 1$) the parameter $\theta$ varies within the limits $-1 \leqslant \theta \leqslant 1$, we obtain $\cos(\pi\theta/2) > 0$, if $-1 < \theta < 1$. From here we get $\lim_{t\to\infty} \left|\exp\left\{t\left(ix - \exp\left\{i\frac{\pi}{2}\theta\right\}\right)\right\}\right| = 0$, if $-1 < \theta < 1$. Thus, the integral in Equation (101) will converge, and, therefore, and the integral $I$ will also converge at $-1 < \theta < 1$.*

*We consider now the cases $\theta = \pm 1$. If $\theta = 1$, then from Equation (102) we obtain*

$$|\exp\{t(ix - \exp\{i\pi/2\})\}| = \cos(t(x-1)). \tag{103}$$

*Substituting this result in Equation (101) we get*

$$|I| \leqslant \int_0^\infty \cos(t(x-1)) dt = \begin{cases} 0, & x \neq 1 \\ \infty, & x = 1. \end{cases} \tag{104}$$

*Similarly, if $\theta = -1$, we have*

$$|I| \leqslant \int_0^\infty \cos(t(x+1)) dt = \begin{cases} 0, & x \neq -1 \\ \infty, & x = -1. \end{cases} \tag{105}$$

*Thus, at $\theta = \pm 1$ the integral $I = 0$ for all $x \neq \pm 1$, and integral $I$ will diverge in the points $x = 1, \theta = 1$ and $x = -1$, $\theta = -1$.*

*Returning to Equation (100) and by calculating the integral directly, we get*

$$g(x,1,\theta) = \frac{1}{\pi}\Re\int_0^\infty \exp\left\{t\left(ix - \exp\left\{i\tfrac{\pi}{2}\theta\right\}\right)\right\}dt = -\frac{1}{\pi}\Re\frac{1}{ix-\exp\{i\pi\theta/2\}}$$
$$= -\frac{1}{\pi}\Re\frac{-ix-\exp\{-i\pi\theta/2\}}{(ix-\exp\{i\pi\theta/2\})(-ix-\exp\{-i\pi\theta/2\})} = \frac{1}{\pi}\frac{\cos(\pi\theta/2)}{x^2-2x\sin(\pi\theta/2)+1}. \quad (106)$$

*From the formula obtained it is clear that if $\theta = \pm 1$*

$$g(x,1,\pm 1) = \frac{\cos(\pi/2)}{\pi(x\mp 1)^2} = \begin{cases} 0, & x \neq \pm 1 \\ \infty, & x = \pm 1. \end{cases} \quad (107)$$

*Thus the behavior of the formula in Equation (106) coincides with the behavior of the integral $I$ in Equations (104) and (105) in the cases $\theta = \pm 1$. Therefore, the formula in Equation (106) is true for any $-1 \leqslant \theta \leqslant 1$. The expression in Equation (107) means that the density $g(x,1,\pm 1)$ is a degenerate distribution in the point $x = \pm 1$. In other words,*

$$g(x,1,\pm 1) = \delta(x \mp 1). \quad (108)$$

Thus, the obtained expression in Equation (106) completely coincides with the one previously obtained in the Theorem 1 the density in Equation (72), and the conclusion presented in this remark is an alternative way of deducing this density.

**Remark 4.** *By a similar method, one can obtain the density value at $x = 0$. Using the first formula in Equation (9) and making a substitution of the integration variable $t^\alpha = \tau$ we get*

$$g(0,\alpha,\theta) = \frac{1}{\pi}\Re\int_0^\infty \hat{g}(t,\alpha,-\theta)dt = \frac{1}{\pi}\Re\int_0^\infty \exp\left\{-t^\alpha \exp\left\{i\tfrac{\pi}{2}\alpha\theta\right\}\right\}dt$$
$$= \frac{1}{\alpha\pi}\Re\int_0^\infty \exp\left\{-\tau\cos\left(\tfrac{\pi}{2}\alpha\theta\right) - i\tau\sin\left(\tfrac{\pi}{2}\alpha\theta\right)\right\}\tau^{1/\alpha-1}d\tau$$
$$= \frac{1}{\alpha\pi}\int_0^\infty \exp\left\{-\tau\cos\left(\tfrac{\pi}{2}\alpha\theta\right)\right\}\cos\left(y\sin\left(\tfrac{\pi}{2}\alpha\theta\right)\right)\tau^{1/\alpha-1}d\tau. \quad (109)$$

*To calculate the integral obtained it is necessary to use the formula (see [45], Section 1.5. the Equation (35))*

$$\int_0^\infty t^{\alpha-1}e^{-ct\cos\beta}\cos(ct\sin\beta)dt = \Gamma(\alpha)c^{-\alpha}\cos(\alpha\beta), \quad c>0,\ \Re\alpha>0,\ -\pi/2<\beta<\pi/2.$$

*It is clear that the integral in Equation (109) completely satisfies the conditions of this integral. Consequently, using it in Equation (109) we get $g(0,\alpha,\theta) = \frac{1}{\pi}\Gamma(1/\alpha+1)\cos(\pi\theta/2)$. The formula obtained coincides completely with the formula in Equation (71).*

We will make another useful remark.

**Remark 5.** *The Theorem 1 formulates an integral representation for the density of a standard strictly stable law. However, it is useful to have a formula that allows one to convert the density of a standard strictly stable law to the density of a strictly stable law with arbitrary $\lambda$. The Property 2 and, in particular, the formula in Equation (8) allows one to obtain such a formula.*

*In fact, in terms of characteristic functions, the formula in Equation (8) will be written as*

$$\hat{g}(t,\alpha,\theta,\lambda) = \hat{g}\left(\lambda^{1/\alpha}t,\alpha,\theta\right). \tag{110}$$

*The density is obtained using the inverse Fourier transform of the characteristic function* $g(x,\alpha,\theta,\lambda) = \frac{1}{2\pi}\int_{-\infty}^{\infty} e^{itx}\hat{g}(t,\alpha,\theta,\lambda)dt$*. Using now the relation in Equation (110) and changing the integration variable* $\lambda^{1/\alpha}t = \tau$ *we obtain the relation for densities.*

$$g(x,\alpha,\theta,\lambda) = \lambda^{-1/\alpha}g(x\lambda^{-1/\alpha},\alpha,\theta). \tag{111}$$

Thus, the Theorem 1 defines an integral representation for the probability density of a standard strictly stable law with the characteristic function in Equation (1). Using this representation, we can obtain an integral representation for the distribution function of the standard strictly stable law. We formulate this result as a corollary to the Theorem 1.

**Corollary 1.** *The distribution function of a stable law* $G(x,\alpha,\theta)$ *with the characteristic function in Equation (1) can be represented as*

1. *If* $\alpha \neq 1$*, then for any* $|\theta| \leqslant \min(1, 2/\alpha - 1)$ *and* $x \neq 0$

$$G(x,\alpha,\theta) = \tfrac{1}{2}(1 - \operatorname{sign}(x)) + \operatorname{sign}(x)G^{(+)}(|x|,\alpha,\theta^*), \tag{112}$$

*where* $\theta^* = \theta \operatorname{sign}(x)$,

$$G^{(+)}(x,\alpha,\theta) = 1 - \frac{(1+\theta)}{4}(1 + \operatorname{sign}(1-\alpha)) + \frac{\operatorname{sign}(1-\alpha)}{\pi}\int_{-\pi\theta/2}^{\pi/2} \exp\left\{-x^{\alpha/(\alpha-1)}U(\varphi,\alpha,\theta)\right\} d\varphi, \quad x > 0, \tag{113}$$

*and* $U(\varphi,\alpha,\theta)$ *is defined by Equation (70).*

2. *If* $\alpha = 1$*, then for any* $-1 \leqslant \theta \leqslant 1$ *and any* $x$

$$G(x,1,\theta) = \frac{1}{2} + \frac{1}{\pi}\arctan\left(\frac{x - \sin(\pi\theta/2)}{\cos(\pi\theta/2)}\right). \tag{114}$$

3. *If* $x = 0$*, then for any admissible* $\alpha$ *and* $\theta$

$$G(0,\alpha,\theta) = (1-\theta)/2. \tag{115}$$

**Proof.** We consider the case $\alpha \neq 1$. It is necessary to obtain the distribution function of a stable law with the characteristic function in Equation (1). For the density of distribution a stable law, one should choose the expression in Equation (89) which defines the density $g(x,\alpha,\theta)$ for $x > 0$. In view of this, we write the distribution function in the form

$$G^{(+)}(x,\alpha,\theta) = 1 - \int_x^{\infty} g(\xi,\alpha,\theta)d\xi, \quad x > 0.$$

Now substituting here Equation (89) we find

$$G^{(+)}(x,\alpha,\theta) = 1 - \frac{\alpha}{\pi|1-\alpha|}\int_{-\pi\theta/2}^{\pi/2} U(\varphi,\alpha,\theta)d\varphi \int_x^{\infty} \xi^{1/(\alpha-1)} \exp\left\{-\xi^{\alpha/(\alpha-1)}U(\varphi,\alpha,\theta)\right\} d\xi$$

$$= \begin{cases} 1 + \dfrac{\text{sign}\,(\alpha-1)}{\pi}\displaystyle\int_{-\pi\theta/2}^{\pi/2}\left(1-\exp\left\{-x^{\alpha/(\alpha-1)}U(\varphi,\alpha,\theta)\right\}\right)d\varphi, & \alpha<1,\ x>0, \\ 1 - \dfrac{\text{sign}\,(\alpha-1)}{\pi}\displaystyle\int_{-\pi\theta/2}^{\pi/2}\exp\left\{-x^{\alpha/(\alpha-1)}U(\varphi,\alpha,\theta)\right\}d\varphi, & \alpha>1,\ x>0. \end{cases}$$

$$= \begin{cases} 1 - \dfrac{(1+\theta)}{2} + \dfrac{1}{\pi}\displaystyle\int_{-\pi\theta/2}^{\pi/2}\exp\left\{-x^{\alpha/(\alpha-1)}U(\varphi,\alpha,\theta)\right\}d\varphi, & \alpha<1,\ x>0, \\ 1 - \dfrac{1}{\pi}\displaystyle\int_{-\pi\theta/2}^{\pi/2}\exp\left\{-x^{\alpha/(\alpha-1)}U(\varphi,\alpha,\theta)\right\}d\varphi, & \alpha>1,\ x>0. \end{cases}$$

If we combine the cases $\alpha < 1$ and $\alpha > 1$ we obtain

$$G^{(+)}(x,\alpha,\theta) = 1 - \frac{(1+\theta)}{4}(1+\text{sign}\,(1-\alpha)) + \frac{\text{sign}\,(1-\alpha)}{\pi}\int_{-\pi\theta/2}^{\pi/2}\exp\left\{-x^{\alpha/(\alpha-1)}U(\varphi,\alpha,\theta)\right\}d\varphi.$$

This formula defines the distribution function of stable law for the case $x > 0$ and $\alpha \neq 1$. The case $x < 0$ is reduced to the case $x > 0$ using the property of an inversion, namely, the formula in Equation (7) for $G(x,\alpha,\theta)$. As a result, we get $G^{(-)}(-x,\alpha,\theta) = 1 - G^{(+)}(x,\alpha,-\theta)$, $x > 0$. This formula gives the distribution function for negative $x$. Combining the formulas for $G^{(+)}(x,\alpha,\theta)$ and $G^{(-)}(-x,\alpha,\theta)$, we obtain the formula in Equation (112) which is true for any $x \neq 0$ and $\alpha \neq 1$.

We now consider the case $\alpha = 1$. According to the definition $G(x,1,\theta) = \int_{-\infty}^{x} g(\xi,1,\theta)d\xi$. Substituting the density (72) here and replacing the variable $\xi - \sin(\pi\theta/2) = \tau$ we obtain

$$G(x,1,\theta) = \frac{\cos(\pi\theta/2)}{\pi}\int_{-\infty}^{x}\frac{d\xi}{\xi^2 - 2\xi\sin(\pi\theta/2)+1}$$

$$= \frac{\cos(\pi\theta/2)}{\pi}\int_{-\infty}^{x-\sin(\pi\theta/2)}\frac{d\tau}{\tau^2+\cos^2(\pi\theta/2)} = \frac{1}{2} + \frac{1}{\pi}\arctan\left(\frac{x-\sin(\pi\theta/2)}{\cos(\pi\theta/2)}\right).$$

Thus, the second item of the corollary has been proved.

To calculate $G(x,\alpha,\theta)$ at $x = 0$ we use the formula in Equation (113). Performing the passage to the limit $x \to 0$ in this expression we get $G(0,\alpha,\theta) = (1-\theta)/2$. Since the formula in Equation (113) is valid for $\alpha \neq 1$, then the result obtained is valid only for $\alpha \neq 1$. To calculate $G(0,1,\theta)$ it is necessary to use (114). Substituting the value $x = 0$ in (114) we obtain $G(0,1,\theta) = (1-\theta)/2$. Thus, the expression (115) is true for any admissible $\alpha$ and $\theta$. □

We make some remarks on the proved corollary.

**Remark 6.** *In the Remark 3 it is emphasized that in the case* $\alpha = 1, \theta = \pm 1$ *the density* $g(x,1,\pm 1)$ *is a degenerate distribution in the point* $x = \pm 1$ *and has the form of Equation (108). Consequently, for the indicated parameter values, the distribution function* $G(x,1,\pm 1)$ *will have the form of the Heaviside function* $G(x,1,\pm 1) = H(x \mp 1)$. *This is directly seen from the form of the distribution function at* $\alpha = 1$. *Indeed, substituting the values* $\theta = \pm 1$ *in Equation (114), we obtain*

$$G(x,1,\pm 1) = \frac{1}{2} + \frac{1}{\pi}\arctan\left(\frac{x \mp \sin(\pi/2)}{\cos(\pi/2)}\right) = \begin{cases} 1, & x > \pm 1, \\ 0, & x < \pm 1 \end{cases} = H(x \mp 1).$$

**Remark 7.** *The proved corollary gives an integral representation for the distribution function of a standard strictly stable law. In order to get the distribution function of a strictly stable law for an arbitrary $\lambda$ it is necessary to use the Remark 5. By definition $G(x,\alpha,\theta,\lambda) = \int_{-\infty}^{x} g(y,\alpha,\theta,\lambda)dy$. Using now the relation in Equation (111) and changing the variable $x\lambda^{-1/\alpha} = \tau$ we arrive at the relation*

$$G(x,\alpha,\theta,\lambda) = G\left(x\lambda^{-1/\alpha},\alpha,\theta\right).$$

It should be noted the density in Equation (72) and distribution function in Equation (114) is not new. These formulas were deduced by V.M. Zolotarev (see § 2.3 in [32]). As we can see, this distribution is the stable law for $\alpha = 1$ and any $-1 \leqslant \theta \leqslant 1$ and it is expressed in terms of elementary functions.

## 4. The Calculation of the Density and Distribution Function of a Stable Law

Integral representations for the probability density and distribution function of a stable law with the characteristic function in Equation (1) were obtained in the Theorem 1 and Corollary 1. These integral representations in Equations (69) and (112) express the probability density and the distribution function in terms of a definite integral. That is why, using the methods of numerical integration, it is possible to calculate the values of these integrals without much difficulty.

In this paper, to calculate definite integrals in Equations (69) and (112) we used the adaptive Gaussian–Kronrod numerical integration algorithm for 31 points. To implement the program for calculating the functions $g(x,\alpha,\theta)$ and $G(x,\alpha,\theta)$ we used the implementation of this algorithm in the gsl library (GNU Scientific Library) of version 1.8 [46]. The calculation results for the functions $g(x,\alpha,\theta)$ and $G(x,\alpha,\theta)$ by the Equations (69) and (112) are given in Figures 9–14.

The figures show the probability density and distribution function for the values of the characteristic exponent $\alpha = 0.3, 0.6, 0.9, 1.2, 1.5, 1.8$ and specified values of the parameter $\theta$ and $\lambda = 1$. It should be noted that the admissible region of the parameter $\theta$ is determined by the inequality $|\theta| \leqslant \min(1, 2/\alpha - 1)$. Thus, $-1 \leqslant \theta \leqslant 1$, if $0 < \alpha \leqslant 1$, and $-(2/\alpha - 1) \leqslant \theta \leqslant 2/\alpha - 1$, if $1 < \alpha \leqslant 2$. It is clear that if $\alpha > 1$ then the admissible region of the parameter $\theta$ narrows and at $\alpha = 2$ the parameter $\theta$ may take a single value $\theta = 0$.

Let us analyze the results presented in more detail. We first consider the case $\alpha < 1$. The results related to this case are given in Figures 9–11. From these figures it can be seen that when $\theta = 1$ the probability density is concentrated on the positive semiaxis. Thus, $g(x,\alpha,1) = 0$, $G(x,\alpha,1) = 0$, if $x < 0$, $\alpha < 1$. Similarly, for the case $\theta = -1$ and $\alpha < 1$, we obtain that the negative semiaxis is the area of concentration of the probability density. Consequently, $g(x,\alpha,-1) = 0$, $G(x,\alpha,-1) = 1$, if $x > 0$, $\alpha < 1$. This result is in complete agreement with remarks 3 and 4 on page 79 of theorem 2.2.3 from the book by [32].

In the introduction, it was noted that in the work [31] (see also [32]) integral representations were obtained for the probability density and distribution function of a stable law with the characteristic function in Equation (5). In order to avoid any confusion the parameters of a stable law with the characteristic function in Equation (5) will be designated as $(\alpha', \beta, \lambda', \gamma)$. For the parameters of a strictly stable law with the characteristic function in Equation (1) we keep the notation $(\alpha, \theta, \lambda)$. The parameters $(\alpha', \beta, \lambda', \gamma)$ are related to the parameters $(\alpha, \theta, \lambda)$ by the relations (see [32,36]): $\alpha' = \alpha$,

$$\begin{aligned} &\theta = \beta K(\alpha')/\alpha', && \lambda = \lambda', && \text{if} \quad \alpha \neq 1, \\ &\theta = (2/\pi)\arctan(2\gamma/\pi), && \lambda = \lambda'\left(\pi^2/4 + \gamma^2\right)^{1/2}, && \text{if} \quad \alpha = 1. \end{aligned}$$

In the case of $\alpha \neq 1$ from the relation for the parameters $\theta$ and $\beta$ we get $\theta = \beta$, if $\alpha < 1$, and $\theta = \beta(1 - 2/\alpha)$, if $\alpha > 1$.

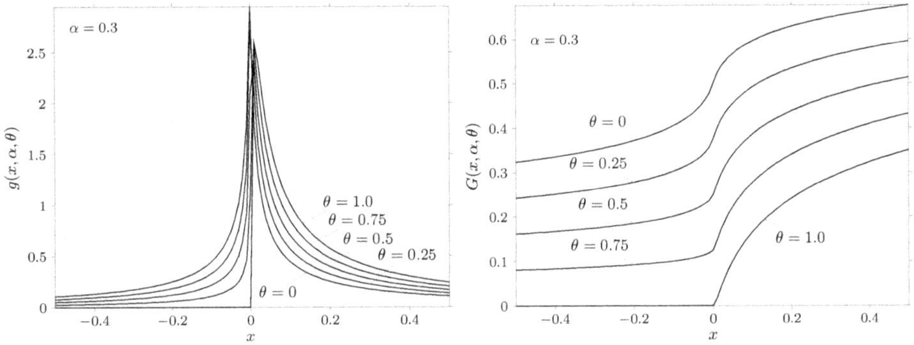


**Figure 9.** The probability density function $g(x, \alpha, \theta)$ (**on the left**) and cumulative distribution function $G(x, \alpha, \theta)$ (**on the right**) of a standard strictly stable law with $\alpha = 0.3$ and specified values $\theta$.

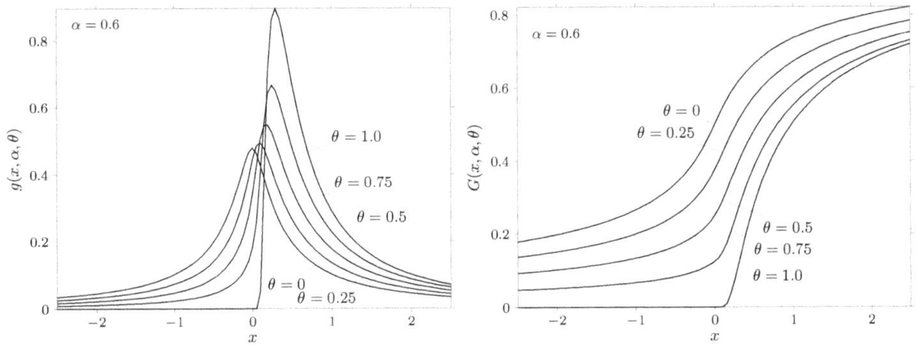


**Figure 10.** The probability density function $g(x, \alpha, \theta)$ (**on the left**) and cumulative distribution function $G(x, \alpha, \theta)$ (**on the right**) of a standard strictly stable law with $\alpha = 0.6$ and specified values $\theta$.

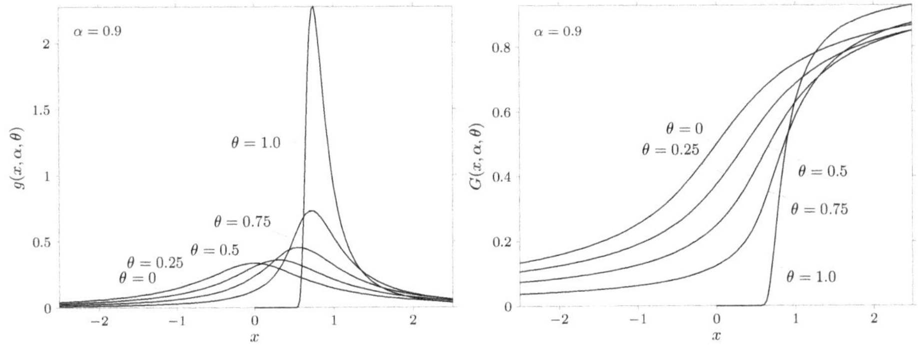


**Figure 11.** The probability density function $g(x, \alpha, \theta)$ (**on the left**) and cumulative distribution function $G(x, \alpha, \theta)$ (**on the right**) of a standard strictly stable law with $\alpha = 0.9$ and specified values $\theta$.

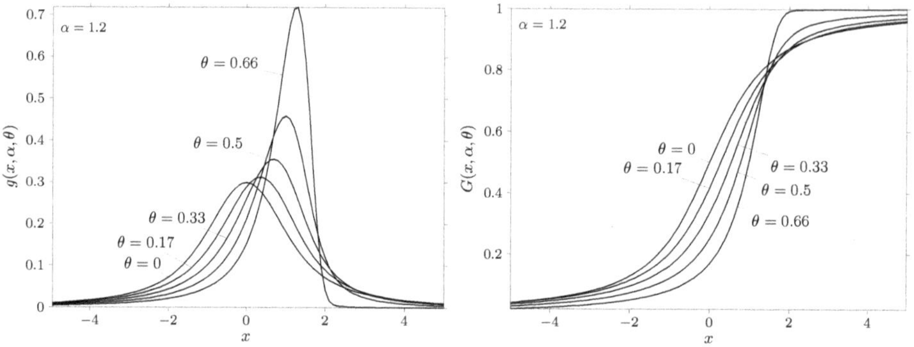


**Figure 12.** The probability density function $g(x, \alpha, \theta)$ (**on the left**) ) and cumulative distribution function $G(x, \alpha, \theta)$ (**on the right**) of a standard strictly stable law with $\alpha = 1.2$ and specified values $\theta$.

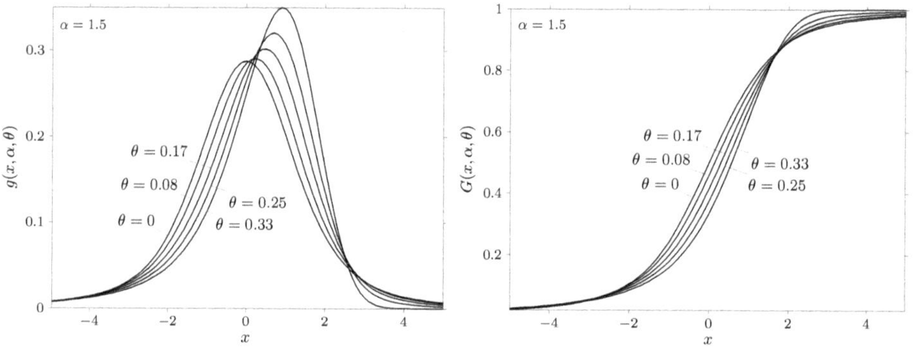


**Figure 13.** The probability density function $g(x, \alpha, \theta)$ (**on the left**) ) and cumulative distribution function $G(x, \alpha, \theta)$ (**on the right**) of a standard strictly stable law with $\alpha = 1.5$ and specified values $\theta$.

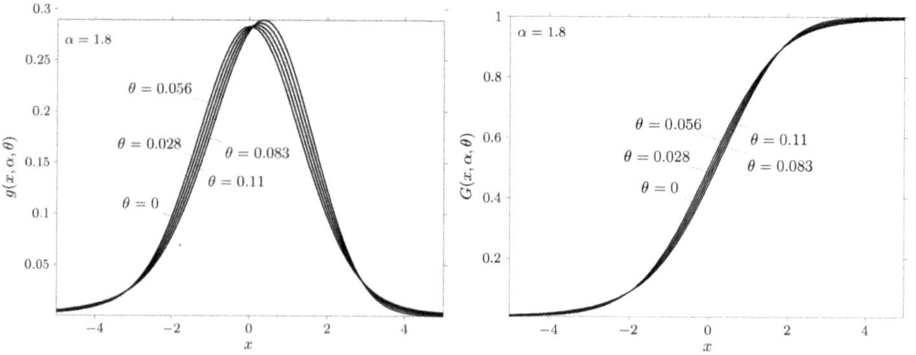


**Figure 14.** The probability density function $g(x, \alpha, \theta)$ (**on the left**) and cumulative distribution function $G(x, \alpha, \theta)$ (**on the right**) of a standard strictly stable law with $\alpha = 1.8$ and specified values $\theta$.

Thus, at $\alpha < 1$ the parameters $\theta$ and $\beta$ coincide. Therefore, the corresponding properties for the probability density and the distribution function of stable laws with characteristic functions in Equations (1) and (5) coincide. The situation slightly changes if $\alpha > 1$. It can be seen from the above relation that, firstly, the admitted region of the parameter $\theta$ narrows in comparison with the admitted region of the parameter $\beta$, secondly, the parameter $\theta$ changes its sign to the opposite with respect to the parameter $\beta$. Since the parameter $\theta$ has the meaning of an asymmetry parameter, then a change of the sign of this parameter when passing from $\alpha < 1$ to $\alpha > 1$ will affect the form of probability density. This is clearly seen in Figures 11 and 12 for densities with extreme values of the parameter $\theta$. Comparing densities for the values $\alpha = 0.9$, $\theta = 1$ given in Figure 11 and density for the values $\alpha = 1.2$, $\theta = 0.66$ given in Figure 12 one can see that these densities are turned into different directions. This fact is a consequence of the fact that the parameter $\theta$ changed the sign compared to the sign of the parameter $\beta$. Thus, for the same sign of the parameter $\theta$ of density $g(x, \alpha, \theta)$ at $\alpha < 1$ and $\alpha > 1$ will be turned into different directions. The reason for this behavior is related to the selected parameter system $(\alpha, \theta, \lambda)$ of the characteristic function in Equation (1). As it was mentioned in the book by Zolotarev V.M. (see page 19 in [32]), distributions from the class of strictly stable laws are continuous in the totality of their parameters in the entire range of their admissible values it is with this choice of parameters.

It should be pointed out that the Theorem 1 and Corollary 1 formulate expressions for probability density and distribution functions for strict stable laws with a scale parameter $\lambda = 1$. To obtain the density and distribution function with an arbitrary value of the scale parameter $\lambda$ it is necessary to use Remarks 5 and 7.

## 5. Conclusions

In this paper, integral representations for the probability density have been obtained (Theorem 1) and distribution function (Corollary 1) of a standard ($\lambda = 1$) strictly stable law with the characteristic function in Equation (1). In the general case $\alpha \neq 1$ and $x \neq 0$ the probability density and distribution function are expressed in terms of a definite integral. In the case $\alpha = 1$ for any $x$ and in the case $x = 0$ for any admissible $\alpha$ and $\theta$ the probability density and distribution function are expressed in terms of elementary functions. Applying the method of numerical integration, the values of the density and distribution function of strictly stable laws with the characteristic function in Equation (1) were calculated. The calculations show that the numerical methods do not have any difficulties in calculating the density and distribution function for the selected parameter values.

However, this does not mean that one can calculate the density and function of distribution for all admissible parameters by using obtained integral representations. Most likely, numerical integration algorithms will have difficulty in calculating the integral for small values $\alpha$, at $\alpha \approx 1$ and for bigger values of $x$. The results of the works in which integral representations for densities of stable laws with characteristic functions in Equations (5) and (4) were investigated testify to this. An integral representation for a stable law with the characteristic function in Equation (5) was obtained in the work [31] (see also § 2.2 in [32], § 4.4 in [36]). In the work [33], it was pointed out that when values of $\alpha$ close to 1 problems arise with the numerical calculation of the integral in this integral representation. An integral representation for the density of a stable law with the characteristic function in Equation (4) was obtained in [33]. In this work, it was emphasized that when calculating the density, calculation difficulties arise at values $0 < |\alpha - 1| < 0.02$ and at values $\alpha$ close to zero. In the works by [38,43] the same problems are mentioned when calculating the integral in the representation obtained in the work [33]. Based on this, it should be expected that, with the above parameter values, calculation difficulties will also arise with the density and distribution functions obtained in the Theorem 1 and Corollary 1. In particular, directly from the expressions in Equations (69) and (113), it can be seen that at $\alpha$ close to 1, but not equal to 1, problems may arise with the numerical calculation of the integral. This is indicated by the exponent $\alpha/(\alpha-1)$. It can be seen that when $\alpha \to 1$ this value increases unlimitedly. Most likely, in this case, one will have to look for other ways of calculating the density and distribution function of a strictly stable law.

In conclusion, we would like to point out that the integral representation of the density $g(x,\alpha,\theta)$ formulated in the Theorem 1 was used to calculate the density in Equation (2). To calculate the improper integral in Equation (2) we used the adaptive quadrature Gaussian–Kornord numerical integration algorithm on 15 points. We used the implementation of this algorithm in the library gsl (GNU Scientific Library) version 1.8 [46]. The calculations performed in some cases show the presence of problems of numerical integration. In particular, at $x$ close to zero, the calculated density behaves like a periodic function. In addition, in some cases, the integration algorithm generates an integration error. All this indicates the need for additional study of the integrand function in Equation (2) and adapting this expression for numerical integration algorithms. It should be noted that the most likely causes of these difficulties may be the ones described above when calculating the density $g(x,\alpha,\theta)$. Therefore, first of all, it is necessary to find a solution to the problems described above. To calculate the density at $x$ close to zero and for bigger values $x$ the most promising approach is to use an expansion of the strictly stable density in the power series. The method described in the article [43] can be used to calculate the density at $\alpha \to 1$. However, the possibility of using this approach requires additional research.

**Funding:** This work was supported by the Ministry of Higher Education and Science of the Russian Federation (project No. 0830-2020-0008).

**Acknowledgments:** The author thanks to M. Yu. Dudikov for translation the article into English.

**Conflicts of Interest:** The author declares no conflict of interest.

## References

1. Kotulski, M. Asymptotic distributions of continuous-time random walks: A probabilistic approach. *J. Stat. Phys.* **1995**, *81*, 777–792. [CrossRef]
2. Kolokoltsov, V.N.; Korolev, V.Y.; Uchaikin, V.V. Fractional Stable Distributions. *J. Math. Sci.* **2001**, *105*, 2569–2576. [CrossRef]
3. Montroll, E.W.; Weiss, G.H. Random Walks on Lattices. II. *J. Math. Phys.* **1965**, *6*, 167. [CrossRef]
4. Scher, H.; Lax, M. Stochastic transport in a disordered solid. I. Theory. *Phys. Rev. B* **1973**, *7*, 4491–4502. [CrossRef]
5. Scher, H.; Lax, M. Stochastic transport in a disordered solid. II. Impurity conduction. *Phys. Rev. B* **1973**, *7*, 4502–4519. [CrossRef]
6. Klafter, J.; Blumen, A.; Shlesinger, M.F. Stochastic pathway to anomalous diffusion. *Phys. Rev. A* **1987**, *35*, 3081–3085. [CrossRef]
7. Metzler, R.; Klafter, J. The random walk's guide to anomalous diffusion: A fractional dynamics approach. *Phys. Rep.* **2000**, *339*, 1–77. [CrossRef]
8. Zaburdaev, V.Y.; Denisov, S.I.; Klafter, J. Lévy walks. *Rev. Mod. Phys.* **2015**, *87*, 483–530. [CrossRef]
9. Uchaikin, V.V. Montroll-Weiss problem, fractional equations, and stable distributions. *Int. J. Theor. Phys.* **2000**, *39*, 2087–2105. [CrossRef]
10. Saenko, V.V. New Approach to Statistical Description of Fluctuating Particle Fluxes. *Plasma Phys. Rep.* **2009**, *35*, 1–13. [CrossRef]
11. Saenko, V.V. Self-similarity of fluctuation particle fluxes in the plasma edge of the stellarator L-2M. *Contrib. Plasma Phys.* **2010**, *50*, 246–251. [CrossRef]
12. Saenko, V.; Saenko, Y. Approximation of Microarray Gene Expression Profiles by the Stable Laws. *Int. J. Environ. Eng.* **2015**, *2*, 98–102.
13. Saenko, V.; Saenko, Y. Application of the fractional stable distributions for approximation of gene expression profiles. *Stat. Appl. Genet. Mol. Biol.* **2015**, *14*, 295–306. [CrossRef] [PubMed]

14. Saenko, V.V. Fractional-Stable Statistics of the Genes Expression in the Next Generation Sequence Results. *Math. Biol. Bioinform.* **2016**, *11*, 278–287. [CrossRef]
15. Ueda, H.R.; Hayashi, S.; Matsuyama, S.; Yomo, T.; Hashimoto, S.; Kay, S.A.; Hogenesch, J.B.; Iino, M. Universality and flexibility in gene expression from bacteria to human. *Proc. Natl. Acad. Sci. USA* **2004**, *101*, 3765–3769. [CrossRef]
16. Hoyle, D.C.; Rattray, M.; Jupp, R.; Brass, A. Making sense of microarray data distributions. *Bioinformatics* **2002**, *18*, 576–84. [CrossRef]
17. Furusawa, C.; Kaneko, K. Zipf's Law in Gene Expression. *Phys. Rev. Lett.* **2003**, *90*, 8–11. [CrossRef]
18. Saenko, V.V. Estimation of the Parameters of Fractional-Stable Laws by the Method of Minimum Distance. *J. Math. Sci.* **2016**, *214*, 101–114. [CrossRef]
19. Schneider, W.R. Stable distributions: Fox function representation and generalization. In *Stochastic Processes in Classical and Quantum Systems*; Albeverio, S., Casati, G., Merlini, D., Eds.; Springer: Berlin/Heidelberg, Germany, 1986; Volume 262, pp. 497–511. [CrossRef]
20. Schneider, W.R. Generalized one-sided stable distributions. In *Stochastic Processes—Mathematics and Physics II. Lecture Notes in Mathematics*; Albeverio, S., Blanchard, P., Streit, L., Eds.; Springer: Berlin/Heidelberg, Germany, 1987; Volume 1250, pp. 269–287. [CrossRef]
21. Hoffmann–Jørgensen, J. Stable Densities. *Theory Probab. Appl.* **1994**, *38*, 350–355. [CrossRef]
22. Zolotarev, V.M. On Representation of Densities of Stable Laws by Special Functions. *Theory Probab. Appl.* **1995**, *39*, 354–362. [CrossRef]
23. Hatzinikitas, A.; Pachos, J.K. One-dimensional stable probability density functions for rational index 0<alpha<=2. *Ann. Phys.* **2008**, *323*, 3000–3019. [CrossRef]
24. Penson, K.A.; Górska, K. Exact and Explicit Probability Densities for One-Sided Lévy Stable Distributions. *Phys. Rev. Lett.* **2010**, *105*, 210604. [CrossRef]
25. Górska, K.; Penson, K.A. Lévy stable two-sided distributions: Exact and explicit densities for asymmetric case. *Phys. Rev. E* **2011**, *83*, 061125. [CrossRef] [PubMed]
26. Pogány, T.K.; Nadarajah, S. Remarks on the Stable $S\alpha$ $(\beta,\gamma,\mu)$ Distribution. *Methodol. Comput. Appl. Probab.* **2015**, *17*, 515–524. [CrossRef]
27. Mittnik, S.; Doganoglu, T.; Chenyao, D. Computing the probability density function of the stable Paretian distribution. *Math. Comput. Model.* **1999**, *29*, 235–240. [CrossRef]
28. Menn, C.; Rachev, S.T. Calibrated FFT-based density approximations for $\alpha$-stable distributions. *Comput. Stat. Data Anal.* **2006**, *50*, 1891–1904. [CrossRef]
29. Nolan, J. An algorithm for evaluating stable densities in Zolotarev's (M) parameterization. *Math. Comput. Model.* **1999**, *29*, 229–233. [CrossRef]
30. Ament, S.; O'Neil, M. Accurate and efficient numerical calculation of stable densities via optimized quadrature and asymptotics. *Stat. Comput.* **2018**, *28*, 171–185. [CrossRef]
31. Zolotarev, V.M. On the representation of stable laws by integrals. *Sel. Transl. Math. Stat. Probab.* **1964**, *4*, 84–88.
32. Zolotarev, V.M. *One-dimensional stable Distributions*; American Mathematical Society: Providence, RI, USA, 1986.
33. Nolan, J.P. Numerical calculation of stable densities and distribution functions. *Commun. Stat. Stoch. Models* **1997**, *13*, 759–774. [CrossRef]
34. Arias-Calluari, K.; Alonso-Marroquin, F.; Harré, M.S. Closed-form solutions for the Lévy-stable distribution. *Phys. Rev. E* **2018**, *98*, 012103. [CrossRef] [PubMed]
35. Bergström, H. On some expansions of stable distribution functions. *Ark. Mat.* **1952**, *2*, 375–378. [CrossRef]
36. Uchaikin, V.V.; Zolotarev, V.M. *Chance and Stability Stable Distributions and Their Applications*; VSP: Utrecht, The Netherlands, 1999; p. 569.
37. Nolan, J.P. Parameterizations and modes of stable distributions. *Stat. Probab. Lett.* **1998**, *38*, 187–195. [CrossRef]
38. Royuela-del Val, J.; Simmross-Wattenberg, F.; Alberola-López, C. Libstable: Fast, parallel, and high-precision computation of $\alpha$-stable distributions in R, C/C++, and MATLAB. *J. Stat. Softw.* **2017**, *78*. [CrossRef]
39. Liang, Y.; Chen, W. A survey on computing Lévy stable distributions and a new MATLAB toolbox. *Signal Process.* **2013**, *93*, 242–251. [CrossRef]

40. Veillette, M. MATLAB Code: Alpha-Stable Distributions. 2008. Available online: http://math.bu.edu/people/mveillet/research.html (accessed on 11 May 2020).
41. Rimmer, R.H.; Nolan, J.P. Stable Distributions in Mathematica. *Math. J.* **2005**, *9*, 776–789.
42. Nolan, J.P. John Nolan's Stable Distribution Page. Available online: http://fs2.american.edu/jpnolan/www/stable/stable.html (accessed on 11 May 2020).
43. Matsui, M.; Takemura, A. Some improvements in numerical evaluation of symmetric stable density and its derivatives. *Commun. Stat. Theory Methods* **2006**, *35*, 149–172. [CrossRef]
44. Bening, V.E.; Korolev, V.Y.; Sukhorukova, T.A.; Gusarov, G.G.; Saenko, V.V.; Uchaikin, V.V.; Kolokoltsov, V.N. Fractionally stable distributions. In *Stochastic Models of Structural Plasma Turbulence*; Korolev, V.Y., Skvortsova, N.N., Eds.; Brill Academic Publishers: Utrecht, The Netherlands, 2006; pp. 175–244. [CrossRef]
45. Bateman, H. *Higher Transcendental Functions*; McGraw-Hill Book Company, Inc.: New York, NY, USA, 1953; Volume 1, p. 302.
46. GSL—GNU Scientific Library. Available online: http://www.gnu.org/software/gsl/ (accessed on 11 May 2020).

mathematics

MDPI

*Article*

# Wavelet Thresholding Risk Estimate for the Model with Random Samples and Correlated Noise

**Oleg Shestakov** [1,2]

[1] Faculty of Computational Mathematics and Cybernetics, M. V. Lomonosov Moscow State University, Moscow 119991, Russia; oshestakov@cs.msu.ru

[2] Institute of Informatics Problems, Federal Research Center "Computer Science and Control" of the Russian Academy of Sciences, Moscow 119333, Russia

Received: 13 February 2020; Accepted: 5 March 2020; Published: 8 March 2020

**Abstract:** Signal de-noising methods based on threshold processing of wavelet decomposition coefficients have become popular due to their simplicity, speed, and ability to adapt to signal functions with spatially inhomogeneous smoothness. The analysis of the errors of these methods is an important practical task, since it makes it possible to evaluate the quality of both methods and equipment used for processing. Sometimes the nature of the signal is such that its samples are recorded at random times. If the sample points form a variational series based on a sample from the uniform distribution on the data registration interval, then the use of the standard threshold processing procedure is adequate. The paper considers a model of a signal that is registered at random times and contains noise with long-term dependence. The asymptotic normality and strong consistency properties of the mean-square thresholding risk estimator are proved. The obtained results make it possible to construct asymptotic confidence intervals for threshold processing errors using only the observed data.

**Keywords:** threshold processing; random samples; long-term dependence; mean-square risk estimate

## 1. Introduction

In digital signal processing tasks, it is often assumed that the recorded signal samples are independent. However, there are many physical processes that demonstrate long-term dependence where correlations between observations decrease rather slowly. For example, long-term dependence is often observed in geophysical processes where it takes the form of long periods of large or small values of observations. Interferences in communication channels demonstrate similar phenomena. Wavelet methods are widely used in the analysis and processing of signals recorded during the study of such processes.

The wavelet decomposition of a function $f(x)$ is a series

$$f(x) = \sum_{j,k\in Z} \langle f, \psi_{j,k}\rangle \psi_{j,k}(x),$$

where $\psi_{j,k}(x) = 2^{j/2}\psi(2^j x - k)$, and $\psi(x)$ is a wavelet function. The indices $j$ and $k$ are called the scale and the shift, respectively. This decomposition provides a time/scale representation of the signal function, that allows one to localise its features. There exist many wavelet functions with various properties.

In practice, a discrete wavelet transform is used. It is a multiplication of a sampled signal vector by an orthogonal matrix defined by the wavelet function $\psi(x)$ (in practice implemented with a fast cascade algorithm [1]). This transform is applied to data, and the threshold processing of the resulting wavelet coefficients is performed [1]. For a model of signal samples with an equispaced grid, these methods were well studied by D. Donoho, I. Johnstone, B. Silverman and others [2–10]. Statistical properties of the mean-square risk estimator were also studied. It is shown that under certain conditions it is strongly consistent and asymptotically normal [11–13].

In some experiments it is not possible to record signal samples at regular intervals [14]. Sometimes registration of samples is made at random times. It was shown by T. Cai and L. Brown [15] that, if the sample points form a variational series based on a sample from the uniform distribution on the data registration interval, then the rate of the mean-square thresholding risk remains, up to a logarithmic factor, equal to the optimal rate in the class of Lipschitz regular functions. A special case of the uniform distribution appears, for example, when considering a Poisson process, and since the conditional distribution of its points on a given time interval, given the number of points, is uniform. These models can arise, for example, in astronomy when considering the stellar intensity. In this paper, it is proven that under some regularity conditions, the statistical properties of the risk estimator also remain the same for both equispaced and random sample grids.

## 2. Long-Term Dependence

Let the signal function $f(x)$ be defined on the segment $[0,1]$ and be uniformly Lipschitz regular with some exponent $\gamma > 0$ and Lipschitz constant $L > 0$: $f \in \mathrm{Lip}(\gamma, L)$. Assume that the samples of $f(x)$ contain additive correlated noise and are recorded at random times that are independent and uniformly distributed on $[0,1]$. Namely, consider the following data model:

$$Y_i = f(x_{(i)}) + e_i, \quad i = 1, \dots, N \ (N = 2^J), \tag{1}$$

where $0 \le x_{(1)} < \dots < x_{(N)} \le 1$ is the variational series based on a sample from the uniform distribution on the segment $[0,1]$, and $\{e_i, i \in \mathbb{Z}\}$ is a stationary Gaussian process with the covariance sequence $r_k = \mathbf{cov}(e_i, e_{i+k})$. We assume that $e_i$ have zero mean and unit variance. We also assume that the noise autocovariance function decreases at the rate of $r_k \sim k^{-\alpha}$, where $0 < \alpha < 1$. This corresponds to the long-term dependence between observations [7].

The observations consist of pairs $(x_{(1)}, Y_1), \dots, (x_{(N)}, Y_N)$, where the distances between the samples are, generally, not equal. It is known that $\mathsf{E}x_{(i)} = i/(N+1)$ (see Lemma 2 in [15]). Along with (1), consider a sample with equal distances between sample points

$$\left(\frac{1}{N+1}, Z_1\right), \dots, \left(\frac{N}{N+1}, Z_N\right). \tag{2}$$

where

$$Z_i = f\left(\frac{i}{N+1}\right) + e_i, \quad i = 1, \dots, N.$$

For the sample (2) threshold processing methods have been developed that effectively suppress the noise and provide an "almost" optimal rate of the mean-square risk [7,8]. The discrete wavelet transform with Meyer wavelets is applied to the sample (2) to obtain a set of empirical wavelet coefficients [1]

$$W_{j,k} = \mu_{j,k} + 2^{\frac{(1-\alpha)(J-j)}{2}} \xi_{j,k}, \ j = 0, \dots, J-1, \ k = 0, \dots, 2^j - 1,$$

where $\mu_{j,k}$ are discrete wavelet coefficients of the sample

$$f\left(\frac{1}{N+1}\right), \dots, f\left(\frac{N}{N+1}\right),$$

and the noise coefficients $\xi_{j,k}$ have the standard normal distribution, but are not independent. The variances of $W_{j,k}$ have the form $\sigma_j^2 = 2^{(1-\alpha)(J-j)}$ [12]. To suppress the noise the coefficients $W_{j,k}$ are processed with the hard thresholding function $\rho_H(y,T) = x\mathbf{1}(|y| > T)$ or the soft thresholding function $\rho_S(y,T) = \mathrm{sgn}(x)\,(|y| - T)_+$ with some threshold $T$, and the estimates $\widehat{W}_{j,k}$ are obtained.

After that, the inverse wavelet transform is performed. The idea of threshold processing is that the wavelet transform provides a "sparse" representation of the useful signal function; i.e., the signal is represented by a relatively small number of modulo large coefficients. To provide a "sparse" representation of a function that is uniformly Lipschitz regular with an exponent $\gamma$, the wavelet function participating in the discrete wavelet transform must have $M$ continuous derivatives ($M \geq \gamma$) and $M$ vanishing moments. It also must decrease fast enough at infinity. It is further assumed that the Meyer wavelets [1] that satisfy all the necessary conditions are used to perform the wavelet transform.

If we apply the discrete wavelet transform to the sample (1), we obtain the set of empirical wavelet coefficients

$$V_{j,k} = \nu_{j,k} + \xi_{j,k},\ j = 0, \dots, J-1,\ k = 0, \dots, 2^j - 1.$$

Here $\nu_{j,k}$ are the coefficients of the discrete wavelet transform of the sample

$$f\left(x_{(1)}\right), \dots, f\left(x_{(N)}\right).$$

In general, $V_{j,k}$ are not equal to $W_{j,k}$, and $\nu_{j,k}$ are not equal to $\mu_{j,k}$. However, one can apply the same thresholding procedure to the coefficients $V_{j,k}$ as to the coefficients $W_{j,k}$ and obtain the estimators $\widehat{V}_{j,k}$. The following sections discuss the properties of the resulting estimators.

## 3. Mean-Square Thresholding Risk

The mean-square thresholding risk for a sample with random grid is defined as

$$R_\nu(f,T) = \sum_{j=0}^{J-1} \sum_{k=0}^{2^j-1} \mathsf{E}(\widehat{V}_{j,k} - \mu_{j,k})^2. \tag{3}$$

We also define the mean-square risk for the equispaced sample as

$$R_\mu(f,T) = \sum_{j=0}^{J-1} \sum_{k=0}^{2^j-1} \mathsf{E}(\widehat{W}_{j,k} - \mu_{j,k})^2.$$

The threshold selection is one of the main problems in threshold processing. For the class $\mathrm{Lip}(\gamma, L)$, the threshold $T_\gamma = \sigma_j\sqrt{\frac{4\alpha\gamma}{2\gamma+\alpha}\ln 2^j}$ (calculated for each $j$) is close to optimal [16]. Using the results of [7] (Theorem 3), we can estimate the rate of $R_\mu(f, T_\gamma)$.

**Theorem 1.** *Let $\alpha > 1/2$ and $f \in \mathrm{Lip}(\gamma, L)$ on the segment $[0,1]$ with $\gamma > (4\alpha-2)^{-1}$. Then for the threshold $T_\gamma$ we have*

$$R_\mu(f, T_\gamma) \leq C 2^{\frac{2\gamma+\alpha-2\alpha\gamma}{2\gamma+\alpha}J} J^{\frac{2\gamma+2\alpha}{2\gamma+\alpha}},$$

*where C is a positive constant.*

Additionally, repeating the arguments of Theorem 1 in [15], it can be shown that similar statement is valid for $R_\nu(f, T_\gamma)$ when $\gamma > \max\{(4\alpha-2)^{-1}, 1/2\}$. Thus, the replacement of equally-spaced samples by random ones does not affect the upper estimate for the rate of the mean-square risk.

## 4. Properties of the Mean-Square Risk Estimate

Since expression (3) explicitly depends on the unknown values of $\mu_{j,k}$, it cannot be calculated in practice. However, it is possible to construct its estimate based only observable data. This estimate is determined by the expression

$$\widehat{R}_\nu(f,T) = \sum_{j=0}^{J-1} \sum_{k=0}^{2^j-1} F[V_{j,k}, T], \tag{4}$$

where $F[V_{j,k}, T] = (V_{j,k}^2 - \sigma^2)\mathbf{1}(|V_{j,k}| \le T) + \sigma^2\mathbf{1}(|V_{j,k}| > T)$ for the hard threshold processing and $F[V_{j,k}, T] = (V_{j,k}^2 - \sigma^2)\mathbf{1}(|V_{j,k}| \le T) + (\sigma^2 + T^2)\mathbf{1}(|V_{j,k}| > T)$ for the soft threshold processing [1,3].

Estimator (4) provides an opportunity to get an idea of the evaluation error for the function $f$, since it can be calculated using only the observable values $V_{j,k}$. The following statement establishes its asymptotic normality, that, in particular, allows constructing asymptotic confidence intervals for the mean-square risk (3).

**Theorem 2.** *Let $f \in \mathrm{Lip}(\gamma, L)$ on the segment $[0,1]$ with $\gamma > \max\{(4\alpha-2)^{-1}, 1/2\}$, $\alpha > 1/2$, and let the Meyer wavelet satisfy the conditions listed above. Then for the hard and soft threshold processing we have*

$$\mathsf{P}\left(\frac{\widehat{R}_\nu(f,T_\gamma) - R_\nu(f,T_\gamma)}{D_J} < x\right) \to \Phi(x) \ \textit{when } J \to \infty,$$

*where $\Phi(x)$ is the distribution function of the standard normal law, $D_J^2 = C_\alpha 2^J$, and the constant $C_\alpha$ depends only on $\alpha$ and the wavelet type.*

**Remark 1.** *In practice, one needs to know the constant $C_\alpha$. Unlike the case of independent observations, this constant depends on the chosen wavelet. The method of calculation of $C_\alpha$ is discussed in [12].*

**Proof.** Let us prove the theorem for the hard threshold processing method. In the case of soft threshold processing, the proof is similar.

Along with $\widehat{R}_\nu(f,T_\gamma)$, consider

$$\widehat{R}_\mu(f,T_\gamma) = \sum_{j=0}^{J-1}\sum_{k=0}^{2^j-1} F[W_{j,k}, T_\gamma]$$

and write the difference $\widehat{R}_\nu(f,T_\gamma) - R_\nu(f,T_\gamma)$ in the form

$$\widehat{R}_\nu(f,T_\gamma) - R_\nu(f,T_\gamma) = \widehat{R}_\mu(f,T_\gamma) - R_\mu(f,T_\gamma) + \widetilde{R},$$

where

$$\widetilde{R} = \widehat{R}_\nu(f,T_\gamma) - \widehat{R}_\mu(f,T_\gamma) - (R_\nu(f,T_\gamma) - R_\mu(f,T_\gamma)).$$

In [12] with the use of the results of [17–19] it is shown that

$$\mathsf{P}\left(\frac{\widehat{R}_\mu(f,T_\gamma) - R_\mu(f,T_\gamma)}{D_J} < x\right) \to \Phi(x) \ \text{when } J \to \infty.$$

Therefore, to prove the theorem, it suffices to show that

$$\frac{\widetilde{R}}{2^{J/2}} \xrightarrow{\mathsf{P}} 0 \ \text{when } J \to \infty.$$

Under the conditions $\gamma > \max\{(4\alpha-2)^{-1}, 1/2\}$ and $\alpha > 1/2$, by virtue of Theorem 1 and a similar statement for $R_\nu(f,T)$, we obtain that

$$\frac{R_\nu(f,T_\gamma) - R_\mu(f,T_\gamma)}{2^{J/2}} \to 0 \ \text{when } J \to \infty.$$

Set

$$j_0 \approx \frac{\alpha}{2\gamma+\alpha}J + \frac{1}{2\gamma+\alpha}\log_2 J.$$

Let us represent $\widehat{R}_\nu(f,T_\gamma) - \widehat{R}_\mu(f,T_\gamma)$ as

$$\widehat{R}_\nu(f,T_\gamma) - \widehat{R}_\mu(f,T_\gamma) = S_1 + S_2,$$

where

$$S_1 = \sum_{j=0}^{j_0-1}\sum_{k=0}^{2^j-1}\left(F[V_{j,k},T_\gamma] - F[W_{j,k},T_\gamma]\right),$$

$$S_2 = \sum_{j=j_0}^{J-1}\sum_{k=0}^{2^j-1}\left(F[V_{j,k},T_\gamma] - F[W_{j,k},T_\gamma]\right).$$

Since for some constant $\tilde{C} > 0$ we have

$$\left|F[V_{j,k},T_\gamma]\right| \le \tilde{C}T_\gamma^2, \quad \left|F[W_{j,k},T_\gamma]\right| \le \tilde{C}T_\gamma^2 \text{ a.s.,} \tag{5}$$

then

$$\frac{S_1}{2^{J/2}} \xrightarrow{\mathsf{P}} 0 \text{ when } J \to \infty.$$

Next

$$\begin{aligned} S_2 = \sum_{j=j_0}^{J-1}\sum_{k=0}^{2^j-1}\left(F[V_{j,k},T_\gamma] - F[W_{j,k},T_\gamma]\right) &= \sum_{j=j_0}^{J-1}\sum_{k=0}^{2^j-1}(V_{j,k}^2 - W_{j,k}^2)+ \\ &+\sum_{j=j_0}^{J-1}\sum_{k=0}^{2^j-1}(W_{j,k}^2 - 2\sigma^2)\mathbf{1}(|V_{j,k}| \le T_\gamma, |W_{j,k}| > T_\gamma)+ \\ &+\sum_{j=j_0}^{J-1}\sum_{k=0}^{2^j-1}(2\sigma^2 - V_{j,k}^2)\mathbf{1}(|V_{j,k}| > T_\gamma, |W_{j,k}| \le T_\gamma)+ \\ &+\sum_{j=j_0}^{J-1}\sum_{k=0}^{2^j-1}(W_{j,k}^2 - V_{j,k}^2)\mathbf{1}(|V_{j,k}| > T_\gamma, |W_{j,k}| > T_\gamma). \end{aligned} \tag{6}$$

Consider the sum $\sum_{j=j_0}^{J-1}\sum_{k=0}^{2^j-1}(V_{j,k}^2 - W_{j,k}^2)$:

$$\sum_{j=j_0}^{J-1}\sum_{k=0}^{2^j-1}(V_{j,k}^2 - W_{j,k}^2) = \sum_{j=j_0}^{J-1}\sum_{k=0}^{2^j-1}(\nu_{j,k}^2 - \mu_{j,k}^2) + 2\sum_{j=j_0}^{J-1}\sum_{k=0}^{2^j-1}\xi_{j,k}(\nu_{j,k} - \mu_{j,k}).$$

Using the results of [12,15,20], it can be shown that the conditional distribution of this sum for fixed $x_i$ is normal with the mean

$$\sum_{j=j_0}^{J-1}\sum_{k=0}^{2^j-1}(\nu_{j,k}^2 - \mu_{j,k}^2)$$

and the variance that is less than

$$\tilde{C}_\alpha \sum_{j=j_0}^{J-1}\sum_{k=0}^{2^j-1}(\nu_{j,k} - \mu_{j,k})^2,$$

where $\tilde{C}_\alpha$ is a positive constant.

Since $f \in \text{Lip}(\gamma, L)$, repeating the arguments of [20], it can be shown that

$$\frac{1}{2^{J/2}}\mathsf{E}_x\left|\sum_{j=j_0}^{J-1}\sum_{k=0}^{2^j-1}(\nu_{j,k}^2 - \mu_{j,k}^2)\right| \to 0,$$

$$\frac{1}{2^{J/2}}\mathsf{E}_x\sum_{j=j_0}^{J-1}\sum_{k=0}^{2^j-1}(\nu_{j,k}-\mu_{j,k})^2 \to 0. \tag{7}$$

Hence, applying the Markov inequality, we obtain

$$\frac{1}{2^{J/2}}\sum_{j=j_0}^{J-1}\sum_{k=0}^{2^j-1}(\nu_{j,k}^2-\mu_{j,k}^2) \xrightarrow{\mathsf{P}} 0,$$

$$\frac{1}{2^{J/2}}\sum_{j=j_0}^{J-1}\sum_{k=0}^{2^j-1}(\nu_{j,k}-\mu_{j,k})^2 \xrightarrow{\mathsf{P}} 0$$

when $J \to \infty$. Thus,

$$\frac{\sum_{j=j_0}^{J-1}\sum_{k=0}^{2^j-1}(V_{j,k}^2-W_{j,k}^2)}{2^{J/2}} \xrightarrow{\mathsf{P}} 0 \text{ when } J \to \infty.$$

The remaining sums in (6) contain indicators where either $|V_{j,k}| > T_\gamma$ or $|W_{j,k}| > T_\gamma$. Repeating the reasoning from [12] and using (7), it can be shown that, when divided by $2^{J/2}$, they also converge to zero in probability. The theorem is proven. □

Theorem 2 provides the possibility to construct asymptotic confidence intervals for the mean-square thresholding risk on the basis of its estimate.

In addition to the asymptotic normality, the estimator (4) also possesses the property of strong consistency.

**Theorem 3.** *Suppose that the conditions of Theorem 2 are satisfied. Then for hard or soft threshold processing, for any $\lambda > 1/2$ we have*

$$\frac{\widehat{R}_\nu(f,T_\gamma)-R_\nu(f,T_\gamma)}{2^{\lambda J}} \to 0 \text{ a.s. when } J \to \infty.$$

Since (5) holds, for fixed $x_i$ the conditional version of Bosq inequality [21] (Theorem 1.3) applies for (4), and the proof of this statement almost completely repeats the proof of the corresponding risk estimator property in [13].

## 5. Discussion

As it has been already mentioned, Theorem 2 provides the possibility to construct asymptotic confidence intervals for the mean-square thresholding risk. For practical purposes, it is desirable to have guaranteed confidence intervals. These intervals could be constructed based on the estimates of the convergence rate in Theorem 2. The estimates should depend on the Lipschitz parameters and parameter $\alpha$. Guaranteed confidence intervals would help to explain how the results of Theorems 2 and 3 affect the error estimation for the finite signal size. We therefore leave the problem of estimation of the rate of convergence and explicit numerical simulation for future work.

The obtained results are applicable to Meyer wavelets. Their advantage is that they possess infinitely many vanishing moments. It simplifies the proof of asymptotic normality in [12]. In view of the results of [8] it is clear that similar conclusions could be obtained with other wavelets that have a large enough number of vanishing moments (e.g., various Daubechies families).

It follows from Theorems 2 and 3 that the statistical properties of the mean-square risk estimator in a model with the uniform random design remain the same as in a model with equispaced samples. Note that this situation is not common. Random times of sample registration can also result in a random sample size. This situation was considered in [22]. In this case, the properties of the model can significantly differ from the properties of the fixed sample size model. For example, the limit distribution of the mean-square risk estimator can be a scale mixture of normal laws, which can have significantly heavier tails than the normal distribution. In particular, this distribution may belong to the class of stable laws, and it is well known that the variances of all stable laws, except the normal one, are infinite (the properties of stable distributions are discussed in detail in the monograph of V. M. Zolotarev [23]; see also [24]).

**Funding:** This research was funded by Russian Science Foundation, project number 18-11-00155.

**Conflicts of Interest:** The author declares no conflict of interest. The funders had no role in the design of the study; in the collection, analyses, or interpretation of data; in the writing of the manuscript, or in the decision to publish the results.

## References

1. Mallat, S. *A Wavelet Tour of Signal Processing*; Academic Press: New York, NY, USA, 1999.
2. Donoho, D.; Johnstone, I.M. Ideal Spatial Adaptation via Wavelet Shrinkage. *Biometrika* **1994**, *81*, 425–455. [CrossRef]
3. Donoho, D.; Johnstone, I.M. Adapting to Unknown Smoothness via Wavelet Shrinkage. *J. Am. Stat. Assoc.* **1995**, *90*, 1200–1224. [CrossRef]
4. Donoho, D.; Johnstone, I.M.; Kerkyacharian, G.; Picard, D. Wavelet shrinkage: Asymptopia? *J. R. Stat. Soc. Ser. B* **1995**, *57*, 301–369. [CrossRef]
5. Marron, J.S.; Adak, S.; Johnstone, I.M.; Neumann, M.H.; Patil, P. Exact risk analysis of wavelet regression. *J. Comput. Graph. Stat.* **1998**, *7*, 278–309.
6. Antoniadis, A.; Fan, J. Regularization of Wavelet Approximations. *J. Am. Stat. Assoc.* **2001**, *96*, 939–967. [CrossRef]
7. Johnstone, I.M.; Silverman, B.W. Wavelet threshold estimates for data with correlated noise. *J. R. Stat. Soc. Ser. B* **1997**, *59*, 319–351. [CrossRef]
8. Johnstone, I.M. Wavelet shrinkage for correlated data and inverse problems adaptivity results. *Stat. Sin.* **1999**, *9*, 51–83.
9. Kudryavtsev, A.A.; Shestakov, O.V. Asymptotic behavior of the threshold minimizing the average probability of error in calculation of wavelet coefficients. *Dokl. Math.* **2016**, *93*, 295–299. [CrossRef]
10. Kudryavtsev, A.A.; Shestakov, O.V. Asymptotically optimal wavelet thresholding in models with non-gaussian noise distributions. *Dokl. Math.* **2016**, *94*, 615–619. [CrossRef]
11. Shestakov, O.V. Asymptotic normality of adaptive wavelet thresholding risk estimation. *Dokl. Math.* **2012**, *86*, 556–558. [CrossRef]
12. Eroshenko, A.A. Statistical Properties of Signal and Image Estimates, Using Threshold Processing of Coefficients in Wavelet Decompositions. Ph.D. Thesis, M. V. Lomonosov Moscow State University, Moscow, Russia, 2015.
13. Shestakov, O.V. Almost everywhere convergence of a wavelet thresholding risk estimate in a model with correlated noise. *Moscow Univ. Comput. Math. Cybern.* **2016**, *40*, 114–117. [CrossRef]
14. Cai, T.; Brown, L. Wavelet Shrinkage for Nonequispaced Samples. *Ann. Stat.* **1998**, *26*, 1783–1799. [CrossRef]
15. Cai, T.; Brown, L. Wavelet Estimation for Samples with Random Uniform Design. *Stat. Probab. Lett.* **1999**, *42*, 313–321. [CrossRef]
16. Jansen, M. *Noise Reduction by Wavelet Thresholding*; Lecture Notes in Statistics; Springer: New York, NY, USA, 2001; Volume 161.
17. Taqqu, M.S. Weak Convergence to Fractional Brownian Motion and to the Rosenblatt Process. *Z. Wahrscheinlichkeitsth. Verw. Geb.* **1975**, *31*, 287–302. [CrossRef]
18. Bradley, R.C. Basic Properties of Strong Mixing Conditions. A Survey and Some Open Questions. *Probab. Surv.* **2005**, *2*, 107–144. [CrossRef]
19. Peligrad, M. On the Asymptotic Normality of Sequences of Weak Dependent Random Variables. *J. Theor. Probab.* **1996**, *9*, 703–715. [CrossRef]
20. Shestakov, O.V. Properties of wavelet estimates of signals recorded at random time points. *Inform. Appl.* **2019**, *13*, 16–21.
21. Bosq, D. *Nonparametric Statistics for Stochastic Processes: Estimation and Prediction*; Springer: New York, NY, USA, 1996.

22. Shestakov, O.V. Convergence of the Distribution of the Threshold Processing Risk Estimate to a Mixture of Normal Laws at a Random Sample Size. *Syst. Means Inform.* **2019**, *29*, 31–38.
23. Zolotarev, V.M. *One-Dimensional Stable Distributions*; AMS: Providence, RI, USA, 1986.
24. Gnedenko, B.V.; Korolev, V.Y. *Random Summation: Limit Theorems and Applications*; CRC Press: Boca Raton, FL, USA, 1996.

MDPI
St. Alban-Anlage 66
4052 Basel
Switzerland
Tel. +41 61 683 77 34
Fax +41 61 302 89 18
www.mdpi.com

*Mathematics* Editorial Office
E-mail: mathematics@mdpi.com
www.mdpi.com/journal/mathematics

www.ingramcontent.com/pod-product-compliance
Lightning Source LLC
LaVergne TN
LVHW070238100526
838202LV00015B/2149